Renormalization Theory

NATO ADVANCED STUDY INSTITUTES SERIES

Proceedings of the Advanced Study Institute Programme, which aims at the dissemination of advanced knowledge and the formation of contacts among scientists from different countries

The series is published by an international board of publishers in conjunction with NATO Scientific Affairs Division

A	Life Sciences	Plenum Publishing Corporation
B	Physics	London and New York
C	Mathematical and Physical Sciences	D. Reidel Publishing Company Dordrecht and Boston
D	Behavioral and Social Sciences	Sijthoff International Publishing Company Leiden
E	Applied Sciences	Noordhoff International Publishing Leiden

Series C – Mathematical and Physical Sciences

Volume 23 – Renormalization Theory

Renormalization Theory

Proceedings of the NATO Advanced Study Institute held at the International School of Mathematical Physics at the 'Ettore Majorana' Centre for Scientific Culture in Erice (Sicily) Italy, 17-31 August, 1975

edited by

G. VELO, *International School of Mathematical Physics, Bologna, Italy*

A.S. WIGHTMAN, *Dept. of Physics, Princeton University, Princeton, N.I., U.S.A.*

D. Reidel Publishing Company

Dordrecht-Holland / Boston-U.S.A.

Published in cooperation with NATO Scientific Affairs Division

ISBN-13: 978-94-010-1492-2 e-ISBN-13: 978-94-010-1490-8
DOI: 10.1007/978-94-010-1490-8

Published by D. Reidel Publishing Company
P.O. Box 17, Dordrecht, Holland

Sold and distributed in the U.S.A., Canada, and Mexico
by D. Reidel Publishing Company, Inc.
Lincoln Building, 160 Old Derby Street, Hingham, Mass. 02043, U.S.A.

TABLE OF CONTENTS

INTRODUCTION

The present volume collects lecture notes from the session of the International School of Mathematical Physics 'Ettore Majorana' on Renormalization Theory that took place in Erice (Sicily), August 17 to August 31, 1975.

The School was a NATO Advanced Study Institute sponsored by the Italian Ministry of Public Education, the Italian Ministry of Scientific and Technological Research, and the Regional Sicilian Government.

Renormalization theory has, by now, acquired forty years of history. The present volume assumes a general acquaintance with the elementary facts of the subject as they might appear in an introductory course in quantum field theory. For more recent significant developments it provides a systematic introduction as well as a detailed discussion of the existing state of knowledge. In particular analytic and dimensional renormalization, normal product technique, and the Bogoliubov-Shirkov-Epstein-Glaser method are treated, with applications to physically important gauge theories. All the preceding deals with perturbative renormalization theory. In recent years there has been an interesting development of non-perturbative renormalization theory in models in space-times of two and three dimensions, with the use of the methods of constructive field theory. Despite the simplicity of these models, the results are of significance because they are exact and answer a number of questions of principle. There are parts of renormalization theory which are not well understood, for instance the renormalization theory of non-renormalizable interactions. The potential physical applications of these interactions are of such importance that the School would not have been complete without some discussion of this topic.

The Editors hope that the book will be useful both to beginners and to those with a long-time interest in renormalization theory.

The Editors wish to thank Ms. S. Bragaglia for tireless and efficient management of the School.

ORIENTATION

A. S. Wightman

Princeton University, Princeton, N.J., U.S.A.

The purpose of this orientation is to provide an introduction to the detailed talks that follow. Renormalization theory is a notoriously complicated and technical subject. As a result it is easy to lose sight of the main themes in the welter of details. In the hope of making the theory more accessible to the earnest student, I shall not flinch from repeating things that all experts know. Furthermore, I hope and expect that many of the points I make will be repeated later; I have chosen them because I believe they are worth repeating.

Gell-Mann-Low Formula and Its Expression as a Functional Integral

For the purposes of renormalization theory, the starting point of Lagrangian field theory is the formula of Gell-Mann and Low expressing the expectation value of a time-ordered product of Heisenberg picture field operators in the physical vacuum, in terms of a ratio of interaction picture vacuum expectation values [1].

$$(\Psi_0, (\prod_j A_j(x_j))_+ \Psi_0) =$$

$$= \frac{(\Phi_0, (\exp[i \int \mathcal{L}_I(\xi) d^4\xi] A_{I1}(x_1) \ldots A_{In}(x_n))_+ \Phi_0)}{(\Phi_0, (\exp[i \int \mathcal{L}_I(\xi) d^4\xi])_+ \Phi_0)} \quad (1)$$

Ψ_0 physical vacuum $\quad A_j$ Heisenberg picture field

Φ_0 interaction picture vacuum $\quad A_{Ij}$ interaction picture field

$(\quad)_+$ time ordered product

G. Velo and A. S. Wightman (eds.), Renormalization Theory, 1-24.

If the exponentials in the numerator and denominator are expanded in series, this expression takes the form from which all perturbative renormalization theory starts

$$(\Psi_0,(\prod_j A_j(x_j))_+\Psi_0) = \frac{\sum_{n=0}^{\infty} \frac{(i)^n}{n!} \int\ldots\int d^4\xi_1\ldots d^4\xi_n(\Phi_0,(\mathcal{L}_I(\xi_1)\ldots\mathcal{L}_I(\xi_n) A_{I1}(x_1)\ldots A_{In}(x_n))_+\Phi_0)}{\sum_{n=0}^{\infty} \frac{(i)^n}{n!} \int\ldots\int d^4\xi_1\ldots d^4\xi_n(\Phi_0,(\mathcal{L}_I(\xi_1)\ldots\mathcal{L}_I(\xi_n))_+\Phi_0)} \quad (2)$$

More than half of what follows will deal with the study of (2) and its transformation into a renormalized perturbation series. But, to begin with, I will follow the development of the non-perturbative form of the theory. Here an alternative expression for (1) in terms of functional integrals , first written down in print by Matthews and Salam [2] following ideas of Feynman, plays a decisive role.

$$(\Psi_0,(\prod_j A_j(x_j))_+\Psi_0) = \frac{\int \prod_j \mathcal{D}A_j \exp[i\int d^4\xi\mathcal{L}(\xi)]\prod_{j=1}^{n} A_j(x_j)}{\int \prod_j \mathcal{D}A_j \exp[i\int d^4\xi\mathcal{L}(\xi)]} \quad (3)$$

where the functional integrals are over all classical field histories, and $\mathcal{L}$ is the Lagrangean of the theory. The fermion classical fields were supposed to be anti-commuting among themselves and commuting with the boson classical fields making the integral even more exotic than it would be in pure boson case in which all classical fields commute. If one doesn't look too closely at the formula, it is fantastic: the solution of non-trivial Lagrangean field theories reduced to quadratures! Much of the recent work on non-perturbative renormalization theory described in the following by Seiler, Feldman, and Fröhlich can be regarded as progress in making a version of (3) that can really be taken literally and used to prove properties of solutions.

Historically, the first serious attempts to make the functional integrals in (3) respectable resulted in setbacks, instructive setbacks however, and hence worth being described. The enthusiasm aroused by formulae such as (3) is evident in the review article <u>Integration in Functional Space and Its Applications in Quantum Physics</u> of I.M. Gelfand and A.M. Yaglom [3] (translated in the first volume (1960) of the Journal of Mathematical Physics although it dates from 1955). Gelfand and Yaglom describe at some length the parallelism first developed by M. Kac between Feynman's expression for the time development of the wave function according to the Schrödinger equation and Wiener's expression for time development according to the diffusion equation.

Formally, the difference between the two is just a factor of i in the exponential:

$$\int \exp[- \frac{1}{4D} \int_{t_0}^{t_1} [\frac{dx(\tau)}{d\tau}]^2 d\tau] \mathcal{D}x \qquad (4)$$

for the diffusion equation

$$\frac{\partial\psi}{\partial t} = D \frac{\partial^2\psi}{\partial x^2} ,$$

$$\int \exp[\frac{im}{2h} \int_{t_0}^{t_1} [\frac{dx}{d\tau}]^2 d\tau] \mathcal{D}x \qquad (5)$$

for the Schrödinger equation

$$i\hbar \frac{\partial\psi}{\partial t} = \frac{\hbar^2\partial^2}{2m\partial x^2} \psi .$$

Thus, going from Schrödinger to the diffusion equation is equivalent to replacing $-i\hbar/2m$ by D. They made clear how natural it is to extend formulae of this type to quantum field theory and thereby to arrive at equation (3).

To make the Feynman history integral rigorous, Gelfand and Yaglom proposed to regard it as the limit of integrals in which Planck's constant h has a negative imaginary part, $-i\delta$. They argued that this gives rise to a complex measure on paths: ..."It is natural that such a complex measure for arbitrary $m > 0$, $\delta > 0$ ... will be just as 'good' as Wiener measure, i.e. it will have just as precise a meaning as measure in the space of continuous functions, and it will allow integration over it of a wide class of functionals including all continuous and bounded functionals. ..." It turned out, alas, that this statement is wrong. It was shown by Cameron [4] that this proposal defines a completely additive complex measure only when $h = 0$, $\delta > 0$, i.e. for the case considered by Wiener. From a practical point of view, this means that one does not have available all the powerful analytical devices of the theory of integration. Instead, one must fashion one's mathematical tools for dealing with these expressions as one goes along. There are those who worked hard along such lines, C. DeWitt [5], for example, but the mainstream of development has taken a different direction. See also [6]. Of course, the 1950's were not without their advances. The beautiful theory of integration on the dual spaces of nuclear spaces [7] probably may be regarded as partly inspired by efforts to make sense of the Feynman history integral.

Euclidean Field Theory and Statistical Mechanics

One of the main preoccupations of the second half of the 1950's was the analyticity of the vacuum expectation values of operator products in space-time variables and of their Fourier transforms in momentum variables. A by-product of this work was the fact that the vacuum expectation value $(\Psi_0, \prod_j A_j(x_j)\Psi_0)$ can be continued analytically to points at which the times are purely imaginary: $i\vec{x}_{10}\ldots i\vec{x}_{n0}$, and the space components real: $\vec{x}_1\ldots\vec{x}_n$, provided that no two points coincide: $(x_{j0}-x_{k0})^2 + (\vec{x}_j-\vec{x}_k) \neq 0$ for $j \neq k$. Such points are called Schwinger points and the corresponding analytically continued vacuum expectation values, Schwinger functions. Schwinger and Nakano pointed out that in Lagrangean field theories the Schwinger functions satisfy equations invariant under the proper Euclidean group in four dimensions, ($\equiv$ inhomogeneous SO(4)), and discussed the possibility of constructing Euclidean field operators whose vacuum expectation values are the Schwinger functions (This requires specification of the singularities at points of coincidence.) [8][9].

Symanzik took up the study of this Euclidean field theory, and almost single-handedly carried it to the stage at which it was ready as a prime tool of constructive quantum field theory [10]. In particular, he recognized that (at least for Bose Fields) the Euclidean version of the formula (3) for vacuum expectation values

$$S(x_1\ldots x_k) = \frac{\int (\prod_j \mathscr{D}A_j)\exp[\int \mathscr{L}(x)d^4x]\prod_{j=1}^{k} A_j(x_j)}{\int (\prod_j \mathscr{D}A_j)\exp[\int \mathscr{L}(x)d^4x]} \qquad (6)$$

is not only analogous to the formula for the correlation function in classical statistical mechanics. It is the correlation function of a certain singular classical system with the denominator playing the role of partition function. Furthermore, although the integrals in (6) are still purely formal as they stand, they are higher dimensional formal analogues of the Wiener integral and therefore provide a possible solution to the difficulties associated with the Feynman history integral mentioned above: it is the Schwinger functions which should be expressible as moments of a measure on the function space of classical fields.

It is, at first sight, surprising that this family of ideas embodied in the work of Symanzik did not arouse greater enthusiasm. Let me venture the following explanation. First, it was only in the early 1960's that the statistical mechanicians finally decided that to establish a sound foundation for their subject, a serious study of the thermodynamic limit was necessary. It was not that previous work did not show an understanding of the role of the thermodynamic limit. What had to be overcome was the psychological barrier created by a tradition in which one evaded any frontal attack on the existence and properties of the limit. A decade of work has resulted in a deep and still evolving theory of the thermodynamic limit, with some completely new methods, for example, the method of correlation inequalities. The moral of this is that to make sense of the formula (6) for the Schwinger functions one must expect to have to introduce some kind of box cutoff and boundary conditions on the integration variables A . Then, provided the other difficulties with (6), such as ultraviolet divergences, are somehow taken care of one must expect to have to cope with at least the same general difficulties as occur in statistical mechanics: both numerator and denominator of (6) diverge as the box becomes large and one has to prove that the ratio approaches a finite limit.

The second part of my explanation is that those working on the foundations of quantum field theory were still pursuing general theorems in axiomatic field theory and although Symanzik could show that every quantum field theory satisfying the axioms has a set of Schwinger functions which are candidates for a Euclidean field theory, he did not have simple conditions on a set of Schwinger functions which would guarantee that they come from an acceptable quantum field theory in Minkowski space. Thus, the general thinkers did not have the satisfaction of knowing that the existence of a non-trivial Euclidean field theory is equivalent to the existence of a non-trivial relativistic quantum field theory in Minkowski space. This situation changed radically when Nelson introduced his theory of Markoff fields [11]. A Markoff field is a Euclidean field with the Markoff property in Euclidean n-dimensional space, $\mathbb{R}^n$. A stochastic process in one dimension has the Markoff property if the history of a random variable up to time t yields no more information about what will happen after than is provided by the knowledge of the random variable <u>at</u> time t . Similarly, in $\mathbb{R}^n$ a stochastic process has the Markoff property if for every open set $\mathcal{O}$, in $\mathbb{R}^n$ knowledge of a random variable in the complement, $\mathcal{O}^c$, of $\mathcal{O}$ yields no more information about the values of a random variable in $\mathcal{O}$ than that obtained from knowledge of the random variable on the boundary of the open set, $\partial\mathcal{O}$. Supplementing the Markoff property with a few other hypotheses, Nelson showed how to recover a uniquely determined quantum field theory in Minkowski space satisfying the usual axioms.

The third and last part of my explanation is that when, in the middle 1960's, the attention of quantum field theorists turned to the construction of solutions of concrete Lagrangean field theories, the remarkable successes initially obtained were mainly based on Hamiltonian field theory rather than on the ideas of Euclidean field theory [12][13][14][15]. It was not until the 1970's that the situation changed.

Two papers, in my opinion, galvanized the constructivists to attention: one by F. Guerra [16] in which the vacuum energy per unit volume in a $P(\phi)_2$ field theory is shown to exist as a simple consequence of Nelson's symmetry (which in turn is an easy consequence of a formula of Euclidean field theory), and the basic work by Osterwalder and Schrader [17] giving necessary and sufficient (as well as convenient and sufficient) conditions on a set of Schwinger functions guaranteeing that they arise from a set of Minkowski space vacuum expectation values satisfying the axioms. As a result of their work, it was clear that in solving relativistic quantum field theories, one may as well work with the corresponding Euclidean field theories.

The above skimpy account of the slow percolation of Euclidean field theory into the minds of students of quantum field theory does not do full justice to the complex history involved. In particular, the hard estimates at the heart of some of the early papers on constructive quantum field theory by Glimm and Jaffe and Nelson are obtained by what nowadays would be called Euclidean techniques. The reader is therefore referred to the introductions of Osterwalder's lectures [18] and Simon's book [19] for more details and more complete references.

Non-perturbative Renormalization Theory In Terms of Euclidean Functional Integrals

In making sense of the Euclidean functional integral (6), it is natural to begin with the free field. For a single scalar free field A in n-dimensional (Euclidean) space-time, (6) reduces to

$$S(x_1 \dots x_n) = \frac{\int \mathcal{D}A \exp\{-\int 1/2[(\nabla A)^2 + m^2A^2]d^nx\} \prod_{j=1} A(x_j)}{\int \mathcal{D}A \exp\{-\int 1/2[(\nabla A)^2 + m^2A^2]d^nx\}} \tag{7}$$

$$= [x_1 \dots x_n]$$

where $[x_1 \dots x_n]$ is the hafnian defined recursively, starting from the two-point symbol by

$$[1\ldots n] = \sum_{j=2}^{n} [1j]\,[\hat{1}\ldots\hat{j}\ldots n] \tag{8}$$

The two-point symbol is

$$[x_j x_k] = S(x_j, x_k) = \frac{1}{(2\pi)^{n/2}} \int d^n k \, \frac{\exp ik\cdot(x_j - x_k)}{[m^2 + k^2]} \tag{9}$$

In probability theory, a precise mathematical meaning is given to the functional integral (7) in terms of the unique Gaussian stochastic process whose expectation value is zero and whose covariance is the two-point function (9). With this definition the formal expression

$$d\mu_0(A) = \exp\{-1/2 \int [(\nabla A)^2 + m^2 A^2] d^n x\} \mathscr{D}A \tag{10}$$

becomes a finite (i.e. $\int d\mu_0(A) < \infty$) measure on a space which can, for convenience, be taken as the space, $\mathscr{S}'$, of tempered distributions. There are good accounts of this construction available in the mathematical literature [20][21], and even one specially prepared for the needs of students of quantum field theory [22], so I will say no more about it, but rather will discuss a few of its consequences.

The first of these has to do with the properties of typical A. (Of course, by the statement "a typical A has property P", I mean "all A's have property P, with the possible exception of a set of A having μ_0-measure zero".) Is the typical A differentiable? continuous ? square integrable? At first sight one might guess from the formal expression (10), that the set of A with

$$\int [(\nabla A)^2 + m^2 A^2] d^n x < \infty \tag{11}$$

would have probability 1 , so a typical A would be square integrable and have square integrable first derivatives. That is false and reconciling oneself to the actual behavior of typical A which is much more singular, is the first hurdle one has to surmount in learning about functional integrals in Euclidean field theory. The easy way to remember what happens is to note that if the Dirichlet form (11) were finite μ_0-almost everywhere, then $\mathscr{D}A$ would exist and be an infinite dimensional analogue of Lebesgue measure, absolutely continuous with respect to μ_0 . Since no infinite dimensional analogue of Lebesgue measure exists, the exponential factor in (10) must be ill-defined and (10) cannot be taken literally. Instead it must be regarded as a formal indication of the construction of a Gaussian measure. As a first step toward a more precise understanding, it is useful to recall that Wiener's treatment of Brownian motion in one dimension is realized on a space of continuous particle paths, but predicts that with probability 1, the paths will fail to have derivatives. Equation (10) reduces to the case considered by Wiener for $n = 1, m = 0$. Although the irregularity of the paths has been known for Brownian motion for more than fifty years the analogous statements for $n \geq 2$ seem to have been worked out precisely only recently for the purposes of constructive field theory [23][24][25][26].

The information available can be summarized as follows.

Irregularity Theorem

For (space time) dimension $n \geq 2$, a typical A is nowhere a signed measure.

Remark A tempered distribution A on $\mathbb{R}^n$ is a signed measure at a point, y, if there exists an open neighborhood of y, U, and a signed measure, μ, on $\mathbb{R}^n$ such that for all infinitely differentiable ϕ with supports in U

$$\langle A,\phi\rangle = \int \phi(x)d\mu(x)$$

Even though a typical A is locally rough, it only requires a little smoothing to become a square integrable function. More precisely,

L^2 Regularity Theorem

Let $P_1 = (-\Delta_{n-1} + 1)^{1/2}$ and $Q = (2 + x^2)^{1/2}$ where Δ_{n-1} is the Laplacean in any $n-1$ of the variables of $\mathbb{R}^n$. Then the set

$$S = P_1^{\,n/2-1+\alpha}\, Q^{n/2}(\log Q)^{\beta}\, L^2(\mathbb{R}^n) \tag{12}$$

has probability 1 for any $\alpha > 0$ and $\beta > 1/2$.

S has probability zero for any $\alpha > 0$ and $\beta \leq 1/2$.

Thus, a typical A needs only to be smoothed in $d - 1$ dimensions by

$$[-\Delta_{n-1} + 1]^{-n/2+1-\alpha} \tag{13}$$

with α arbitrarily small, to become a locally square integrable function. The factor $[2+x^2]^{-n/2}\,[\log[2+x^2]^{1/2}]^{-\beta}$ suffices for any $\beta > 1/2$ to make it square integrable at infinity.

There is a local version of this in terms of sup norms rather than L^2 norms of which I quote a part

Local Regularity Theorem

Let $0 < \alpha \leq 1/2$ then the set of A for which

$$P_1^{1-d/2\ \alpha}\ A \tag{14}$$

is a locally Hölder continuous function for all exponents $\alpha' < \alpha$ has probability 1 .

Thus smoothing with $P_1^{1-d/2-\alpha}$, with α no matter how small, suffices to make the typical A continuous.

These statements do not look much like what one finds in the physics literature on functional integration but they are facts of life that have to be faced if one is to make sense of (7). The roughness of typical A (according to the measure μ_0) determines whether the integral is well defined or not, and if not what kind of dependence on the ultraviolet cutoff the cutoff integrand will have. The theorems have straightforward generalizations to the case of several Bose fields.

My second remark about the Euclidean free field theory concerns the relation between problems of renormalizability with and without the use of perturbation theory, when interaction is introduced. The standard approach to this is to introduce cutoffs and cutoff-dependent counterterms in such a way that the functional integral is well defined and, in the limit as the cutoffs are removed, converges. This amounts to replacing the exponential

$$\exp[\int d^nx\, \mathcal{L}_I(A(x))]$$

by

$$\exp[\int d^nx\ g(x)\mathcal{L}_I^{(K)}(A^{(K)}(x))]$$

where $A^{(K)}$ is the field A with its high frequency components amputated $\mathcal{L}_I^{(K)}$ is $\mathcal{L}_I$ with the inclusion of counterterms and g is a function of compact support on $\mathbb{R}^n$ describing the Euclidean space-time box (not to be confused with the analogous Minkowski space-time box).

If one examines the proofs of convergence as $K \to \infty$, discussed below one sees that they deal only indirectly with the properties of regularity and irregularity of the typical Euclidean field described above. It seems technically more convenient instead to control the cutoff dependence of integrals over the cutoff measures; such a technique also seems more likely to be generalizable to renormalizable but not superrenormalizable theories.

In perturbative renormalization theory the criterion of renormalizability is obtained by expanding the exponential in its power series and examining the convergence of the terms of the resulting series. For each term the functional integral can be done explicitly and yields the familiar expressions of perturbation theory. On the other hand, in a non-perturbative approach the functional integral cannot, in general, be done explicitly and the criteria of renormalizability must be extracted from the mathematical properties of the integrand itself. At first sight, it looks hopeless to see in the unexpanded expression the subtle cancellations of cutoff dependent terms that forty years of work have taught us to make in the perturbative expansion. Nevertheless, it is one of the achievements of constructive quantum field theory to carry it through in several models as you will hear from E. Seiler, J. Feldman, and J.Fröhlich. I will preface what they have to say only with some general remarks about strategy.

It seems sensible to separate the ultraviolet divergence problem from the thermodynamic limit problem. That means removing the ultraviolet cutoff first and then letting the box become large: $g \to 1$. This procedure maximizes the parallelism with non-relativistic statistical mechanics where there is no analogue of the first step. Experience with models in constructive field theory is so far consistent with this procedure.

As far as ultraviolet divergences are concerned, the three most thoroughly studied cases are $P(\phi)_2$, Y_2, and ϕ^4_3. According to perturbation theoretic analysis, there are no divergences in $P(\phi)_2$ provided one remembers to Wick order the powers of the field. (In the Euclidean world, Wick ordering is usually defined not as an "ordering" but as a subtraction process involving vacuum expectation values of lower degree products of fields. Nevertheless, by custom it is still called Wick ordering, since by the introduction of appropriate annihilation and creation operators it can also be defined by ordering.) Thus, the candidate for the exponential in the integrand of (7) in the limit of no ultraviolet cutoff is

$$\exp[- \int g(x) :P(A(x)): d^2x] \qquad (15)$$

The new feature of the situation as compared with perturbation theory is that this function is unbounded in spite of the requirements $g(x) \geq 0$ and $P(\xi) \geq 0$, because the Wick ordering has the effect of making $:P(A(x)):$ unbounded below. Thus even if one has verified that (15) exists for μ_0 almost all A, one has the non-trivial task of proving it integrable i.e. proving

$$Z(g) = \int d\mu_0(A)\exp[-\int g(x):P(A(x)):d^2x] < \infty \qquad (16)$$

I refer you to Nelson's Erice lectures of 1973 for proofs of these statements.

I should also remind you that there is a class of models involving entire functions for which Nelson's argument is unnecessary because $-:\mathcal{L}_I(A(x)):$ is automatically bounded below. Notable among these is Albeverio and Hoegh-Krohn's exponential model [27].

The next most complicated case is Y_2, the Yukawa interaction of a two-component fermion field with a scalar or pseudoscalar field, which will be discussed here by Erhard Seiler. The first complication is the necessity of a Euclidean theory of fermions. Osterwalder and Schrader [28] gave a workable solution of that problem. If the solution has not aroused universal enthusiasm, it is probably because no one has seen how to use the resulting formalism to make direct estimates of the behaviour of the Schwinger functions. What can be done and was done by Seiler [29], is to follow the Euclidean version of a suggestion of Matthews and Salam [1], and carry out the functional integral over the fermion variables. One is then left with a functional integral over the boson fields. Explicitly, the (still formal!) expression for the Schwinger function of n fermion fields, n adjoints of fermion fields, and ℓ boson fields

$$S(x_1\ldots x_n, y_1\ldots y_n; z_1\ldots z_\ell) =$$

$$[Z(g)]^{-1}\int d\mu_0(A)\det_{rs}\{S(x_r,y_s;A)\}\prod_{t=1}^{\ell}A(z_t)\det_2(1-K) \qquad (17)$$

where

$$Z(g) = \int d\mu_0(A)\det_2(1-K(A)) \qquad (18)$$

and K is the integral operator whose kernel is

$$K(x,y;A) = S(x,y)\Gamma A(y)g(y) \qquad (19)$$

$\Gamma = 1$ or γ_5 and $S(x_r,y_s;A)$ is the Schwinger function of the fermion field in the presence of the field A . By definition

$$\det_2(1-K) = \det((1-K)\exp K) \tag{20}$$

is the ordinary Fredholm determinant with all terms containing tr K deleted.

The remarkable feature of this model is that the ultraviolet divergences are all contained in the determinant, and can be explicitly separated and cancelled. To see this one may write the determinant as an exponential and expand the exponent as a series in powers of K .

$$\ell n \det_2(1-K) = \sum_{r=2}^{\infty} \frac{1}{r} \operatorname{tr} K^r \tag{21}$$

The $r^{\underline{th}}$ term of the sum corresponds to the Feymman diagram

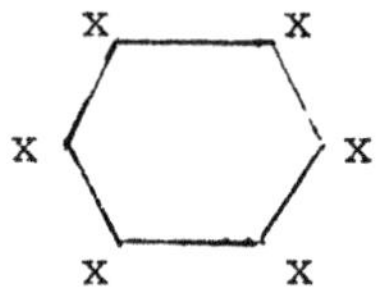

r vertices

Only the trace tr K =  diverges and it can be rearranged

 = :[ $_{\text{ren}}$]: +  + $(\delta m^2):A^2:$ (22)

The last two terms are precisely of the form that will be cancelled if vacuum energy and mass renormalization counterterms are added to the Lagrangean. The resulting expression can be regarded as defining the renormalized determinant

$$\det_{\text{ren}}(1-K)$$

It is non-trivial to show that this quantity is non-negative and integrable but those are among Seiler's results. Thus, in this model the process of renormalization of the ultraviolet divergences can be regarded as giving rise to a new boson measure, $d\mu$, absolutely continuous with respect to $d\mu_0$.

$$d\mu(A) = \det_{\text{ren}}(1-K(A))d\mu_0(A) \tag{23}$$

Replacing $d\mu_0$ by $d\mu$ corresponds to including the effects of virtual fermion anti-fermion pairs.

To pass to the thermodynamic limit in the ultraviolet renormalized functional integral requires new information about the properties of the renormalized determinant, as you will hear from Seiler.

The models $P(\phi)_2$ and Y_2 have one feature in common which they do not share with $(\phi^4)_3$. The process of renormalization gives rise to a new measure absolutely continuous with respect to μ_0,

$$\exp[-\int g(x)d^2x:P(A(x):]d\mu_0(A) \tag{24}$$

and

$$\det_{ren}(1-K(A))d\mu_0(A) \tag{25}$$

respectively, On the other hand, as you will hear from Joel Feldman

$$\begin{aligned}\exp[-\lambda \int d^3x\ g(x):A^{(K)4}:(x) \\ - \int d^3x:A^{(K)2}:(x)\delta m_K^2(\lambda,x) \\ - E_{0K}(\lambda,g)]\end{aligned} \tag{26}$$

does <u>not</u> converge μ_0-almost everywhere. There is a limiting measure μ but the convergence takes place only in the weak sense

$$\int f(A)d\mu^{(K)}(A) \rightarrow \int f(A)d\mu(A) \tag{27}$$

for a certain set of continuous functions, f. Here $d\mu^{(K)}$ is (26) multiplied by $d\mu_0$. The following simple argument which I learned from I. Herbst shows why this happens [30].

Suppose the function (26) converges μ_0-almost everywhere as $K \rightarrow \infty$ for $\lambda = \lambda_1$, for $\lambda = \lambda_1 + \delta\lambda$ and for $\lambda = \delta\lambda$, and that there is a subset B of non-zero μ_0-measure where the limits are non-vanishing. Then

$$\begin{aligned}\frac{d\mu^{(K)}_{\lambda_1+\delta\lambda}(A)}{d\mu^{(K)}_{\lambda_1}(A)} &= \exp\{[-\delta\lambda \int :A^{(K)4}:(x)g(x)d^3x] \\ &- [\int ((\delta m^2)_K(\lambda_1+\delta\lambda) - (\delta m^2)_K(\lambda_1)) :A^2:(x)g(x)d^3x] \\ &- [E_{0K}(\lambda_1+\delta\lambda) - E_{0K}(\lambda_1)]\} =\end{aligned}$$

$$= \frac{d\mu_{\delta\lambda}^{(K)}(A)}{d\mu_0(A)} \exp\{-\int [(\delta m^2)_K(\lambda_1+\delta\lambda) - (\delta m^2)_K(\lambda_1) - (\delta m^2)_K(\delta\lambda)] :A^2:(x)g(x)d^3x - [E_{0K}(\lambda_1+\delta\lambda) - E_{0K}(\lambda_1) - E_{0K}(\delta\lambda)]\} \quad (28)$$

for A in B. Now both of the last two square brackets diverge as $K \to \infty$, because $\delta m^2(\lambda)$ is quadratic in λ while $E_{0K}(\lambda)$ contains both quadratic and cubic terms in λ. Thus we have a contradiction. (The analogous expressions are finite in $P(\phi)_2$. In Y_2 the relevant coupling constant is the square of the Yukawa coupling constant and both $(\delta m^2)_K$ and E_{0K} are linear, so the terms cancel.

The moral of this story is that the absolute continuity of $d\mu$ with respect to $d\mu_0$ which holds for $P(\phi)_2$ and Y_2 is a special gift which should not be expected in general models.

That $d\mu^{(K)} \to d\mu$ in the above-described weak sense is proved by one of the most extraordinarily difficult and subtle arguments ever invented in quantum field theory [31]. You will hear more about it from Feldman.

Before leaving Herbst's argument I would like to make one last point about it. I have sketched it for the case in which $g(x)$ is of compact support so that interaction takes place only in a finite volume. In the thermodynamic limit, $g \to 1$, and the ground state energies $E_{0K}(\lambda)$ become proportional to the volume of the support of g. Thus, even if the mass renormalizations $(\delta m^2)_K(\lambda)$ have finite limits as $K \to \infty$, the right-hand side of (28) will not have a limit different from 0 or ∞ unless the energy per unit volume is a linear function of the coupling constant. Here is yet another reason for the inequivalence of interacting measures. It occurs even in the simplest cases [32].

As J. Fröhlich will relate in his lectures, there have been other remarkable developments in the theory of quantum field theory models in two-dimensional space-time which are interesting both in themselves and for the lessons they present for higher dimensional theories.

First, S. Coleman [33] discovered a connection between the solutions of the massless sine-Gordon equation in two-dimensional space-time i.e. the theory of a neutral scalar field, ϕ, satisfying

$$\Box\phi(x) + \frac{\alpha_0}{\beta} :\sin \beta\phi(x): = 0$$

and the massive Thirring model, the theory of a two-component field satisfying the anti-commutation relations and the wave equation

$$(i\gamma^\mu\partial_\mu - m)\psi = gj^\mu\gamma_\mu\psi$$

where

$$j^\mu(x) = :\psi^+\gamma^\mu\psi: \quad .$$

The connection is

$$:\psi^+(x)\psi(x): = -\frac{\alpha_0}{\beta^2 m} : \cos \beta\phi:$$

$$:\psi^+(x)\gamma^\mu\psi(x): = -\frac{\beta}{2\pi}\varepsilon^{\mu\nu}\partial_\nu\phi$$

$$\frac{4\pi}{\beta^2} = 1 + g/\pi$$

The model illustrates an aspect of relativistic local quantum theory whose significance was first brought out by Haag and Kastler [34] over a decade ago: the occurrence of superselection sectors arising from inequivalent representations of the quasi-local algebra of observables. For the sine-Gordon equation functions of the field ϕ applied on the vacuum sector generate the vacuum sector. However, the resulting Hilbert space of states cannot define a complete theory, because in the vacuum sector there occur fermion anti-fermion pairs whose full description requires the introduction of sectors containing arbitrary numbers of fermions and anti-fermions. In those other sectors the representations of the quasi-local algebra of observables generated by the field ϕ are not unitarily equivalent to the representation occurring in the vacuum sector. It is really remarkable that this phenomenon of the general theory should be so neatly illustrated in Lagrangean field theory.

Coleman's discovery is intimately related to a series of new results on the significance for quantum field theory of solitary wave solutions of classical non-linear wave equations: solitons. I will not attempt to describe these developments but instead refer you to Fröhlich's lectures. Looked at from the perspective of the agonizingly long period during which our principal guide to the predictions of Lagrangean field theories was renormalized perturbation theory, these new non-perturbative insights are the first indications of what must be an immense richness of structure in the solutions.

What the moral of these developments is for renormalization theory is at the moment unclear, and worth investigating for the light it might throw on the next logical step in non-perturbative renormalization theory: the treatment of renormalizable but not super-renormalizable theories.

Perturbative Renormalization Theory: Natural History

One of the first steps in natural history is to establish a classificatory nomenclature. I will do this for perturbative renormalization theory, but in so doing, I want to tell stories with a moral for the earnest student: Renormalization theory has a history of egregious errors by distinguished savants. It has a justified reputation for perversity; a method that works up to 13th order in the perturbation series fails in the 14th order. Arguments that sound plausible often dissolve into mush when examined closely. The worst that can happen often happens. The prudent student would do well to distinguish sharply between what has been proved and what has been made plausible, and in general he should watch out!

My first cautionary tale has to do with the early days of renormalization theory. When F.J. Dyson analyzed the renormalization theory of the S-matrix for the quantum electrodynamics of spin one-half particles in his two great papers of 1948-9 [35][36], he laid the foundations for most later work on the subject, but his treatment of one phenomenon, overlapping divergences was incomplete. Among the methods offered to clarify the situation, that of J. Ward [37] seemed outstandingly simple, so much so that it was adopted in Jauch and Rohrlich's standard textbook [38]. Several years later Mills and Yang noticed that unless further refinements are introduced the method does not work for the photon self energy [39]. The lowest order for which the trouble manifests itself is the fourteenth e.g. in the graph

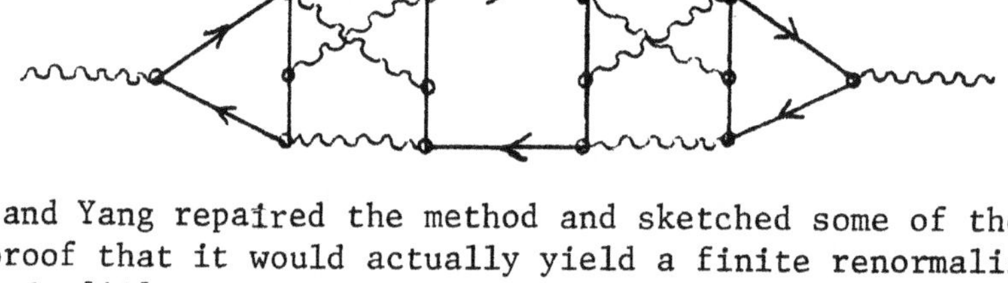

Mills and Yang repaired the method and sketched some of the steps in a proof that it would actually yield a finite renormalized amplitude [40]. An innocent, reading the textbook of Jauch and Rohrlich, would never suspect such refinements are necessary.

Another attempt to cope with the overlapping divergences was made by Salam [41]. I will not describe it, if for no other reason than that I never have succeeded in understanding it. Salam and Matthews commenting on this and related work somewhat later [42], remarked "...The difficulty, as in all this work, is to find a notation which is both concise and intelligible to at least two people of whom one may be the author". The belief is widespread that when Salam's work is combined with later significant work by S. Weinberg [43], the result should be a mathematically coherent version of renormalization theory. At least that is what one reads in the text book of Bjorken and Drell for quantum electrodynamics [44], and in the work of R. Johnson [45] and the lectures of K. Symanzik for meson theories [46], so apparently the Matthews-Salam criterion has been satisfied. I only wish they had spelled it out a little for the peasants.

Another foundation for renormalization theory with a rather different starting point was put forward by Stueckelberg and Green [47]. It was refounded and brought to a certain stage of completion in the standard text book of Bogoliubov and Shirkov [48]. The mathematical nut that had to be cracked is in the paper of Bogoliubov and Parasiuk [49] (amazingly, not quoted in the English translation of Bogoliubov and Shirkov). This paper introduces a systematic combinatorial and analytic scheme for overcoming the overlapping divergence problem. This paper is very important for later developments. Unfortunately, it was found by K. Hepp [50] that Theorem 4 of the paper is false, and that consequently the proof of the main result is incomplete as it stands. However, Hepp showed that theorem 4 is not essential to derive the main result and he could fill all the gaps. Thus, it is appropriate to introduce the initials BPH to stand for the renormalization method described in [49] and [50]. So far as I know it was the first version of renormalization theory on a mathematically sound basis.

An alternative method is ANALYTIC renormalization. The ideas here go back to Marcel Riesz' generalization of the Riemann-Liouville definition of fractional differentiation to treat the fundamental solutions of the relativistic wave equation [51]. Riesz proposed to his physicist colleagues that they try to use his technique to obtain finite answers in the quantum theory of fields. After some initial successes, it was found that the method did not work [52]. It was pointed out by G. Källén that if instead of trying to use the method to calculate the values of quantities like self-energies one calculates Green's functions, the Riesz method should yield a convenient method of doing renormalization calculations. Although this was illustrated by a calculation of lowest-order vacuum polarization [53] no systematic effort was made to deal with more complex situations. The same applies to later work by Bollini , Giambiagi, and Dominguez who took advantage of the elegant mathematical tools offered by the Gelfand treatise on generalized functions to show that analytic renormalization works for bubble graphs [54]. A version of the method for general graphs was worked out by E. Speer [55]. Analytic renormalization so defined was later shown to be equivalent to BPH normalization in a sense that they predict the same answers up to a finite renormalization [56].

It is characteristic of analytic renormalization that it introduces a regularized Feynman amplitude in which every particle propagator is replaced by a regularized propagator depending on a complex variable, λ. The resulting regularized Feynman amplitude is then meromorphic in $\lambda_1 \dots \lambda_L$ where L is the number of lines in the graph associated with the amplitude. On the other hand, in DIMENSIONAL RENORMALIZATION a single complex variable, ν, is introduced which can be interpreted as an interpolation and extension to complex numbers of the integer-valued variable, n, giving the dimension of space-time [57]. Dimensional renormalization is again equivalent to BPH and therefore to analytic renormalization. However, it has the remarkable property that it preserves gauge invariance insofar as that is possible (i.e. modulo the anomalies discussed in the lectures of John Lowenstein and Peter Breitenlohner). Among methods which introduce regularized Feynman amplitudes, this property makes dimensional renormalization the "best".

There are other renormalization methods which do not introduce regularized Feynman amplitudes but instead define the renormalized Feynman amplitude directly. One of these about which you will hear much from John Lowenstein is BPHZ. Here the B, P, and H stand for the same names as before but Z is for Zimmermann. Zimmermann's method has evolved out of an extension of BPH to deal with Green's functions involving not only the basic fields of a Lagrangean theory but also renormalized currents formed from those fields. However, in the process he developed an expression for the renormalized Feynman amplitudes (usually referred to as Zimmermann's forest formula) in which renormalization is an operation that replaces the momentum space integrand by a renormalized integrand without resort to any regularization procedure. A somewhat tricky point is that the counterterms involved are more numerous than in BPH renormalization because a slightly altered definition of the notion of subdiagram is used. BPHZ renormalization has been shown to be a good language in which to discuss gauge theories as you will hear from Lowenstein, Becchi and Stora.

There are two other renormalization methods which have in common with the BPHZ method that no regularization is used:

STEINMANN'S method and the E(pstein) G(laser) method. Steinmann's method is a recursive procedure which produces in addition to renormalized Green's functions renormalized generalized retarded and advanced functions [58]. For lack of time, there will be no lectures on it here. The EG method is somewhat similar in technique to Steinmann's [59]. You will hear about it in detail from H.Epstein. An important application of the EG method to quantum electrodynamics by Blanchard and Seneor [60] will be described by Seneor.

This completes my natural history of perturbative renormalization procedures. The main thing to be remembered about them is that they are all equivalent up to a finite renormalization, so that which one is used is a matter of technical convenience. The convenience or inconvenience involved can be considerable if, as happens, a renormalization procedure does not preserve a desired invariance. The lecture notes contain proofs of equivalence for some pairs of procedures and references to proofs for most of the others. The discussion is based on the elegant axiomatization of renormalization by Hepp [61].

Extension of the Class of Perturbatively Renormalizable Theories; Gauge Theories; Non-Renormalizable Theories

Two of the most important developments of renormalization theory in the last few years have been the treatment of theories with spontaneously broken symmetry and the discovery of classes of renormalizable gauge theories. The general ideas involved have been so extensively discussed that they need no further comment from me [62][63]. You will hear in detail about some important aspects of the subject in the lectures of Becchi and Stora. However, I cannot resist remarking on one irony involved. The underlying mechanism which makes possible the perturbative renormalizability of gauge theories is the cancellation of contributions from different diagrams. The conspiracy of signs that makes these cancellations possible typically involves indefinite metric in Hilbert space and non-physical states. Yet when these features turned up in the Gupta-Bleuler formalism for quantum electrodynamics a quarter of a century ago they were widely regarded as 1) odious, 2) to be avoided at all costs by special choice of gauge. Now they turn out to be keys to such guiding physical ideas as asymptotic freedom and quark confinement.

No general discussion of renormalization theory can be complete without some examination of the status of non-renormalizable theories. The main general question involved, a perennial one, is whether the conventional distinction between renormalizable and non-renormalizable theories is an artifact produced by reliance on the techniques of perturbation theory. As you will hear from Pohlmeyer, Symanzik and he have done exploratory work which seems to indicate a parallelism between infrared problems in $(\phi^4)_{4-\varepsilon}$ and ultraviolet (non-renormalizable) problems in $(\phi^4)_{4+\varepsilon}$. There are preliminary indications that in both cases an appropriate resummation of the perturbation series leads to a renormalized theory with only a finite number of arbitrary parameters. It is an intriguing prospect.

Even from this brief survey, it should be evident that renormalization theory is a subject with plenty of open problems. We are still looking forward to the day when cancellations of infinities for four-dimensional theories can be carried out directly on the non-perturbative expression for the solution of the theories in terms of functional integrals. Nothing has turned up so far that would indicate anything fundamentally wrong with that aspiration.

REFERENCES

[1] M. Gell-Mann and F. Low, Phys. Rev. 84 (1951) 350-4.

[2] P.T. Matthews and A. Salam, Nuovo Cimento 2 (1955) 120-134.

[3] I.M. Gelfand and A.M. Yaglom Integration in Functional Spaces and its Applications in Quantum Physics, Jour. Math. Phys. 1 (1960) 48-69.

[4] R.H. Cameron, Comm. in Pure and Applied Math (1960) 126-140.

[5] C. DeWitt-Morette, Comm. in Math. Phys. 28 (1972) 47-67 and 37 (1974) 63-81.

[6] S. Albeverio and R. Hoegh-Krohn, Mathematical Theory of Feyman Path Integrals Oslo preprint 1974.

[7] I.M. Gelfand and N.Ya. Vilenkin, Generalized Functions, Vol. 4, Academic Press 1964.

[8] T. Nakano, Prog. Theoret Phys. 21 (1959) 241-59.

[9] J. Schwinger, Proc. Nat. Acad. Sci USA 44 (1958) 956-65.

[10] K. Symanzik, Euclidean Quantum Field Theory, pp. 153-226 in Local Quantum Theory, R. Jost Ed. Academic Press N.Y. (1969). This review contains references to earlier papers on Euclidean field theory.

[11] E. Nelson, Jour. Funct. Analys. 12 (1973) 97-112.

[12] E. Nelson, A Quartic Interaction in Two Dimensions pp. 69-73 in Mathematical Theory of Elementary Particles Ed. R. Goodman, I. Segal MIT Press, 1966.

[13] J. Glimm and A. Jaffe, Phys. Rev. 176 (1968) 1945-51.

[14] J. Glimm and A. Jaffe, Ann. Math. 91 (1970) 362-401.

[15] J. Glimm and A. Jaffe, Acta. Math. 125 (1970) 203-261.

[16] F. Guerra, Phys. Rev. Letts. 28 (1972) 1213-14.

[17] K. Osterwalder and R. Schrader, Comm. in Math. Phys. 31 (1973) 83-112 and 42 (1975) 281-305.

[18] K. Osterwalder, Euclidean Green's Functions and Wightman Distributions, pp. 71-93 in Constructive Quantum Field Theory Lecture Notes in Phys. Vol. 25 Springer (1973).

[19] B. Simon The $P(\phi)_2$ Euclidean (Quantum) Field Theory, Princeton Univ. Press (1974).

[20] L. Breiman, Probability, Addison-Wesley, 1968 .

[21] T. Hida, Stationary Stochastic Processes, Princeton University Press, 1970 .

[22] M. Reed, Functional Analysis and Probability Theory pp. 2-93 in Constructive Quantum Field Theory Lecture Notes in Phys. Vol. 25 Springer 1973 .

[23] J. Cannon, Comm. in Math Phys. 35 (1974) 215-33.

[24] P. Collela and O. Lanford, Sample Field Behavior for the Free Markov Random Field pp. 44-70 in Constructive Quantum Field Theory Lecture Notes in Phys. 25 Springer 1973 .

[25] M. Reed and L. Rosen, Comm. in Math Phys. 36 (1974) 123-132.

[26] B. Simon and J. Rosen, Duke Math. Jour. 42 (1975) 51-55.

[27] S. Albeverio and R. Hoegh-Krohn, Jour. Funct. Anal. 16 (1974) 39-82.

[28] K. Osterwalder and R. Schrader, Helv. Phys. Acta. 46 (1973) 277-301.

[29] E. Seiler, Comm. in Math Phys. 42 (1975) 163-182.

[30] I. Herbst, private communication.

[31] J. Glimm and A. Jaffe, Fortschritte der Physik 21 (1973) 327-376.

[32] A. Klein, Zeits. Wahrsch. Verw. Geb. 28 (1974) 323-34.

[33] S. Coleman, Phys. Rev. D11 (1975) 2088-97.

[34] R. Haag and D. Kastler, Jour. Math. Phys. 5 (1964) 848-61.

[35] F.J.Dyson, Phys. Rev. 75 (1949) 486-502.

[36] F.J. Dyson, Phys. Rev. 75 (1949) 1736-55.

[37] J. Ward, Proc. Phys. Soc. Lond. A64 (1951) 54-6.

[38] J.M. Jauch and F. Rohrlich, Theory of Photons and Electrons, Addison Wesley (1955) Chapters 9 and 10.

[39] T.T. Wu, Phys. Rev. 125 (1962) 1436-50 esp. p 1439.

[40] R.L. Mills and C.N. Yang, Prog. Theoret. Phys. Supp. 37 (1966) 507-511.

[41] A. Salam, Phys. Rev. 82 (1951) 217-227 and 84 (1951) 426-431.

[42] P.T. Matthews and A. Salam, Rev. Mod. Phys. 23 (1951) 311-4.

[43] S. Weinberg, Phys. Rev. 118 (1960) 838- 49

[44] J.D. Bjorken and S. Drell, Relativistic Quantum Fields, McGraw Hill (1965) chapter 19.

[45] R. Johnson, J. Math. Phys. 11 (1970) 2161-5,

[46] K. Symanzik, Green's Functions Method and Renormalization of Renormalizable Quantum Field Theories, pp. 485-517 in Lecture on High Energy Physics II, Hercegnovi, 1961, Ed. B. Jakšić.

[47] E.C.G. Stueckelberg and T.A. Green, Helv.Phys. Acta. 24 (1951) 153-174.

[48] N. Bogoliubov and D. Shirkov, Introduction to the Theory of Quantized Fields, New York Interscience, 1959 .

[49] N. Bogoliubov and O. Parasiuk, Acta. Math. 97 (1957) 227-266. See also O. Parasiuk, Ukrainskii Math. Jour. 12 (1960) 287-307.

[50] K. Hepp, Comm. in Math Phys. 2 (1966) 301-26.

[51] M. Riesz, Acta. Math. 81 (1948) 1-223 and earlier references quoted there.

[52] S.B. Nilsson, Arkiv Fysik 1 (1950) 369-423 and earlier work quoted there.

[53] E. Karlson, Arkiv für Fysik 7 (1953) 221-37.

[54] C. Bollini, J. Giambiagi and A. Dominguez, Nuovo Cim. 31 (1964) 550-61.

[55] E. Speer, Generalized Feynman Amplitudes, Annals of Math. Study No. 62 Princeton Press, 1969 .

[56] K. Hepp, Comm. in Math. Phys. 14 (1969) 67-9.

[57] G. 't Hooft and M. Veltman, Nuclear Physics B44 (1972) 189-213.

[58] O. Steinmann, Perturbation Expansions in Axiomatic Field Theory Lecture Notes in Physics 11 J. Springer (1971).

[59] H. Epstein and V. Glaser, Ann. Inst. H. Poincaré 19 (1973) 211-95.

[60] P. Blanchard and R. Seneor, Ann. Inst. H. Poincaré, to appear.

[61] K. Hepp, Renormalization Theory pp. 429-500 in Statistical Mechanics and Quantum Field Theory Ed. C. DeWitt and R. Stora Gordon and Breach, 1971 .

[62] S. Coleman, Secret Symmetry: An Introduction to Spontaneous Symmetry Breakdown and Gauge Fields in Proceedings of the 1973 School Ettore Majorana.

[63] J. Bernstein, Rev. Mod. Phys. 46 (1974) 7-48.

DIMENSIONAL AND ANALYTIC RENORMALIZATION

Eugene R. Speer*

Department of Mathematics
Rutgers University
New Brunswick, New Jersey

I. INTRODUCTION

Our goal in these lectures is to define time ordered vacuum expectation values of products of field polynomials:

$$\langle 0|T(Y_1(x_1) \ldots Y_m(x_m))|0\rangle, \tag{1.1}$$

where $Y_i(x_i)$ is a polynomial in certain (interacting) field operators evaluated at the space time point $x_i \in R^4$. The Gell-Mann Low formula expresses (1.1) as a formal power series whose terms are also time-ordered vacuum expectation values, similar to (1.1), but involving free rather than interacting fields; since we are not interested in the actual convergence of this series, it sufficies to define (1.1) when only free fields occur. To do this, we expand the time ordered product as a finite sum of normal products (Wick's theorem). Correspondingly, (1.1) is expressed as a sum of Feynman amplitudes, each associated with a Feynman graph. These amplitudes are given as divergent integrals over momentum variables, and it is the general task of renormalization theory to assign appropriate finite values (which are in fact distributions in the variables $x_1, \ldots x_m$) to these integrals. The first part of these lectures will be devoted to this problem; once it is solved, we will have expressed (1.1) as a formal power series whose coefficients are well-defined distributions. We will then investigate properties of this power series.

* Partially supported by NSF Grant #MPS74-05783 A01.

G. Velo and A. S. Wightman (eds.), Renormalization Theory, 25-93.

We will discuss two renormalization methods, analytic renormalization and dimensional renormalization. The distinctive idea in both cases is to replace the divergent integral mentioned above by an analytic function of certain regularizing parameters, which we denote for the moment by $z = (z_1, \ldots z_k)$. This regularized amplitude is chosen so that at a certain value z_0 of the parameters it is formally equal to the divergent integral we wish to renormalize. However, the equality is only formal; the regularized amplitude will in fact have a singularity (usually a pole) at z_0. Renormalization is then accomplished by discarding the pole and thus extracting a finite part.

When the amplitude arises from a sufficiently complicated graph, it is impossible to take z to be a single complex variable and simply to discard the singular part at the pole. At this stage there are two alternative procedures possible, exemplified by analytic and dimensional renormalization.

In analytic renormalization [1,2], the amplitude is a function of several regularizing parameters. The analytic structure in these variables is sufficiently rich that a renormalized amplitude may be defined by the application of a universal operator, independent of the graph in question, to the regularized amplitude. This formal simplicity of the method makes it particularly well suited for discussing such properties as unitarity [3], analyticity [4], and the equivalence of various renormalization schemes. There is a variant of dimensional renormalization which uses a similar scheme [5,6], but we will not discuss it in these lectures.

In dimensional renormalization [7,8,9], a single regularizing parameter, the complex dimension, is used. A more complicated subtraction procedure is then necessary; it involves recursive subtractions associated with various subgraphs of the main graph. The method is of value because the renormalized amplitudes maintain certain invariance properties which are formally possessed by the original divergent integrals. In these notes this will be exemplified by the equations of motion studied in Chapter 5.

Before beginning a study of these renormalization methods for a general Feynman graph, we will illustrate them in a simple situation.

Example: Consider the vacuum expectation value

$$\langle 0|T[\phi^2(x_1)\phi^2(x_2)]|0\rangle, \tag{1.2}$$

when $\phi(x)$ is a free scalar field. Wick's theorem applied to (1.2) produces a sum of amplitudes; one of the corresponding graphs is shown in Figure 1.

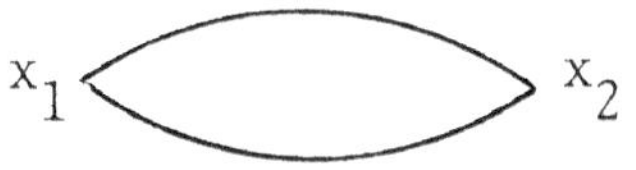

Figure 1

We will shortly give general rules for writing down the amplitude associated with such a graph, but for the moment, we simply state that the amplitude for Fig. 1 is a distribution in x_1, x_2 whose Fourier transform T has the formal expression $(p_1, p_2 \in R^4)$

$$T(p_1,p_2) = \delta(p_1+p_2) \int_{\mathbb{R}^4} \frac{d^4k}{(m^2-k^2)(m^2-(p-k)^2)} . \tag{1.3}$$

Here m is the mass of the particles associated with the field ϕ. (1.3) is formal because the integral is divergent (for large values of k it behaves like $\int d^4k\, |k|^{-4}$).

We are going to modify (1.3) in two ways. First, we assume a world of n, rather than 4, dimensions, so that p_1, p_2, and k all lie in $\mathbb{R}^n$; secondly, we introduce complex powers λ_1, λ_2 for the two factors in the denominator. Thus (ignoring the δ-function) we must study the integral

$$I(\lambda_1,\lambda_2,n) = \int_{\mathbb{R}^n} \frac{d^nk}{(m^2-k^2)^{\lambda_1}(m^2-(p_1-k)^2)^{\lambda_2}} . \tag{1.4}$$

At large k (1.4) looks like $\int \frac{d^nk}{|k|^{2(\lambda_1+\lambda_2)}}$; thus it will converge (ignoring for the moment the vanishing denominators) if $\mathrm{Re}(\lambda_1+\lambda_2) > n/2$. In this region we may combine the denominators by the Feynman formula

$$a^{-\lambda_1} b^{-\lambda_2} = \frac{\Gamma(\lambda_1+\lambda_2)}{\Gamma(\lambda_1)\Gamma(\lambda_2)} \int_0^1 x^{\lambda_1 - 1}(1-x)^{\lambda_2 - 1}(xa+(1-x)b)^{-\lambda_1-\lambda_2} dx,$$

then interchange the order of integration and evaluate the k integral explicity. The result is

$$I(\lambda_1,\lambda_2,n) = \frac{\Gamma(\lambda_1+\lambda_2-n/2)}{\Gamma(\lambda_1)\Gamma(\lambda_2)} \int_0^1 x^{\lambda_1-1}(1-x)^{\lambda_2-1}[m^2-x(1-x)p^2]^{\frac{n}{2}-\lambda_1-\lambda_2} dx \tag{1.5}$$

Observe now that in (1.5) we may consider n as a complex variable rather than an integer. Moreover, the integral in (1.5) is absolutely convergent if $\lambda_1 > 0$, $\lambda_2 > 0$, not just in the original region $\mathrm{Re}(\lambda_1+\lambda_2) > n$; in particular, (1.5) defines I for all complex λ_1,λ_2,n in a neighborhood of the physically relevant point $\lambda_1 = \lambda_2 = 1$, $n = 4$. (Recall the principle of analytic continuation, which guarantees that for integer n (1.5) defines the unique analytic function which agrees with (2.4) for $\mathrm{Re}(\lambda_1+\lambda_2) > n/2$.) However, I is singular at the physical point because the gamma function has a pole at zero.

We now discuss the dimensional and analytic renormalization of this integral. For dimensional renormalization we set $\lambda_1 = \lambda_2 = 1$; (1.5) then defines a function of the complex dimension n alone:

$$I(1,1,n) = \Gamma(2-\tfrac{n}{2}) \int_0^1 [m^2-x(1-x)p^2]^{\frac{n}{2}-2} dx. \tag{1.6}$$

To obtain a renormalized amplitude we expand (1.6) in a Laurent series around the physical point $n = 4$, drop the pole, and take the constant term as our value. Thus $I(1,1,n) =$

$$\{-\frac{2}{n-4} + \gamma + O(n-4)\}\{1 + \frac{n-4}{2}\int_0^1 \log[m^2-x(1-x)p^2]dx + O((n-4)^2)\}$$

and the renormalized amplitude is $-\{\gamma + \int_0^1 \log[m^2-x(1-x)p^2]dx\}$. This is physically acceptable, as we will discuss later.

In the analytic renormalization of (1.5) we set $n = 4$ at the beginning, giving us a function $I(\lambda_1,\lambda_2,4)$ with a pole on the surface $\lambda_1+\lambda_2 = 2$. Again we must discard the pole to obtain a finite value at $\lambda_1 = \lambda_2 = 1$. In general it will not suffice to set $\lambda_1 = \lambda_2 = \lambda$ and extract the finite part of a Laurent series in λ (although this would work in our present example). Instead, we proceed as follows: (i) take the finite part in a Laurent series in λ_1 around the point $\lambda_1 = 1$; (ii) take the finite part of a Laurent series for the result around $\lambda_2 = 1$; (iii) repeat this procedure, taking first λ_2 and then λ_1; (iv) average the two results. This iterated, symmetrized Laurent series prescription will be justified in Chapter 3.

II. REGULARIZATION

A. Integration in Complex Dimension

In the example treated above, the complex dimension was introduced by first calculating the integral in arbitrary integer dimension n, then reinterpreting any occurrence of n in the answer as a complex variable. This procedure may be generalized to an arbitrary Feynman graph [8-11]. In these lectures, however, we will use an extension of the original method of t'Hooft, giving a direct definition of integration in complex dimension.

We are thus led to the following problem. Let $f(k)$ be a function of $k = (k_1,\dots k_m)$ with $k_i = (k_i^0, \vec{k}_i) \in \mathbb{R}^n$, and let f be invariant under orthogonal transformations of the space components $\vec{k}_i$. (We do not require Lorentz invariance since we will need to introduce non-Lorentz invariant quantities at intermediate stages of calculation). We want to define

$$F(k_1,\dots k_M) = \int \dots \int d^\nu k_{M+1} \dots d^\nu k_m\, f(k), \tag{2.1}$$

where ν is a complex dimension. The original dimension n is presumably not relevant for this definition; so we introduce a space of functions similar to f but not associated with any space-time dimension.

Definition 2.2. Let U_m denote the set of positive symmetric semi-definite $m\times m$ real matrices, let $X_m = \mathbb{R}^n \times U_m$, and let $\mathcal{Q}_m$ denote all measurable complex-valued functions on X_m. For

$n \geq 0$, define $\phi_{mn}: (\mathbb{R}^n)^m \to X_m$, $\phi_{mn}(k_1,\dots k_m) = ((k_1^0,\dots k_m^0), \vec{k}_i\cdot\vec{k}_j)$, for $k_i = (k_i^0, \vec{k}) \in \mathbb{R}^n = \mathbb{R}\times\mathbb{R}^{n-1}$.

Remark: If $n \geq m+1$, the map ϕ_{mn} is onto. In this case, if $f \in \mathcal{Q}_m$, then $f \circ \phi_{mn}$ is a function of the m vectors $k_1,\dots k_m$ which is invariant under rotations in the space-like components. This is technically not true if $n \leq m$ since then the image of ϕ_{mn} (i.e. $\{(y,S) \mid \text{rank } S = n-1\}$) has measure zero, and we cannot necessarily restrict f to this set. However, $f \circ \phi_{mn}$ will be well defined if, for example, f is continuous at almost every point of this image, which will generally be true for us. We will frequently abuse notation and write $f(k)$ for $f(\phi_{mn}(k))$, $k \in \mathbb{R}^{nm}$.

Now consider the integral (2.1) in integer dimension n:

$$\int\cdots\int d^n k_{M+1}\cdots d^n k_m\, f(k) \tag{2.2}$$

for $f \in \mathcal{Q}_m$, $k \in \mathbb{R}^{nm}$. (2.2) defines a function of $(k_1,\dots k_M) \in \mathbb{R}^{nM}$, but since the function is rotation invariant, we may view it as an element of $\mathcal{Q}_M$. As such, it is determined once it is known for all values of k_i^0 and $\vec{k}_i \cdot \vec{k}_j$, i.e. since $\phi_{M,M+1}$ is onto X_M, once it is known for $k_i \in \mathbb{R}^{M+1}$, $i \leq M$. Thus, (2.2) defines an element $F \in \mathcal{Q}_M$ which may be computed as follows: for $x \in X_M$, choose $k \in (\mathbb{R}^{M+1})^M$ with $\phi_{M,M+1}(k) = x$, and, regarding k_i as an element of $\mathbb{R}^n$ whose last $n-(M+1)$ coefficients vanish, define

$$F(x) = \int\cdots\int d^n k_{M+1}\cdots d^n k_m\, f(\phi_{mn}(k)). \tag{2.3}$$

The integral over angle variables in the last $n-(M+1)$ components of $k_{M+1},\dots k_m$ may be done using the next lemma, with $h = m-M$ and $s = n-(M+1)$:

Lemma 2.2. Let $g(k_1,\dots k_h) = G(k_i\cdot k_j)$ be a function of the Euclidean scalar products $k_i\cdot k_j$, $k_i \in \mathbb{R}^s$, and suppose $g \in L^1(\mathbb{R}^{sh})$. Then for $s \geq h$,

$$I \equiv \int_{\mathbb{R}^{nr}} g(k)\, d^s k_1 \dots d^s k_h$$

$$= N(s,h) \int_{S\geq 0} \prod_{i\leq j} dS_{ij}\, (\det S)^{\frac{s-h-1}{2}}\, G(S_{ij}) \qquad (2.4)$$

$$= M(s,h) \int_{\mathbb{R}^{h^2}} d^h\kappa_1 \dots d^h\kappa_h\, |\det \kappa|^{s-h}\, G(\kappa_i\cdot\kappa_j), \qquad (2.5)$$

with (2.4) and (2.5) absolutely convergent. Here S is a symmetric $h\times h$ matrix, κ the $h\times h$ matrix formed by the h vectors $\{\kappa_i\}$, and

$$N(s,h) = \prod_{j=1}^{h} \{\pi^{\frac{1}{2}(s-j+1)}/\Gamma[\tfrac{1}{2}(s-j+1)]\}, \qquad (2.6)$$

$$M(s,h) = N(s,h)/N(h,h).$$

Proof: We introduce new variables $\{\alpha_{ij}|h\geq i\geq j\geq 1,\ \alpha_{ij}\in\mathbb{R},\ \alpha_{ii}\geq 0\}$ and $\{u^i\in\mathbb{R}^s|1\leq i\leq h,\ u^i\cdot u^j = \delta_{ij}\}$ by $k_i = \sum_{j\leq i}\alpha_{ij}u^j$. Note that u^i lies in the $(s-i)$-sphere $\{u|u\cdot u^j = 0,\ j<i;\ u\cdot u = 1\}$; if $\sigma^{(s-i)}$ is the restriction of Lebesque measure to this sphere,

$$d^s k_1 \dots d^s k_h = \prod_i \alpha_{ii}^{(s-i)}\, d\alpha_i\, d\sigma^{(s-i)} \prod_{j<i} d\alpha_{ij}.$$

Since $k_i\cdot k_j = \sum_{r\leq\min(i,j)} \alpha_{ir}\alpha_{jr}$ is independent of $u_1,\dots u_h$ the integrals over the u variables give the areas of the spheres: $\int d\,\sigma^{(k)} = 2\pi^{\frac{k+1}{2}}/\Gamma[\frac{1}{2}(k+1)]$. If we introduce the variables $S_{ij} = k_i\cdot k_j$ $(i\leq j)$, we have Jacobean $\frac{\partial S}{\partial\alpha} = 2^h \prod_{i=1}^{h}\alpha_{ii}^{(h-i+1)}$ and an easy matrix manipulation shows that $\det S = \prod_{i=1}^{h}\alpha_{ii}^2$. (2.4) follows immediately; the integral converges absolutely by Fubini's Theorem.

To prove (2.5), we note that

$|\det \kappa| = |\det \kappa\, \kappa^T|^{\frac{1}{2}} = (\det [\kappa_i \cdot \kappa_j])^{\frac{1}{2}}$ so that the integral in (2.5) is of the form of the original integral I. A priori, we do not know that (2.5) is absolutely convergent. However, we may make the same variable changes as above and regard (2.5) as an iterated integral, integrating first over the variables $\{u_i\}$. These integrations are absolutely convergent, and the result is (2.4). Since (2.4) is absolutely convergent, Fubini's theorem implies that (2.5) is also, and that they are equal.

The importance of this result is that in (2.4) and (2.5), s may be taken as complex. Thus we may make

Definition 2.3. Suppose that $f \in \mathcal{Q}_m$, that $M+h = m$, and that $\nu \geq m+1$. We write vectors $k \in \mathbb{R}^{m+1}$ as $k = (K,\kappa)$, with $K = (K^0,\vec{K}) \in \mathbb{R}^{M+1}$ and $\kappa \in \mathbb{R}^h$. Then

$$\int \cdots \int f\, d^\nu k_{M+1} \cdots d^\nu k_m$$

is that element of $\mathcal{Q}_M$ such that

$$\left(\int \cdots \int f\, d^\nu k_{M+1} \cdots d^\nu k_m\right)(\phi_{M,M+1}(K))$$

$$= M(\nu-(M+1),h) \int d^{M+1}K_{M+1} \cdots d^{M+1}K_m d^h\kappa_{M+1} \cdots d^h\kappa_m \tag{2.7}$$

$$\times |\det \kappa|^{\nu-(m+1)} f(\phi_{m,m}(k)),$$

whenever the integral converges absolutely.

Remark. This definition (2.7) splits the ν-dimensional integral into one of dimension $M+1$ and one of dimension $\nu-(M+1)$; the latter evaluated using Lemma 2.2. The number $M+1$ was chosen since $r = M$ is the minimum value of r for which $\phi_{M,r+1}$ is onto. However, this choice is essentially arbitrary, as the next lemma shows. The proof, which is by direct induction on r, will be omitted.

Lemma 2.4. Suppose that $f \in \mathcal{Q}_m$, that $\nu \geq m+1$, and that $\int f \prod_{M+1}^{m} d^\nu k_i$ is defined. Then for $r \geq M$, $\nu \geq r+h+1$, and

$K_i \in \mathbb{R}^{r+1}$, $(\int f\, d^\nu k_{M+1} \cdots d^\nu k_m)(\phi_{M,r+1}(K))$

$$= M(\nu-(r+1),h) \int \prod_{i=M+1}^{m} d^{r+1}K_i d^h\kappa_i \;|\det \kappa|^{\nu-(r+h+1)} f(\phi_{m,r+h+1}(k)).$$

Remark: The restriction $r \geq M$ in Lemma 2.4 arises again from the difficulty in restricting f to $\operatorname{Im} \phi_{mn}$ for $n \leq m$. If this does not arise, (2.8) will hold for all $r \geq 0$. In particular, this implies that, if the function $\bar{f}$ in Def. 2.3 is independent of, say, k_1, the same value of $f\, d^\nu k_{M+1} \cdots d^\nu k_m$ is obtained if we consider f to belong to $\mathcal{Q}_{M-1}$.

We give a result which justifies linear variable changes and a sort of "Fubini's theorem" for these integrals.

Lemma 2.5. If $k_1', \ldots k_m' \in \mathbb{R}^{m+1}$ are defined by

$$k_i' = \begin{cases} k_i, & 1 \leq i \leq M, \\ \sum_{j=M+1}^{m} A_{ij} k_j + \sum_{j=1}^{M} B_{ij} k_j, & M < i \leq m, \end{cases} \tag{2.8}$$

then for $f \in \mathcal{Q}_M$,

$$\int f(k') \sum_{j=M+1}^{m} d^\nu k_j' = |\det A|^\nu \int f(k'(k)) \prod_{j=M+1}^{m} d^\nu k_j. \tag{2.9}$$

Proof: This follows directly by writing out both sides of (2.9) according to Def. 2.3. Note that we are again abusing notation above. A technically correct version could be stated as follows: (2.8) implicitly defines a map $\psi: U_m \to U_m$ by $\psi(\phi_{m,m+1}(k) = \phi_{m,m+1}(k')$; (2.9) may then be written

$$\int (f \circ \psi) \prod_{M+1}^{m} d^\nu k = |\det A|^\nu \int f \prod_{M+1}^{m} d^\nu k.$$

Lemma 2.6. If $f \in \mathcal{Q}_m$, $\nu \geq m+1$, and M, $H \geq 0$ with $M+H \leq m$, then

$$\int f\, d^\nu k_{M+1} \cdots d^\nu k_m = \int \left(\int f\, d^\nu k_{M+H+1} \cdots d^\nu k_m\right) d^\nu k_{M+1} \cdots d^\nu k_{M+H},$$

whenever either side is defined.

Proof: Let $h = m-M$, as usual. Then

$$\int f\, d^\nu k_{M+1} \cdots d^\nu k_m = M(\nu-(M+1),h) \int f \prod_{i=M+1}^{m} d^{M+1}K_i d^h \kappa_i |\det \kappa|^{\nu-m-1} \tag{2.10}$$

We integrate over κ_j, $M+H < j \le m$, holding the other variables fixed, and write $\kappa_j = \kappa_j' + \kappa_j''$, where κ_j'' is in the span of $\kappa_{M+1} \ldots \kappa_H$ and κ_j' is perpendicular to this span. $\det \kappa$ may be evaluated in a basis appropriate to this decomposition to give

$$|\det \kappa| = (\det T)^{\frac{1}{2}} |\det \kappa'|$$

with $T = (T_{ij})$, $T_{ij} = \kappa_i \cdot \kappa_j$, $M < i,j \le M+H$. Thus (2.10) becomes

$$M(\nu-(M+1),h) \int [\det T]^{\frac{1}{2}(\nu-(m+1))} \prod_{j=M+1}^{M+H} d^{M+1}K_j\, d^h \kappa_j$$
$$\times \left\{ \int f \prod_{j+H+M+1} d^{M+1}K_i d^H \kappa_i'\, d^{m-(M+H)} \kappa_i'' \, |\det \kappa_i'|^{\nu-(m+1)} \right\}. \tag{2.11}$$

By Definition 2.3 the quantity in brackets is

$$M(\nu-(M+H+1),\, m-(M+H))^{-1} \int f \prod_{i=M+H+1}^{m} d^\nu k_i,$$

so that (2.11) becomes

$$\frac{M(\nu-(m+1),\, h)}{M(\nu-(m+1),\, m-(M+H))} \int \left(\int f \prod_{M+H+1}^{m} d^\nu k_i\right) (\det T)^{\frac{1}{2}(\nu-(m+1))} \prod d^{M+1}K_j d^h \kappa_j \tag{2.12}$$

(2.12) is nearly the desired form. Using Lemma 2.2, the integral over $\kappa_j \in \mathbb{R}^h$, $M < j \le M+H$, is converted to an integral over $\kappa_j \in \mathbb{R}^H$, and the result is

$$\frac{M(\nu-(M+1),h)M(h,H)}{M(\nu-(M+H+1),m-(M+H))} \int \left(\int \left(f \prod_{M+H+1}^{m} d^{\nu} k_i\right) \Pi d^{M+1} K_j d^{H} \kappa_j \; |\det \kappa|^{\nu-(M+H)-1}\right).$$

The ratio of constants is in fact $M(\nu-(M+1),H)$, from (2.6).

We will need the evaluation of one specific complex dimensional integral--that of a Gaussian.

Lemma 2.7. Let A be an $h\times h$ real positive definite matrix, B an $M\times h$ real matrix. For $\beta \in C$ with $\mathrm{Im}\,\beta > 0$, and $k_1, k_2 \in \mathbb{R}^{m+1}$, define

$$\langle k_1, k_2\rangle = \beta\, k_1^0 k_2^0 - \bar{\beta} \sum_{i=1}^{m} k_1^i k_2^i. \tag{2.13}$$

Then

$$\int d^{\nu} k_{M+1} \cdots d^{\nu} k_m \exp i\{\sum_{i,j=M+1}^{m} A_{ij}\langle k_i, k_j\rangle + 2\sum_{i=1}^{m}\sum_{j=M+1}^{m} B_{ij}\langle k_i, k_j\rangle\}$$

$$= (-i\beta)^{-\frac{1}{2}h}(i\bar{\beta})^{-\frac{1}{2}h(\nu-1)}\pi^{\frac{1}{2}h\nu}(\det A)^{-\frac{1}{2}\nu} \exp -i \sum_{i,j=1}^{m} (BA^{-1}B^T)_{ij}\langle k_i, k_j\rangle; \tag{2.14}$$

it is understood that complex powers z^a in (2.14) are defined with $-\pi < \arg z < \pi$.

Proof: We prove (2.14) with $\beta = i\gamma$ and $\gamma > 0$; it then follows for general β by analytic continuation. We make a change of integration variable in (2.14) to complete the square in the exponential:

$$k_i' = \sum_{j=M+1}^{m} (\gamma A)^{\frac{1}{2}}_{ij}\left(k_j + \sum_{s=M+1}^{m}\sum_{r=1}^{m} A^{-1}_{js} B_{sr} k_r\right),$$

for $M+1 \le i \le m$. An application of Lemma 2.5 then reduces the problem to proving that

$$\int \prod_{M+1}^{m} d^{\nu} k_i' \, e^{-\Sigma ||k_i'||^2} = \pi^{\frac{1}{2}h\nu},$$

and by Lemma 2.6 this will follow from

$$\int d^{\nu} k \, e^{-||k||^2} = \pi^{\frac{1}{2}\nu}. \tag{2.15}$$

According to Def. 2.3 the left-hand side of (2.15) is

$$M(\nu-1,1)\int_{\mathbb{R}} dK \int_{\mathbb{R}} d\kappa |\kappa|^{\nu-2} e^{-(K^2+\kappa^2)}$$

and (2.15) is immediate.

B. Feynman Amplitudes

In this section we will consider scalar Feynman amplitudes; those involving vector or spinor fields, or derivatives of fields, will be discussed later. Consider then a finite collection of scalar fields $\{\phi_\alpha\}$. They are described by a free Lagrangian L_0, quadratic in the fields and their first derivatives, which we take to be

$$L_0 = \tfrac{1}{2}\sum_\alpha (\partial_\mu\phi_\alpha\partial^\mu\phi_\alpha - m_\alpha^2\phi_\alpha^2); \tag{2.16}$$

thus the fields satisfy free field equations

$$0 = \frac{\partial L_0}{\partial\phi_\alpha} - \partial_\mu \frac{\partial L_0}{\partial(\partial_\mu\phi_\alpha)} = -(\partial^2 + m_\alpha^2)\phi. \tag{2.17}$$

The mass m_α is positive. Let $\{Y_i(x)\,|\,i=1,\ldots m\}$ be polynomials in these fields:

$$Y_i(x) = \sum_{\underline{i}} c_{\underline{i}} \prod_\alpha \phi_\alpha^{\,i_\alpha}(x),$$

where $\underline{i} = (i_\alpha)$ is a multi-index. Eventually, we wish to define time-ordered vacuum expectation values

$$\langle 0|T[Y_1(x_1)\ \ldots\ Y_m(x_m)]|0\rangle \tag{2.18}$$

as distributions in $S'(\mathbb{R}^{4m})$; for the moment our goal is to define a regularized version of (2.18).

We require multi-linearity in the polynomials, so we may assume that each Y_i is in fact a monomial. Then (2.18) is expanded as a sum of terms indexed by Feynman graphs. (A general discussion of graph-theoretical terms is given in the Appendix.)

One such graph is generated by each partitioning of the set of factors in $\prod_{i=1}^{m} Y_i(x_i)$ into pairs containing two fields with the same index; the graph has vertices labelled $V_1, \ldots V_m$, each with one external line, and, for each pair $\{\phi_\alpha(x_i), \phi_\alpha(x_j)\}$, an internal line joining V_i to V_j. We do not exclude the possibility of loop lines $(i=j)$; because of this the field products we define will not correspond to the usual normal products.

Consider now a specific graph G. We let $L(=L(G))$ denote the set of lines of G, give each line an arbitrary orientation, and define the <u>incidence matrix</u> e of G by

$$e_i^\ell = \begin{cases} -1, & \text{if } \ell \text{ is } \underline{\text{into}}\ V_i, \\ 1, & \text{if } \ell \text{ is } \underline{\text{out of}}\ V_i, \\ 0, & \text{if neither or both.} \end{cases}$$

Let $\Delta_m(x) \in S'(\mathbb{R}^4)$ be given in terms of its Fourier transform as follows:

$$\Delta_m(x) = (2\pi)^{-4} \int d^4q\, e^{iq\cdot x}\, \tilde{\Delta}_m(q)$$

$$\tilde{\Delta}_m(q) = \frac{-i}{m^2-q^2-i0} \equiv \lim_{\varepsilon\to 0+} \frac{-i}{(m^2+\vec{q}^{\,2})(1-i\varepsilon)-q_0^2(1+i\varepsilon)} \tag{2.19}$$

Then for the graph G, the amplitude T_G is formally

$$T_G(x) = \prod_{\ell\in L} \Delta_{m_\ell}\Big(\sum_i e_i^\ell x_i\Big) \tag{2.20}$$

where we write $m_\ell = m_\alpha$ whenever ℓ arises from pairing fields of type α. Finally, the amplitude (2.18) is

$$\sum_G T_G(x),$$

the sum taken over all G as above.

Continuing in a formal vein, we may calculate the Fourier transform of the product (2.20), assuming that G is connected:

$$\tilde{T}_G(p) \equiv \int d^4x_1 \ldots d^4x_m \, e^{-i\sum_{i=1}^{m} p_i x_i} \, T_G(x)$$

$$= (-i)^L (2\pi)^{-4h} [(2\pi)^4 \delta(\sum_{i=1}^{m} p_i)] \int d^4k_1 \ldots d^4k_h \prod_L (m_\ell^2 - q_\ell^2 - i0)^{-1}. \tag{2.21}$$

Here $L = |L|$ is the number of lines in the graph, $h = L-m+1$ is the number of loops, and

$$q_\ell(p,k) = \sum_{i=1}^{m-1} a_{\ell i} p_i + \sum_{j=1}^{h} b_{\ell j} k_j. \tag{2.22}$$

(2.22) expresses the momentum q_ℓ in the line ℓ in terms of the external momenta p and the loop momenta k. Because of the δ-function in (2.21), it suffices to specify the value of the integral as a function of $p_1, \ldots p_{m-1}$. We have made this explicit in (2.22).

The matrices a and b are arbitrary except for the requirement that the determinant of the L×L matrix [a b] be 1, and that $\sum_\ell e_i^\ell a_{\ell j} = \delta_{ij}$ $(1 \leq i,j \leq m-1)$ and $\sum_\ell e_i^\ell b_{\ell j} = 0$ $(1 \leq i \leq m-1,\ 1 \leq j \leq h)$. Such matrices are usually found by choosing a set T of lines of G, called a tree, which connects all vertices but contains no loops. Then the requirements that $a_i^\ell = 0$, $\ell \notin T$, and $b_j^{\ell_i} = \delta_{ij}$, for $(\ell_1, \ldots \ell_h)$ an enumeration of the lines not in T, determine a and b completely.

Now the integral in (2.21) is not, in general, convergent. There are two sources of difficulty. The first is that the denominators become infinite when $q_\ell^2 = m_\ell^2$. This is not a serious problem since the i0 tells us how to handle it; specifically, we reintroduce the ε dependence of (2.19) and remove it when all calculations are finished. The second difficulty is the fundamental one of ultraviolet divergence. When the loop momenta k in (2.21) are large, say order of magnitude β, the integral looks like $\int \beta^{-2L} \beta^{4h-1} d\beta$, which is divergent if $4h-2L \geq 0$.

To avoid this divergence we introduce a regularized integral, depending on certain complex parameters $\{\lambda_\ell | \ell \in L\}$. We do this by replacing the propagator $(q_\ell^2 - m_\ell^2 - i0)^{-1}$ by $(q_\ell^2 - m_\ell^2 - i0)^{-\lambda_\ell}$; if $\mathrm{Re}\,\lambda_\ell$ is large this function decreases rapidly for large q_ℓ. We will also introduce a complex dimension as an additional regularization. The δ function in (2.21), which does not generalize readily to complex dimension, will be omitted from our definition; it must eventually be replaced to obtain the correct amplitude in dimension 4.

Definition 2.8. The regularized Feynman amplitude for the graph G is the element $F_G(\lambda,\nu,\varepsilon) \in \mathcal{Q}_{m-1}$ defined by

$$F_G(\lambda,\nu,\varepsilon) = z^{\omega}(-i)^L(2\pi)^{-h\nu} \int d^\nu k_1 \ldots d^\nu k_h \prod_L \tilde{\Delta}_\ell(\lambda_\ell,\varepsilon)(q_\ell) \tag{2.23}$$

where $\tilde{\Delta}_\ell(\lambda_\ell,\varepsilon)(q_\ell) = (m_\ell^2(1-i\varepsilon) - \langle q_\ell,q_\ell\rangle)^{-\lambda_\ell} \in \mathcal{Q}_L$, $\langle q_\ell,q_\ell\rangle = q_0^2(1+i\varepsilon) - \vec{q}^{\,2}(1-i\varepsilon)$, as in (2.13), and $\omega = \sum_L (\lambda_\ell - 1) - \frac{1}{2}h(\nu-4)$.

Remarks: 1. We recall from Def. 2.3 the full interpretation of (2.23). For $\xi \in X_{m-1}$, choose a point $p \in \mathbb{R}^m$ with $\phi_{m-1,m-1}(p) = \xi$. For $k \in \mathbb{R}^{m+h}$ write $k = (K,\kappa)$ with $K \in \mathbb{R}^m$ and $\kappa \in \mathbb{R}^h$, and define $q_\ell = \Sigma\, a_{\ell i} p_\ell + \Sigma\, b_{\ell j} k_j \in \mathbb{R}^{m+h}$. Then

$$\begin{aligned} F_G(\lambda,\nu,\varepsilon)(\xi) = {} & (-i)^L(2\pi)^{-\frac{1}{2}h\nu} M(\nu-m,h) \int \prod d^m K_i d^h \kappa_i \times \\ & \times |\det \kappa|^{\nu-(m+h)} \prod_L (m_\ell^2(1-i\varepsilon) - q_\ell,q_\ell\rangle)^{-\lambda_\ell}. \end{aligned} \tag{2.24}$$

2. The integral (2.24) is absolutely convergent if $\mathrm{Re}\,\nu \geq m+h$ and if $\mathrm{Re}\,\lambda_\ell$ is sufficiently large. Explicitly, it is sufficient that the superficial divergence $\nu h(H) - 2\sum_{\ell \in H} \lambda_\ell$ have negative real part for each subgraph H of G. This can be proved by Weinberg's theorem; in the next section we will give a proof based on Feynman parameterization of the integral.

3. Again, it may be argued directly that for λ,ν as

above, F is an analytic function of λ and ν. In the next section we will see that F is actually a meromorphic function of λ,ν for all $\lambda,\nu \in \mathbb{C}^{L+1}$.

4. Comparing (2.21) and (2.23) yields the relation

$$T_G(p) = (2\pi)^4 \, \delta(\sum_1^m p_i) \; F_G(\lambda^0,4,0)(\phi_{m-1,3}(p_1,\ldots p_{m-1}))$$

with $\lambda^0 = (1,1,\ldots 1)$. Since F_G is in general singular at $(\lambda^0,4)$, this equality is only formal. Removal of the singularity in order to define T_G is discussed in Chapter 3.

5. The variable z in (2.23) is a parameter with dimension $(\text{mass})^2$; its introduction guarantees that F_G has the same dimension for all ν,λ. The effect on renormalization will be discussed in Chapter 6.

C. Analytic Continuation

In this section we introduce the Feynman parametric representation of the regularized integral, which allows a discussion of analytic properties. The basic trick is to write each Feynman denominator in (2.24) as an integral superposition of exponentials, so that the complex dimensional k integrations may be done explicitly. F_G is then expressed as an integral over Feynman parameters, and scaling transformations on these parameters lead to analytic continuation in the parameters (λ,ν).

Thus we write

$$\tilde{\Delta}_\ell(\lambda_\ell,\varepsilon) = \frac{e^{\lambda_\ell \pi i}}{\Gamma(\lambda_\ell)} \int_0^\infty d\alpha_\ell \alpha_\ell^{\lambda_\ell - 1} \exp i\alpha_\ell[\langle q_\ell,q_\ell\rangle - m_\ell^2(1-i\varepsilon)]. \qquad (2.25)$$

This integral converges absolutely for all q_ℓ as long as $\varepsilon > 0$ and $\text{Re}\,\lambda_\ell > 0$. We insert (2.25) into (2.23) and interchange the k and α integrations. The resulting momentum integral is

$$\int d^\nu k_1 \ldots d^\nu k_h \exp i\{\sum_{\ell\in L} \alpha_\ell[\langle q_\ell,q_\ell\rangle - m_\ell^2(1-i\varepsilon)]\}. \qquad (2.26)$$

(2.26) may be evaluated via Lemma 2.7, applied with $\beta = 1+i\varepsilon$, $A_{ij} = \sum_\ell b_{\ell j}\, b_{\ell j}\, \alpha_\ell$, and $B_{ij} = \sum_\ell a_{\ell i}\, b_{\ell j}\, \alpha_\ell$. We write

$$(-i\beta)^{-h/2}(i\bar{\beta})^{-h(\nu-1)/2} = i^h i^{-h\nu/2} c(\varepsilon),$$

where $c(\varepsilon) = (1+i\varepsilon)^{-h/2}(1-i\varepsilon)^{-h(\nu-1)/2}$ with phase chosen so that $\lim_{\varepsilon\to 0} c(\varepsilon) = 1$. Thus, if the interchange of integrations described above is justified, we have from Lemma 2.7:

$$F_G(\lambda,\nu,\varepsilon)(p) = i^h \prod_L \left(\frac{-i\, e^{\lambda_\ell \pi i}}{\Gamma(\lambda_\ell)}\right) (4\pi i)^{-\frac{1}{2}h\nu}\, z^\omega\, c(\varepsilon) \times$$

$$\int_0^\infty \cdots \int_0^\infty \prod_L \alpha_\ell^{\lambda_\ell - 1} d\alpha_\ell\, d(\alpha)^{-\frac{1}{2}\nu} \exp i[D^*_\varepsilon(\alpha,p) - \sum_L m_\ell^2(1-i\varepsilon)\, \alpha_\ell] \tag{2.27}$$

with

$$d(\alpha) = \det A,$$

$$D^*_\varepsilon(\alpha,p) = \sum_{i,j=1}^{m-1} D^*_{ij}(\alpha) \langle p_i, p_j\rangle$$

$$D^*_{ij}(\alpha) = -(BA^{-1}B^T)_{ij} + \sum_\ell \alpha_\ell a_{\ell i} a_{\ell j}.$$

We will first study the integral (2.27), then return to its justification.

Definition 2.9. A tree T in the graph G is a set of lines connecting all vertices but containing no loops. A 2-tree T_2 is a tree with one line omitted; such a 2-tree naturally separates the vertices into two disjoint non-empty sets.

Lemma 2.10. The functions d and D^* are given by

$$d(\alpha) = \sum_T \prod_{\ell \notin T} \alpha_\ell,$$

the sum taken over all trees in G, and

$$D^*_{ij}(\alpha) = d(\alpha)^{-1} D_{ij}(\alpha)$$

$$D_{ij}(\alpha) = \sum_{T_2} \prod_{\ell \notin T_2} \alpha_\ell,$$

the sum running over all 2-trees T_2 such that V_i and V_j are separated from V_m.

Proof: Omitted; see [2,12,13].

The scaling transformations necessary to analyze the integral (2.27) are indexed by certain families of subgraphs of G.

Definition 2.11. A singularity family (s-family) is a maximal family $\mathcal{E}$ of non-overlapping subgraphs of G, each 2-connected (see Appendix) or consisting of a single line, such that, for $H_1, H_2 \in \mathcal{E}$, either $H_1 \subset H_2$, $H_2 \subset H_1$, or $L(H_1) \cap L(H_2) = \emptyset$, and such that no union of two or more disjoint elements of $\mathcal{E}$ is 2-connected. We note that maximality implies that every 2-connected component of G is in $\mathcal{E}$.

Every s-family contains L graphs. For a fixed s-family $\mathcal{E}$ we introduce variables $t_H \in \mathbb{R}$, $H \in \mathcal{E}$, and define the Feynman parameters $\{\alpha_\ell\}$ as functions of $\{t_H\}$ by

$$\alpha_\ell = \prod_{\ell \in H} t_H. \tag{2.28}$$

Let $\mathcal{D}_\mathcal{E}$ denote the set of α's obtained in this way, for $t_H \geq 0$, $H \in \mathcal{E}$, and $t_{H'} \leq 1$ if H' is not a maximal element of $\mathcal{E}$.

Lemma 2.12. a) The union of the sets $\mathcal{D}_\mathcal{E}$ is $\{\alpha | \alpha_\ell \geq 0\}$; the intersection of any two distinct $\mathcal{D}_\mathcal{E}$ has measure 0.

b) Under the substitution (2.28),

$$d(\alpha) = \prod_{H \in G} t_H^{h(H)} E(t)$$

$$D_{ij}(\alpha) = \prod_{H \in G} t_H^{h(H)} F_{ij}(t)$$

with $E(t)$, $F_{ij}(t)$ polynomials, and $E(t) > 0$ and bounded away from 0 for $t_H \geq 0$.

Proof: Omitted, see [2,13].

Now according to Lemma 2.12, the integral in (2.27) may be written as a sum of integrals over the regions $\mathcal{D}_E$, i.e., the right hand side is

$$\sum_{E} F_E(\lambda,\nu,\varepsilon),$$

with

$$F_E(\lambda,\nu,\varepsilon) = i^h c(\varepsilon)(4\pi i)^{-\frac{1}{2}h\nu} z^{\omega} \prod_{L} \left(\frac{-i\, e^{\lambda_\ell \pi i}}{\Gamma(\lambda_\ell)}\right) \int \prod_{H\in E} dt_H\, t_H^{\omega_H - \frac{1}{2}\mu_H - 1}\, E(t)^{-\nu/2}$$

$$\times \exp i[\sum_{ij} E(t)^{-1} F_{ij}(t) \langle p_i, p_j\rangle - (1-i\varepsilon) \sum \alpha_\ell m_\ell^2]. \tag{2.29}$$

Here $\omega_H = \sum_{\ell\in L(H)} (\lambda_\ell - 1) - \frac{1}{2}h(\nu-4)$, and $\mu_H = 4h(H) - 2L(H)$ is the superficial divergence of H.

Lemma 2.13. The integral (2.29) is absolutely convergent, and defines a holomorphic function of λ,ν, if $\mathrm{Re}\,\omega_H > \frac{1}{2}\mu_H$ for every $H \in E$. Moreover, for such values, (2.24) converges absolutely, (2.27) is correct, and thus

$$F_G = \sum_{E} F_E. \tag{2.30}$$

Proof sketch: The exponential factor in (2.29) is easily seen to be bounded by

$$\prod_{H\in E_{max}} e^{-\varepsilon m_0 t_H},$$

where $E_{max} \subset E$ consists of maximal subgraphs, and m_0 is the minimum mass on any line of G. Since $E(t) > a > 0$, the convergence statement is clear. Using this, and the Lebesque dominated convergence theorem, one can justify a differentiation under the integral with respect to λ and ν, proving analyticity. Finally, (2.24) is shown to converge and (2.27) is justified by showing that the integral over k and α from which (2.27) was obtained is absolutely convergent. For replacing the integrand by its absolute value, we may still do the k integrations explicitly; the resulting α integral is similar to (2.27) and can be shown to be convergent in the same way.

Theorem 2.14. The regularized amplitude $F_G(\lambda,\nu,\varepsilon)$ extends to a meromorphic function of λ,ν, with simple poles on the hyperplanes

$$\sum_{\ell\in H} \lambda_\ell - \tfrac{1}{2}\nu h(H) = 0,-1,-2, \ldots \tag{2.31}$$

where H is a 2-connected subgraph of G.

Proof: According to Lemma 2.11 we may write $F_G = \sum_{E} F_E$, with F_E given by (2.29). We may regard each factor $t_H^{\omega_H - \frac{1}{2}\mu_H - 1}$ as a distribution; it is known [14] that $T(\phi) = \int_0 t^{\gamma-1}\phi(t)dt$, ϕ infinitely differentiable, defines a function meromorphic in γ, with simple poles at $\gamma = 0,-1,-2,\ldots$. [The analytic continuation is carried out by adding and subtracting from ϕ its finite Taylor series of some order with center 0.] This leads to the desired conclusion; note that if $H \in E$ is not 2-connected but rather consists of a single line ℓ, the corresponding poles in the integral are at $\lambda_\ell = 0,-1,-2,$ and are cancelled by the zeros of the factor $\Gamma(\lambda_\ell)^{-1}$ in (2.29).

Finally, we state the standard result on the existence of the $\varepsilon \to 0$ limit of the regularized amplitude. For $M \geq 0$, $k > 0$ $(P_1,\ldots P_k) \in (\mathbb{R}^M)^k$, and $1 \leq i_1 < \ldots < i_k \leq m$, define $p(P) \in (\mathbb{R}^M)^m$ by $p_{i_j}(P) = P_j$, $p_i = 0$ if $i \neq i_j$ for some j; then $F_G(\lambda,\nu,\varepsilon)(\phi_{m,M}(p(P)))$ is the regularized Feynman amplitude for G in a space of M dimensions, with some external momenta set to 0.

Theorem 2.15. The limit

$$\lim_{\varepsilon\to 0} \delta\left(\sum_1^k P_i\right) F_G(\lambda,\nu,\varepsilon)(\phi_{m,M}(p(P)))$$

exists in the sense of tempered distributions, i.e., in $S'((R^M)^k)$, at every point not on any hyperplane (2.31).

Proof: Omitted [2,15].

III. RENORMALIZATION

We wish to obtain a finite value for F_G at $\lambda_\ell = 1$, $\nu = 4$, by "subtracting" the pole in some way. We begin by discussing analytic renormalization, and then use the results to study the structure of the dimensional method.

A. Analytic Renormalization

Consider first the amplitude $F_G(\lambda,4,\varepsilon)$. This is a meromorphic function of the λ_ℓ, with simple poles on various planes $\{\lambda | \sum_{\ell\in L'} (\lambda_\ell - 1) = 0\}$ passing through $\underline{\lambda}^0$. We may define a finite part by repeatedly taking the finite part of a Laurent series in each variable separately; since the answer depends on the order in which we proceed, we must average over all possible orders. Recall that the finite part of the Laurent series for a function f(z), analytic in an annulus $r_1 < |z| < r_2$, is

$$g(z) = \frac{1}{2\pi i} \int_{|\zeta|=r} \frac{f(\zeta)}{\zeta - z}\, d\zeta, \tag{3.1}$$

where $r_1 < r < r_2$ and $|z| < r$. By iterating (3.1), we derive the following formula: let $r_1, r_2, \ldots$ be positive numbers (chosen so that $r_j > \sum_{i=1}^{j-1} r_i$, and let C_i be the contour $\{\zeta |\ |\zeta - 1| = r_i\}$. Then the finite part of the iterated Laurent series for $F_G(\lambda,4,\varepsilon)$ is

$$g(\mu) = \frac{1}{(2\pi i)^L} \int_{C_1} d\lambda_1 \cdots \int_{C_L} d\lambda_L \; \frac{F_G(\lambda,4,\varepsilon)}{\Pi(\lambda_\ell - \mu_\ell)} \tag{3.2}$$

defined for $|\mu_\ell - 1| < r_1$. (3.2) must then be averaged over possible orders.

The resulting formula is adequate for a discussion of analytic renormalization. However, in applying the results to dimensional renormalization, we will need a simple generalization which includes the ν dependence. For fixed ν, with $\nu \approx 4$, the various singular surfaces of $F_G(\lambda,\nu,\varepsilon)$ are displaced slightly and no longer pass through λ^0, and we cannot renormalize by taking an iterated series at this point. However, we may directly adapt the formula (3.2).

Definition 3.1. Let $f(\lambda,\nu)$ be meromorphic in a neighborhood of $(\lambda^0,4)$, with possible simple poles on a finite number of subvarieties of the form

$$\sum_{\ell\in L'} (\lambda_\ell - 1) - h(\nu-4) = 0. \qquad (3.3)$$

Set $h_0 = \sup h$, the supremum taken over all singular varieties (3.3) for f, and choose $r_0, r_1, \ldots r_L$ with $r_j > \sum_{i=0}^{j-1} r_i$ and such that, aside from the singularities (3.3), f is analytic in $\{(\lambda,\nu) \,|\, |\lambda_\ell - 1| < 2r_L,\ |\nu-4| < r_0/h_0\}$. Then for $|\mu_\ell - 1| < r_0$ and $|\nu-4| < r_0/h_0$,

$$(\mathcal{V}f)(\underline{\mu},\nu) = \frac{(2\pi i)^{-L}}{L!} \sum_{\sigma\in S_L} \int_{C_{\sigma(1)}} d\lambda_1 \cdots \int_{C_{\sigma(L)}} d\lambda_L \frac{f(\lambda,\nu)}{\Pi(\lambda_\ell - \mu_\ell)} .$$

Definition 3.2. Let $F_G(\lambda,\nu,\varepsilon)$ be the regularized amplitude for the Feynman graph G (Definition 2.8). Then the analytically renormalized amplitude for G is

$$\mathcal{W}_G(p) = \lim_{\varepsilon\to 0} (2\pi)^4\, \delta\Big(\sum_1^m p_i\Big)\, \mathcal{V}F_G(\lambda^0,4,\varepsilon)(\phi_{m-1,4}(p)).$$

Remarks: 1. The ν dependence plays no role at all in this definition; when using analytic renormalization, one may work in 4 dimensions throughout.

2. As in Theorem 2.14, some external momenta may be set to zero.

The definition of a renormalization operation such as $\mathcal{W}$ must of course be justified as physically reasonable. One approach to this problem is to prove that $\mathcal{W}$ satisfies certain axioms (unitarity, causality, etc.) which Hepp [3] has shown to characterize physically acceptable renormalizations; a complete proof is given in [3,13]. Here we take an alternate tack by describing the additive structure of $\mathcal{W}$. It is this structure which shows that a renormalization may be implemented by (infinite) counter terms in the Lagrangian of a theory, and thus leads to the usual mass, charge, and field strength renormalizations.

Definition 3.3. Suppose that H is a 1PI subgraph of G with external line set $E(H)$ (see Appendix). A vertex part for H, $X_H(\lambda,\nu)$, is an element of $\mathcal{Q}_{|E(H)|-1}$ which is a polynomial in the variables p^0_i and s_{ij} $(i,j \in E(H)-\{k\}$, with k some distinguished external line) with coefficients meromorphic functions of ν and λ_ℓ, $\ell \in L(H)$.

Suppose now that for each 1PI subgraph H of G we have defined a vertex part X_H. Let A be a set of (totally) disjoint 1PI subgraphs of G, and let $\{q_\ell | \ell \in L(G/A)\}$ and $\{k_i | i=1,\ldots h(G/A)\}$ be line and loop variables for the quotient graph G/A (see Appendix), as in (2.24). Then define

$$F^X_{G:A}(\lambda,\nu,\varepsilon) = z^{\omega'}(-i)^{L'}(2\pi)^{-h'\nu}\int d^\nu k_1 \ldots d^\nu k_{h'} \times$$

$$\prod_{L(G/A)} \tilde{\Delta}_\ell(\lambda_\ell,\varepsilon)(q_\ell) \prod_{H\in A} X_H(\lambda,\nu)(P(k,p)). \tag{3.4}$$

In (3.4), for $i \in E(H)$,

$$P_i(k,p) = \begin{cases} p_i & \text{if } i \in E(G), \\ (-1)^k q_\ell & \text{if } i = (k,\ell), \end{cases}$$

and primed variables refer to G/A. $F^X_{G:A}$ is the amplitude for the generalized graph in which the subgraphs H play the role of vertices. Then the additive structure of the renormalization operator $\mathcal{W}$ is described by

Theorem 3.4. There exists a set of vertex parts X such that

$$VF_G = \sum_{A\in A(G)} F^X_{G:A} , \qquad (3.5)$$

where $A(G) = \{\{H_1,\ldots H_k\} \mid H_i$ pairwise disjoint 1PI subgraphs of $G\}$.

Remark. (3.5) contains subtractions for all 1PI subgraphs. The alternate formulation in which subtractions appear only for generalized vertices is discussed in [13,16]. The two are equivalent, but it is the above form which we will use in the next section. For more details of the proof, see [16].

Proof sketch: For H any subgraph of G, let V_H be the operator, as in Def. 3.1, which extracts a finite part at $\nu = 4$, $\lambda_\ell = 1$ for $\ell \in L(H)$. Let $S(H) = \sum_{H'\subset H} (-1)^{|L(H)-L(H')|} V_{H'}$,

so that $V_H = \sum_{H'\subset H} S(H)$; for a 1PI subgraph H we may define

$$X_H = S(H)F_H$$

and easily verify that X_H is a vertex part. If $f_i(\lambda,\nu)$, for $i=1,2$, depends only on those λ_ℓ with $\ell \in L(H_i)$, and $L(H_1) \cap L(H_2) = \emptyset$, then one may prove that

$$S(H)\ f_1 f_2 = \begin{cases} 0, & \text{if } H \not\subset H_1 \cup H_2 \\ S(H\cap H_1)f_1 S(H\cap H_2)f_2, & \text{otherwise.} \end{cases} \qquad (3.6)$$

Moreover, $S(H)f = 0$ unless f is singular on a set of hyperplanes

$$\sum_{\ell\in L(H_i)} (\lambda_\ell - 1) - \tfrac{1}{2}h(H_i)(\nu-4) = 0$$

with $L(H) = L(H_i)$, so that $S(H)F_G = 0$ unless each component of H is 1PI; such an H is naturally identified with an element $A \in A(G)$. Then

$$\mathcal{V}F_G = \sum_H S(H)\, F_G = \sum_H S(H)\{z^{\omega'}(-i)^{L'}(2\pi)^{-h'\nu}\int d^\nu k_1 \ldots d^\nu k_{h'}$$

$$\times \prod_{L(G/A)} \tilde{\Delta}_\ell \prod_{H'\in A} F_{H'}\}$$

(as in (3.4)), using Lemma 2.6, and $\mathcal{V}F_G = \sum_A F^{\mathcal{X}}_{G:A}$, using (3.6).

B. Dimensional Renormalization

The definition of dimensional renormalization does not use the λ dependence of the regularized amplitudes, although we will use it in our proof that we have a renormalization; thus we initially set $\lambda_\ell = 1$, for all ℓ. [It should be emphasized that in our approach in these lectures the λ variables were essential at an intermediate stage, i.e., to give a convergent and well-defined integral (2.24) representing the regularized amplitude, which could then be analytically continued to $\lambda = \lambda^0$. However, the same amplitude $F_G(\lambda^0,\nu,\varepsilon)$ may be obtained by other means, e.g., intermediate formal manipulations.] For simple graphs such as that treated in the introduction, renormalization is accomplished by subtracting the pole which is present at $\nu = 4$; for multi-loop graphs a recursive procedure, involving subtractions for subgraphs, is necessary.

Definition 3.5. If $f(\nu)$ has an isolated singularity at $\nu = 4$, let Kf be the singular part, defined by

$$Kf(\nu) = \int_{|\nu'-4|=r} \frac{f(\nu')}{\nu'-\nu}\, d\nu'$$

for $|\nu-4| > r$.

Definition 3.6. Let H be a 1PI subgraph of G. Suppose inductively that we have defined vertex parts $\mathcal{Y}_J$ for each 1PI $J \subset H$. Then define

$$\mathcal{Y}_H(\nu) = -K \sum_{A\in A_0(H)} F^{\mathcal{Y}}_{H:A}(\lambda^0,\nu,\varepsilon), \tag{3.7}$$

where $A_0(H) = \{A\in A(H)\,|\,A\neq\{H\}\}$. We will prove shortly that $\mathcal{Y}_H$

is a vertex part; thus we may define $\mathcal{D}_G \in S'(\mathbb{R}^{4m})$ by

$$\mathcal{D}_G(p) = \lim_{\substack{\nu\to 4\\ \varepsilon\to 0}} (2\pi)^4 \delta(\sum_1^m p_i) \sum_{A\in A(G)} F^{\mathcal{Y}}_{G:A}(\lambda^0,\nu,\varepsilon). \tag{3.8}$$

$\mathcal{D}_G$ is the <u>dimensionally renormalized amplitude</u> for G.

<u>Remark</u>: The $A = \emptyset$ term in (3.8) is the unrenormalized amplitude; other terms with $A \neq \{G\}$ represent subtractions for subgraphs. If G is 1PI, the sum in (3.8) may be written as

$$(1-K) \sum_{A\in A_0(G)} F^{\mathcal{Y}}_{G:A} ,$$

using (3.7) and the relation $F^{\mathcal{Y}}_{G:\{G\}} = \mathcal{Y}_G$. Thus the basic renormalization is again subtraction of a pole.

We now prove that $\mathcal{Y}_H$ is in fact a vertex part for H; this is the justification of the definition as a renormalization.

<u>Lemma 3.7.</u> $\mathcal{Y}_H$ is a vertex part for H.

<u>Proof</u>: We need a slightly stronger form of Theorem 3.4. Let $\{\hat{X}\}$ be a set of <u>finite</u> vertex parts for G, i.e., vertex parts regular at $\lambda = \lambda^0$, $\nu = 4$, and fix $A \in A(G)$. Then there are vertex parts $X_H(A,\hat{X})$, defined for those H such that, for each $H' \in A$, $H \cap H' = \emptyset$ or $H \supset H'$ (and in that case depending only on H and $\{H' \in A | H \supset H'\}$) so that

$$VF^{\hat{X}}_{G:A} = \sum_{B>A} F^{X(A,\hat{X})}_{G:B} (\lambda,\nu,\varepsilon); \tag{3.9}$$

here $B > A$ means that $B \in A(G)$ and that each $H \in A$ is contained in some $H' \in B$.

We now define finite vertex parts $\hat{X}_J(\nu)$ for 1PI $J \subsetneq H$ by

$$\hat{X}_J(\nu) = -(1-K) \sum_{A\in A_0(J)} X_J(A,\hat{X})(\lambda^0,\nu). \tag{3.10}$$

Let us assume inductively that for 1PI $J \subsetneq H$

$$Y_J(\nu) = \sum_{A\in A(J)} X_J(A,\hat{X})(\lambda^0,\nu), \tag{3.11}$$

i.e., from (3.10) and using $X_J(\{J\},\hat{X}) = \hat{X}_J$,

$$Y_J(\nu) = K \sum_{A_0(J)} X_J(A,\hat{X})(\lambda^0,\nu)$$

$$= K \sum_{A(J)} X_J(A,\hat{X})(\lambda^0,\nu). \tag{3.12}$$

Since (3.12) is clearly a vertex part, it suffices to verify (3.12) for $J = H$. Now

$$Y_H(\nu) = -K \sum_{A\in A_0(H)} F^{Y}_{H:A}(\lambda^0,\nu)$$

$$= -K \sum_{A\in A_0(H)} \sum_{B<A} F^{X(B,\hat{X})}_{H:A} (\lambda^0,\nu)$$

(using (3.11) and linearity in vertex parts)

$$= -K\{ \sum_{B\in A(G)} (\sum_{A>B} F^{X(B,\hat{X})}_{H:A} (\lambda^0,\nu)) - \sum_{B\in A(G)} X_H(B,\hat{X})(\lambda^0,\nu)\}$$

where we have used $F^{X}_{H:\{H\}} = X_H$. But the first term is regular at $\nu = 4$ from (3.9), and hence annihilated by K, so that we have verified (3.12).

IV. VECTOR AND SPINOR FIELDS

To date we have considered only normal products involving polynomials in scalar fields. When vector fields, spinor fields, or the derivatives of scalar fields are introduced, a more complicated formalism is necessary. This is because the Feynman integrals which formally define the amplitudes for various graphs no longer depend only on invariant inner products of momenta, but also involve directly the momentum vectors p^μ, as well as the Dirac matrices γ^μ and the metric tensor $g_{\mu\kappa}$. The proper interpretation of such quantities in complex dimension is delicate. Essentially, it is true that they may be treated as formal quantities which are to be manipulated according to obvious rules, e.g., $p^{\mu}p_{\mu} = p^2$, $\gamma^{\mu}\gamma^{\nu} + \gamma^{\nu}\gamma^{\mu} = g^{\mu\nu}$, $g^{\mu\nu}p_{\kappa} = p^{\mu}$, $g^{\mu}{}_{\mu} = \nu$ [6,11]. In this chapter we want to give a precise mathematical structure into which such formal manipulations may be imbedded.

A. Tensors

We recall from Def. 2.1 the space $\mathcal{Q}_m$ of measurable functions on $X_m = \mathbb{R}^m \times \{(S_{ij}) \mid S \geq 0, \quad i,j = 1,\dots m\}$. The elements of $\mathcal{Q}_m$ are scalar functions but do not include any dependence on the complex dimension; such dependence is implicitly present when tensor quantities are involved, due to the relation $g^{\mu}{}_{\mu} = \nu$. We therefore define the scalar quantities in our theory as follows.

Definition 4.1. Let R_m denote the space of measurable functions $f(x,\nu)$ on $X_m \times \mathbb{C}$. As above, if $k \in (\mathbb{R}^M)^m$, we frequently write $f(k,\nu)$ rather than $f(\phi_{mM}(k),\nu)$. We let $P_m \subset R_m$ be those functions polynomial in k.

R_m has the structure of an algebra over the complex numbers, i.e., it is a complex vector space with multiplication of any two elements defined. As such, it is a ring [i.e., we may add, subtract and multiply elements], and we may use the standard mathematical concept of a R_m module [17]; for any ring R, an R-module M is a set of elements for which addition, subtraction, and multiplication by elements of R are defined and such obvious rules as $f(a+b) = fa+fb$, $f(ga) = (fg)a$ $[a,b \in M,\ f,g \in R]$ are satisfied. For any ring R, one particularly simple R-module is the free R-module of rank s, $\bigoplus_{i=1}^{s} R$, defined as s-tuples $(a_1,\dots a_s)$ of elements of R with component-wise operations.

Definition 4.2. For R_m as above, define the R_m-module M_m by

$$M_m = \bigoplus_{i=1}^{m} R.$$

If M_m is defined by m-tuples as above, let $\underline{k}_i$ = $(0,0,\ldots 1, 0,\ldots 0)$, with the "1" in the ith place; every element a of M_m may thus be written uniquely as

$$a = \sum_{1}^{m} f_i \underline{k}_i \tag{4.1}$$

for some $f_1, \ldots f_m \in R_m$.

<u>Remark</u>: Elements of M_m represent the <u>vector</u> quantities of our theory. That is, just as any $f \in R_m$ gives a function $\phi^*_{mM} f$ of $k \in \mathbb{R}^{Mm}$, with $(\phi^* f)(k,\nu) = f(\phi_{mM}(k),\nu)$, so the element (4.1) of M_m gives a vector field $\phi^*_{mM}(a)$ on $\mathbb{R}^{Mm}$ with

$$[\phi^*_{mM}(a)]^\mu(k) = \sum_{i=1}^{m} \phi^*_{mM} f_i(k,\nu)(k_i)^\mu, \quad \mu = 0,1,\ldots M-1.$$

The natural way to generate tensors from the R-module M is to take repeated tensor products of M with itself. Recall that if M and N are R-modules, the tensor product $M \otimes N$ $(=M \otimes_R N)$ denotes all finite linear combinations, with coefficients in R, of expressions of the form $a \otimes b$ $[a \in M, b \in N]$, with the understanding that

$$(fa) \otimes b = f(a \otimes b) = a \otimes fb;$$

$$(a_1+a_2) \otimes b = a_1 \otimes b + a_2 \otimes b$$

$$a \otimes (b_1+b_2) = a \otimes b_1 + a \otimes b_2.$$

We let $M^r = M \otimes \ldots \otimes M$ (r factors); by convention $M^0 = R$, $M^1 = M$. The direct sum

$$T(M) = \bigoplus_{r=0}^{\infty} M^r$$

is called the <u>tensor algebra</u> of M; $T(M)$ is an R-module and in addition has a multiplicative structure $\otimes$, defined by $(a_1 \otimes \ldots \otimes a_r) \otimes (b_1 \otimes \ldots \otimes b_s) = a_1 \otimes \ldots \otimes b_s$.

In our case it does not suffice to consider the tensor algebra of M_m, since we will encounter the tensor $g^{\mu\kappa}$ as well as tensors $\underline{k}_i^{\mu}\, \underline{k}_j^{\kappa}$. We therefore give

Definition 4.3. Let R be a commutative ring, M an R-module. For $r \geq 0$ and $k \leq [\frac{r}{2}]$, let I_{rk} be the set of all partitions of $\{1,\ldots r\}$ into $k+1$ subsets $\{X_1,\ldots X_k,\psi\}$, with $|X_i| = 2$. The module $W(M)$ is defined by $W(M) = \bigoplus_{r=0}^{\infty} W^r(M)$, with $W^0(M) = R$, $W^1(M) = M$, and for $r \geq 2$,

$$W^r(M) = \bigoplus_{k=0}^{[\frac{r}{2}]} \bigoplus_{\alpha \in I_{rk}} [M^{(r-2k)}]_\alpha . \tag{4.2}$$

In (4.2) the second direct sum is of disjoint copies of the same module $M^{(r-2k)}$.

Definition 4.4. W_m is the R_m module $W(M_m)$; $W^r{}_m = W^r(M_m)$.

Remark. a) We introduce the following general notation. Let $(\mu_1,\mu_2,\ldots)$ be a sequence of distinct indices. Then if $\alpha = \{X_1,\ldots X_k,\psi\} \in I_{rk}$, with $X_i = \{a_i,b_i\}$ and $\psi = \{c_1,\ldots c_{r-2k}\}$, and if $x_1,\ldots x_{r-2k} \in M$, we dénote the element $x_1 \otimes \ldots \otimes x_{r-2k}$ in the summand M_α of (4.2) by

$$g^{\mu_{a_1}\mu_{b_1}} \ldots g^{\mu_{a_k}\mu_{b_k}} x_1^{\mu_{c_1}} \ldots x_{r-2k}^{\mu_{c_{r-2k}}}, \tag{4.3}$$

or by the similar expression produced by any permutation of these symbols, together with a change from $g^{\mu_a\mu_b}$ to $g^{\mu_b\mu_a}$.

b) In particular, W_m^r is a free R-module with basis consisting of all expressions

$$g^{\kappa_1\lambda_1} \ldots g^{\kappa_k\lambda_k} \underline{k}_{i_1}^{\xi_1} \ldots \underline{k}_{i_s}^{\xi_s} \tag{4.4}$$

where $2k+s = r$ and $\{\kappa_1,\dots \kappa_k, \lambda_1,\dots \lambda_k, \xi_1,\dots \xi_s\} = \{\mu_1,\dots \mu_r\}$. These expressions are precisely the symbolic quantities we have set out to define; i.e., each element w of W^r_m defines a tensor field $\phi^*_{mM}(W)$ of rank r (depending parametrically on ν) over $\mathbb{R}^{Mm}$, for any M.

There are three additional algebraic structures on $W(M)$ which we will need: (i) $W(M)$ is an algebra (compare $T(M)$): the element (4.3) in W^r and a similar element in $W^{r'}$ have product $g^{\mu_{a_1}\mu_{b_1}} \dots g^{\mu_{a_1'+r}\mu_{b_1'+r}} \dots x_1^{\mu_{c_1}} \dots x_2'^{\mu_{c_1'+r}} \dots$.

(ii) There is an R-linear representation π of S_r (the symmetric group on r letters) on W^r, where for $\sigma \in S_r$, $\pi(\sigma)$ takes (4.3) into

$$g^{\mu_{\sigma(a_1)}\mu_{\sigma(a_2)}} \dots x_1^{\mu_{\sigma(c_1)}} \dots .$$

(iii) Suppose that M is equipped with an inner product, i.e. an R-bilinear function $\langle x,y\rangle$ defined for $x,y \in M$ with values in R. Then for $1 \leq i < j \leq r$, the <u>contraction</u> map

$$C_{ij}: W^r \to W^{r-2}$$

is defined by the conditions

$$C_{12}(x^{\mu_1} y^{\mu_2}) = \langle x,y\rangle,$$

$$C_{12}(x^{\mu_1} g^{\mu_2\mu_3}) = x^{\mu_3}, \tag{4.5}$$

$$C_{12}(g^{\mu_1\mu_2}) = \nu,$$

$$C_{\sigma(i)\sigma(j)}\ \pi(\sigma) = \pi(\sigma')\ C_{ij}$$

for $\sigma \in S_r$ and σ' the corresponding element in S_{r-2}, and

$$C_{ij}(a\ b) = (C_{ij}a)b$$

if $a \in W^r$, $b \in W^s$ and $1 \leq i < j \leq r$.

Remark. a) The equation (4.5) above needs comment: the right-hand side is an element of M, which would normally be labelled with μ_1, the first member in the index set. Since our index set is arbitrary, however, we agree to make the notational convention that a contraction C_{ij} also changes the index set by omitting μ_i and μ_j. In this way the standard manipulations of tensor calculus hold for our quantities.

b) There is a natural scalar product on $W^r(M)$, given by

$$\langle a,b\rangle = (C_{1,r+1} \cdots C_{r,2r})(ab).$$

c) The above structures exist in W_m. The first two are clear; to define contractions, we introduce a scalar product into M_m: for fixed $\varepsilon > 0$ and $x = (y,S) \in X_m = (\mathbb{R}^m \times U_m)$, define

$$\langle \underline{k}_i, \underline{k}_j\rangle\ (x,\nu) = (1+i\varepsilon)\ y_i y_j - (1-i\varepsilon)\ S_{ij},$$

corresponding to the scalar product introduced in Def. 2.8. We note that consistency requires

$$\overset{*}{\phi}_{mM}(g)^{\kappa_1\kappa_2} = \begin{cases} i+i\varepsilon, & \kappa_1 = \kappa_2 = 0 \\ 1-i\varepsilon, & 0 < \kappa_1 = \kappa_2 \leq M-1 \\ 0, & \kappa_1 \neq \kappa_2. \end{cases}$$

To complete our discussion of complex dimensional tensors we must discuss integration, i.e., for $w \in W_m^r$, we want to define

$$\int w d^\nu k_{M+1} \cdots d^\nu k_m$$

as an element of W_M^r. Let $M' = M_M \otimes_{R_M} R_m$ (note that R_m is an

R_M-module), i.e., elements of M' are linear combinations of $k_1, \ldots \underline{k}_M$ whose coefficients depend on $k_1, \ldots k_m$. Then M' is an R_m-module and $W^r(M') = W^r_M \otimes_{R_M} R_m$ is in a canonical way an R_m submodule of W^r_m; let E^r_M denote orthogonal projection onto this submodule. [E^r_M exists because, if $\{w_i\}$ is our standard basis (4.4) for W^r_M, the matrix $A_{ij} = \langle w_i, w_j \rangle$ with elements in R_M is invertible, and we may set $E^r_M(w) = \Sigma A^{-1}{}_{ij} \langle w_i, w \rangle w_j$ for $w \in W^r_m$.] Now $f \longmapsto \int f \prod_{M+1}^{m} d^\nu k_i$ is an R_M-linear map of R_m into R_M, and thus extends to a natural R_M linear map (for which we use the same notation) of $W^r_M \otimes_{R_M} R_m$ into W^r_M:

$$\int \sum_i w_i f_i \prod d^\nu k = \sum_i w_i \int f_i \prod d^\nu k,$$

for $w_i \in W^r_M$, $f_i \in R_m$. Finally we have

<u>Definition 4.5.</u> For $w \in W^r_m$,

$$\int w \, d^\nu k_{M+1} \ldots d^\nu k_m = \int (E^r_M w) \, d^\nu k_{M+1} \ldots d^\nu k_m .$$

<u>Example:</u> Take $m = 2$, and let $w \in W^2_2$ be

$$w = \underline{k}^{\mu_1} \underline{k}^{\mu_2} e^{i\langle k-p, k-p\rangle};$$

we wish to calculate $\int w \, d^\nu k$. In this case the projection E^2_1 is given by

$$E^2_1(\underline{k}^{\mu_1} \underline{k}^{\mu_2}) = \frac{1}{\nu-1} \{[\langle p,p\rangle\langle k,k\rangle - \langle p,k\rangle^2]\langle p,p\rangle^{-1} g^{\mu_1\mu_2}$$

$$- [\langle p,p\rangle\langle k,k\rangle - \nu\langle p\cdot k\rangle^2]\langle p,p\rangle^{-2} \underline{p}^{\mu_1}\underline{p}^{\mu_2}\}.$$

We use the formulae

$$\int e^{\,i<k-p,k-p>}\, d^\nu k = \pi^{\frac{1}{2}\nu}\, c(\varepsilon)$$

$$\int <p,k>^2\, c^{i<k-p,k-p>}\, d^\nu k = \pi^{\frac{1}{2}\nu}\, c(\varepsilon)(<p,p>^2 + \frac{i}{2} <p,p>)$$

$$\int <k,k>\, e^{i<k-p,k-p>}\, d^\nu k = \pi^{\frac{1}{2}\nu}\, c(\varepsilon)(\frac{i\nu}{2} + <p,p>)$$

where $c(\varepsilon) = (1+i\varepsilon)^{-\frac{1}{2}}(1-i\varepsilon)^{-(\nu-1)/2}$; the first is from Lemma 2.6 and the others from similar calculations. Thus

$$\int w\, d^\nu k = \pi^{\frac{1}{2}\nu} c(\varepsilon)[g^{\mu_1\mu_2} - \underline{p}^{\mu_1}\underline{p}^{\mu_2}],$$

which is easily seen to be correct in any integer dimension $\nu = n$.

Remark. This simple example leads to a calculation sufficiently complicated to suggest that Def. 4.5 is not directly suited for practical calculations. Rather, one may derive standard formulae usually used to reduce Feynman amplitudes to α-space integrals [7,11]. We will not work through the details here; rather, we emphasize that Def. 4.5 has the advantage of generality, and is useful for deriving properties of the integration, as in the next theorem.

Theorem 4.6. a) For $w \in W_m^r$ and $1 \leq i < j \leq r$,

$$\int (C_{ij}\, w)\, d^\nu k_{M+1} \ldots d^\nu k_m = C_{ij} \{\int w\, d^\nu k_{M+1} \ldots d^\nu k_m\}. \tag{4.6}$$

b) For $w' \in W_M^{r'}$ and $w \in W_m^r$,

$$w' \int w\, d^\nu k_{M+1} \ldots d^\nu k_m = \int w'w\, d^\nu k_{M+1} \ldots d^\nu k_m. \tag{4.7}$$

Proof: a) Let $E = E_M^r$. It suffices to show that $C_{ij}\, Ew = EC_{ij}w$; to verify this, we take the inner product with an arbitrary element v of W_M^{r-2}:

$$\begin{aligned}
<v, C_{ij} E w> &= <g^{\mu_i\mu_j} v, E w> \\
&= <g^{\mu_i\mu_j} v, w> \\
&= <v, C_{ij} w> \\
&= <v, E C_{ij} w>
\end{aligned}$$

since $g^{\mu_i\mu_j} v \in W_M^r$ and E is orthogonal.

b) Similar.

Finally, consider the subring R_{am} of R_m consisting of all $f(x,\nu)$ which, for almost all x, are analytic in a fixed neighborhood of $\nu = 4$ except for a possible pole at $\nu = 4$. There is a corresponding module W_{am}, an R_{am}-submodule of W_m, and we can make

Definition 4.7. K: $W_{am} \longrightarrow W_{am}$ is defined to act by Def. 3.5 on the coefficients of tensors expressed in the standard basis (4.4).

Remark. K does not commute with contractions of the form $C_{ij}\, g^{\mu_i\mu_j}$. For example,

$$(1-K)\, C_{12}[g^{\mu_1\mu_2}/(\nu-4)] = (1-K)[\nu/(\nu-4)] = 1$$

$$C_{12}(1-K)[g^{\mu_1\mu_2}/(\nu-4)] = 0.$$

This is one source of anomalies in dimensional renormalization.

K does commute with contractions if they do not produce the trace of a metric tensor; that is, if A and B are disjoint subsets of $\{1,\ldots r\}$, $W_m^r(A|B) \subset W_m^r$ the submodule consisting of terms in (4.2) such that α does not contain a pair of indices, one from each of A and B, and $i_1, \ldots i_p \in A$, $j_1, \ldots j_p \in B$ distinct indices, then $K \Pi C_{i_s j_s} w = \Pi C_{i_s j_s} Kw$ for $w \in W_m^r(A|B)$. Now integration can produce metric tensors, but we have

Lemma 4.8. Let $k = (k_1, \ldots k_L)$, $p = (p_1, \ldots p_M)$, $q = (q_1, \ldots q_N)$, and $m = L+M+N$. If $w'(k,p) \in W_m^{r'}$, and $w''(k,q) \in W_m^{r''}$, then $\int w'w'' \Pi d^\nu p \, \Pi d^\nu q \in W_m^{r'+r''}(1,\ldots r'|r'+1,\ldots r'+r'')$, i.e. K and contractions C_{ij} with $1 \leq i \leq r < j \leq r+r'$ commute on this integral.

Proof: Omitted.

B. Spinor Amplitudes

We will discuss briefly the generalizations of the formulae of the previous section which are necessary for the inclusion of spinor quantities. (We do not discuss γ_5.) The goal is to introduce an algebraic structure in which expressions involving γ-matrices occur and in which such formulae as $\gamma^\mu\gamma^\nu + \gamma^\nu\gamma^\mu = 2g^{\mu\nu}$ have a natural interpretation. It is clear that we need a bigraded R_m-module:

$$W = \bigoplus_{r,s=0}^{\infty} W_m^{rs} ,$$

where as before r denotes the rank of an element of W as a Lorentz tensor, and s denotes the number of free pairs of spinor indices, i.e., for a Feynman amplitude, the number of open spinor lines in the graph. Of course,

$$W_m^{r0} \equiv W_m^r .$$

where the right hand side was defined in Section A.

We now construct the modules W_m^{r1}. W_m^{01} is an algebra; we write $W_m^{01} = A_m$. A_m must contain all the quantities $\not{k}_i$ as well as their products; thus we define it to be the Clifford algebra of M_m:

$$A = A_m = W_m^{01} = C(M_m),$$

according to

Definition 4.9. For M a module with inner product over the commutative ring R_0, the Clifford algebra $C(M)$ is defined to be $T(M)/T^\circ(M)$, where $T^\circ(M)$ is the subalgebra of $T(M)$ generated by all elements $a\otimes b + b\otimes a - 2\langle a,b\rangle$, $a,b \in M$. The natural injection $M \longrightarrow C(M)$ is denoted by $a \longmapsto \not{a}$; thus $\not{a}\,\not{b} + \not{b}\,\not{a} = 2\langle a,b\rangle$, $a,b \in M$.

The module W_m^{11} must contain the vectors $\underline{k}_i$, $i = 1,\dots m$ (equivalently, $\underline{k}_i^{\ \mu}$ or $\underline{k}_i^{\ \mu_1}$) as well as a vector $\gamma \equiv \gamma^\mu$ which, on specialization to integer dimension n, gives the usual n-vector of Dirac matrices. Thus we define

$$W^* = W_m^{10} \otimes_{R_m} A_m \oplus A_m \otimes_{R_m} A_m$$

and let γ denote the element $0 \oplus 1 \otimes 1$. W^* is a two-sided module over the algebra A; the significance of the tensor product with respect to R_m in the second summand, rather than with respect to A, is that elements of A cannot move through γ, i.e., $\not{a}\,\gamma = \not{a} \otimes 1 \neq 1 \otimes \not{a} = \gamma\,\not{a}$. Finally

$$W_m^{11} = W^* / W^{**},$$

where W^{**} is the two-sided A_m submodule of W^* generated by all elements $\not{a}\,\gamma + \gamma\,\not{a} - a \otimes 1$, $a \in M_m$.

We want to construct the other modules W_m^{r1}, $r > 1$, as in Def. 4.2, i.e., by taking tensor products of things in W_m^{11} and then adding metric tensors. A new complication arises, however. Consider the formula

$$\gamma^{\mu_1}\gamma^{\mu_2} + \gamma^{\mu_2}\gamma^{\mu_1} = 2g^{\mu_1\mu_2}$$

which should certainly hold, in some sense, in W_m^{21}. But how are we to distinguish $\gamma^{\mu_1}\gamma^{\mu_2}$ from $\gamma^{\mu_2}\gamma^{\mu_1}$? Both would seem to be simply $\gamma \otimes \gamma$. Our next definition makes this distinction by brute force, taking $\gamma^{\mu_1}\gamma^{\mu_2}$ to be $\gamma \otimes \gamma$ (recall that μ_1, μ_2 came from an ordered index set) and adding $\gamma^{\mu_2}\gamma^{\mu_1}$ as an extra element.

Remark. Another way to look at the above dilemma is to recall the natural representation π of S_2 on $M_m \otimes_{R_m} M_m$, given by $\pi(12)(a\otimes b) = b\otimes a$. In fixed integer dimension M such a rep may also be defined on the equivalent of $W^{11} \otimes_A W^{11}$, since in that case W^{11} is the tensor product of an R_m module (of vector fields on R^{Mm}) with a finite dimensional Clifford algebra, and the rep on $W^{11} \otimes W^{11}$ simply permutes the vector fields in the first factors. But there is no corresponding A-linear rep on $W^{11} \otimes_A W^{11}$ here. For we would have

$$\pi(12)[\not{a}\ \gamma \otimes \gamma] = \not{a}\,\pi(12)[\gamma \otimes \gamma] = \not{a}[\gamma \otimes \gamma],$$

but also

$$\pi(12)[\not{a}\ \gamma \otimes \gamma] = \gamma \otimes \not{a}\ \gamma = \gamma\ \not{a} \otimes \gamma,$$

and these two expressions are not the same.

The next definition is the proper generalization of Def. 4.2 for a non-commutative ring.

Definition 4.10. Let A be a (possibly) non-commutative ring, with center R, and let M be a 2-sided A module with $fa = af$ for $a \in M$ and $f \in R$. Define $W^*(M) = \bigoplus_{r=0}^{\infty} W^{*r}(M)$ as a graded A module by

$$W^{*r}(M) = \bigoplus_{k=0}^{[r/2]} \bigoplus_{\alpha \in I_{rk}} \bigoplus_{\sigma \in S(\psi)} (M \otimes_A M \otimes_A \cdots \otimes_A M)_{\alpha,\sigma}$$

where $\alpha = \{\chi_1, \ldots \chi_k, \psi\}$, $S(\psi)$ is the symmetric group on ψ, and this is a direct sum of disjoint copies of the tensor product of M with itself $r-2k$ times. For

$\alpha = \{\chi_1, \ldots \chi_k, \psi\}$, with $\chi_i = \{a_i, b_i\}$ and $\psi = \{c_1, \ldots c_{r-2k}\}$,

and $x_1, \ldots x_{r-2k} \in M$, we denote the element $x_1 \otimes \ldots \otimes x_{r-2k}$ in the (α,σ) summand by

$$g^{\mu_{a_1}\mu_{b_1}} \ldots x_1^{\mu_{\sigma(c_1)}} \ldots x_{r-2k}^{\mu_{\sigma(c_{r-2k})}}; \tag{4.11}$$

the order in which the x factors are written <u>is</u> important. We define a representation π of S_r on W^{*r} so that, for $\tau \in S_r$, $\pi(\tau)$ takes the element (4.11) into

$$g^{\mu_{\tau(a_1)}\mu_{\tau(b_1)}} \ldots x_1^{\mu_{\tau(\sigma(c_1))}} \ldots x_{r-2k}^{\mu_{\tau(\sigma(c_{r-2k}))}} \tag{4.12}$$

Note that (4.12) is an element of $(M \otimes \ldots \otimes M)_{\alpha',\sigma'}$, with $\alpha' = \{\tau(\chi_1) \ldots \tau(\chi_k), \tau(\psi)\}$ and $\sigma'\colon \tau(\psi) \to \tau(\psi)$ given by $\sigma'(c_i') = \tau\sigma(c_i)$. We also make W^* into an algebra by defining the product of (4.11) with a similar (primed) element to be

$$g^{\mu_{a_1}\mu_{b_1}} \ldots g^{\mu_{a_1'+r}\mu_{b_1'+r}} \ldots x_1^{\mu_{\sigma(c_1)}} \ldots (x_1')^{\mu_{\sigma(c_1')+r}} \ldots,$$

an element of $(M \otimes \ldots M)_{\alpha''\alpha''}$, where

$$\alpha'' = \{\{a_1,b_1\}, \ldots \{a_k'+r, b_k'+r\}, \{c_1, \ldots, c_{r'-2k}'+r\}$$

and $\sigma''(c_i) = \sigma(c_i)$; $\sigma''(c_{j+r}) = \sigma'(c_j')+r$. Now let $M_0 \subset M$ be the R-submodule on which the two actions of A agree, i.e., $M_0 = \{x \in M | ax = xa$ for all $a \in A\}$, let W^{**} be the A-subalgebra of W^* generated by elements

$$x_1^{\mu_1} x_2^{\mu_2} - x_1^{\mu_2} x_2^{\mu_1} \in W^{*2}$$

for $x_1, x_2 \in M_0$, and invariant under π, and let $W^+(M) = W^*(M)/W^{**}(M)$.

We apply Def. 4.10 to our problem, taking $A = A_m$, $R = R_m$, and $M = W_m^{11}$. Let W^{++} be the π-invariant subalgebra of W^+ generated by

$$\gamma^{\mu_1}\gamma^{\mu_2} + \gamma^{\mu_2}\gamma^{\mu_1} - 2g^{\mu_1\mu_2}.$$

Then we define

$$W_m^{r1} = (W^+)^r/(W^{++})^r.$$

Again there is a natural contraction map $C_{ij}: W_m^{rs} \longrightarrow W_m^{r-2,s}$, acting as before on $\underline{k}_i$ and g and with

$$C_{12}\,\gamma^{\mu_1}\gamma^{\mu_2} = \nu$$

$$C_{12}\,\gamma^{\mu_1}\underline{k}_i^{\mu_2} = \not{k}_i$$

$$C_{12}\,\gamma^{\mu_1}g^{\mu_2\mu_3} = \gamma^{\mu_3}.$$

There is also a map $\omega: W_m^{r1} \longrightarrow W_m^{r0}$, the <u>trace</u>, which we will not discuss in detail.

We also omit a detailed construction of W_m^{rs} for $s \geq 2$. The only new complication is illustrated for $s = 2$: W_m^{02} must contain, along with the obvious summand $W_m^{01} \otimes_{R_m} W_m^{01}$, new terms corresponding to contraction of γ matrices from different spinor lines: $\gamma_1^{\mu}\,\gamma_{2\mu}$. We also omit discussion of integration, which is completely similar to the tensor case.

V. PROPERTIES OF DIMENSIONALLY RENORMALIZED PRODUCTS

A. Free Fields

We now return briefly to the definition of Feynman amplitudes. Again we consider free fields; for notational simplicity we consider a collection of scalar fields $\{\phi_\alpha\}$, although vector and spinor fields could easily be introduced. Covariant quantities will occur in Feynman amplitudes because we will consider vacuum expectation values of field derivatives. The fields are described by a Lagrangian L_0, i.e., a covariant homogeneous polynomial of degree two in the fields and their first derivatives. The field equations

$$\frac{\partial L}{\partial \phi_\alpha} - \partial_\mu \frac{\partial L}{\partial(\partial_\mu \phi_\alpha)} = 0 \tag{5.1}$$

have the form

$$\sum_\beta K_{\alpha\beta}\, \phi_\beta = 0 \tag{5.2}$$

with K a second order differential operator. The propagators are then the Greens functions for (5.2); if we denote them by $\Delta_{\alpha\beta}$, we have

$$\sum_\beta K_{\alpha\beta}\, \Delta_{\beta\gamma}(x) = i\, \delta_{\alpha\gamma}\delta(x). \tag{5.3}$$

In momentum space each propagator may be expanded in the form

$$\tilde{\Delta}_{\alpha\beta}(p) = \sum_j (-i)\, Z^j_{\alpha\beta}(p)(m_j^2 - p^2 - i0)^{-1}. \tag{5.4}$$

Equation (5.3) then becomes

$$\sum_\beta \tilde{K}_{\alpha\beta}\, \tilde{\Delta}_{\beta\gamma} = i\, \delta_{\alpha\gamma} \tag{5.5}$$

with $\tilde{K}$ now a covariant polynomial in p.

Now let $\{Y_i | i=1,\ldots m\}$ be a collection of formal monomials in the fields and their derivatives; such a monomial Y is completely specified by non-negative integers $N(\alpha,r)$ giving the number of occurrences of the rth order derivative $(r \geq 0)$ of ϕ_α in Y, and an integer $k \geq 0$ giving the number of metric tensor factors. We consider vacuum expectation values of linear combinations (with complex coefficients) of expressions

$$C\,Y_1(x_1)\ldots Y_m(x_m). \tag{5.6}$$

Here C denotes a formal contraction of some Lorentz indices. A complete mathematical specification of such quantities may be given (compare the definitions of Lam [18]); it must recognize the equivalence of contraction of different derivative indices on the same field, etc., and must include an ordering of the uncontracted indices in order, for example, to distinguish the two terms in $\partial^{\mu_1}\phi_\alpha\,\partial^{\mu_2}\phi_\beta + \partial^{\mu_2}\phi_\alpha\,\partial^{\mu_1}\phi_\beta$. We omit a more thorough discussion; there are no logical pitfalls.

An ambiguity remains in the contraction of a metric tensor with itself. We choose to have a 1-1 correspondence between linear combinations of our forms (5.6) and ordinary covariant field polynomials; this forbids ν-dependent coefficients and thus we agree that any trace of a metric tensor implicit in (5.6) be given value 4 before regularization.

TOVEV's of these linear combinations are defined by linearity; an elementary TOVEV

$$\langle 0|T[C\,Y_1(x_1)\ldots Y_m(x_m)|0\rangle$$

is defined as follows:

(a) Using Wick's theorem, each term is expanded as a sum over Feynman graphs. The propagator for a line ℓ arising from the contraction of $(\partial^{\mu_1}\ldots\partial^{\mu_k}\phi_\alpha)$ with $(\partial^{\kappa_1}\ldots\partial^{\kappa_j}\phi_\beta)$ is, in momentum space,

$$\tilde{\Delta}_\ell(q) = (i\underline{q})^{\mu_1}\ldots(i\underline{q})^{\mu_k}(-i\underline{q})^{\kappa_1}\ldots(-i\underline{q})^{\kappa_j}\,\tilde{\Delta}_{\alpha\beta}(q) \tag{5.7}$$

[we are assuming that the line representing the contraction is oriented from ϕ_α to ϕ_β]. (5.7) is an element of W_1^{k+j}, in the notation of Chapter 4.

(b) Using (5.4), we define

$$\tilde{\Delta}_\ell(\lambda_\ell,\varepsilon)(q) = \sum_j (-i)\, Z_{\ell j}(q)\,(m_j^2(1-i\varepsilon)-\langle q_\ell,q_\ell\rangle)^{-\lambda_\ell},$$

with

$$Z_{\ell j} = \prod_{r=1}^{k} (i\underline{q})^{\mu_r} \prod_{s=1}^{j} (-i\underline{q})^{\kappa_s} Z^j_{\alpha\beta}(q);$$

again, $\tilde{\Delta}_\ell \in W_1^{k+j}$. If the graph G is connected, we define the regularized amplitude by

$$F_G(\lambda,\nu,\varepsilon) = z^\omega(2\pi)^{-h\nu} \int \prod_1^h d^\nu k_i \; C \prod_L \tilde{\Delta}_\ell(\lambda_\ell,\varepsilon)(q_\ell). \tag{5.8}$$

Here $q_\ell = q_\ell(p,k)$ as in (2.22), the contraction operator C is now a product of contraction operators C_{ij} in W_L, and the integration is in the sense of Def. 4.5.

(c) Renormalization is as in Def. 3.6:

$$Y_H(\nu) = -K \sum_{A\in A_0(H)} F^Y_{H:A}(\lambda^0,\nu,\varepsilon);$$

$$\mathcal{D}_G(p) = \lim_{\substack{\nu\to 4\\ \varepsilon\to 0}} (2\pi)^4\, \delta\Big(\sum_1^m p_i\Big) \sum_{A\in A(G)} \phi^*_{m-1,4}[F^Y_{G:A}(\lambda^0,\nu,\varepsilon)].$$

The amplitude $F^Y_{H:A}$ contains all contractions over two indices associated with internal lines of H.

There is one remark about the above procedure: the expansion (5.4) of the propagator is not unique. [For example, one can always add

$$\frac{(m_1^2-p^2)}{(m_1^2-p^2-i0)} - \frac{(m_2^2-p^2)}{(m_2^2-p^2-i0)}.]$$

This means that the regularized amplitude $F_G(\lambda,\nu,\varepsilon)$ is not uniquely defined. One could attempt to impose conditions on (5.4) to make these expansions unique; however, since we are primarily interested in dimensional renormalization, we will simply prove

Lemma 5.1. If G is a Feynman graph for which each line ℓ has propagator

$$\tilde{\Delta}_\ell(p) = \sum_{i=1}^{r} (-i)\, Z_{\ell i}(p)(m_{\ell i}^2 - p^2 - i0)^{-1}, \tag{5.9}$$

the dimensionally regularized and renormalized amplitudes $F_G(\lambda^0,\nu,\varepsilon)$ and $\mathcal{D}_G$, defined using the propagator

$$\tilde{\Delta}_\ell(\lambda_\ell)(p) = \sum_{i=1}^{r} (-i)\, Z_{\ell i}(p)(m_{\ell i}^2 - p^2 - i0)^{-\lambda_\ell}$$

are independent of the choice of expansion (5.8).

Proof: Fix a line ℓ, and let F_G, F'_G be two regularized amplitudes defined using two different forms $\Delta_\ell(\lambda_\ell)$, $\Delta'_\ell(\lambda_\ell)$. Suppose first that the endpoints of ℓ are distinct. Then in the defining integral of $F_G(\lambda,\nu,\varepsilon)$ (5.8) we can, by choosing $\mathrm{Re}\,\lambda_{\ell'}\,(\ell'\neq\ell)$ sufficiently large, set $\lambda_\ell = 1$ without destroying absolute covergence; comparison of the F_G and F'_G integrals shows that $F_G|_{\lambda_\ell=1} = F'_G|_{\lambda_\ell=1}$; this proves the statement for F_G. The same proof shows that $F^y_{G:A}$ is independent of the choice of expansion (5.9) for the propagators on lines not contained in an element of A, and from this one sees by induction that $\mathcal{D}_G$ is independent of these choices.

If ℓ is a loop line, we modify the argument slightly. Let H be a graph obtained from G by adding an extra vertex x' and line ℓ', as in Figure 5.1. As propagator for ℓ' we take $\Delta_{\ell'} = 1$; this has an expansion of the form (5.9) as

$$\Delta'_\ell = \frac{(M^2-p^2)}{(M^2-p^2-i0)}$$

for some fixed $M > 0$; thus by the case considered above,

$$F_H|_{\lambda_\ell=\lambda_{\ell'}=1}$$

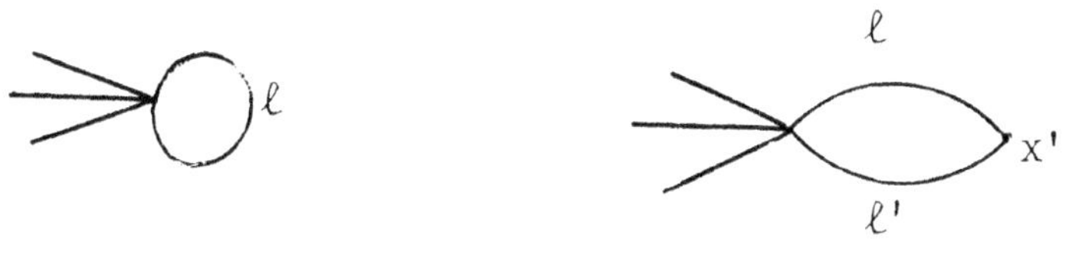

Fig. 5.1.

is independent of the choice of the form of propagator for Δ_ℓ. But

$$F_H\Big|_{\lambda_{\ell'}=1,\ p'=0} = F_G$$

and hence $F_G\Big|_{\lambda_\ell=1}$ is also independent.

The important properties of time-ordered vacuum expectation values defined using dimensional renormalization are stated in the next two theorems.

Theorem 5.2. (Differentiation):

(a) $$\frac{\partial}{\partial x_{i\mu}} \langle 0|T[C\ Y_1(x_1)\dots\ Y_n(x_m)]|0\rangle$$
$$= \langle 0|T[C\ Y_1(x_1)\dots\ (\partial\ Y_i(x_i)/\partial x_{i\mu})\dots]|0\rangle \qquad (5.10)$$

where $\partial Y_i/\partial x$ is the new covariant polynomial obtained by formal differentiation.

(b) Equation (5.10) also holds if the index μ is contracted with some index μ_i on the left, and the contraction of μ with μ_i is included with the operator C on the right.

(c) The Fourier transform of (5.10) can be restricted by $p_{i\mu} = 0$, and gives 0. Formally,

$$\int d^4x_i \frac{\partial}{\partial x_{i\mu}} \langle 0|T[C\ Y_1\dots\ Y_m]|0\rangle = 0.$$

Proof: (a) We expand $<0|T[C\,Y_1\dots Y_m]|0>$ as a sum of amplitudes for various graphs, and let $F_G(\lambda,\nu,\varepsilon)$ be the regularized amplitude for one of these. Then in W_m,

$$i\,\underline{p}_i\,F_G = i\,\underline{p}_i z^{\omega}(2\pi)^{-h\nu}\int d^{\nu}k_1\dots d^{\nu}k_h\,\Pi\,\tilde{\Delta}_{\ell}(\lambda_{\ell},\varepsilon)(q_{\ell})$$

$$= z^{\omega}(2\pi)^{-h\nu}\int d^{\nu}k_1\dots d^{\nu}k_h(-\sum_{\ell} e_i^{\ell}(i\,\underline{q}_{\ell})\Pi\,\tilde{\Delta}_{\ell}),$$

and the last expression is precisely a term in the expansion of the right hand side of (5.10). This equality persists through the renormalization steps; on specialization to 4 dimensions it becomes the Fourier transform of (5.10).

(b) This obvious-looking result is mentioned explicitly because, as remarked in Section 4, contraction does not in general commute with renormalization. Here, however, we cannot be contracting a metric tensor with itself; the desired commutation therefore holds.

(c) The Fourier transform of $<0|T[C\,Y_1\dots Y_m]|0>$ can be restricted to $p_{i\mu} = 0$ by Lemma 2.14; since we obtain the Fourier transform of (5.10) by multiplication by p_i^{μ}, the result is obvious.

The statement of the next theorem is simplified by the following notation. Let

$$D^{\mu}_{rs} = \partial^{r}/\partial x_{s\mu_1}\dots\partial x_{s\mu_r};$$

we suppress the index s when only one variable occurs. For a monomial $Y(x)$ containing N factors of the rth derivative of ϕ_{α}, written symbolically

$$Y(x) = \prod_{i=1}^{N}[(D_r^{\mu_i}\phi_{\alpha})(x)]\hat{Y}(x),$$

we define

$$\frac{\partial Y}{\partial(D_r^{\kappa}\phi_\alpha)} = \sum_{i=1}^{N} [(\sum_{\sigma\in S_r} \prod_{k=1}^{r} \delta_{\eta_{\sigma(k)}}^{\mu_{ik}}) \prod_{j\neq i} (D_r^{\mu_j}\phi_\alpha)\hat{Y}], \tag{5.11}$$

and in general

$$\frac{\bar{\partial}\, Y}{\bar{\partial}\,\phi_\alpha} = \sum_{r=0}^{\infty} (-1)^r D_r^{\mu} \frac{\partial Y}{\partial(D_r^{\kappa}\phi_\alpha)}$$

(5.11) is still symbolic, but its interpretation in a precise definition of covariant polynomials (see discussion following (5.6)) is clear.

Theorem 5.3. Let $\{Y_i(x) \mid i=1,\dots m\}$ be covariant polynomials as above, with

$$Y_1(x) = \left(\frac{\bar{\partial}\, L_0}{\bar{\partial}\,\phi_\alpha}\right) Y(x).$$

Then

$$\langle 0|T[C\, Y_1(x_1)\dots Y_m(x_m)]|0\rangle = i \sum_{i=2}^{m} \sum_{r,s=0}^{\infty} \binom{r+s}{r}$$
$$D_{ri}^{\mu}\{\delta(x_1-x_i)\langle 0|T[C\, Y(x_i)\, Y_2(x_2)\dots (D_{si}^{\kappa}(\frac{\partial\, Y_i}{\partial\, D_r^{\mu} D_s^{\kappa}\,\phi_\alpha})(x_i)) \tag{5.12}$$
$$\dots Y_m(x_m)]|0\rangle$$

Remark 5.4. (a) This is the equation of motion for ϕ_α. Similar equations have been derived using BPHZ renormalization [19], and (5.10) is a special case of the action principle discussed using BPHZ by Lam [20]. [10] and [11] discuss (5.12) and its generalizations in dimensional renormalization; the important advantage, as we will see, is that (5.12) holds with arbitrary factor $Y(x)$; in BPHZ, anomalies occur if $Y(x)$ is degree two or higher in the fields.

(b) It is understood in (5.12) that the contractions over κ are made before renormalization, while those over μ are ordinary 4-dimensional sums. Compare Theorem 5.2 (b).

(c) It should be noticed in the following proof that two choices made earlier are crucial to the validity of (5.12). First, in expanding time-ordered products via Wick's theorem, we permitted Feynman graphs containing loop lines. Second, our subtraction procedure allowed for subtractions for all 1PI subgraphs (the usual procedure in momentum space), not just for subgraphs generalized vertices (as in BPH renormalization).

(d) The notation in (5.12) could be simplified by introducing the functional derivative

$$\frac{\delta\, Y(x_i)}{\delta\, \phi_\alpha(x\,)} = \sum_{r,s=0}^{\infty} \binom{r+s}{r} D^\mu_{ri}\{\delta(x-x_i)\, D^\kappa_{si} \frac{\partial\, Y}{\partial\, D^\mu_r\, D^\kappa_s\, \phi_\alpha}\},$$

(5.12) then becomes

$$\frac{\bar{\partial}\, L_0}{\bar{\partial}\, \phi_\alpha}(x) = i\, \frac{\delta}{\delta\, \phi_\alpha(x)}\, .$$

We have avoided this notation; its use would have to be accompanied by a prescription for taking the δ function and D^μ_{ri} operator outside the TOVEV in (5.12).

(e) In typical applications of (5.12), $Y_2, \ldots Y_M$ are field operators and we integrate over $x_{M+1} \ldots x_m$. In this case, by Theorem 5.2 (c), only the $r = 0$ terms contribute:

$$\int d^4x_{M+1} \ldots d^4x_m \, \langle 0|T[C\; Y_1(x_1) \ldots Y_m(x_m)|0\rangle$$

$$= i \sum_{i=2}^{m} \delta(x_1 - x_i) \langle 0|T[C\; Y(x_1) \ldots (\bar{\partial}\, Y_i/\bar{\partial}\, \phi_\alpha) \ldots]|0\rangle .$$

We may use the functional derivative notation of (d) without ambiguity in this case.

<u>Proof of Theorem 5.3</u>: By linearity, we may assume that $Y(x)$, $Y_2(x), \ldots Y_m(x)$ are monomials in the fields and their derivatives. Consider then a particular graph G in the Wick's theorem expansion of the right hand side of (5.12). Since $\bar{\partial}\, L_0/\bar{\partial}\, \phi_\alpha$ is homogeneous of degree 1 in the fields, we may think of this factor as being joined by a line ℓ to some field $D^\eta_{ni}\, \phi_\beta$ occurring in $Y_i(x_i)$. Since $\bar{\partial}\, L_0/\bar{\partial}\, \phi_\alpha$ has the form

(5.2), (5.5) implies that ℓ has propagator

$$\tilde{\Delta}_\ell(q_\ell) = i\delta_{\alpha\beta}(-i\,\underline{q}_\ell)^{\eta_1}\ldots(-i\,\underline{q}_\ell)^{\eta_n} \qquad (5.13)$$

Now there are two cases to be distinguished.

Case 1: $i = 1$. In this case ℓ is a loop joining V_1 to itself; we will show that the renormalized amplitude $\mathcal{D}_G$ vanishes. For in the integral for $\mathcal{F}_G$:

$$\mathcal{F}_G(\lambda^0,\nu,\varepsilon) = z^\omega(2\pi)^{-h\nu}\int d^\nu k_1\ldots d^\nu k_h\,\Pi\,\tilde{\Delta}_{\ell'}(q_{\ell'}) \qquad (5.14)$$

we may choose (say) k_h so that $q_\ell = k_h$ and $q_{\ell'}$, $\ell \neq \ell'$, is independent of k_h. By Lemma 2.6 we may treat (5.14) as an iterated integral, so that $\mathcal{F}_G = 0$ if

$$\int d^\nu k\,\underline{k}^{\eta_1}\ldots\underline{k}^{\eta_n} = 0. \qquad (5.15)$$

According to our general prescriptions, (5.15) is evaluated as

$$\lim_{\lambda\to 0}\int d^\nu k\,\underline{k}^{\eta_1}\ldots\underline{k}^{\eta_n}\,(m^2(1-i\varepsilon) - \langle k,k\rangle)^{-\lambda}.$$

We exponentiate the propagator as in (2.25); the factor $\Gamma(\lambda)$ in the denominator is not cancelled by any pole produced during the k integration (all such poles depend on ν); hence the integral vanishes for $\lambda = 0$. Thus $\mathcal{F}_G = 0$. Now

$$\mathcal{D}_G = \lim_{\nu\to 4}(2\pi)^4\,\delta(\Sigma\,p_i)(1-K)\sum_A \phi^*(\mathcal{F}^{\mathcal{Y}}_{G:A});$$

if $\ell \notin H$ for any $H \in A$, $\mathcal{F}_{G:A}$ vanishes by the above argument; if $\ell \in H$, then $\mathcal{Y}_H = 0$ by an inductive step; hence $\mathcal{D}_G = 0$.

Case 2. $i \neq 1$. We have isolated a particular factor $D^\eta_{ni}\,\phi_\beta$ of Y_i; suppose $Y_i = (D^\eta_{ni}\,\phi_\beta)\hat{Y}_i$. Then

$$\frac{\partial Y_i}{\partial (D_r^\mu D_s^\kappa \phi_\alpha)} = \delta_{r+s,n} \sum_{\sigma\in S_n} \delta_{\mu_1}^{\eta_{\sigma(1)}} \cdots \delta_{\kappa_1}^{\eta_{\sigma(r+1)}} \cdots \hat{Y}_i$$

$$+ \text{(other terms)},$$

so that the right hand side of (5.12) contains a term

$$\sum_{B\subset\{1,\dots n\}} \left(\prod_{j\in B} \partial/\partial x_{i\eta_j}\right) \delta(x_1 - x_i) \langle 0|T[C\, Y(x_i)\dots$$

$$\prod_{j\notin B} \left(\frac{\partial}{\partial x_{i\eta_j}}\right) \hat{Y}_i \dots]|0\rangle. \tag{5.16}$$

When the vacuum expectation value appearing in (5.16) is expanded via Wick's theorem, there will be a graph $\tilde{G}$ in which precisely the same contractions are made as in the graph G considered originally, so that $\tilde{G} = G/\{\ell\}$. The regularized amplitude will be

$$F_{\tilde{G}}(\lambda,\nu,\varepsilon) = \sum_B \prod_{j\in B} (-i\, \underline{p}_i^{\eta_j})\, z^\omega\, (2\pi)^{-h\nu} \int d^\nu k_1 \dots d^\nu k_h$$

$$\prod_{j\notin B} (i\underline{Q}^{\eta_j}) \prod_{L(\tilde{G})} \tilde{\Delta}_{\ell'}(\lambda_{\ell'},\varepsilon)(q_{\ell'}). \tag{5.17}$$

Here $\underline{Q}$ is the total momentum incident on V_i from those internal linear of $\tilde{G}$ which arise from contractions of fields in $\hat{Y}_i$:

$$\underline{Q} = \sum_{\ell'\in L(\tilde{G})} e_i^{\ell'}\, \underline{q}_{\ell'},$$

with e the incidence matrix of G. But since $\underline{q}_\ell = \underline{p}_j - \underline{Q}$, (5.17) is the amplitude for G, with λ_ℓ set to one, so that

$$F_G(\lambda^0,\nu,\varepsilon) = F_{\tilde{G}}(\lambda^0,\nu,\varepsilon). \tag{5.18}$$

We must now extend (5.18) to equality of renormalized amplitudes, proving (5.12).

The equality of $\mathcal{D}_G$ and $\mathcal{D}_{\tilde{G}}$ is not obvious because the counterterm structure used to define them does not appear to be the same. In fact, because certain counterterms for $\mathcal{D}_G$ vanish, the two structures are equivalent; it is here that our decisions to use all 1PI graphs for counterterms and to allow loop lines become important. The desired equality follows by taking $G = H$ in

Lemma 5.5. If H is a connected subgraph of G which contains ℓ, and $\tilde{H} = H/\{\ell\}$, then $\mathcal{D}_H = \mathcal{D}_{\tilde{H}}$; if H is 1PI, then also $\mathcal{Y}_H = \mathcal{Y}_{\tilde{H}}$.

Proof: Let

$$P_H(\nu,\varepsilon) = \sum_{A \in A_0(H)} F^{\mathcal{Y}}_{H:A}(\lambda^0,\nu,\varepsilon); \tag{5.19}$$

it suffices to show $P_H = P_{\tilde{H}}$. If H has no proper 1PI subgraphs, then neither does $\tilde{H}$, and

$$P_H(\nu,\varepsilon) = F_H(\lambda^0,\nu,\varepsilon) = F_{\tilde{H}}(\lambda^0,\nu,\varepsilon) = P_{\tilde{H}}(\nu,\varepsilon),$$

by (5.17). Thus we may assume by induction that $\mathcal{Y}_J = \mathcal{Y}_{\tilde{J}}$ for all 1PI subgraphs J of H which contain ℓ.

Now fix $A = \{J_1,\dots J_k\} \in A_0(H)$ and consider the three possible cases:

Case 1: $\ell \in L(J_i)$ for some i. Let $\hat{A} = \{J_1,\dots,\tilde{J}_i,\dots J_k\}$; then $\hat{A} \in A_0(\tilde{H})$ and

$$F^{\mathcal{Y}}_{H:A}(\lambda^0,\nu,\varepsilon) = F^{\mathcal{Y}}_{\tilde{H}:\hat{A}}(\lambda^0,\nu,\varepsilon) \tag{5.20}$$

using the induction assumption $\mathcal{Y}_{J_i} = \mathcal{Y}_{\tilde{J}_i}$ and the same argument which led to (5.18).

Case 2: $\ell \notin L(\bigcup J_i)$, with ℓ a loop line in G/A. Here $F^{\mathcal{Y}}_{H:A} = 0$ by (5.15).

Case 3: $\ell \notin L(\bigcup J_i)$, with ℓ not a loop line in G/A. Let

$\hat{A}$ be the set of components of the subgraph of $\tilde{H}$ which contains the same lines as $\bigcup J_i$. Either $\hat{A} = A$, in which case (5.20) holds by the argument of (5.18), or, if ℓ joins two elements J_1 and J_2 of A, $\hat{A} = \{J, J_3, J_4, \ldots\}$ with $L(J) = L(J_1) \cup L(J_2)$. In this case, from Lemma 5.6 (below), $Y(J) = Y(J_1)Y(J_2)$, and again (5.20) holds.

Now an easy verification shows that the families $\hat{A}$ produced in cases 1 and 3 are distinct and exhaust $A_0(\tilde{H})$. Summing (5.20) over all $A \in A_0(H)$, we find $P_H = P_{\tilde{H}}$.

The different cases possible are illustrated in Fig. 5.2. Note that $\tilde{H}$ contains a loop line even though H does not and that, in the second example under Case 3, a subtraction corresponding to two generalized vertices in H becomes a subtraction for a subgraph of $\tilde{H}$ which is not a generalized vertex. It is such graphs which force the choices mentioned in Remark 5.4 (c).

We finally verify that the vertex parts Y of dimensional renormalization have the factorization property needed above.

Lemma 5.6. Let H be a 1PI graph with two 2-connected components H_1, H_2. Then

$$Y_H = Y_{H_1} Y_{H_2}. \tag{5.21}$$

Proof: By induction we may assume that (5.21) holds for all 1PI subgraphs J of H such that $J = J_1 \cup J_2$, $J_i \subset H_i$, and either $J_1 \neq H_1$ or $J_2 \neq H_2$. (It is easiest to start the induction with either J_1 or J_2 a trivial graph, with vertex part 1). Now any $A \in A(H)$ generates a unique pair (A_1, A_2), with $A_i \in A(H_i)$; by the induction assumption above and "Fubini's theorem" (Lemma 2.6),

$$F^{Y}_{H:A} = F^{Y}_{H_1:A_1} F^{Y}_{H_2:A_2}$$

if $A \neq \{H\}$. But then

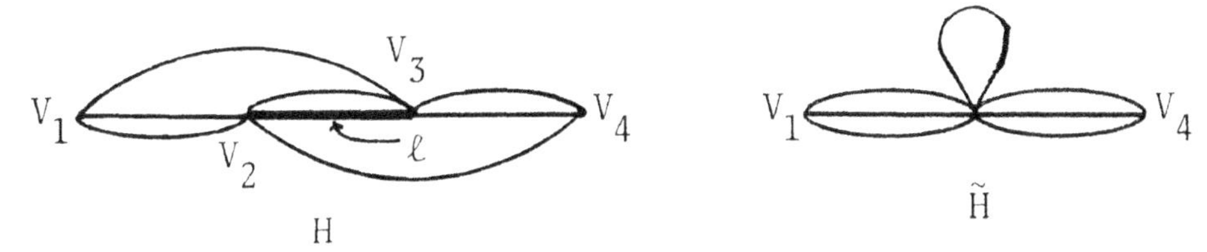

Fig. 5.2a Graph and quotient

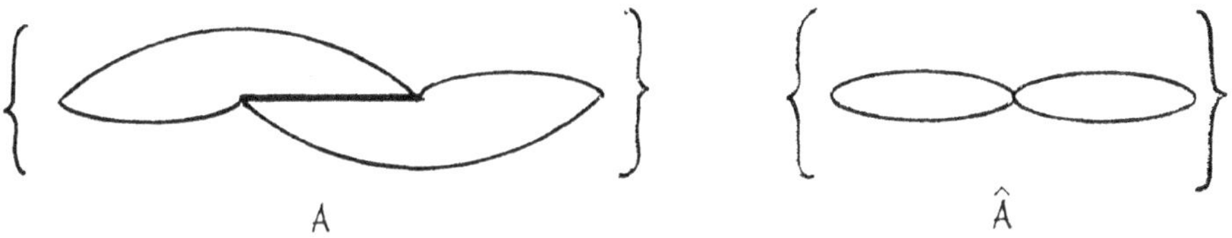

Fig. 5.2b Case 1

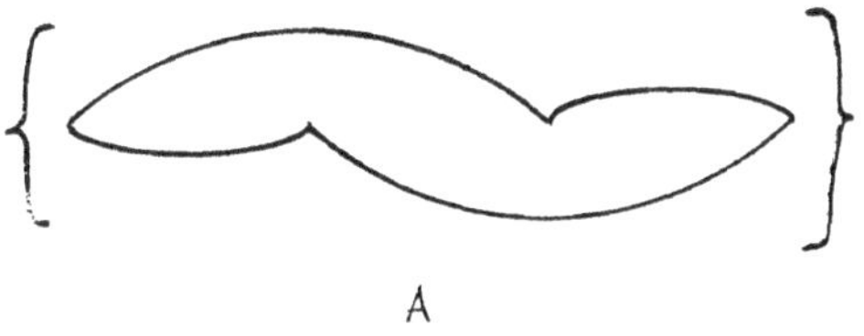

$F^{y}_{H:A} = 0$

Fig. 5.2c Case 2

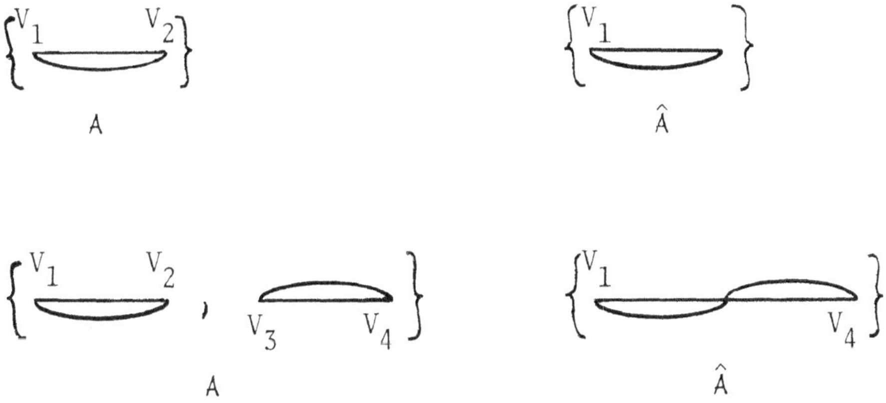

Fig. 5.2d Case 3

$$P_H = \sum_{A\in A_0(H)} F^y_{H:A}$$

$$= \sum_{\substack{A_1\in A_0(H_1)\\ A_2\in A_0(H_2)}} F^y_{H_1:A_1} F^y_{H_2:A_2} + \sum_{A_1\in A_0(H_1)} F^y_{H_1:A_1} y_{H_2}$$

$$+ y_{H_1} \sum_{A_2\in A_0(H_2)} F^y_{H_2:A_2}$$

$$= P_{H_1} P_{H_2} + P_{H_1} y_{H_2} + y_{H_1} P_{H_2}$$

$$= (1-K)P_{H_1} (1-K)P_{H_2} - y_{H_1} y_{H_2}, \qquad (5.22)$$

since $y_{H_i} = -KP_{H_i}$. Now $y_H = -KP_H$; the first term in (5.22) is analytic at $\nu = 4$ and hence annihilated by K; the second is a product of factors with only negative terms in their Laurent series, hence preserved by K (Lemma 4.8 is needed here to ensure that contractions present in the amplitude for H do not produce ν-dependence in this term).

B. Interacting Fields

We describe briefly the standard derivation of properties of time-ordered vacuum expectation values of interacting fields from those for free fields. Consider interacting fields $\{\Phi_\alpha\}$ corresponding to the free fields of the previous section, governed by a Lagrangian

$$L = L_0 + L_I.$$

Here L_0 is the Lagrangian function of section A, and L_I is a formal power series in one or more coupling constants, all terms being at least first order. The Gell-Man Low formula,

$$<0|T[Y_1(x_1)\dots Y_m(x_m)]|0> =$$

$$<0|T[\overline{Y}_1(x_1)\dots \overline{Y}_m(x_m)\, e^{i\int \overline{L}_I(y)dy}]|0>_0$$

($\langle\,\rangle_0 \equiv$ excluding vacuum-vacuum parts), where $\{Y_i\}$ are covariant polynomials in the interacting fields and $\{\overline{Y}_i\}$ the same polynomials in the free fields, defines the vacuum expectation value on the left as a formal power series in the coupling constants, whose coefficients are the vacuum expectation values of products of free fields defined earlier. Specifically, the right hand side is

$$\sum_{n=0}^{\infty} i^n \int dy_1 \ldots dy_n \, \langle 0|T[\overline{Y}_1(x_1)\ldots \overline{Y}_m(x_m)\overline{L}_I(y_1)\ldots\overline{L}_I(y_n)]|0\rangle_0$$

and the y integrations are interpreted in the sense of Theorem 5.2.

The differentiation properties described by Theorem 5.2 carry over to the interacting case without change. Moreover, the interacting fields satisfy the correct equations of motion derived from L. For in the sense of formal power series,

$$\frac{\delta}{\delta\,\phi_\alpha(x)}\, e^{i\int L_I(y)dy} = i\,\frac{\overline{\partial}\,L_I}{\overline{\partial}\,\phi_\alpha}(x)\; e^{i\int L_I(y)dy} .$$

(5.12) thus implies that

$$\frac{\overline{\partial}\,L}{\overline{\partial}\,\Phi_\alpha}(x) = i\,\frac{\delta}{\delta\,\Phi_\alpha(x)}$$

in the shorthand notation of Remark 5.4 (d).

VI. THE ϕ^4 THEORY

In order to illustrate some particular aspects of dimensional renormalization, we consider the well-known theory of a single scalar field with quartic self-interaction. We follow Lowenstein and Zimmermann [19,20] in our general approach. The Lagrangian is

$$L = \frac{1}{2}(1+b)\partial_\mu\phi\partial^\mu\phi - \frac{1}{2}(m^2-a)\phi^2 - \frac{1}{4!}(g-c)\phi^4 \equiv L_0 + L_I$$

$$L_0 = \frac{1}{2}\partial_\mu\phi\partial^\mu\phi - \frac{1}{2}m^2\phi^2$$

where g is the coupling constant and a,b,c are power series in g chosen to normalize the two and four point functions. Specifically, if Γ_N is the N point vertex function (defined for $N = 2$ as the negative inverse of the Green's function and for $N > 2$ by restricting the expansion of

$$<0|T[\phi(x_1)\dots \phi(x_n)]|0>$$

to 1PI graphs and eliminating $\delta(\Sigma p_i)$ and propagators associated with external lines), then

$$\Gamma_2 (p,-p) \Big|_{p^2=m^2} = 0 \tag{6.1}$$

$$\Gamma_2(p,-p) \Big|_{p^2=\mu^2} = i(\mu^2-m^2) \tag{6.2}$$

$$\Gamma_4 \Big|_{\text{symmetry point}} = -ig. \tag{6.3}$$

It can be seen from the above that time-ordered vacuum expectation values in our theory depend parametrically on m^2, μ^2, g, and z, the mass parameter used to normalize the complex dimensional integrals. The normalization conditions guarantee that the Green's functions of the fields are in fact independent of z; however, vacuum expectation values of polynomials will be z-dependent. a, b, and c also depend on z; this has an effect on the derivation of the Callen-Symanzik equation, which we discuss shortly.

We note one minor difference between the BPHZ and dimensionally renormalized theories: there is a first order contribution to the two point function in the dimensional theory, from the diagram of Fig. 6.1.

Fig. 6.1.

This means that a contains a first order term also.

One may derive Callen-Symanzik and renormalization group equations much as for BPHZ renormalization [19], a discussion is given in [11]. If Δ_1, Δ_2, and Δ_3 represent the differential vertex operations of inserting an $\frac{i}{2}\phi^2$ vertex, an $\frac{i}{2}\phi_\mu\phi^\mu$ vertex, and an $\frac{i}{4!}\phi^4$ vertex, respectively, Lowenstein's action principle implies that

$$\frac{\partial\Gamma_N}{\partial m^2} = [(\frac{\partial a^2}{\partial m^2} - 1)\Delta_1 + \frac{\partial b}{\partial m^2}\Delta_2 + \frac{\partial c}{\partial m^2}\Delta_3]\Gamma_n \tag{6.4}$$

$$\frac{\partial\Gamma_N}{\partial\mu^2} = [\frac{\partial a}{\partial\mu^2}\Delta_1 + \frac{\partial b}{\partial\mu^2}\Delta_2 + \frac{\partial c}{\partial\mu^2}\Delta_3]\Gamma_N \tag{6.5}$$

$$\frac{\partial\Gamma_N}{\partial g} = [\frac{\partial a}{\partial g}\Delta_1 + \frac{\partial b}{\partial g}\Delta_2 + (\frac{\partial c}{\partial g} - 1)\Delta_3]\Gamma_N \tag{6.6}$$

$$N\Gamma_N = [2(a-m^2)\Delta_1 + 2(1+b)\Delta_2 + 4(c-g)\Delta_3]\Gamma_N. \tag{6.7}$$

The structure of (6.4) - (6.7) implies the existence of a linear relation among the four quantities on the left hand side:

$$(\mu^2\frac{\partial}{\partial\mu^2} + A\,m^2\frac{\partial}{\partial m^2} + B\frac{\partial}{\partial g} - CN)\Gamma_N = 0; \tag{6.8}$$

where A, B, and C are functions of the parameters. The normalization condition (6.1) implies that $A = 0$; (6.8) is then the renormalization group equation for the theory.

The Callen-Symanzik equation is a different linear combination of (6.4)-(6.7), with the form

$$(m^2\frac{\partial}{\partial m^2} + \mu^2\frac{\partial}{\partial\mu^2} + \beta\frac{\partial}{\partial g} - \gamma N)\Gamma_N = \alpha\, m^2\Delta_1\Gamma_N. \tag{6.9}$$

For (6.9) to hold, β and γ must satisfy

$$m^2\frac{\partial b}{\partial m^2} + \mu^2\frac{\partial b}{\partial\mu^2} + \beta\frac{\partial b}{\partial g} - 2\gamma(1+b) = 0,$$

$$m^2\frac{\partial c}{\partial m^2} + \mu^2\frac{\partial c}{\partial\mu^2} + \beta(\frac{\partial c}{\partial g} - 1) - 4\gamma(c-g) = 0;$$

the relevant determinant

$$\begin{vmatrix} \frac{\partial b}{\partial g} & -2(1+b) \\ (\frac{\partial c}{\partial g} - 1) & -4(c-g) \end{vmatrix}$$

is -2 in zero order, so the equations have solutions as formal power series in g. Then

$$\alpha = m^2(\frac{\partial a}{\partial m^2} - 1)+\mu^2 \frac{\partial a}{\partial \mu^2} + \beta \frac{\partial a}{\partial g} - 2\gamma(a-m^2).$$

It is instructive to compare this derivation with the BPHZ treatment. In that case one may insert either a "hard" ϕ^2 vertex $\Delta_1 = \int N_4(\phi^2)$ or a "soft" vertex $\Delta_0 = \int N_2(\phi^2)$ and the four equations (6.4)-(6.7) are supplemented with a fifth, the Zimmermann identify, expressing Δ_0 in terms of $\Delta_1,\Delta_2,\Delta_3$. It is the soft insertion Δ_0 which appears on the right hand side of the Callen-Symanzik equation. In our case Δ_1 is soft; there is no hard ϕ^2 insertion available. Moreover, in the BPHZ version, a, b, and c are homogeneous functions, of degree 1, 0, and 0, respectively, of μ^2 and m^2. This is not true in dimensional renormalization; an attempt to use it will trivialize the Callen-Symanzik equation. The reason is the parameter z; Γ_N is independent of z, as mentioned earlier, but a, b, and c are not. In fact, they are homogeneous (of the above degrees) in μ^2, m^2, and z.

As a final illustration of dimensional renormalization in ϕ^4 theory, we discuss the energy-momentum tensor and its relation to scaling. The "new, improved" Callen-Coleman-Jackiw energy-momentum tensor [21] is

$$\Theta^{\mu\kappa} = (1+b)\ \partial^\mu\phi\partial^\kappa\phi-g^{\mu\kappa}L - \frac{(1+b)}{6}\ (\partial^\mu\partial^\kappa-g^{\mu\kappa}\partial^2)\phi^2.$$

We first consider the vacuum expectation values

$$<0|T[\Theta^{\mu\kappa}(x)\ X]|0>$$

where

$$X = \prod_{i=1}^{m} \phi(x_i).$$

By Theorem 5.2,

$$\frac{\partial}{\partial x^\mu} <0|T[\Theta^{\mu\kappa}(x)X]|0> = <0|T[\partial_\mu \Theta^{\mu\kappa}(x)\ X]|0>$$

$$= <0|T[\{(1+b)\partial^2\phi + (m^2-a)\phi^2 + \frac{1}{3!}(g-c)\phi^3\}\partial^\kappa\phi\ X]0>$$

$$= <0|T[(-\frac{\overline{\partial L}}{\partial\phi})\ \partial^\kappa\phi\ X]|0>, \tag{6.10}$$

and so by Theorem 5.3

$$\frac{\partial}{\partial x^\mu} <0|T[\Theta^{\mu\kappa}\ X]|0> = \\ -i \sum_{j=1}^{m} \delta(x-x_j)\ <0|T[\partial^\kappa\phi(x_j)\phi(x_1)\ldots\widehat{\phi(x_j)}\ldots]|0>. \tag{6.11}$$

This is the Ward identity for the energy momentum tensor.

To investigate scaling we will need to consider more general vacuum expectation values than were discussed in Section 5, specifically those of the form

$$<0|T[C\ P(x)\ Y_1(x_1)\ldots\ Y_m(x_m)|0> \tag{6.12}$$

where C is a contraction operator as in (5.6) and $P(x)$ is a covariant polynomial in $x_1, \ldots x_m$. We do this in an obvious way: if F is the regularized amplitude for some graph in the expression of $<0|T[Y_1\ldots Y_m]|0>$, the corresponding amplitude for (6.12) is

$$C\ P(i\frac{\partial}{\partial p})\ F,$$

where the partial derivatives are given an obvious definition on elements of the algebras W^r_m. Renormalization proceeds as before. The only new feature is that part (b) of Theorem 5.2 [contraction of the index for a total derivative of a vacuum expectation value] does not hold, since, for example,

$\partial^\mu x^\kappa = g^{\mu\kappa}$.

We now give an incorrect "derivation" of asymptotic scale invariance. First, using the relation $g^\mu{}_\mu = 4$, we find the trace of Θ:

$$\Theta^\mu{}_\mu = (1+b)\partial_\mu\phi\partial^\mu\phi - 4[\tfrac{1}{2}(1+b)\partial_\mu\phi\partial^\mu\phi - \tfrac{1}{2}(m^2-a^2)\phi^2$$

$$- \frac{1}{4!}(g-c)\phi^4] + (1+b)[\phi\partial^2\phi + \partial_\mu\phi\partial^\mu\phi]$$

$$= \phi(-\frac{\overline{\partial L}}{\partial\phi}) + (m^2-a^2)\phi^2. \tag{6.13}$$

Then from (6.10) and (6.13),

$$\partial_\mu(x_\kappa\,\Theta^{\mu\kappa}) = (x_\kappa\partial^\kappa\phi+\phi)(-\frac{\partial L}{\partial\phi}) + (m^2-a^2)\phi^2$$

and

$$0 = \int d^4x\,\partial_\mu <0|T[x_\kappa\Theta^{\mu\kappa}(x)\ X]|0>$$

$$= \sum_{j=1}^{m} (-i) <0|T[\phi(x_1)\dots\{x_{j\kappa}\partial^\kappa+1\}\phi(x_j)\dots\ \phi(x_m)|0>$$

$$+ (m^2-a^2) <0|T[\phi^2(x)X]|0>. \tag{6.14}$$

Because ϕ^2 is a soft insertion, (6.14), if true, would express asymptotic scale invariance of dimension one.

The error in this calculation, with our conventions, is in the inappropriate application of Theorem 5.2 (b) in (6.14). The use of $g^\mu{}_\mu = 4$ in (6.13) is correct, according to the definition established in Chapter 5 for defining TOVEV's containing expressions such as $\Theta^\mu{}_\mu$. In (6.14), however, we will encounter the trace of the metric tensor in one of the algebras W^2_L of Chapter 4; the resulting factor of ν will not commute with renormalization.

Comparison with the BPHZ treatment [19], in which an

equation like (6.14) holds but with a hard $N_4(\phi^2)$ on the right, shows what seems to be a general principle. In dimensional renormalization, anomalies occur because a naive 4-dimensional calculation does not generalize to ν dimensions; in BPHZ, the anomalies occur because normal products of inappropriate degree appear. A calculus for treating these anomalies systematically in the dimensional methods, similar to the use of oversubtracted normal products, would be desirable.

VII. THEORIES WITH ZERO MASS PARTICLES

A. General Considerations

We now consider field theories in which some particles have mass zero, i.e., we allow propagators of the form

$$\Delta(p) = -i\, Z(p)\, (-p^2 - i0)^{-1},$$

and ask to what extent we may carry through the dimensional renormalization program of the previous chapters.

Let G be a connected Feynman graph; for each line $\ell \in L(G)$ we suppose given a propagator

$$\Delta_\ell(q_\ell) = -i\, Z_\ell(q_\ell)(m_\ell^2 - q_\ell^2 - i0)^{-1},$$

where now $m_\ell \geq 0$. From our previous Def. 2.7, it is natural to attempt to define a regularized amplitude for G by

$$F_G(\lambda,\nu,\varepsilon) = z^{\omega}(-i)^{\ell}(2\pi)^{-h\nu} \int d^\nu k_1 \ldots d^\nu k_h \times \prod_L (m_\ell^2(1-i\varepsilon) - \langle q_\ell, q_\ell\rangle)^{-\lambda_\ell} Z_\ell(q_\ell). \tag{7.1}$$

This definition is justified by our next theorem. We use the following standard terminology: a set of momenta $p_1, \ldots p_m \in R^{M+1}$ is (Euclidean) non-singular if, for any non-empty subset $A \subset \{1, \ldots m\}$,

$$\sum_{j=0}^{M} \left(\sum_{i\in A} p_i^j\right)^2 \neq 0.$$

Correspondingly, a point $x = (y, S) \in X_m = R^m \times U_m$ is nonsingular if

$$\left(\sum_{i\in A} y_i\right)^2 + \sum_{i,j\in A} S_{ij} \neq 0.$$

Theorem 7.1. Let G be a connected Feynman graph with no detachable subgraphs (see Appendix). Then there is a nonempty open set $\Omega \subset \mathbb{C}^{L+1}$ such that, for $p_1, \ldots p_m$ nonsingular and $(\lambda,\nu) \in \Omega$, (7.1) converges absolutely. For these momentum values, $F_G(\lambda,\nu,\varepsilon)$ may be analytically continued to a meromorphic function on $\mathbb{C}^{L+1}$.

We will discuss the proof of this theorem below. For a graph G with no detachable subgraphs, we thus have defined a regularized amplitude for G. When detachable subgraphs exist, (7.1) will not converge for any value of the regularizing parameters. As an example, the amplitude for a massless loop line is given by

$$\int d^\nu k \, (k^2)^{-\lambda}$$

which diverges at 0 or infinity for any choice of λ,ν.

Definition 7.2. For any Feynman graph G, the amplitude $F_G(\lambda,\nu,\varepsilon)$ is given by (7.1) if G has no detachable subgraphs; otherwise,

$$F_G(\lambda,\nu,\varepsilon) = 0. \tag{7.2}$$

Remark. (7.2) is motivated by more than its simplicity. We want the equations of motion to continue to hold in a massless theory; recall that the crucial condition was that the amplitude for a line with propagator 1 be equal to the amplitude for the contracted graph. Now detachable subgraphs can arise from contractions in a graph G without such pieces (see Fig. 5.2). Consider such a G, with ℓ the line for which $\Delta_\ell = 1$; this propagator is treated by defining

$$\Delta_\ell(\lambda_\ell,\varepsilon) = (m^2-q_\ell^2-i\varepsilon)^{-\lambda_\ell} \tag{7.3}$$

during regularization, and then setting $\lambda_\ell = 0$ (see Chapter 5). In α-space, exponentiation of the propagator (7.3) introduces a factor $\Gamma(\lambda_\ell)^{-1}$ ((2.25)), but the resulting α-space integral (corresponding to (2.27)) is not singular at $\lambda_\ell = 0$. [This follows from the analysis of singularities given in [22]; in the terminology of that paper, the graph consisting only of the line ℓ is not saturated and hence does not generate a singularity.] Thus $F_G|_{\lambda_\ell=0} = 0$; since this should be the regularized amplitude for the contracted graph, (7.2) is justified.

Proof of Theorem 7.1 (Sketch): Again we exponentiate propagators and obtain an α-space representation of (7.1). The behavior of such integrals for scalar propagators is analyzed in [22], and the extension to the general situation is straightforward.

Suppose that $Z_\ell(q_\ell)$ is a polynomial of degree r_ℓ, all terms being of degree at least $s_\ell (\leq r_\ell)$. For any subgraph H of G let

$$\pi_H = h(H)\,\frac{\nu}{2} - \sum_{\ell\in L(H)} \lambda_\ell + \left[\sum_{\ell\in L(H)} \frac{r_\ell}{2}\right]$$

(with [ ] the greatest integer function) and, for any quotient graph $Q = G/S$, where S contains all massive lines of G and connects all external vertices, let

$$\pi_Q = h(Q)\,\frac{\nu}{2} - \sum_{\ell\in L(Q)} \lambda_\ell + \left[1 + \sum_{\ell\in L(Q)} \frac{s_\ell}{2}\right].$$

From [22] one can show that the α-space integral is absolutely convergent if $\pi_H < 0$ for all H, and $\pi_Q > 0$ for all Q, and that the set of (λ,ν) satisfying these conditions is not empty. An argument as in Section 2 then implies the convergence of the momentum space integral (7.1). The analytic continuation is also carried out in [22]; there are series of ultraviolet poles on varieties

$$\pi_H = 0,\ 1,\ 2\ \ldots\ , \tag{7.4}$$

and of infrared poles on varieties

$$\pi_Q = 0, -1, -2, \dots \tag{7.5}$$

for certain subgraphs H and quotient graphs Q.

Having defined regularized amplitudes, we may use Def. 3.6 to obtain renormalized amplitudes (the discussion above also defines regularized amplitudes $F^{y}_{H:A}$). However, the proof that y_H as given by (3.5) is a polynomial essentially assumed that we were subtracting ultraviolet poles; the physical interpretation of subtracting an infrared pole is not clear. We therefore make

Definition 7.3. Suppose that G is an infrared convergent Feynman graph, i.e. $\pi_Q(\lambda^0,4) > 0$ for all quotients as above of G; and that corresponding infrared convergence holds for all amplitudes $F^{y}_{G:A}$. Then the renormalized amplitude for G is defined as in Def. 3.6.

Remark. (a) The properties of renormalized amplitudes discussed in Section 5 carry over to these new amplitudes without change.

(b) It must be emphasized that we have defined the amplitude, for $\varepsilon > 0$, only for non-singular momenta. In some cases, it is possible to show that

$$\int \mathcal{D}_G(\varepsilon)(p)\, f_1(p_1)\dots f_m(p_m)\, d^4p_1\dots d^4p_m,$$

with $f_i \in S(\mathbb{R}^4)$, converges absolutely; a result of Lowenstein and myself [23] then implies the existence of

$$\lim_{\varepsilon\to 0} \mathcal{D}_G(\varepsilon)$$

as a tempered distribution in the external momenta.

B. Massless ϕ^4 Theory

One theory to which this renormalization applies is the massless ϕ^4 theory, with Lagrangian

$$L = \frac{1}{2}(1+b)\partial_\mu\phi\partial^\mu\phi - \frac{1}{4!}(g-c)\phi^4 \equiv L_0 + L_I$$

$$L_0 = \frac{1}{2}\partial_\mu\phi\partial^\mu\phi.$$

Consider a graph G occurring in the Gell-Mann Low expansion of a Green's function

$$<0|T[\phi(x_1)\ldots\phi(x_m)]|0>. \qquad (7.6)$$

G will have vertices as in Fig. 7.2; note that the lines in Fig. 7.2 (b) have $s_\ell = 1$. Let $Q = G/S$ be a quotient

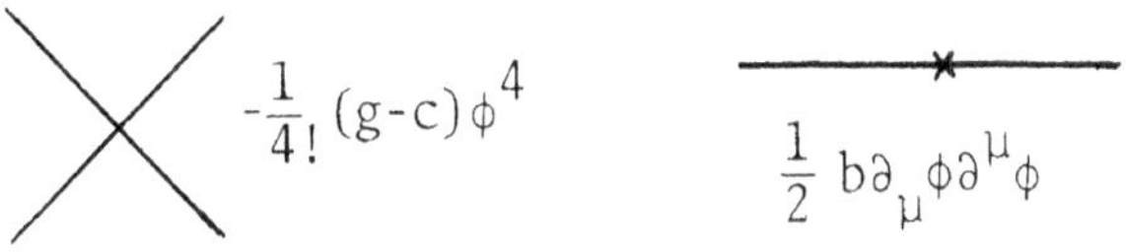

(a) (b)

Fig. 7.2.

graph as above, i.e., assume that S connects all external vertices of G. We must verify that $\pi_Q > 0$ at $(\lambda^0,4)$. Actually, [22] implies that we must check this only for subgraphs S which are connected and connect all external vertices; thus Q has one vertex corresponding to S with some number E of lines incident, m_1 vertices as in Fig. 7.1 (a), and m_2 as in Fig. 7.1 (b), a total of $L = 2m_1+m_2+E/2$ lines. Then

$$\pi_Q(\lambda^0,4) = 2(L-(m_1+m_2))-L+m_2 = E/2,$$

so the infrared convergence conditions hold.

According to Def. 7.3 we must also verify infrared convergence for the amplitudes $F^{y}_{G:A}$, with $A \in A(G)$ and $\{y_H\}$ the vertex parts defined inductively. Now $y_H \neq 0$ for two types of subgraph G; those with 4 external lines, for which the superficial divergence is zero and hence y_H is independent of momentum, and those with 2 external lines, for which the superficial divergence is two and hence y_H is quadratic in the momentum. Since there is no mass parameter in the theory except

z, the parameter used to normalize complex dimensional integrals, and since it follows from Chapter 3 that y_H does not depend on z, y_H must in fact be proportional to p^2. Thus only vertices as in Fig. 7.2 appear in calculating $F^y_{H:A}$, and the above analysis implies infrared convergence here also.

Thus Def. 7.3 leads to renormalized Green's functions as functions defined for almost all momenta. The constants b and c in the Lagrangian may be used to normalize the two and four point functions at some Euclidean momentum value.

The ϕ^4 theory is one in which the results of [23] may be used to prove existence of Green's functions as tempered distributions in external momenta. We will not discuss details here.

APPENDIX

Discussion at the school suggested the value of collecting graph-theoretical definitions in a single place. We attempt to do so here; the terminology introduced should be a good approximation to that used by most speakers.

A graph G is an ordered sextuple (V,L,E,i_1,i_2,g) where V, L and E are finite sets called the vertices, internal lines, and external lines of G, respectively. i_1 and i_2 are maps $i_k\colon L \to V$, and g is a map $g\colon E \to V$; $i_1(\ell)$ and $i_2(\ell)$ are the initial and final vertices of the line ℓ, and $g(e)$ is the unique vertex on which the external line e is incident. G is trivial if $L = \emptyset$. The empty graph is the unique graph with $V = L = E = \emptyset$.

A subgraph H of G $(H \subset G)$ is a graph $H = \{V',L',E',i_1',i_2',g'\}$ such that $L' \subset L$, $i_1(L') \cup i_2(L') \subset V' \subset V$, $i_k' = i_k|_{L'}$ $(k = 1,2)$,

$$E' = \bigcup_{k=1}^{2} [\{k\}\times(i_k^{-1}(V')-L')] \cup g^{-1}(V'),$$

and $g'((k,\ell)) = i_k(\ell)$ $(k = 1,2)$, $g'(e) = g(e)$ for $e \in E$. Thus the external lines incident on a vertex in H are the original incident lines together with "fragments" of internal lines from G not contained in H. Note that specifying V' and L' completely specifies H. A generalized vertex of G

is a subgraph H containing all lines of G connecting a pair of its vertices, i.e., for which $L' = \bigcap_{k=1}^{2} i_k^{-1}(V')$.

Given two subgraphs H_1, H_2 of G, their union $H_1 \cup H_2$ is the subgraph with vertex set $V_1 \cup V_2$ and line set $L_1 \cup L_2$; their intersection is defined similarly. Note that if $V_1 \subset V_2$ and $L_1 \subset L_2$, then $H_1 \subset H_2$ in the sense above. H_1 and H_2 are non-overlapping if $H_1 \subset H_2$ or $H_2 \subset H_1$, or if $H_1 \cap H_2$ is the empty graph.

A loop line ℓ is one for which $i_1(\ell) = i_2(\ell)$; when no confusion can arise, we also use this term for the subgraph composed of ℓ and its vertex. If v,v' are distinct vertices of G, a path from v to v' is a sequence $v = v_0, v_1, \ldots v_n = v'$ of distinct vertices and a sequence $\ell_1, \ldots \ell_n$ of lines with $\{i_1(\ell_j), i_2(\ell_j)\} = \{v_j, v_{j-1}\}$. G is connected if any distinct vertices may be joined by a path, and is one particle irreducible (1PI) if, for any line ℓ, distinct vertices may be joined by a path not containing ℓ. G is 2-connected if it is a loop line or if it contains no loop lines and is 1PI and if, for any distinct vertices v, v', and v'', there is a path from v to v' which does not contain v''. (Note that 2-connected $\Rightarrow$ 1PI $\Rightarrow$ connected.) For any G, the connected (resp. 1PI, 2-connected) components of G are the maximal subgraphs which are connected (resp. 1PI, 2-connected). A graph is the union of its connected components and is the union of its 1PI or 2-connected components plus some subgraphs containing a single line; a 1PI graph is the union of its 2-connected components.

A detachable subgraph of a graph G is either a loop line (together with its vertex) or a connected subgraph H with distinguished vertex $v_0 \in V'$ such that $V - V' \neq \emptyset$, $V' - \{v_0\} \neq \emptyset$, $g^{-1}(V'-\{v_0\}) = \emptyset$, and for $v \in V'-\{v_0\}$ and $v' \in V-V'$, any path from v to v' goes through v_0. A tadpole is a detachable subgraph for which $|i_1'^{-1}(v_0) \cup i_2'^{-1}(v_0)| = 1$; the term may also refer to the subgraph obtained from this by deleting v_0 and the single line incident on it (the "tail"), i.e., to a subgraph with precisely one external line which is internal in G.

Let H be a subgraph of G with connected components $H_1, \ldots H_k$. The <u>reduced</u>, <u>contracted</u>, or <u>quotient</u> graph $G/H \equiv G/H_1 \ldots H_k \equiv G/\{H_1, \ldots H_k\}$ has vertex set $V'' = \{V_1, \ldots V_k\} \cup \{v | v \notin V_k, \quad k\}$, where V_k is the vertex set of H_k. There is a natural map $\pi: V \to V''$ with $\pi(v) = V_k$ if $v \in V_k$, and $\pi(v) = v$ otherwise; then G/H has $L'' = L - L'$, $E'' = E$, and $i''_k = \pi \circ i_k$ $(k = 1,2)$, $g'' = \pi \circ g$. If $G \supset K \supset H$, there is a natural interpretation of K/H as a subgraph of G/H and a natural isomorphism of (G/H)/(K/H) with G/K; such isomorphisms are generally used without explicit mention.

ACKNOWLEDGMENT

I would like to thank Lynn Braun for her excellent typing of these notes.

BIBLIOGRAPHY

[1] E. Speer, J. Math. Phys. 9, 1404 (1968).
[2] E. Speer, Generalized Feynman Amplitudes, Princeton University Press (1969).
[3] K. Hepp, in Statistical Mechanics and Quantum Field Theory, ed. DeWitt and Stora, Gordon and Breach, New York, 1971.
[4] C. Chandler, Commun. Math. Phys. 19, 169 (1968).
[5] J. Ashmore, Commun. Math. Phys. 29, 177 (1973).
[6] E. Speer, Commun. Math. Phys. 37, 83 (1974).
[7] G. t'Hooft and M. Veltman, Nucl. Phys. B 44, 189 (1972).
[8] E. Speer, J. Math. Phys. 15, 1 (1974).
[9] H. DeVega and F. Schaposnik, J. Math. Phys. 15, 1998 (1974).
[10] J. Collins, "Normal Products in Dimensional Regularization," DAMTP preprint.
[11] P. Breitenlohner and D. Maison, "Dimensional Renormalization and the Action Principle," Max Planck Institut preprint.
[12] N. Nakanishi, Graph Theory and Feynman Integrals, Gordon and Breach, New York, 1971.
[13] E. Speer, "Analytic Renormalization," University of Maryland lectures, 1972.
[14] I. Gel'fand and G. Shilov, Generalized Functions, Vol. I, Academic Press, New York, 1964.
[15] K. Hepp, Commun. Math. Phys. 2, 301 (1966).
[16] E. Speer, Commun, Math. Phys. 23, 23 (1971) and 25, 336 (1972).
[17] S. MacLane, Homology, Springer-Verlag, New York, 1967.
[18] Y. Lam, Phys. Rev. D6, 2145 (1972).
[19] J. Lowenstein, "Normal Product Methods in Renormalized Perturbation Theory," University of Maryland lectures, 1972.
[20] W. Zimmermann, Commun. Math. Phys. 11, 1 (1968).
[21] C. G. Callen, S. Coleman, R. Jackiw, Ann. Phys. 59, 42 (1970).
[22] E. Speer, Ann. Inst. H. Poincare, XXIII, 1 (1975).
[23] J. Lowenstein and E. Speer, "Distributional Limits of Renormalized Feynman Integrals with Zero Mass Denominators," to appear in Commun. Math. Phys.

BPHZ RENORMALIZATION

John H. Lowenstein*

Department of Physics
New York University

1. INTRODUCTION

Composite fields --- local, covariant fields which are formally products of the "elementary" fields of a given theory --- have played an important role in the theoretical developments of recent years, and promise to do so for many years to come. The areas of investigation in which composite fields arise in an essential way include:

(a) field equations, in which the sources of "elementary" fields must be composite if there is to be non-trivial interaction;

(b) current algebra, in which symmetry currents appear as bilinear expressions in some "elementary" fermion fields;

(c) operator product expansions of products of fields at different points;

(d) interpolating fields for particles in theories whose dynamics is expressed in terms of a small number of "elementary" fields (e.g. quark fields);

(e) non-linear symmetry transformations of fields, such as supergauge transformations, Slavnov transformations in non-Abelian gauge theories and chiral transformations in the non-linear sigma model.

---

* Research supported in part by the National Science Foundation Grant No. MPS74-21778.

G. Velo and A. S. Wightman (eds.), Renormalization Theory, 95–160.

The first person to study composite fields systematically in renormalized perturbation theory was Zimmermann (in his 1970 Brandeis Summer School lectures [1]). He exploited an important aspect of the renormalization program of Bogoliubov [2] and Hepp [3] which even today is not fully appreciated by practitioners of perturbative field theory, namely: the Green functions of local composite fields (normal products) can be defined in perturbation theory by essentially the same methods used to define the Green functions of elementary renormalized fields. This holds true for all of the most common approaches to renormalization, including the traditional regularized counterterm approach. In some formulations, the normal-product Green functions have properties which are particularly convenient for studying, to all orders in perturbation theory, conserved and non-conserved currents and their Ward identities, and more generally any local or global symmetry of the Green functions. In this respect, I have found the normal-product Green functions defined by Zimmermann's momentum-space subtraction prescription [1,4] to be particularly favorable, although for particular models other methods may be more efficient (for example, dimensional renormalization in the pure Yang-Mills model [5,6]).

As in other renormalization schemes, Zimmermann's approach is based ultimately on the Gell-Mann-Low formula

$$\langle T \prod_i N[Q_i(x_i)] \rangle$$

$$(1.1) \qquad = \frac{\langle T \prod_i N[Q_i^{(0)}(x_i)] \exp i \int dy\, N[\mathcal{L}_I^{(0)}(y)] \rangle_0}{\langle T \exp i \int dy\, N[\mathcal{L}_I^{(0)}(y)] \rangle_0}$$

which expresses, formally, the Green functions of local products of renormalized fields, to every finite order in perturbation theory, in terms of the Green functions of products of free fields. As is well known, the right-hand side of (1.1) can be expanded (formally), using Wick's theorem, to obtain the usual sum over Feynman diagrams. Each diagram corresponds, in momentum space, to a product of delta functions, free two-point functions and formal integrals

$$\int d^{4m}k\, I_\Gamma(p,k) ,$$

$$\Gamma = \text{one-particle-irreducible (1PI) diagram}$$

$$(1.2) \qquad p = p_1 p_2 \,\ldots\ldots\, p_n = \text{external 4-momenta of } \Gamma$$

$$k = k_1 k_2 \,\ldots\ldots\, k_m = \text{internal 4-momenta of } \Gamma$$

The main point of Zimmermann's method is to replace (1.2) by the subtracted integral

(1.3)
$$\int d^{4m}k\, R_\Gamma(p,k)$$

$$R_\Gamma(p,k) = I_\Gamma(p,k) - \text{subtractions} = (1 - t_p^{\delta(\Gamma)})\,\bar{R}_\Gamma(p,k) \;,$$

where $t_p^{\delta(\gamma)}$ denotes the Taylor series to order $\delta(\gamma)$ about vanishing external momentum (the determination of $\delta(\gamma)$ will be discussed in Section 3; it is an integer at least as large as 4m plus the degree of $I_\Gamma(p,k)$ considered as a rational function of k), and $\bar{R}_\Gamma(p,k)$ includes all subtractions corresponding to formally divergent 1PI subgraphs properly contained in Γ .

For any formal product $Q^{(0)}$ of free fields $\phi_\alpha^{(0)}$, $\alpha=1,2,$ ..., and their derivatives of arbitrary order, Zimmermann's subtraction method [1] allows one to define a sequence of normal products

$$N_\delta[Q(x)] \quad , \quad \delta = \dim Q + \text{non-negative integer} \;,$$

where dim Q is the "naive" or "canonical" dimension of Q (counting 1 for each boson field, 3/2 for each fermion field, 1 for each derivative). Written in terms of those normal products, the Gell-Mann-Low formula (1.1) for composite-field Green functions in a theory with effective interaction Lagrangian density $\mathcal{L}_{\text{eff}\,I}$ takes the form

$$\langle T \prod_i N_{\delta_i}[Q_i(x_i)] \rangle$$

(1.4)
$$= \langle T \prod_i N_{\delta_i}[Q_i^{(0)}(x_i)] \exp i \int dy\, N_4[\mathcal{L}^{(0)}_{\text{eff}\,I}(y)] \rangle_0$$

Note that the denominator of the right-hand member of (1.1) is equal to unity with the assumption of N_4 interaction terms.

I shall now list a number of convenient properties of Zimmermann's normal products. For a more complete discussion, the reader is referred to Section 4 below and to Ref. [7]. Throughout the following discussion, ϕ_α denotes an elementary field and Q(x) is a formal product of elementary fields and their derivatives.

(i) Derivative rules [8]

Suppose Q has the factorization $Q = \prod_{i=1}^{a} A_i$; then,

$$\partial_x^\mu \langle T N_\delta[Q(x)]\, X \rangle = \langle T N_{\delta+1}[\partial^\mu Q(x)]\, X \rangle$$

$$= \sum_j \langle T N_{\delta+1}[(\prod_{i<j} A_i)\, \partial^\mu A_j\, (\prod_{k>j} A_j)]\, X \rangle$$

(ii) Field equations [1,8-10]

Let P be a polynomial in the elementary fields and their derivatives of arbitrary order; then ($d_\alpha \equiv \dim \phi_\alpha$)

$$\langle T\, N_{\rho+4-d_\alpha}\Big[P(x)\Big(\frac{\partial \mathcal{L}_{eff}}{\partial \phi_\alpha}(x) - \frac{\partial}{\partial x^\mu}\frac{\partial \mathcal{L}_{eff}}{\partial(\partial_\mu \phi_\alpha)}\Big)\Big] \prod_j \phi_{\beta_j}(y_j)\rangle$$

$$= i \langle T\, N_\rho[P(x)] \frac{\delta}{\delta \phi_\alpha(x)} \prod_j \phi_{\beta_j}(y_j)\rangle$$

+ normal product corrections (= 0 if P is linear in ϕ_γ)

(iii) Normalization conditions [1].

The vertex function (the tilde indicates Fourier transformation)

$$\langle T\, N_\delta[Q(0)] \prod_{i=1}^{n} \tilde{\phi}_{\alpha_i}(p_i) \prod_{j=1}^{m} N_{\delta_j}[\tilde{Q}_j(q_j)] \rangle^{p}$$

is defined by a sum over 1PI diagrams whose external lines correspond to the fields ϕ_{α_i} , i=1,2,...,n , with naive dimensions d_{α_i} . Let D_p^μ be a partial derivative, with respect to the p variables, of order $|\mu|$. Then

$$D_p^\mu \langle T\, N_\delta[Q(0)] \prod_{i=1}^{n} \tilde{\phi}_{\alpha_i}(p_i) \rangle^{p}$$

= contribution from trivial (one-vertex) graphs only

$$\text{if} \quad |\mu| + \sum_{i=1}^{n} d_{\alpha_i} \leq \delta \quad .$$

Note that the trivial-graph contribution vanishes unless Q contains n factors and $|\mu|$ derivatives.

(iv) Zimmermann identities [1].

$$\langle T\big(N_\delta[Q(x)] - N_{q}[Q(x)]\big) \prod_i \phi_{\alpha_i}(y_i)\rangle$$

$$= \sum_i r_i \langle T\, N_\delta[M_i(x)] \prod_i \phi_{\alpha_i}(y_i)\rangle \quad ,$$

where the M_i are field monomials of naive dimension $\leq \delta$ and the coefficients r_i all vanish in zeroth order.

(v) Action principle for field transformations [9].

Consider the infinitesimal field transformation

$$\phi_\alpha \rightarrow \phi_\alpha + \delta\phi_\alpha$$

$$\delta\phi_\alpha = N_{\rho_\alpha}[P_\alpha] \qquad P_\alpha = \text{polynomial in fields}$$

Then

$$\int dx < T\, \delta N_4[\mathcal{L}_{eff}(x)] \prod_j \phi_{\alpha_j}(y_j) >$$

$$= \int dy < T\, \delta\phi_\alpha(y) \frac{\delta}{\delta\phi_\alpha(y)} \prod_j \phi_{\alpha_j}(y_j) >$$

+ normal-product corrections (= 0 for P linear in ϕ_γ)

where $\delta N_4[\mathcal{L}_{eff}(x)] = \int dy \sum_\alpha N_{\rho_\alpha+4-d_\alpha}[P_\alpha(y)\, \delta\mathcal{L}_{eff}(x)/\delta\phi_\alpha(y)]$.

Note that (v) is actually a simple consequence of (ii).

(vi) Action principle for parametric derivatives [11].

Suppose that $L_{eff} = L_o + L_{eff\ I}$, where L_o is the Lagrangian for the free fields $\phi_\alpha^{(0)}$ of a theory and $L_{eff\ I}$ is the effective interaction Lagrangian appearing in (1.4), has the form

$$\mathcal{L}_{eff} = \sum_i c_i M_i$$

where M_i are formal field products of naive dimension ≤ 4, and the coefficients c_i are functions of parameters ξ_r, $r = 1,2,\ldots$. Then

where
$$\frac{\partial}{\partial\xi_r} < T \prod_j \phi_{\alpha_j}(x_j) > = i \sum_i (\partial c_i/\partial\xi_r) \int dx < T\, N_4[M_i(x)] \prod_j \phi_{\alpha_j}(x_j) >$$

Note that for ξ_r not appearing in L_o, (vi) is a simple consequence of (1.4).

(vii) Counting identities [11].

$$-N_\alpha < T \prod_i \phi_{\alpha_i}(x_i) > = i \int dx < T\, N_4\left[\phi_\alpha(x) \frac{\delta}{\delta\phi_\alpha(x)} \int dy\, \mathcal{L}_{eff}(y)\right] \times \prod_i \phi_{\alpha_i}(x_i) >$$

where

$$N_\alpha = \sum_i \delta_{\alpha\alpha_i}.$$

Note that (vii) is a simple consequence of (v), with $P_\alpha = \phi_\alpha$. There is a corresponding identity for vertex functions, with N_α replaced by $-N_\alpha$.

To show how properties (i-vii), taken together, comprise a powerful "normal product algorithm , let us present, without full details, the derivations of the Ward identity of the axial-vector current and the Callan-Symanzik equation in quantum electrodynamics, whose covariant-gauge effective Lagrangian, corresponding to "intermediate" normalization conditions, is

$$\mathcal{L}_{eff} = -\frac{1}{4}(\partial_\mu A_\nu - \partial_\nu A_\mu)^2 - \frac{1}{2\alpha}(\partial_\mu A^\mu)^2 + i\bar{\Psi}(\partial_\mu - ieA_\mu)\psi - (M-c)\bar{\Psi}\psi \tag{1.5}$$

where the finite mass counterterm c is determined recursively to insure the mass-shell vanishing of the fermion's inverse propagator. For more details, as well as for references to earlier work on the axial-current Ward identity and the Callan-Symanzik equation, the reader is urged to consult Refs.[12, 13].

<u>Ward identity of the axial-vector current [12-16].</u> The Green functions of the gauge-invariant axial-vector current are given, up to a finite normalization factor, by

$$\langle T\, N_3[\bar{\psi}\gamma_\mu\gamma^5\psi(x)]\, X\rangle$$

Applying (i) and linearity(notation: $\frac{\hat{\partial}}{\hat{\partial}\phi} \equiv \frac{\partial}{\partial\phi} - \frac{\partial}{\partial x^\mu}\frac{\partial}{\partial(\partial_\mu\phi)}$)

$$\begin{aligned} &\partial_x^\mu \langle T N_3[\bar{\psi}\gamma_\mu\gamma^5\psi(x)]\, X\rangle \\ &= \langle T N_4[\bar{\psi}\overleftarrow{\partial\!\!\!/}\gamma^5\psi(x) - \bar{\psi}\gamma^5\overrightarrow{\partial\!\!\!/}\psi(x)]\, X\rangle \\ &= -i\langle T N_4\Big[\frac{\hat{\partial}\mathcal{L}_{eff}}{\hat{\partial}\psi}\gamma^5\psi(x) - \bar{\psi}\gamma^5\frac{\hat{\partial}\mathcal{L}_{eff}}{\hat{\partial}\bar{\psi}}(x) - 2(M-c)\bar{\psi}\gamma^5\psi(x)\Big]X\rangle \end{aligned} \tag{1.6}$$

Thus, by (ii),

$$\begin{aligned} &\partial_x^\mu \langle T N_3[\bar{\psi}\gamma_\mu\gamma^5\psi(x)]\, X\rangle \\ &= -\langle T\Big[(\gamma^5\psi(x))\frac{\delta}{\delta\psi(x)} + (\bar{\psi}(x)\gamma^5)\frac{\delta}{\delta\bar{\psi}(x)}\Big]X\rangle \\ &\quad + 2i(M-c)\langle T N_4[\bar{\psi}\gamma^5\psi]\, X\rangle \end{aligned} \tag{1.7}$$

The last term on the right-hand side can now be re-written in terms of minimally subtracted normal products using the Zimmermann identity (iv). In particular($\tilde{F}_{\mu\nu} \equiv \epsilon_{\mu\nu\rho\sigma}F^{\rho\sigma}$, $F_{\mu\nu} \equiv \partial_\mu A_\nu - \partial_\nu A_\mu$),

$$\begin{aligned} N_4[\bar{\psi}\gamma^5\psi] = N_3[\bar{\psi}\gamma^5\psi] &+ r\, N_4[F_{\mu\nu}\tilde{F}^{\mu\nu}] \\ &+ s\, N_4[\partial^\mu(\bar{\psi}\gamma_\mu\gamma^5\psi)] . \end{aligned} \tag{1.8}$$

The "anomaly" r can be evaluated by application of the normalization conditions of the two-photon vertex functions at zero momentum. From (iii), we see that of the N_4 terms in (1.8), only $r\, N_4[F_{\mu\nu}\tilde{F}^{\mu\nu}]$ contributes when D^{μ}_{p} is a second-order derivative. Moreover, that contribution comes entirely from the diagram with a single vertex. Carrying out the computation of r and substituting (1.8) into (1.7), one obtains, finally,

$$\partial^{\mu}_{x}\big(1-2i(M-c)s\big)\langle T\, N_3[\bar{\Psi}\gamma_{\mu}\gamma^5\Psi(x)]\, X\rangle$$

(1.9)
$$= 2i(M-c)\langle T\, N_3[\bar{\Psi}\gamma^5\Psi(x)]\, X\rangle$$
$$-\langle T\Big[(\gamma^5\Psi(x))\frac{\delta}{\delta\Psi(x)} + (\bar{\Psi}(x)\gamma^5)\frac{\delta}{\delta\bar{\Psi}(x)}\Big] X\rangle$$
$$+\, 2i(M-c)\, r\, \langle T\, N_4[F_{\mu\nu}\tilde{F}^{\mu\nu}(x)]\, X\rangle\ ,$$

with

(1.10) $$8r\,\epsilon_{\varkappa\lambda\rho\sigma} = \frac{\partial}{\partial p^{\varkappa}}\frac{\partial}{\partial q^{\lambda}}\langle T\, N_3[\bar{\Psi}\gamma^5\Psi(0)]\,\tilde{A}_{\nu}(p)\,\tilde{A}_{\sigma}(q)\rangle^{p}\Big|_{\substack{p=0\\ q=0}}.$$

Callan-Symanzik equation. [17,18,13] We define integrated normal products (also known as vertex insertions or differential vertex operations [11])

$$\Delta_j \equiv i\int dx\, N_{\delta_j}[Q_j(x)] \quad , \qquad j = 0,1,2,3,4$$

(1.11)
$$Q_0 = Q_1 = \bar{\Psi}\Psi$$
$$Q_2 = i\,\bar{\Psi}\gamma^{\mu}(\partial_{\mu} - ieA_{\mu})\Psi$$
$$Q_3 = \frac{1}{4}(\partial_{\mu}A_{\nu} - \partial_{\nu}A_{\mu})^2$$
$$\delta_0 = 3\ ,\quad \delta_1 = \delta_2 = \delta_3 = 4$$

by

(1.12) $$\Delta_i G^{(B,F)}_{\mu}(x,y,z) = i\int dt\,\langle T\, N_{\delta_i}[Q_i(t)]\prod_j^{B} A_{\mu_j}(x_j)\prod_k^{F/2}\Psi(y_k)\cdot\prod_{\ell}^{F/2}\bar{\Psi}(z_{\ell})\rangle$$

Writing also

(1.13) $$G^{(B,F)}_{\mu}(x,y,z) = \langle T\prod_j^{B} A_{\mu_j}(x_j)\prod_k^{F/2}\Psi(y_k)\prod_{\ell}^{F/2}\bar{\Psi}(z_{\ell})\rangle$$

we have, upon application of (iv) (with additional gauge invariance arguments [7,12,13], (vi) and (vii), the following identities

$$\begin{aligned}
\Delta_0 G_\mu^{(B,F)} &= (\Delta_1 + u\Delta_2 + v\Delta_3)\, G_\mu^{(B,F)} \\
-F\, G_\mu^{(B,F)} &= (2(c-M)\Delta_1 + 2\Delta_2)\, G_\mu^{(B,F)} \\
\left(-B - e\frac{\partial}{\partial e} + 2\alpha\frac{\partial}{\partial\alpha}\right) G_\mu^{(B,F)} &= \left(\left(2\alpha\frac{\partial c}{\partial\alpha} - e\frac{\partial c}{\partial e}\right)\Delta_1 - 2\Delta_3\right) G_\mu^{(B,F)} \\
M\frac{\partial}{\partial M} G_\mu^{(B,F)} &= (c-M)\Delta_1 G_\mu^{(B,F)} .
\end{aligned} \tag{1.14}$$

Simple algebra implies that there must be a linear relation among the four quantities on the left-hand side of (1.14). We write this relation (Callan-Symanzik equation) as

$$\left(M\frac{\partial}{\partial M} + \beta\left(e\frac{\partial}{\partial e} - 2\alpha\frac{\partial}{\partial\alpha} + B\right) + \gamma F\right) G_\mu^{(B,F)} = \alpha M \Delta_0 G_\mu^{(B,F)} \tag{1.15}$$

(there is a corresponding relation for the vertex functions, with $B \to -B$, $F \to -F$) where α, β, γ are the perturbative solutions (which obviously exist) of

$$\begin{aligned}
-M\alpha + \left(e\frac{\partial c}{\partial e} - 2\alpha\frac{\partial c}{\partial\alpha}\right)\beta + (c-M)\gamma &= -(c-M) \\
-u\alpha \qquad\qquad\qquad - 2\gamma &= 0 \\
-v\alpha \qquad + 2\beta \qquad\qquad &= 0
\end{aligned} \tag{1.16}$$

Equations (1.16) are obtained by substituting (1.14) into (1.15) and equating to zero the coefficient of each Δ_i, $i = 1,2,3$.

The remainder of these lecture notes are organized as follows. In Sections 2 and 3 we shall discuss how ultraviolet and infrared divergences arise in momentum-space Feynman integrals, and how the former may be removed by systematic subtractions. In Section 3 Zimmermann's "forest formula" will be discussed and his convergence theorem stated. A sketch of the convergence proof, suitably generalized to take into account the vanishing photon mass in QED, will be presented later, in Section 6. In Section 4, the properties (i-iv) of normal products listed above will be stated in a more complete manner and their derivations sketched. Gauge invariance in QED will be discussed in Section 5, based on these normal product properties. Finally, the formulation of zero-mass theories, and in particular of the pure Yang-Mills model, will be treated in Section 7.

2. ULTRAVIOLET AND INFRARED DIVERGENCES

The first task of any renormalization scheme for composite fields is to give meaning to Green functions of normal products of free fields,

(2.1) $$\langle T \prod_i N[Q_i^{(0)}(x_i)] \rangle_0$$

where each $Q_i^{(0)}$ is a formal product of free fields $\phi_\alpha^{(0)}$ and their derivatives. The simplest prescription for free-field Green functions is that of Wick [19], according to which (2.1) can be expanded as a sum of products of the two-point functions

(2.2) $$\langle T \phi_\alpha^{(0)}(x) \phi_\beta^{(0)}(y) \rangle_0$$

with pairings of factors in the same $Q_i^{(0)}$ forbidden. As is well known, each term in the Wick expansion corresponds to a Feynman diagram (graph) whose vertices correspond to the normal products and whose lines correspond to the two-point functions (propagators) linking them. If Γ is a connected graph, the Fourier transform of the term in the Wick expansion represented by Γ has the form

(2.3) $$(2\pi)^4 \delta\left(\sum_{i=1}^{n+1} q_i(p)\right) \lim_{\epsilon \to 0} \int \frac{d^{4m}k}{(2\pi)^{4m}} I_{\Gamma\epsilon}(p,k)$$

$k = k_1 \ldots k_m$ = basis for internal momenta of Γ

$p = p_1 \ldots p_n$ = basis for external momenta of Γ

q_i = momentum entering at i^{th} external vertex

(2.4)

$\prod_V$ = product over vertices V of Γ

$\prod_L$ = product over internal lines L of Γ

ℓ_L = momentum flowing through line L

$\{\ell_{vi}\}$ = set of momenta of internal and external lines meeting at V

P_V = polynomial in components of ℓ_{vi}

$$\Delta_{L\varepsilon}(\ell_L) = \frac{P_L(\ell_L)}{\prod_j [\ell_L^2 - m_{Lj}^2 + i\varepsilon(\vec{\ell}_L^{\,2} + m_{Lj}^2)]}\,, \qquad P_L = \text{polynomial}$$

In this lecture we shall study under what circumstances the integral in (2.3) converges absolutely. Here we shall encounter well known divergence problems whose removal (or avoidance) is the main focus of renormalization theory.

The particular form of the ε term in the denominator factors of $\Delta_{L\varepsilon}$ is perhaps unfamiliar to the reader. It is a form which is particularly well suited to the study of absolute convergence, due to the inequalities [20]

$$(2.5)\quad \frac{1}{\sqrt{1+\varepsilon^2}}\,\frac{1}{\ell_0^2+\vec{\ell}^{\,2}+m^2} \le \frac{1}{|\ell^2-m^2+i\varepsilon(\vec{\ell}^{\,2}+m^2)|} \le \sqrt{1+\frac{4}{\varepsilon^2}}\;\frac{1}{\ell_0^2+\vec{\ell}^{\,2}+m^2}\,.$$

Thus, the absolute convergence of $\int dk I_{\Gamma\varepsilon}(p,k)$ is equivalent to that of $\int dk I^E_{\Gamma\varepsilon}(p,k)$ =

$$(2.6)\qquad \int d^{4m}k \prod_V P_V \prod_L \frac{P_L}{\prod_j(\ell_{L0}^2+\vec{\ell}_L^{\,2}+m_{Lj}^2)}\,.$$

It is clear that divergences of (2.6), with all denominator factors positive semi-definite, can come from only two sources:

<u>Ultraviolet (UV) Divergence</u>: when $I^E_{\Gamma\varepsilon}$ does not fall off fast enough as one goes to infinity in some directions in k-space.

<u>Infrared (IR) Divergence</u>: when $I^E_{\Gamma\varepsilon}$ has a non-integrable singularity for some finite k (possible only if some of the masses m_{Lj} vanish).

To see how UV and IR divergences arise in practice, we consider two simple examples.

<u>Example 1</u>: The Feynman integral

$$\int \frac{d^4k}{(2\pi)^4}\,\frac{i^2}{[k^2-m^2+i\varepsilon(\vec{k}^{\,2}+m^2)][(p-k)^2-m^2+i\varepsilon((\vec{p}-\vec{k})^2+m^2)]}$$

corresponding to the graph  , converges absolutely if and only if

Table

Integration Variables		Formal Convergence or Divergence	
u("small")	v("large")	UV ($v\to\infty$)	IR ($u\to 0$) $p\neq 0$
k	q-k q-p ℓ	logarithmic divergence	convergence
q-k	k q-p ℓ	logarithmic divergence	convergence
q-p	k q-k ℓ	logarithmic divergence	convergence
ℓ'	k q-k q-p ℓ	convergence	convergence
k , q-k k , q-p k , ℓ q-k , q-p q-k , ℓ q-p , ℓ ℓ , ℓ'	——	——	convergence
——	k , q-k k , q-p k , ℓ q-k , q-p q-k , ℓ q-p , ℓ ℓ' , ℓ	quadratic divergence	——

$$\int d^4k \frac{1}{[k_\varepsilon^2 + m^2][(p-k)_\varepsilon^2 + m^2]} \qquad k_\varepsilon^2 = k_0^2 + \vec{k}^2, \text{ etc.}$$

converges absolutely.

For $k \to \infty$, $I_{\Gamma\varepsilon}^{E} \sim 1/(k_\varepsilon^2)^2 \to$ log. divergence.

For $m \neq 0$, $I_{\Gamma\varepsilon}^{E}$ has no singularity for finite k .

For $m = 0$, $p \neq 0$, $I_{\Gamma\varepsilon}^{E}$ has integrable singularities at $k = 0$ and $k = p$.

For $m = 0$, $p = 0$, $I_{\Gamma\varepsilon}^{E}$ has a non-integrable singularity (log. divergence) at $k = 0$.

Example 2: The Feynman integral

$$\int \frac{d^4k}{(2\pi)^4} \frac{d^4q}{(2\pi)^4} \frac{i^3}{[k^2 + i\epsilon\vec{k}^2][(q-k)^2 + i\epsilon(\vec{q}-\vec{k})^2][(p-q)^2 + i\epsilon(\vec{p}-\vec{q})^2]}$$

corresponding to  with massless propagators, converges absolutely if and only if

$$\int d^4k\, d^4q \frac{1}{k_\varepsilon^2 \; (q-k)_\varepsilon^2 \; (p-q)_\varepsilon^2}$$

converges absolutely. The UV and IR divergence or convergence of representative sub-integrals is presented in the accompanying table. There generic linear combinations of k , q and p ($\neq k$, $q-k$, $p-q$) are denoted by ℓ and ℓ' .

In the above examples, a systematic study of possible UV and IR divergences was made (for fixed external momentum), based on the assumption that they could all be found by considering all possible choices of four-vector integration variables $u_1 \ldots u_a \;\; v_1 \ldots v_{m-a}$ with Jacobian $\partial(u,v)/\partial(k) \neq 0$ and simply counting powers in numerator and denominator for

(UV) $v \to \infty$ almost all fixed u

(IR) $u \to \infty$ almost all fixed v

This assumption of the validity of naive power counting in fact holds, not only for this graph, but for general Feynman integrals. Thus, to prove absolute convergence of $\int dk I_{\Gamma\varepsilon}^{E}$, and hence of $\int dk I_{\Gamma\varepsilon}$, it is sufficient to show that for all choices u , v of integration variables, the cut-off integral

$$\int\limits_{u_{iE}^2 < r^2 < v_{jE}^2} du\, dv\, I_{rE}^{E}$$

is convergent by simple power counting.

To state this result more precisely, we introduce the <u>upper</u> and <u>lower degrees</u> of a rational function $F(x,y)$, $x = x_1 \ldots x_f$ $y = y_1 \ldots y_g$. We define

$$\overline{\deg}_{x|y} F = \bar{\nu} \quad , \quad \underline{\deg}_{x|y} F = \underline{\nu}$$

if for almost all x and y ,

$$\lim_{\lambda \to \infty} \lambda^{-\bar{\nu}} F(\lambda x, y) \neq 0, \infty \quad ,$$

$$\lim_{\lambda \to 0} \lambda^{-\underline{\nu}} F(\lambda x, y) \neq 0, \infty \quad .$$

These degrees satisfy the following properties:

$$\overline{\deg}_{x|y} \prod_i F_i = \sum_i \overline{\deg}_{x|y} F_i$$

$$\underline{\deg}_{x|y} \prod_i F_i = \sum_i \underline{\deg}_{x|y} F_i$$

(2.7)

$$\overline{\deg}_{x|y} \sum_i F_i \leq \max\{\overline{\deg}_{x|y} F_i\}$$

$$\underline{\deg}_{x|y} \sum_i F_i \geq \min\{\underline{\deg}_{x|y} F_i\}$$

with equality in the last two relations if the F_i are linearly independent monomials. If F is a <u>homogeneous</u> function of degree n , then we have in addition

(2.8) $$\overline{\deg}_{x|y} F + \underline{\deg}_{y|x} = n$$

The power counting theorem [21] may then be stated as follows (in a form which can also be used to study the subtracted Feynman integrals to be introduced in the next lecture).

<u>Theorem.</u> <u>Let $R(p,k)$ be a function of four-vector variables $p = p_1 \ldots p_n$ and $k = k_1 \ldots k_m$ of the form</u>

$$\mathcal{R}(p,k) = \frac{P(p,k)}{\prod_i \left(\ell_i^2 - m_i^2 + i\epsilon(\vec{\ell}_i^{\,2} + m_i^2)\right)} \tag{2.9}$$

where $P(p,k)$ is a polynomial in the components of the p_i and k_j and each ℓ_i is an element of $\mathbf{L}$, the space linear forms

$$\sum_i \alpha_i k_i + \sum_j \beta_j p_j$$

The integral

$$\int d^{4m}k \; \mathcal{R}(p,k)$$

converges absolutely if for every choice of $u_1 \ldots u_a$ $v_1 \ldots v_b$, $a+b = m$, such that the Jacobian $\partial(u,v)/\partial(k)$ is non-vanishing, the following UV and IR convergence conditions are satisfied:

$$\overline{\deg}_{v|u} \mathcal{R}(p, k(u,v,p)) + 4b < 0 \tag{2.10}$$

$$\underline{\deg}_{u|v} \mathcal{R}(p, k(u,v,p)) + 4a > 0 \,. \tag{2.11}$$

Let us now study how to check in a systematic fashion the UV and IR power counting conditions for a Feynman integral $\int dk I_{\Gamma\varepsilon}(p,k)$ corresponding to a one-particle-irreducible graph Γ (Feynman integrals of connected or disconnected graphs may be written as products of such integrals). Here it is convenient to introduce the concept of the augmented graph $\hat{\Gamma}$ formed from Γ by drawing (if possible) auxiliary lines starting at the external vertices of Γ and ending at a new vertex, V_∞. Such a graph correctly represents the linear relations among momenta (internal and external) for generic values of k and p. The vertex V_∞ represents the overall conservation of momentum. (There are, of course, exceptional values of the external momenta for which there are additional relations.) For a tadpole (one external line) or a vacuum bubble (no external lines) we simply define $\hat{\Gamma} \equiv \Gamma$.

Let us suppose that a set of integration variables has been given. Define

Γ_v = the subgraph of Γ spanned by lines whose momenta are variable with respect to v (i.e. lines L such that L flows momentum $\ell_L = \alpha \cdot u + \beta \cdot v + \gamma \cdot p$ with $\beta \neq 0$)

Γ_{vp} = the subgraph of $\hat{\Gamma}$ spanned by lines whose momenta are variable with respect to v or p ($\beta \neq 0$ or $\gamma \neq 0$)

Clearly UV convergence can be checked by counting powers in Γ_v, (dimension of phase space + ~~degree~~ of integrand) whereas IR convergence can be verified by counting powers in the reduced graph Γ/Γ_{vp} (dimension of phase space + degree of integrand). The following fact allows one to state a convenient sufficient condition for the absolute convergence of $\int dk I_{\Gamma\varepsilon}$: the connected components of Γ_v and $\hat{\Gamma}_{vp}$ are one-particle irreducible. (The trivial verification of this is left as an exercise for the reader.) Thus, the integral

$$\int d^{4m}k I_{\Gamma\varepsilon}$$

converges absolutely if

(a) for every 1PI subgraph γ of Γ, the integral $\int d^{4m(\gamma)}k^\gamma I_{\gamma\varepsilon}(p^\gamma,k^\gamma)$ is UV convergent by simple power counting, i.e.

$$d(\gamma) = \overline{\deg}\, I_\gamma + 4m(\gamma) < 0 \tag{2.12}$$

where (denoting the sets of vertices and internal lines of γ by $V(\gamma)$ and $L(\gamma)$, respectively),

$$\overline{\deg}\, I_\gamma = \sum_{v\in\mathcal{V}(\gamma)} \overline{\deg}\, P_v + \sum_{L\in\mathcal{L}(\gamma)} \overline{\deg}\, \Delta_L \quad , \tag{2.13}$$

(b) for every reduced graph $\Lambda = \hat{\Gamma}/\{\gamma_i\}$, $\gamma_1\dots\gamma_c$ disjoint non-trivial 1PI subgraphs of $\hat{\Gamma}$, with all auxiliary lines contained in γ_1 , the integral

$$\int d^{4m(\Lambda)}k^\Lambda I_{\Lambda\varepsilon}(p^\Lambda,k^\Lambda)$$

is IR convergent by simple power counting, i.e.

$$r(\Lambda) = \underline{\deg}\, I_\Lambda + 4m(\Lambda) > 0 \tag{2.14}$$

where

$$\underline{\deg}\, I_\Lambda = \sum_{v\in\mathcal{V}(\gamma)} \underline{\deg}\, P_v + \sum_{L\in\mathcal{L}(\gamma)} \underline{\deg}\, \Delta_L \, . \tag{2.15}$$

The UV and IR superficial divergences, $d(\gamma)$ and $r(\Lambda)$, may be put in more usable form by introducing

$\bar{d}_\alpha$ = UV dimension of $\phi_\alpha^{(0)}$

$\underline{d}_\alpha$ = IR dimension of $\phi_\alpha^{(0)}$

$$\nu_\alpha(V) \;=\; \text{no. of } \alpha\text{-type line endings at } V$$

$$d_V \;=\; \sum_\alpha \nu_\alpha(V)\bar{d}_\alpha + \overline{\deg}\, P_V$$

$$r_V \;=\; \sum_\alpha \nu_\alpha(V)\underline{d}_\alpha + \underline{\deg}\, P_V$$

Then one can easily show (denoting the set of external lines of γ by $\mathcal{E}(\gamma)$)

(2.15) $$d(\gamma) = 4 - \sum_{L\in\mathcal{E}(\gamma)} \bar{d}_{\alpha(L)} + \sum_{V\in\mathcal{V}(\gamma)} (d_V - 4) \quad ,$$

(2.16) $$r(\Lambda) = 4 + \sum_{V\in\Lambda} (r_V - 4) \quad .$$

In a theory with interaction, in which a diagram may have an arbitrary number of internal interaction vertices, one clearly wants to restrict oneself to such vertices with $d_V \leq 4$, $r_V \geq 4$. But then, in a theory with fermions and bosons with canonical dimensions, one still may have UV trouble from 1PI subgraphs with 2 , 3 and 4 external lines and IR trouble from those with 2 or 3 external massless lines.

3. MOMENTUM-SPACE SUBTRACTIONS

Momentum-space subtraction schemes for avoiding divergences are grounded in some simple properties of the Taylor series. In particular, suppose that $F(x,y)$ is a homogeneous rational function of $x = x_1 \ldots x_a$, and $y = y_1 \ldots y_b$ of degree n which is regular at $y = 0$. Then

(3.1) $$\underline{\deg}_{y|x} (1 - t_y^d) F(x,y) \geq d+1$$

and, since

(3.2) $$\underline{\deg}_{y|x} F + \overline{\deg}_{x|y} F = n \quad ,$$

we have

(3.3) $$\overline{\deg}_{x|y} (1 - t_y^d) F \leq n - d - 1 \quad .$$

To see how this property can be applied to Feynman integrals, consider $\int d^4k I(p,k,m) \equiv$

(3.4) $$\int d^4k \frac{1}{[k^2 - m^2 + i\epsilon(\vec{k}^2 + m^2)][(p-k)^2 - m^2 + i\epsilon((\vec{p}-\vec{k})^2 + m^2)]}$$

corresponding to the diagram

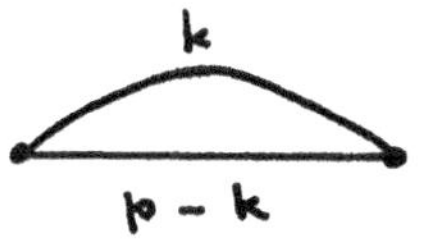

As pointed out in Section 2, the integral is logarithmically UV divergent. Now let us observe that $I(p,k,m)$ is a homogeneous rational function of degree -4, and hence, by (3.3)

$$\deg_{km|p}(1-t_p^0)I \leq -5 .$$

We now observe that since m^2 in a denominator factor always is accompanied by a non-zero k^2 term,

$$(3.5) \quad \deg_k(1-t_p^0)I \leq \deg_{km|p}(1-t_p^0)I \leq -5$$

and so

$$\int d^4k\,(1-t_p^0)I(p,k,m)$$

$$(3.6) \quad = \int d^4k\left[I(p,k,m) - I(0,k,m)\right]$$

is UV convergent. More generally, if the Feynman integrand $I(p,k,m)$ is a homogeneous function of $p_1 \ldots p_n$, $k_1 \ldots k_m$, $m_1 \ldots m_\ell$ such that

$$(3.7) \quad \deg_{pkm} I = d-4m$$

we have

$$(3.8) \quad \deg_k(1-t_p^d)I \leq -4m-1 .$$

This is of course not sufficient to establish the UV convergence of $\int dkI(p,k,m)$, since there may be divergent sub-integrals.

Let us now return to our simple one-loop example to examine briefly the question of IR convergence (for $p \neq 0$). We noted in Section 2 that there is no IR problem for the unsubtracted integral (3.4), even for $m = 0$, unless p is also zero. But this is precisely the form of our subtraction term. By subtracting at zero momentum, to avoid a UV divergence, we have introduced an IR divergence where there was none before. The remedy, at least in this example, is obvious: either subtract at some non-zero momentum point, or subtract at some non-zero value of the mass. The latter alternative turns out to be simpler, and we shall adopt it here and in what follows. In the present case, we use a subtracted integral

$$\int d^4k \; R(p,k,M) \tag{3.9}$$

where

$$R(p,k,M) = I(p,k,0) - I(p,k,M) .$$

The integrand in (3.9) now satisfies both UV and IR convergence conditions.

To generalize the prescription (3.9), we note that $R(p,k;M)$ can be reexpressed using a Taylor operator with respect to p and an auxiliary variable s as

$$R(p,k,M) = (1 - t^0_{ps}) \cdot \tag{3.10}$$

$$\cdot \left[\frac{1}{[k^2 - (1-s)^2M^2 + i\epsilon(\vec{k}^2 + (1-s)^2M^2)][(p-k)^2 - (1-s)^2M^2 + i\epsilon((\vec{p}-\vec{k})^2 + (1-s)^2M^2)]} \right]$$

More generally, if F is a homogeneous function of p, k, M and sM of degree $d-4m$, then

$$\deg_k (1 - t^d_{ps}) I \leq -4m - 1 . \tag{3.11}$$

The assignment of mass terms $(1-s)^2M^2$ to all denominator factors with $m^2 = 0$ and the use of the Taylor operator t^d_{ps} gets rid of the artificial IR problem caused by our special way of cancelling UV divergences, namely by subtraction at $p = 0$. It does not, of course, obviate the more fundamental problem, cited in the previous lecture, of possible IR divergences due to self-energy and 3-point vertex parts with zero-mass external lines. That problem will be dealt with in the final section. Until then, we shall not have occasion to resort to the "s-trick".

For a primitively divergent graph (no properly contained superficially divergent subgraphs) with UV superficial divergence d, relation (3.8) shows that application of the Taylor remainder operator $(1 - t^d_p)$ is sufficient to render the Feynman integral UV convergent. For a diagram Γ with only non-overlapping divergent 1PI subgraphs this procedure may be applied iteratively, starting with the smallest subgraphs and proceeding to successively larger ones until all superficial divergences have been eliminated. This is essentially the subtraction method introduced by Dyson [22]. Using somewhat vague symbolic notation, the procedure may be summarized by the following recipe:

Dyson's Method (Shorthand Notation)

Replace the integrand I_Γ by

$$R_\Gamma = \prod_{\gamma \in \hat{U}} (1 - t_{p^\gamma}^{d(\gamma)}) I_\Gamma \tag{3.12}$$

where $\hat{U}$ = set of all 1PI $\gamma \subseteq \Gamma$ with $d(\gamma) \geq 0$

$\lambda \subset \gamma \Longrightarrow (1 - t_{p^\lambda}^{d(\lambda)})$ stands to the right of $(1 - t_{p^\gamma}^{d(\gamma)})$.

The formula (3.12) has no well defined meaning for cases in which Γ contains overlapping renormalization parts (1PI γ with $d(\gamma) \geq 0$). However, a simple formal manipulation of (3.12), namely the expansion of the product of $(1 - t_{p^\gamma}^{d(\gamma)})$ factors into a sum of "monomials" in the Taylor operators, yields an expression which can be so generalized, namely

Zimmermann's Forest Formula (Shorthand Notation)

Replace the integrand I_Γ by

$$R_\Gamma = \sum_{U \in \mathcal{F}_\Gamma} \prod_{\gamma \in U} (-t_{p^\gamma}^{d(\gamma)}) I_\Gamma \tag{3.13}$$

where F_Γ is the set of a forests (families of non-trivial, non-overlapping 1PI subgraphs) of Γ

$\lambda \subset \gamma \Longrightarrow t_{p^\lambda}^{d(\lambda)}$ stands to the right of $t_{p^\gamma}^{d(\gamma)}$

$t_{p^\gamma}^{d(\gamma)} \equiv 0$ if $d(\gamma) < 0$.

Let us now turn to the problem of making precise formulas (3.12) and (3.13). The first step is to specify a standard routing of momentum through a 1PI graph Γ and its subgraphs (this is not the only method for routing momentum, but it is the simplest). Let $p = p_1 \ldots p_n$ be a basis for the external momenta of Γ and let $k = k_1 \ldots k_n$ be a basis for the internal momenta (integration variables) of Γ, with each k_i representing the momentum circulating in a loop C_i of Γ. Then the momentum $\ell_{ab\nu}$ flowing through $L_{ab\nu}$, the νth line connecting vertex V_b to V_a, is written

$$\ell_{ab\nu} = q_{ab\nu}(p) + k_{ab\nu}(k) \tag{3.14}$$

where

$$k_{ab\nu} = \sum_i \epsilon_{ab\nu i} k_i$$

$$\epsilon_{ab\nu i} = \begin{cases} 1 & \text{if } L_{ab\nu} \in C_i \text{ (incl. orientation)} \\ -1 & \text{if } L_{ba\nu} \in C_i \quad \text{"} \quad \text{"} \\ 0 & \text{otherwise} \end{cases}$$

and $q_{ab\nu}(p)$ is determined by "Kirchhoff's Laws" of circuit theory,

$$\forall\, V_a \in \mathcal{V}(\Gamma),\ \sum_{b,\nu} q_{ab\nu} + q_a = 0 \tag{3.15}$$

$$\forall \text{ loop } C \text{ of } \Gamma,\ \sum_{L_{ab\nu} \in C} r_{ab\nu}\, q_{ab\nu} = 0 , \tag{3.16}$$

where the "resistances" $r_{ab\nu} = r_{ba\nu}$ may be chosen in any convenient manner (including $r_{ab\nu} = 0$ or ∞), except that closed loops of zero-resistance or infinite-resistance lines are not permitted (it is clearly advantageous to maximize the number of 0- and ∞- resistances).

In addition to (3.14), we shall also need an expression for the momentum flow through $L_{ab\nu}$ considered as a line of $\gamma \subset \Gamma$. If $p^\gamma = p_1^\gamma \ldots p_{n(\gamma)}^\gamma$ is a basis for the external momenta of γ, we write

$$\ell_{ab\nu}^\gamma = q_{ab\nu}^\gamma(p^\gamma) + k_{ab\nu}^\gamma(k) \tag{3.17}$$

where $q_{ab\nu}^\gamma$ is determined by "Kirchhoff's Laws" restricted to γ,

$$\forall\, V_a \in \mathcal{V}(\gamma),\ \sum_{b,\nu} q_{ab\nu}^\gamma + q_a^\gamma = 0 \tag{3.18}$$

$$\forall \text{ loop } C \text{ of } \gamma,\ \sum_{L_{ab\nu} \in C} r_{ab\nu}\, q_{ab\nu}^\gamma = 0 , \tag{3.19}$$

and $k_{ab\nu}^\gamma$ is determined by the identity

$$q_{ab\nu}(p) + k_{ab\nu}(k) = q_{ab\nu}^\gamma(p^\gamma(p,k)) + k_{ab\nu}^\gamma \tag{3.20}$$

with $p^\gamma(p,k)$ determined by momentum conservation at the vertices of γ . It is left as an exercise for the reader to prove that $k^\gamma_{ab\nu}$ defined this way is in fact a function only of k .

Example

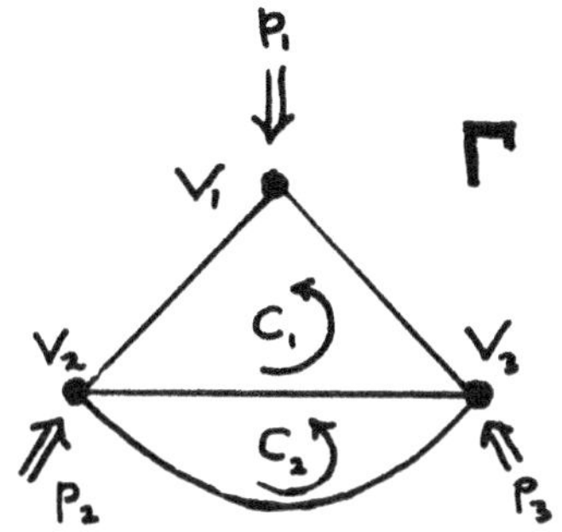

with 1PI subgraph

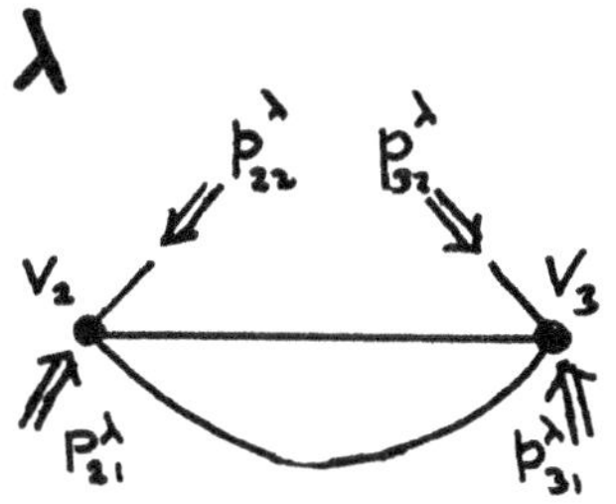

Assigning "resistance" 0 to L_{13} and L_{232} and 1 to L_{12} and L_{231}, the flow of external momentum is given by

$$q_{12} = q_{231} = q^\lambda_{231} = 0$$

$$q_{13} = -p_1 \quad , \quad q_{232} = -p_2 \quad , \quad q^\lambda_{232} = -(p^\lambda_{21} + p^\lambda_{22})$$

Assigning loop momenta k_1 and k_2 to C_1 and C_2 , the flow of internal momentum is given by

$$k_{13} = k_{21} = k_1 \quad , \quad k_{322} = k_2 \quad , \quad k_{321} = k_1 - k_2$$

$$k^\lambda_{322} = k_2 - k_1 = k^\lambda_{23} \quad .$$

We are now in a position to write down mathematically precise expressions for the subtracted integrands corresponding to

(3.21) $$I_{\Gamma G} = \prod_{V_a \in \Gamma} P_a(\ell_{ab\sigma}) \prod_{L_{ab\nu} \in \Gamma} \Delta_{ab\nu}(\ell_{ab\nu})$$

Specifically,

(3.22) $$R_{\Gamma G} = S_\Gamma \prod_{\gamma \in \hat{U}} (1 - t^{d(\gamma)}_{p^\gamma}) S_\gamma I_{\Gamma G}(\hat{U})$$ (non-overlapping case)

(3.23) $$R_{\Gamma G} = S_\Gamma \sum_{U \in \mathcal{F}_\Gamma} \prod_{\gamma \in U} (-t^{d(\gamma)}_{p^\gamma} S_\gamma) I_{\Gamma G}(U)$$ (general case)

where, for any Γ-forest U ,

$$I_{\Gamma\varepsilon}(U) = \prod_{V_a \in \mathcal{V}(\Gamma)} P_a(\ell_{ab\sigma}^{\gamma(V_a)}) \prod_{L_{ab\nu} \in \mathcal{L}(\Gamma)} \Delta_{ab\nu}(\ell_{ab\nu}^{\gamma(L_{ab\nu})})$$

$\gamma(V_a)$ = smallest element of $U \cup \{\Gamma\}$ containing V_a

$\gamma(L_{ab\nu})$ = smallest element of $U \cup \{\Gamma\}$ containing $L_{ab\nu}$

and the substitution operator S_γ is defined by

$$[S f](p^\gamma, k) = f(p^\lambda(p^\gamma, k), k)$$

for an arbitrary function f of p^λ, k. The order of the Taylor operators in (3.22) and (3.23) is that indicated in (3.12) and (3.13) respectively. It is not difficult to show that in the non-overlapping case, formulas (3.22) amd (3.23) give the same result.

Example Let us apply (3.23) to obtain $R\Gamma$ for the two-loop graph whose momentum routing was derived above. The relevant forests are ϕ, $\{\lambda\}$, $\{\Gamma\}$, $\{\Gamma,\lambda\}$ (note: $d(\Gamma) = d(\lambda) = 0$), with

$$I_\Gamma(\emptyset) = I_\Gamma(\{\Gamma\}) = \Delta_F(k_1)\Delta_F(p_1-k_1)\Delta_F(p_2+k_2)\Delta_F(k_1-k_2)$$

$$I_\Gamma(\{\lambda\}) = I_\Gamma(\{\Gamma,\lambda\}) = \Delta_F(k_1)\Delta_F(p_1-k_1)\Delta_F(p_{21}^\lambda+p_{22}^\lambda+k_2-k_3)\times$$
$$\times\,\Delta_F(k_1-k_2)$$

Thus,

$$R_\Gamma = I_\Gamma(\emptyset) - t_p^\circ I_\Gamma(\{\Gamma\}) - S_\Gamma t_{p^\lambda}^\circ I_\Gamma(\{\lambda\}) + t_p^\circ S_\Gamma t_{p^\lambda}^\circ I_\Gamma(\{\Gamma,\lambda\})$$

$$\begin{aligned} &= \Delta_F(k_1)\Delta_F(p_1-k_1)\Delta_F(p_2+k_2)\Delta_F(k_1-k_2) \\ &\quad - \Delta_F(k_1)^2\Delta_F(k_2)\Delta_F(k_1-k_2) \\ &\quad - \Delta_F(k_1)\Delta_F(p_1-k_2)\Delta_F(k_2-k_1)^2 + \Delta_F(k_1)^2\Delta_F(k_2-k_1)^2 \end{aligned}$$

We conclude this section with a statement of Zimmermann's convergence theorem [4,20]: the integral

(3.24) $$J_\varepsilon(p) = \int d^{4m}k\, R_{\Gamma\varepsilon}(p,k) \qquad \text{all } m_{L_j} > 0,\ \varepsilon > 0$$

is absolutely convergent, and when $\varepsilon \to 0$,

$$J_\varepsilon(p) \longrightarrow J(p) \, , \tag{3.25}$$

where $J(p_1 \ldots p_n)$ is a Lorentz covariant tempered distribution and the limit (3.25) is in the sense of tempered distributions.

Although convergence proof of Refs. [4] and [20] applies only to the case where all masses are non zero, the same subtraction prescription may be used in QED with its massless photon. The convergence proof for this and closely related models, as well as all purely massive renormalizable models, is sketched in Section 5 below.

It was pointed out in Section 1 that one of the advantages of Zimmermann's subtraction method is the possibility of making oversubtractions. This corresponds simply to the fact [4] that Zimmermann's convergence theorem remains valid if (3.23) is replaced by

$$R_{\Gamma\varepsilon} = S_\Gamma \sum_{U \in \mathcal{F}_\Gamma} \prod_{\gamma \in U} \left(-t_{p^\gamma}^{\delta(\gamma)} S_\gamma\right) I_{\Gamma\varepsilon}(U) \tag{3.26}$$

where $\delta(\gamma) = d(\gamma)$ + non-negative integer, provided that the degree function $\delta(\gamma)$ satisfies the following property for reduced subdiagrams $\gamma/\lambda_1 \ldots \lambda_c$:

$$\delta(\gamma/\lambda_1 \ldots \lambda_c) \equiv \delta(\gamma) - \sum_i^c \delta(\lambda_i) \geq d(\gamma/\lambda_1 \ldots \lambda_c) \, . \tag{3.27}$$

It is in the more general form (3.26) that we shall use the forest formula in the remainder of these lectures.

4. NORMAL PRODUCTS AND THEIR BASIC PROPERTIES

The connected Green functions of free-field normal products

$$\langle T \prod_i N_{\delta_i} [Q_i(x_i)] \rangle^c \tag{4.1}$$

(we suppress the superscript (0) used previously for products of free fields) are uniquely defined by the Feynman diagram expansion and Zimmermann's forest formula (3.26) once one has specified a rule for calculating subtraction degrees $\delta(\gamma)$ from a knowledge of the formal products Q_i and indices δ_i . Each Q_i is of the form

$$\prod_r D^{\mu_{ir}} \phi_{\alpha_{ir}}$$

where D^μ is a partial derivative of order $|\mu|$ and the subscript α labels the various free fields. Most commonly δ_i is an integer or half-integer at least as large as the naive UV dimension of Q_i ,

$$\text{(4.2)} \qquad \dim Q_i = \sum_r (\bar{d}_{\alpha_{ir}} + |\mu_{ir}|)$$

where $\bar{d}_\alpha$ denotes the UV dimension of ϕ_α (1 for canonical boson, 3/2 for canonical fermion fields), and $\delta(\gamma)$ is computed according to

$$\text{(4.3)} \qquad \delta(\gamma) = 4 + \sum_{V_i \in \mathcal{V}(\gamma)} (\delta_i - 4) - \sum_{L \in \mathcal{E}(\gamma)} \bar{d}_{\alpha(L)} .$$

Occasionally one has need of more intricate subtraction schemes in which case δ_i may stand for a sequence of numbers and the rule specifying $\delta(\gamma)$ will be more complicated than (4.3) (see, e.g., [9]). In these lectures, unless otherwise stated, the symbol $N_\delta[Q(x)]$ will denote an "ordinary" ("isotropic") normal product with δ an integer or half-integer and $\delta(\gamma)$ given by (4.3).

We now turn to the derivation of four important properties of normal-product Green functions (already mentionèd in abbreviated fashion in Section 1): derivative rules, small-momentum behavior of vertex functions (normalization conditions), Zimmermann identities and field equations.

Derivative Rules

We wish to establish the identities

(notation: $P = \prod_r A_r \quad , \quad A_r = D^{\mu_r} \phi_{\alpha_r}$)

$$\text{(4.4)} \qquad \partial_x^\mu \langle T N_\delta[P(x)] \prod_i^n N_{\eta_i}[Q_i(y_i)] \rangle^c$$

$$= \langle T N_{\delta+1} [\partial_\mu P(x)] \prod_i^n N_{\eta_i} [Q_i(y_i)] \rangle^c \quad =$$

$$= \sum_s \langle T N_{\delta+1}[\prod_{r<s} A_r (\partial^\mu A_s) \prod_{t>s} A_t] \prod_i^n N_{\eta_i}[Q_i(y_i)] \rangle^c$$

To derive (4.4), we expand the left-hand side as a sum over diagrams, and then write the contribution of each diagram as a sum over forests

$$\partial_x^\mu \langle T N_\delta[P(x)] \prod_i^n N_{\eta_i}[Q_i(y_i)] \rangle^c$$

(4.5) $$= \int \frac{d^4p_1}{(2\pi)^4} \cdots \frac{d^4p_n}{(2\pi)^4}\, e^{i\sum_{\ell=1}^n p_\ell(x-y_\ell)} \times$$

$$\times\, i q^\mu \langle T N_\delta[P(0)] \prod_i^n N_{\eta_i}[\tilde{Q}_i(p_i)] \rangle^c$$

where $q^\mu = \sum_{\ell=1}^n p_\ell$ and the tilde denotes Fourier transformation;

(4.6) $$q^\mu \langle T N_\delta[P(0)] \prod_i N_{\eta_i} [\tilde{Q}_i(p_i)] \rangle^c$$

$$= \sum_\Gamma \sigma(\Gamma) \lim_{\epsilon \to 0} \int d^{4m}k (q^\mu) R_{\Gamma\epsilon}(p_1 \cdots p_n ; k_1 \cdots k_m)$$

where $\sigma(\Gamma)$ is a combinatorial factor;

(4.7) $$q^\mu R_{\Gamma\epsilon}(p,k) = q^\mu \sum_{U\in\mathcal{F}_\Gamma} S_\Gamma \prod_{\gamma\in U} (-t_{p^\gamma}^{\delta(\gamma)} S_\gamma) I_{\Gamma\epsilon}(U) .$$

We now commute the q^μ factor through the product of $(-t_{p^\gamma}^{\delta(\gamma)} S_\gamma)$ operators for each forest U . If γ is an element of U containing $V(P)$, the vertex corresponding to $N_\delta[P(x)]$, then q is an external momentum of γ , which in the variables appropriate to γ we write q^γ . For given U , let $\gamma(P;U)$ be the smallest element of U containing $V(P)$. Using

(4.8) $$q^{\gamma\mu} t_{p^\gamma}^{\delta(\gamma)} = t_{p^\gamma}^{\delta(\gamma)+1} q^{\gamma\mu}$$

we obtain

$$q^\mu R_{\Gamma\epsilon}(p,k) = \sum_{U\in\mathcal{F}_\Gamma} S_\Gamma \prod_{\gamma\in U} (-t_{p^\gamma}^{\delta'(\gamma)} S_\gamma)\, q^{\gamma(M;U)\mu} I_{\Gamma\epsilon}(U)$$

$$\delta'(\gamma) = \begin{cases} \delta(\gamma)+1 & \text{if} \quad V(P) \in \mathcal{V}(\gamma) \\ \delta(\gamma) & \text{if} \quad V(P) \notin \mathcal{V}(\gamma) \end{cases} \tag{4.9}$$

But this is just the subtracted integrand contributing to

$$\langle T\, N_{\delta+1}[\partial^\mu P(0)] \prod_i N_{\eta_i}[\tilde{Q}_i(p_i)] \rangle^c$$

This establishes the first equality of (4.4). To obtain the second, we simply write $q^{\gamma(M;U)}$ in (4.9) as $-\sum_L \ell_L^{\gamma(M;U)}$ (sum over all lines L, internal and external, of γ which meet at $V(P)$).

Small-Momentum Behavior of Vertex Functions (Normalization Conditions

As a consequence of the subtraction formula (3.26), the momentum-space vertex functions have the following behavior for small values of their arguments:

$$\begin{aligned} &\langle T\, N_\delta[P(0)] \prod_{i=1}^{a} N_{\eta_i}[\tilde{Q}_i(q_i)] \prod_{j=1}^{b} \tilde{\phi}_{\alpha_j}(r_j) \rangle^p \\ &= \delta_{a0}\, T(r_1 \ldots r_b) + \Delta(q,r) , \end{aligned} \tag{4.10}$$

where $T(r_1 \ldots r_b)$ is the contribution of trivial (one-vertex) graphs and

$$D^\mu_{qr}\, \Delta(q,r) \Big|_{q=0=r} = 0 \tag{4.11}$$

for arbitrary derivative D^μ_{qr} of order

$$|\mu| \leq \delta + \sum_{i=1}^{a} (\eta_i - 4) - \sum_{j=1}^{b} \bar{d}_{\alpha_j}$$

To establish (4.11), we note that for any non-trivial 1PI Γ appearing in the Feynman diagram expansion of $\Delta(q,r)$, the subtracted integrand is of the form

(4.12) $$R_{\Gamma\varepsilon}(q,r,k) = (1 - t_{qr}^{\delta(\Gamma)})\,\bar{R}_{\Gamma\varepsilon}(q,r,k)$$

with $$\bar{R}_{\Gamma\varepsilon} = \sum_{\substack{U\in\mathcal{F}_\Gamma \\ \Gamma\notin U}} S_\Gamma \prod_{\gamma\in U} (-t_{p^\gamma}^{\delta(\gamma)} S_\gamma)\, I_{\Gamma\varepsilon}(U)$$

$$\delta(\Gamma) = \delta + \sum_{i=1}^{a} (\eta_i - 4) - \sum_{j=1}^{b} \bar{d}_{\alpha_j}$$

Hence

(4.13) $$\underline{\deg}_{qr}\, R_{\Gamma\varepsilon}(q,r,k) > \delta(\Gamma)$$

Using the fact that for all masses nonzero, the vanishing of $D^{\mu}_{qr}\Delta_{\varepsilon}(q,r)$ for all positive ε implies the vanishing of $D^{\mu}_{qr}\Delta(q,r)$ for $\varepsilon = 0$, (4.13) gives immediately the desired (4.11).

Zimmermann Identities

The Zimmermann Identities are relations between normal products $N_\delta[P(x)]$ and $N_\phi[P(x)]$ with $\delta > \phi \geq \overline{\dim}\, P$. $N_\delta[P(x)]$ is assumed to be an "ordinary" (isotropic) normal product, with δ an integer or half-integer. On the other hand, we allow $N_\phi[P(x)]$ to be anisotropic. Given a diagram with a vertex $V(P)$ corresponding to $P(x)$, let us denote by $\delta(\gamma)$ and $\phi(\gamma)$ the subtraction degrees associated with the respective assignments δ and ϕ to $V(P)$. Then the inequality $\delta > \phi$ means that

$\delta(\gamma) \geq \phi(\gamma)$ for all 1PI γ containing $V(P)$

$\delta(\gamma) > \phi(\gamma)$ for at least one γ.

From the forest formula (3.26), we have

(4.14) $$\langle T\Big(N_\phi[P(0)] - N_\delta[P(0)]\Big)\prod_{i=1}^{n} N_{\eta_i}[\tilde{Q}_i(p_i)]\rangle^c$$
$$= \sum_\Gamma \sigma(\Gamma) \lim_{\varepsilon\to 0} \int d^{4m}k\, \Big(R^{(\phi)}_{\Gamma\varepsilon}(p,k) - R^{(\delta)}_{\Gamma\varepsilon}(p,k)\Big)$$

where

(4.15) $$R^{(\varphi)}_{\Gamma\varepsilon} - R^{(\delta)}_{\Gamma\varepsilon} = \sum_{U\in\mathcal{F}_\Gamma} S_\Gamma \left(\prod_{\gamma\in U} (-t_{p^\gamma}^{\varphi(\gamma)} S_\gamma) - \prod_{\gamma\in U} (-t_{p^\gamma}^{\delta(\gamma)} S_\gamma) \right) I_{\Gamma\varepsilon}(U)$$

$$\sigma(\Gamma) = \text{combinatorial factor}$$

(4.16) $$\left.\begin{aligned} \delta(\gamma) &= \delta + \sum_{V_i\in\mathcal{V}(\gamma)} (\eta_i - 4) - \sum_{L\in\mathcal{E}(\gamma)} \bar{d}_{\alpha(L)} \\ \varphi(\gamma) &= \varphi_\gamma + \sum_{V_i\in\mathcal{V}(\gamma)} (\eta_i - 4) - \sum_{L\in\mathcal{E}(\gamma)} \bar{d}_{\alpha(L)} \end{aligned}\right\} \text{ if } V(P)\in\mathcal{V}(\gamma)$$

$$\delta(\gamma) = \varphi(\gamma) = 4 + \sum_{V_i\in\mathcal{V}(\gamma)} (\eta_i - 4) - \sum_{L\in\mathcal{E}(\gamma)} \bar{d}_{\alpha(L)} \qquad \text{if } V(P)\notin\mathcal{V}(\gamma).$$

In (4.16), ϕ_γ is an integer or half-integer, at least as large as $\overline{\dim}\ P$, which depends in general on the structure of Γ and γ.

Writing

$$t_{p^\gamma}^{\delta(\gamma)} = (t_{p^\gamma}^{\delta(\gamma)} - t_{p^\gamma}^{\varphi(\gamma)}) + t_{p^\gamma}^{\varphi(\gamma)}$$

we can expand $\prod_{\gamma\in U} (-t_{p^\gamma}^{\delta(\gamma)} S_\gamma)$, grouping terms according to the smallest τ of U which contains $V(P)$ and is assigned $t_{p^\tau}^{\delta(\tau)} - t_{p^\tau}^{\phi(\tau)}$ rather than $t_{p^\tau}^{\phi(\tau)}$. Hence

(4.17) $$\sum_{U\in\mathcal{F}_\Gamma} \left(\prod_{\gamma\in U} (-t_{p^\gamma}^{\varphi(\gamma)} S_\gamma) - \prod_{\gamma\in U} (-t_{p^\gamma}^{\delta(\gamma)} S_\gamma) \right)$$

$$= \sum_{U\in\mathcal{F}_\Gamma} \sum_{\substack{\tau\in U \\ V(P)\in\mathcal{V}(\tau)}} \left[\prod_{\substack{\gamma\in U \\ \gamma\supset\tau \\ \text{or } \gamma\cap\tau=\emptyset}} (-t_{p^\gamma}^{\delta(\gamma)} S_\gamma)(t_{p^\tau}^{\delta(\tau)} - t_{p^\tau}^{\varphi(\tau)}) S_\tau \prod_{\substack{\lambda\in U \\ \lambda\subset\tau}} (-t_{p^\lambda}^{\varphi(\lambda)}) \right] =$$

$$= \sum_{\substack{\tau \\ V(P)\in\mathcal{U}(\tau)}} \sum_{\substack{U\in\mathcal{F}_\Gamma \\ \tau\in U}} \left[\prod_{\substack{\gamma\in U \\ \gamma\supset\tau \text{ or} \\ \gamma\cap\tau=\emptyset}} (-t_{p^\gamma}^{\delta(\gamma)} S_\gamma)(t_{p^\tau}^{\delta(\tau)} - t_{p^\tau}^{\varphi(\tau)}) S_\tau (1 - t_{p^\tau}^{\varphi(\tau)}) \prod_{\substack{\lambda\in U \\ \lambda\subset\tau}} (-t_{p^\lambda}^{\varphi(\lambda)} S_\lambda) \right]$$

In commuting all $t_{p^\gamma}^{\phi(\gamma)} S_\gamma$ factors, $\gamma\cap\tau = \emptyset$, to the left, we have used the fact that for such γ, $\phi(\gamma) = \delta(\gamma)$.

For fixed τ, a forest containing τ can be written

$$U = U_1 \cup \{\tau\} \cup U_2 \quad , \quad U_1 \in \mathcal{M}(\tau) , \; U_2 \in \mathcal{F}_\tau$$

where

$$\mathcal{M}(\tau) = \{ U_1 : \gamma\in U_1 \Rightarrow \gamma\supset\tau \text{ or } \gamma\cap\tau = \emptyset \}$$

Note that there is a one-to-one correspondence between elements of $M(\tau)$ and those of $F_{\Gamma/\tau}$. Moreover,

$$(4.18) \qquad I_{\Gamma\epsilon}(U) = I_{\Gamma/\tau\,\epsilon}(U_1)\, I_{\tau\epsilon}(\{\tau\}\cup U_2) \; .$$

To write the difference of Taylor operators in (4.17) more explicitly, it is convenient to use two labels, q^τ and r^τ, for external momenta p^τ of τ, with q^τ associated with the normal-product vertices and r^τ associated with the external lines of τ. Then

$$(4.19) \qquad t_{p^\tau}^{\delta(\tau)} - t_{p^\tau}^{\varphi(\tau)} = \sum_{\substack{\mu,\nu \\ |\mu|+|\nu| \geq \varphi(\tau)}}^{|\mu|+|\nu|\leq\delta(\tau)} \frac{1}{\mu!\,\nu!}\, r_\nu^\tau\, q_\mu^\tau\, D_{q^\tau}^\mu D_{r^\tau}^\nu \Big|_{q^\tau = 0 = r^\tau}$$

where, in highly abbreviated notation,

$$p^\lambda = p_1^{\lambda_{11}} \cdots p_1^{\lambda_{1a_1}}\, p_2^{\lambda_{21}} \cdots p_2^{\lambda_{2a_2}} \cdots\cdots p_n^{\lambda_{n1}} \cdots p_n^{\lambda_{na_n}}$$

$$\lambda! = a_1!\, a_2! \cdots a_n!$$

$$|\lambda| = \sum_{i=1}^n a_i \qquad \text{etc.}$$

Substituting (4.18) and (4.19) into (4.17), one obtains for the difference of subtracted integrands the expression

$$\sum_{\substack{\tau \\ V(\rho)\in\mathcal{V}(\tau)}} \sum_{\substack{\mu,\nu \\ |\mu|+|\nu|\geq\phi(\tau)+1}}^{|\mu|+|\nu|\leq\delta(\tau)} \frac{1}{\mu!\,\nu!} \left[D^{\mu}_{q^{\tau}} D^{\nu}_{r^{\tau}} R^{(\phi)}_{\tau\epsilon}(q^{\tau}, r^{\tau}, k) \right]_{q^{\tau}=0=r^{\tau}} \times \tag{4.20}$$

$$\times\; q^{\tau}_{\mu}\, R^{(\delta)}_{\Gamma/\tau\,\epsilon\,\nu}(q, r, k)$$

where $R^{(\delta)}_{\Gamma/\tau\epsilon}$ is the subtracted integrand for the reduced diagram Γ/τ, with the reduced τ vertex assigned factor r^{τ}_{ν} and subtraction degree

$$\delta + \sum_{V_i\in\mathcal{V}(\tau)} (\eta_i - 4) - |\mu| \ .$$

Note that because of the subtractions of $R^{(\phi)}_{\tau\epsilon}$, the restriction on the μ,ν summation to $|\mu|+(\nu) \geq \phi(\tau) + 1$ is superfluous (see discussion of the small-momentum behavior of vertex functions earlier in this section).

The double summation over Γ and τ in (4.14) with (4.20) is equivalent to the following three-step summation:

(i) for fixed sequence $\alpha = \alpha_1, \alpha_2, \ldots \alpha_{|\alpha|}$ of positive integers and fixed non-empty subset β of $\{1,\ldots,n\}$, sum over all 1PI diagrams $\tau(\alpha,\beta)$ with vertices V_i, $i \in \beta$ and external lines corresponding to fields ϕ_{α_j}, $j = 1\ldots,|\alpha|$;

(ii) with α and β still fixed, sum over all connected diagrams with vertices V_j, $j \notin \beta$ and $V(\alpha)$, a special vertex whose line-endings are labeled by α;

(iii) sum over all α and β such that

$$\delta(\alpha,\beta) \equiv \delta + \sum_{i\in\beta} (\eta_i - 4) - \sum_{\alpha} \bar{d}_{\alpha_i} \geq 0 \ .$$

Moreover, the combinatorial factor $\sigma(\Gamma)$ may be split up as $\sigma(\Gamma/\tau)\,\frac{1}{|\alpha|!}\,\sigma(\tau)$ and the k-integrations and ϵ limit in (4.14) performed. One obtains, finally,

$$\langle T \left(N_\varphi[P(x)] - N_\delta[P(x)] \right) \prod_{i=1}^{n} N_{\eta_i}[Q_i(y_i)] \rangle^c$$

$$(4.21) \quad = \sum_{\beta \subseteq \{1,2,\cdots n\}} \sum_{\substack{\alpha = \{\alpha_i\} \\ \delta(\alpha,\beta) \geq 0}} \frac{1}{|\alpha|!} \sum_{\mu,\nu}^{|\mu|+|\nu| \leq \delta(\alpha,\beta)} \frac{(i)^{|\mu|-|\nu|}}{\mu!\,\nu!} D^{\nu}_{y(\beta)} \delta(x - y(\beta)) \times$$

$$\times \left[D_{q(\beta)\mu} D_{r(\alpha)\nu} \langle T N_\varphi[P(0)] \prod_{i\in\beta} N_{\eta_i}[\tilde{Q}_i(q_i) \prod_{k=1}^{|\alpha|} \tilde{\phi}_{\alpha_{|\alpha|-k+1}} (r_{|\alpha|-k+1}) \rangle^{p} \right]_{\substack{q(\beta)=0 \\ r(\alpha)=0}}$$

$$\times \langle T N_{\delta(\beta)-|\mu|} \left[\prod_{j=1}^{|\alpha|} D_x^{\mu_j} \phi_{\alpha_j}(x) \right] \prod_{i\notin\beta} N_{\eta_i}[Q_i(y_i)] \rangle^c$$

where

$$\delta(\beta) = \delta + \sum_{i\in\beta} (\eta_i - 4)$$

$$y(\beta) = y_{\beta_1}, \cdots y_{\beta_{|\beta|}}$$

$$q(\beta) = q_{\beta_1}, \cdots q_{\beta_{|\beta|}}$$

$$r(\alpha) = r_{\alpha_1}, \cdots r_{\alpha_{|\alpha|}}$$

$$\delta(x - y(\beta)) = \prod_{i=1}^{|\beta|} \delta(x - y_{\beta_i})$$

Field Equations

We wish to derive the quantum analogue of the classical free-field Euler-Lagrange equations,

$$(4.22) \quad \frac{\hat{\partial} \mathcal{L}_0}{\hat{\partial} \phi_\alpha}(x) = \frac{\partial \mathcal{L}_0}{\partial \phi_\alpha}(x) - \frac{\partial}{\partial x^\mu} \frac{\partial \mathcal{L}_0}{\partial(\partial_\mu \phi_\alpha)}(x) = 0$$

For the two-point function, this is simply a matter of the definition of the time-ordered Green function,

$$(4.23)\quad \langle T \frac{\hat\partial \mathcal{L}_0}{\hat\partial \phi_\alpha}(x)\, \phi_\beta(y)\rangle = i\,\delta_{\alpha\beta}\,\delta(x-y)$$

Equation (4.23) is readily generalized, using the amputation formula (a simple consequence of Wick's theorem)

$$\langle T\, \phi_\alpha(x) \prod_{i=1}^{n} N_{\eta_i}[Q_i(y_i)]\rangle^c$$

$$(4.24)\quad = \sum_{j,\beta} (-1)^{\sigma_j} \int dz\, \langle T\, \phi_\alpha(x)\, \phi_\beta(z)\rangle \times$$

$$\times \langle T\, N_{\eta_j - d_\beta}\left[\frac{\delta Q_j(y_j)}{\delta \phi_\beta(z)}\right] \prod_{i\neq j}^{n} N_{\eta_i}[Q_i(y_i)]\rangle^c ,$$

where $(-1)^{\sigma_j}$ is the sign of the fermion-field permutation, to obtain

$$(4.25)\quad \langle T \frac{\hat\partial \mathcal{L}_0}{\hat\partial \phi_\alpha}(x) \prod_{i=1}^{n} N_{\eta_i}[Q_i(y_i)]\rangle^c$$

$$= i \sum_{j=1}^{n} \langle T\, N_{\eta_j - d_\alpha}\left[\frac{\delta Q_j(y_j)}{\delta \phi_\alpha(x)}\right] \prod_{i\neq j}^{n} N_{\eta_i}[Q_i(y_i)]\rangle^c$$

In (4.24-25) the functional derivative is defined by

(4.26)

$$\frac{\delta}{\delta\phi_\alpha(x)} \prod_i D_{y_i}^{\mu_i} \phi_{\beta_i}(y_i) = \sum_j \delta_{\alpha\beta_j}\, D_{y_j}^{\mu_j} \delta(x-y_j)\, (-1)^{\sigma_j} \prod_{i\neq j} \phi_{\beta_i}(y_i)$$

where

$$\sigma_j = \begin{cases} 0 \text{ if } \phi_\alpha \text{ is a boson field} \\ \text{no. of fermionic } \phi_{\beta_i},\ i < j,\ \text{if } \phi_\alpha \text{ is a fermion field.} \end{cases}$$

When $\frac{\hat\partial L_0}{\hat\partial \phi_\alpha}$ appears inside a normal product multiplied by a field monomial $P(x)$, one also obtains field equations, analogous to (4.25), for the Green functions. Here the derivation is complicated by the possibility that graphically the line associated with $\hat\partial L_0/\hat\partial\phi_\alpha$ may be the external line of a 1PI subgraph requiring subtractions. To be more precise, suppose $\hat\partial L_0/\hat\partial\phi_\alpha =$

$\sum_i c_i D^{\mu_i} \phi_{\alpha_i}$. Let us consider "elementary" fields $\chi_{\alpha i}$ (which could be some of the original ϕ_α) , assigned (not necessarily canonical) power-counting dimension $4 - \bar{d}\alpha_i$, with two-point functions

$$(4.27) \qquad \langle T \chi_{\alpha_i}(x) \phi_\beta(y) \rangle = \langle T D^{\mu_i} \phi_{\alpha_i}(x) \phi_\beta(y) \rangle \ .$$

It is possible to equate $N_\delta[P\chi_{\alpha i}]$ with a normal product of $PD^{\mu_i}\phi_{\alpha_i}$, but one whose diagrams are subtracted in an anisotropic manner. Specifically, in the notation of [10] ,

$$(4.28) \qquad N_\delta[P\chi_{\alpha_i}] = N_\delta[\{P\} D^\mu \phi_{\alpha_i}]$$

where in computing $\delta(\gamma)$ for 1PI subgraph γ the corresponding vertex is assigned degree δ if the line associated with ϕ_{α_i} is internal to γ , and degree $\delta + \overline{\dim}\, D^{\mu_i}\phi_{\alpha_i} - \overline{\dim}\, \chi_{\alpha_i} \leq \delta$ if the line is external. Clearly there is a Zimmermann identity relating $N_\delta[PD^{\mu_i}\phi_{\alpha_i}]$ to $N_\delta[\{P\}D^{\mu_i}\phi_{\alpha_i}]$ and hence to $N_\delta[P\chi_{\alpha i}]$. Summing over all terms of $\hat{\partial} L_0/\hat{\partial}\phi_\alpha$, we obtain a Zimmermann identity (of the form (4.21)) relating $N_\delta\left[P\frac{\hat{\partial}\mathcal{L}_0}{\hat{\partial}\phi_\alpha}\right]$ to

$$N_\delta[P\chi_\alpha]$$

where χ_α is a field assigned dimension $4 - \bar{d}\alpha$ and two-point function

$$(4.29) \qquad \langle T \chi_\alpha(x) \phi_\beta(y) \rangle = i\,\delta_{\alpha\beta}\,\delta(x-y) \ .$$

To obtain a relation analogous to (4.25), we must now establish a correspondence between

$$\langle T N_\delta[P\chi_\alpha(x)] \prod_{i=1}^n N_{\eta_i}[Q_i(y_i)] \rangle^c \quad \text{and}$$

$$i \sum_j \langle T N_{\delta+\eta_j-4}\left[P(x)\frac{\delta Q_j(y_j)}{\delta\phi_\alpha(x)}\right] \prod_{i\neq j} N_{\eta_i}[Q_i(y_i)] \rangle (-1)^{\sigma_j}$$

We first note that by Wick's theorem,

$$\langle T N_\delta[P\chi_\alpha(x)] \prod_{i=1}^{n} N_{\eta_i}[Q_i(y_i)] \rangle^c$$

may be written as a sum of contributions, each of which corresponds to the contraction of χ_α with a factor of a given Q_j, plus a contribution from graphs with detachable one-line loops. The latter fortunately vanishes due to cancellations of terms in the forest formula [9].

Now consider the term corresponding to the contraction of χ_α with Q_j (see Fig. 4.1). We let V_1, V_2 and L denote the $P\chi_\alpha$- and Q_j-vertices and the χ_α line connecting them, respectively. The unsubtracted integrand is clearly the same as if V_1 and V_2 were fused in a single $N_{\delta+\eta_j-4}[P\, \delta Q_j/\delta\phi_\alpha]$ vertex. Here the normal-product subtraction degree has been chosen to give the correct number of subtractions for 1PI subgraphs containing the line L, or else containing neither V_1 nor V_2. For other subgraphs containing V_1 or V_2, the number of subtractions could be different in Case I ($P\chi_\alpha$ and Q_j vertices linked by a χ_α-line) than in Case II (single $P\,\delta Q_j/\delta\phi_\alpha$ vertex). We would like to relate Cases I and II (illustrated in Fig. 4.1) by applying the Zimmermann identity, but this is not immediately possible. A given diagram in Case I corresponds uniquely to a diagram in Case II whose unsubtracted integrand is identical (recall: the χ_α propagator is trivial); nevertheless, the forest structures are different in the two cases and so the Zimmermann identity is not directly applicable.

To resolve this difficulty, (our treatment is essentially that of [6]) let us first catalogue those forests which are present in one case but not in the other. First of all, there are some forests in Case I containing a subgraph γ with $V_1, V_2 \in \mathcal{V}(\gamma)$, $L \notin \mathcal{L}(\gamma)$. These clearly have no counterpart in Case II. However, such forests come in pairs: one containing both γ and $\gamma' = \gamma \cup \{L\}$ (smallest subgraph containing γ and L), the other containing γ but not γ'. The subtraction terms in the forest formula exactly cancel, since

$$(1 - t_{p^{\gamma'}}^{\delta(\gamma')}) S_{\gamma'} (-t_{p^\gamma}^{\delta(\gamma)} S_\gamma) = 0$$

The remaining Case I forests are of the form

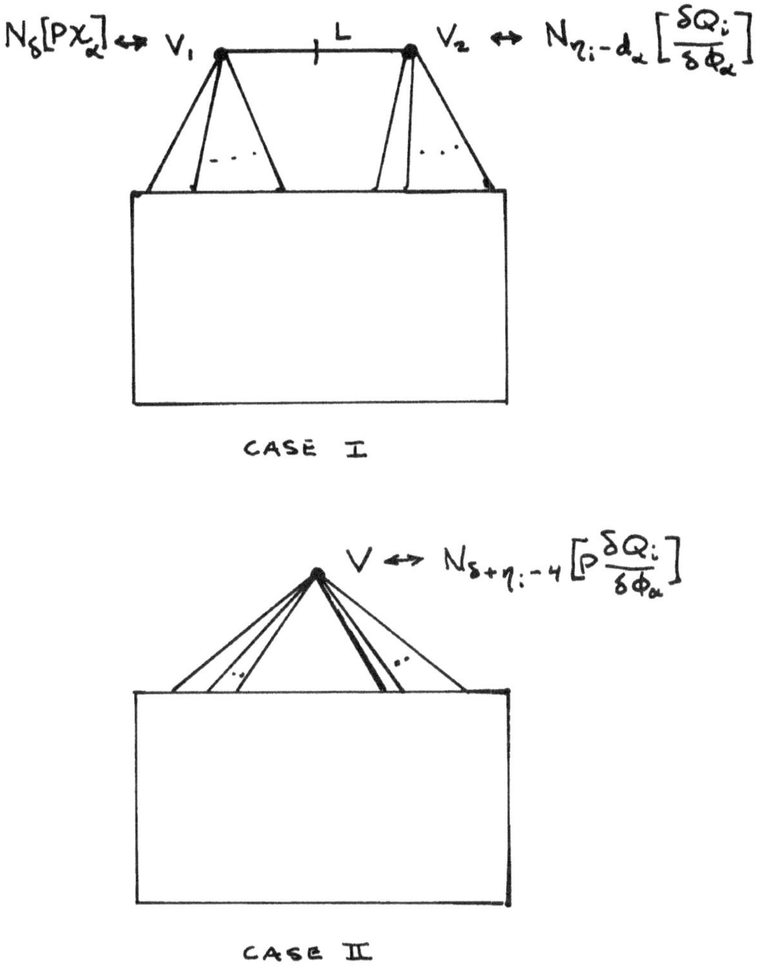

Fig. 4.1

The derivation of normal-product field equations consists mainly in relating the subtracted integrals for Case I, shown schematically in the upper figure (the vertical bar across line L is a reminder that L is assigned a trivial propagator) and Case II, shown in the lower figure.

$$U_I = \{\gamma_{1k}\} \cup \{\gamma_{2\ell}\} \cup \{\rho_m\}$$

$$V_1 \in \gamma_{1k} \;,\quad V_2 \notin \gamma_{1k} \qquad k=1,2,\ldots$$

$$V_1 \notin \gamma_{2\ell} \;,\quad V_2 \in \gamma_{2\ell} \qquad \ell=1,2,\ldots$$

$$V_1 , V_2 \in \rho_m \;\; \underline{\text{or}} \;\; V_1 , V_2 \notin \rho_m \;. \qquad m=1,2,\ldots$$

Each such Case I forest corresponds to a subset $F_{II}(U_I)$ of Case II forests consisting of those which, when the vertex V is split into V_1 and V_2, yield precisely U_I. If $\{\gamma_{1k}\} = \emptyset$ or $\{\gamma_{2\ell}\} = \emptyset$, there is only one such forest; if $\{\gamma_{1k}\} \neq \emptyset$, $\{\gamma_{2\ell}\} \neq \emptyset$, each U_{II} $F_{II}(U_I)$ may be generated recursively in the following manner (notational conventions:

$$\gamma_{1\ell} \subset \gamma_{1m} \quad \text{for} \quad \ell > m$$

$$\gamma_{2\ell} \subset \gamma_{2m} \quad \text{for} \quad \ell > m$$

$$\hat{\gamma} = \text{graph formed from } \gamma \text{ by contracting } L$$

$$\widehat{\gamma_{1j}\,\gamma_{2k}} = \text{graph formed from } \gamma_{ij} , \gamma_{2k} \text{ by contracting } L \text{).}$$

In the first step, let $\widehat{\gamma_{11}\,\gamma_{21}} \in U_{II}$ (forced by non-overlapping property). Now suppose that in the $(n-1)$st step γ_{1k} and $\gamma_{2\ell}$ have been paired to form $\widehat{\gamma_{1k}\,\gamma_{2\ell}}$ and there remain unpaired γ_{ij} and γ_{2m}, $j > k$, $m > \ell$. Then in the nth step let

$$\widehat{\gamma_{1k}\,\gamma_{2(\ell+1)}} \;\; \underline{\text{or}} \;\; \widehat{\gamma_{1(k+1)}\,\gamma_{2\ell}} \;\; \underline{\text{or}} \;\; \widehat{\gamma_{1(k+1)}\,\gamma_{2(\ell+1)}} \in U_{II}$$

(again forced by non-overlapping property). Proceed until $\{\gamma_{1k}\}$ or $\{\gamma_{2\ell}\}$ is exhausted. Finally, let the remaining γ_{1k} (or $\gamma_{2\ell}$), as well as ρ_m, become $\hat{\gamma}_{1k}$ (or $\hat{\gamma}_{2\ell}$) and $\hat{\rho}_m \in U_{II}$.

The subtracted integrand $R^{\delta_I}_{T\in I}$ is defined with the forest structure of Case I (subscript) and the degree function of Case I (superscript); $R^{\delta_{II}}_{T\in II}$ is defined analogously for Case II. We now wish to show that there exists a new degree function ϕ_I appropriate to the forest structure of Case II such that

$$\varphi_{I} \leq \delta_{II}$$

(4.30) $$R^{\varphi_I}_{\Gamma\in II} = R^{\delta_I}_{\Gamma\in I} \; .$$

Then we shall be in a position to apply the Zimmermann identity.

To obtain (4.30), we show that the contribution of the U_I term to $R^{\delta_I}_{\Gamma\in I}$ is precisely equal to the sum of $F_{II}(U_I)$ terms to $R^{\phi_I}_{\Gamma\in II}$. This is obvious for U_I with $\{\gamma_{1k}\}$ or $\{\gamma_{2\ell}\}$ empty. For $\{\gamma_{1k}\}$ and $\{\gamma_{2\ell}\}$ both non-empty, the contribution to $R^{\delta_I}_{\Gamma\in I}$ has the form (notation: $t^{\delta_I}_{\gamma} \equiv t^{\delta_I(\gamma)}_{p^\gamma} S_\gamma$)

(4.31) $$\cdots(-t^{\delta_I}_{\gamma_{11}})(-t^{\delta_I}_{\gamma_{21}})\cdots\prod_{j>1}\left[(-t^{\delta_I}_{\gamma_{1j}})\cdots\right]\prod_{k>1}\left[(-t^{\delta_I}_{\gamma_{2k}})\cdots\right] I_{\Gamma\epsilon}(U_I)$$

where the dots stand for appropriately placed $t^{\delta_I}_{\rho_m}$.

We now apply the identity

(4.32) $$\begin{aligned} t^{\delta_I}_{\gamma_{11}} t^{\delta_I}_{\gamma_{21}} &= t^{\varphi_I}_{\widehat{\gamma_{11}\gamma_{21}}}\, t^{\delta_I}_{\gamma_{11}} t^{\delta_I}_{\gamma_{21}} \\ &= -t^{\varphi_I}_{\widehat{\gamma_{11}\gamma_{21}}}\left(1 - t^{\delta_I}_{\gamma_{11}} - t^{\delta_I}_{\gamma_{21}}\right) \end{aligned}$$

with

$$\varphi_I(\widehat{\gamma_{1j}\gamma_{2k}}) = \delta_I(\gamma_{1j}) + \delta_I(\gamma_{2k}) \; .$$

Expression (4.31) then becomes

(4.33) $$\begin{aligned} \Big\{ &\cdots(-t^{\varphi_I}_{\widehat{\gamma_{11}\gamma_{21}}})\cdots(-t^{\delta_I}_{\gamma_{12}})(-t^{\delta_I}_{\gamma_{22}})\cdots\prod_{j>2}\left[(-t^{\delta_I}_{\gamma_{1j}})\cdots\right]\prod_{k>2}\left[(-t^{\delta_I}_{\gamma_{2k}})\cdots\right] \\ &+\cdots(-t^{\varphi_I}_{\widehat{\gamma_{11}\gamma_{21}}})\cdots(-t^{\delta_I}_{\gamma_{11}})(-t^{\delta_I}_{\gamma_{22}})\cdots\prod_{j>1}\left[(-t^{\delta_I}_{\gamma_{1j}})\cdots\right]\prod_{k>2}\left[(-t^{\delta_I}_{\gamma_{2k}})\cdots\right] \\ &+ \end{aligned}$$

$$+\cdots(-t^{\phi_I}_{\widehat{\gamma_{11}\gamma_{21}}})\cdots(-t^{\delta_I}_{\gamma_{12}})(-t^{\delta_I}_{\gamma_{21}})\cdots$$

$$\cdots\prod_{j>2}\left[(-t^{\delta_I}_{\gamma_{1j}})\cdots\right]\prod_{k>1}\left[(-t^{\delta_I}_{\gamma_{2k}})\cdots\right]\Big\}\, I_{\Gamma\varepsilon}(U_I).$$

This procedure can now be iterated until one runs out of $t^{\delta_I}_{\gamma_{1j}}$'s or $t^{\delta_I}_{\gamma_{2k}}$'s to pair up. In this way one constructs all the terms corresponding to forests of $F_{II}(U_I)$ contributing to $R^{\phi_I}_{\Gamma\varepsilon II}$ (compare the recursive contruction of $F_{II}(U_I)$).

With the question of forest structure taken care of, one can now write down a Zimmermann identity relating the Green functions of $N_\delta[P\chi_\alpha]$ to those of $N_{\delta+\eta_j-4}[P\,\frac{\delta Q_j}{\delta\phi_\alpha}]$. Combining this identity with that relating $N_\delta[P\chi_\alpha]$ to $N_\delta[P\,\frac{\partial L_0}{\partial\phi_\alpha}]$, one obtains a field equation of the form

$$\langle T\, N_\delta[P\frac{\hat\partial \mathcal{L}_0}{\hat\partial\phi_\alpha}(x)]\prod_{i=1}^{n} N_{\eta_i}[Q_i(y_i)]\rangle^c$$

$$= i\sum_{j=1}^{n}(-1)^{\sigma_j}\langle T\, N_{\delta+\eta_j-4}[P_{(x)}\frac{\delta Q_j(y_j)}{\delta\phi_\alpha(x)}]\prod_{i\neq j}N_{\eta_i}[Q_i(y_i)]\rangle^c$$

(4.34)

$$+\sum_{\beta\subseteq\{1,\cdots,n\}}\ \sum_{\substack{\alpha=\{\alpha_i\}\\ \delta(\alpha,\beta)\geq 0}}\ \sum_{\mu,\nu}^{|\mu|+|\nu|\leq\delta(\alpha,\beta)} C^{\alpha\beta}_{\mu\nu}\, D^{\nu}_{y(\beta)}\,\delta(x-y(\beta))\ \times$$

$$\times\langle T\, N_{\delta(\beta)-|\mu|}[\prod_{j=1}^{|\alpha|}D^{\mu_j}\phi_{\alpha_j}(x)]\prod_{i\notin\beta}N_{\eta_i}[Q_i(y_i)]\rangle^c$$

The coefficients $C^{\alpha\beta}_{\mu\nu}$ may be calculated if needed, from the details of the above derivation.

Normal Products of Interacting Fields

The normal-products of renormalized fields in a theory with interaction are specified by the Gell-Mann-Low formula (1.4). The derivative rules, normalization condition, Zimmermann identities and field equations appropriate to those Green functions can be readily derived from the analogous free-field properties (this is left as an exercise for the reader). The various identities have the same form as in the free-field case, with L_0 being replaced by the full effective Lagrangian L_{eff} in the field equations.

For Green functions involving only one normal product which is not an elementary field or a derivative thereof, Eqs. (4.21) and (4.34) lead to particularly simple forms (stated in Section 1) for the Zimmermann identities and field equations. For further discussion of these properties and applications to various models the reader is referred to Ref. [7].

5. WARD IDENTITIES IN QUANTUM ELECTRODYNAMICS

Quantum electrodynamics, oddly enough, has only recently been seen to be treatable in a simple and direct fashion within the BPHZ framework. Although there is no essential infrared problem associated with defining Green functions for QED, provided one normalizes away from the fermion's mass shell (P. Blanchard and R. Seneer [23] provided a rigorous proof of this within the Epstein-Glaser framework), the same is not obviously true of contributions of individual diagrams when subtracted according to the prescription of Zimmermann. For example, consider the diagram

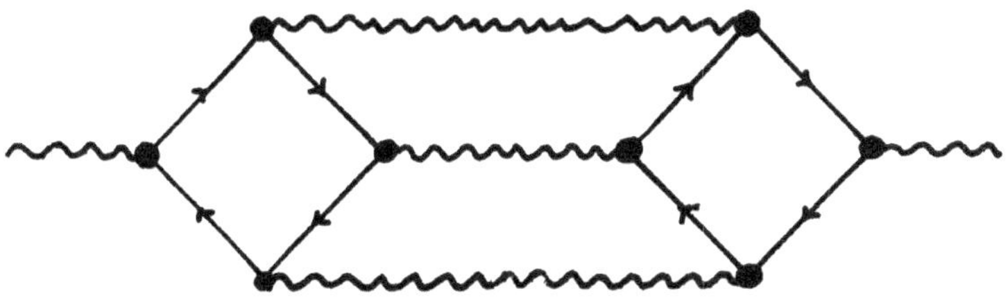

The corresponding Feynman integral requires Taylor subtractions up to second order. Are the second order subtractions IR finite? This is far from clear (recall that for $\varepsilon > 0$ one cannot use arguments based on Lorentz invariance or gauge invariance). One might argue that since the box subgraphs are once-subtracted at zero external momentum, there will be enough momentum powers in the numerator to avoid IR problems. But this argument clearly breaks down if all of the external momentum p is routed through only one of the intermediate photon lines, and this can always be arranged by a suitable change of integration variables (Recall: the IR power counting criterion must be satisfied for all choices of integration variables). It turns out that the subtractions of four-photon vertex parts do save the day, but the proof is definitely nontrivial, thanks to the presence of overlaps (of which there are plenty already in the simple example just considered). This proof will be postponed to the next section, where it will be shown that the Green functions of QED are, graph by graph, well defined as tempered distributions. In this section we shall apply the "normal product algorithm" discussed in Sections 1 and 4 to derive the most important structural properties (Ward identities) of these Green functions.

The connected time-ordered functions of QED are specified, to every finite order in e, by the Gell-Mann-Low formula (1.4) with effective Lagrangian

$$\begin{aligned} \mathcal{L}_{eff} &= \mathcal{L}_0 + \mathcal{L}_{eff\,I} \\ \mathcal{L}_0 &= -\tfrac{1}{4}(\partial_\mu A_\nu - \partial_\nu A_\mu)^2 - \tfrac{1}{2\alpha}(\partial_\mu A^\mu)^2 + \bar{\Psi}(i\not{\partial} - M)\Psi \\ \mathcal{L}_{eff\,I} &= e\bar{\Psi}\gamma_\mu\Psi A^\mu + c\bar{\Psi}\Psi \end{aligned} \tag{5.1}$$

The B-photon, F-fermion vertex functions, $\Gamma^{(B,F)}_{\mu_1\ldots\mu_B}$, constructed this way have the low momentum behavior ("intermediate" normalization conditions)

$$\begin{aligned} \Gamma^{(2,0)}_{\mu\nu}(k,-k) &= i(g_{\mu\nu}k^2 - k_\mu k_\nu) + \frac{i}{\alpha}k_\mu k_\nu + O(k^4) \\ \Gamma^{(0,2)}(p,-p) &= i(\not{p} - M + c) + O(p^2) \\ \Gamma^{(1,2)}_\mu(-p-q,p,q) &= ie\gamma_\mu + O(p,q) \\ \Gamma^{(4,0)}_{\mu_1\mu_2\mu_3\mu_4}(k_1\cdots k_4) &= O(k^2) \end{aligned} \tag{5.2}$$

The constant c is chosen recursively to enforce

$$\Gamma^{(0,2)}(p,-p)\Big|_{\not{p}=M} = 0 \tag{5.3}$$

From (5.1) and the results of Section 4, we have the following field equations (notation:

$$X = \hat{\prod_{i=1}} N_{\eta_i}[Q_i(y_i)] \quad , \quad X_j = \hat{\prod_{i\neq j}} (-1)^{\sigma_j} N_{\eta_i}[Q_i(y_i)] \ ):$$

(5.4)

$$\langle T\, N_{\frac{5}{2}}\left[\frac{\hat{\partial}\mathcal{L}_{eff}}{\hat{\partial}\bar{\Psi}}(x)\right] X \rangle^c = i \sum_j \langle T\, N_{\eta_j - \frac{3}{2}}\left[\frac{\delta Q_j(y_j)}{\delta \bar{\Psi}(x)}\right] X_j \rangle^c$$

$$\langle T\, N_{\frac{5}{2}}\left[\frac{\hat{\partial}\mathcal{L}_{eff}}{\hat{\partial}\Psi}(x)\right] X \rangle^c = i \sum_j \langle T\, N_{\eta_j - \frac{3}{2}}\left[\frac{\delta Q_j(y_j)}{\delta \Psi(x)}\right] X_j \rangle^c$$

$$\langle T\, N_3\left[\frac{\hat{\partial}\mathcal{L}_{eff}}{\hat{\partial}A_\mu}(x)\right] X \rangle^c = i \sum_j \langle T\, N_{\eta_j - 1}\left[\frac{\delta Q_j(y_j)}{\delta A_\mu(x)}\right] X_j \rangle^c$$

where

(5.5)

$$\frac{\hat{\partial}\mathcal{L}_{eff}}{\hat{\partial}\bar{\Psi}} = i(\not{\partial} - ie\not{A})\Psi - (M-c)\Psi$$

$$\frac{\hat{\partial}\mathcal{L}_{eff}}{\hat{\partial}\Psi} = i\bar{\Psi}(\overleftarrow{\not{\partial}} + ie\not{A}) + (M-c)\bar{\Psi}$$

$$\frac{\hat{\partial}\mathcal{L}_{eff}}{\hat{\partial}A_\mu} = \left[(g_{\mu\nu}\partial^2 - \partial_\mu\partial_\nu) + \frac{1}{\alpha}\partial_\mu\partial_\nu\right]A^\nu + e\bar{\Psi}\gamma_\mu\Psi$$

Similarly,

(6.6)

$$\langle T\, N_4[\bar{\Psi}_k \frac{\hat{\partial}\mathcal{L}_{eff}}{\hat{\partial}\bar{\Psi}_\ell}(x) X\rangle^c = i\sum \langle T\, N_{\eta_j}\left[\bar{\Psi}_k(x)\frac{\delta Q_j(y_j)}{\delta\bar{\Psi}_\ell(x)}\right] X_j \rangle^c$$

$$\langle T\, N_4\left[\frac{\hat{\partial}\mathcal{L}_{eff}}{\hat{\partial}\Psi_\ell}\Psi_k(x)\right] X \rangle^c = \frac{1}{i}\sum_j \langle T\, N_{\eta_j}\left[\Psi_k(x)\frac{\delta Q_j(y_j)}{\delta\Psi_\ell(x)}\right] X_j \rangle^c$$

$$\langle T\, N_4\left[A_\nu \frac{\hat{\partial}\mathcal{L}_{eff}}{\hat{\partial}A^\mu}(x)\right] X \rangle^c = i\sum_j \langle T\, N_{\eta_j}\left[A_\nu(x)\frac{\delta Q_j(y_j)}{\delta A^\mu(x)}\right] X_j \rangle^c$$

We now derive the fundamental Ward identity of QED, that corresponding to the formal invariance of L_{eff} under restricted (infinitesimal) gauge transformations of the second kind,

$$\begin{aligned} \delta(\omega)A_\mu(x) &= \partial_\mu\omega(x) \\ \delta(\omega)\psi(x) &= ie\,\omega(x)\psi(x) \\ \delta(\omega)\bar\psi(x) &= -ie\,\omega(x)\bar\psi(x) \end{aligned} \tag{5.7}$$

$$\omega(x) = \omega^*(x)$$

$$\partial^2\omega(x) = 0 \quad .$$

Taking the divergence of the A_μ equation in (5.4), we obtain

$$\begin{aligned} &\frac{1}{\alpha}\langle T\,\partial^2\partial_\mu A^\mu(x)\,X\rangle^c \\ &= -e\,\partial_x^\mu\langle T\,N_3[\bar\psi\gamma_\mu\psi(x)]\,X\rangle^c \\ &\quad + i\sum_j \partial_x^\mu\langle T\,N_{\eta-1}\Big[\frac{\delta Q_j(y_j)}{\delta A^\mu(x)}\Big]\,X_j\rangle^c \end{aligned} \tag{5.8}$$

Applying the derivative rules and field equations (5.6), the first term on the right-hand side becomes (this is just the Ward identity for the vector current)

$$\begin{aligned} &e\langle T\,N_4[\bar\psi\overleftarrow{\partial\!\!\!/}\psi(x) + \bar\psi\overrightarrow{\partial\!\!\!/}\psi(x)]\,X\rangle^c \\ &= -ie\langle T\,N_4\Big[\bar\psi\frac{\hat\partial\mathcal{L}_{eff}}{\hat\partial\bar\psi}(x) + \frac{\hat\partial\mathcal{L}_{eff}}{\hat\partial\psi}\psi(x)\Big]\,X\rangle^c \\ &= e\sum_j\langle T\,N_{\eta_j}\Big[\Big(\bar\psi(x)\frac{\delta}{\delta\bar\psi(x)} - \psi(x)\frac{\delta}{\delta\psi(x)}\Big)Q_j(y_j)\Big]\,X_j\rangle^c . \end{aligned} \tag{5.9}$$

Hence,

$$\begin{aligned} &\frac{1}{\alpha}\langle T\,\partial^2\partial_\mu A^\mu(x)\,X\rangle \\ &= -i\sum_j\langle T\,N_{\eta_j}\Big[\Big(-\partial_x^\mu\frac{\delta}{\delta A^\mu(x)} + ie\,\psi(x)\frac{\delta}{\delta\psi(x)} - ie\,\bar\psi(x)\frac{\delta}{\delta\bar\psi(x)}\Big)Q_j(y_j)\Big]\times \\ &\qquad \times X_j\rangle^c , \end{aligned} \tag{5.10}$$

and, for arbitrary test function $\omega(x)$,

(5.11) $$\frac{i}{\alpha}\int dx\, \omega(x) < T\, \partial^2 \partial_\mu A^\mu(x)\, X >$$

$$= \sum_j < T\, N_{\eta_j}[\delta(\omega)\, Q_j(y_j)]\, X_j >$$

It is tempting to generalize (5.11) to include $\omega(x)$ satisfying $\partial^2\omega = 0$; however, this does not make the left-hand side vanish due to the boundary term in the integration by parts. Thus the restricted invariance of the effective section is not reflected in a true invariance of the Green functions. More fruitful is to replace $\omega(x)$ by

$$\Delta_y(x) = \alpha\, D_F(x-y)$$

--- (5.11) then remains true in the distributional sense --- to obtain what we would today call the Slavnov identity,

(5.12) $$< T\, \partial_\mu A^\mu(x)\, X > = \sum_j < T\, N_\eta[\delta(\Delta_x)\, Q_j(y_j)]\, X_j > .$$

The graphical structure of this identity is depicted in Fig. 5.1.

The implications of (5.10) for the vertex functions may be found by introducing the generating functionals

(5.13)
$$G\{J,\xi,\bar{\xi}\} = < T \exp i\int dx [J^\mu(x) A_\mu(x) + \bar{\xi}(x)\psi(x) + \bar{\psi}(x)\xi(x)] >^c$$

$$\Gamma\{K,\eta,\bar{\eta}\} = < T \exp \int dx\, [K^\mu(x) A_\mu(x) + \bar{\eta}(x)\psi(x) + \bar{\psi}(x)\eta(x)] >^p$$

related by

(5.14)
$$K_\mu(x) = \frac{1}{i}\frac{\delta G}{\delta J^\mu(x)} \qquad J_\mu(x) = i\,\frac{\delta \Gamma}{\delta K^\mu(x)}$$

$$\eta(x) = \frac{1}{i}\frac{\delta G}{\delta \bar{\xi}(x)} \qquad \xi(x) = i\,\frac{\delta \Gamma}{\delta \bar{\eta}(x)}$$

$$\bar{\eta}(x) = \frac{1}{i}\, G\frac{\overleftarrow{\delta}}{\delta \xi(x)} \qquad \bar{\xi}(x) = i\,\Gamma\frac{\overleftarrow{\delta}}{\delta \eta(x)}$$

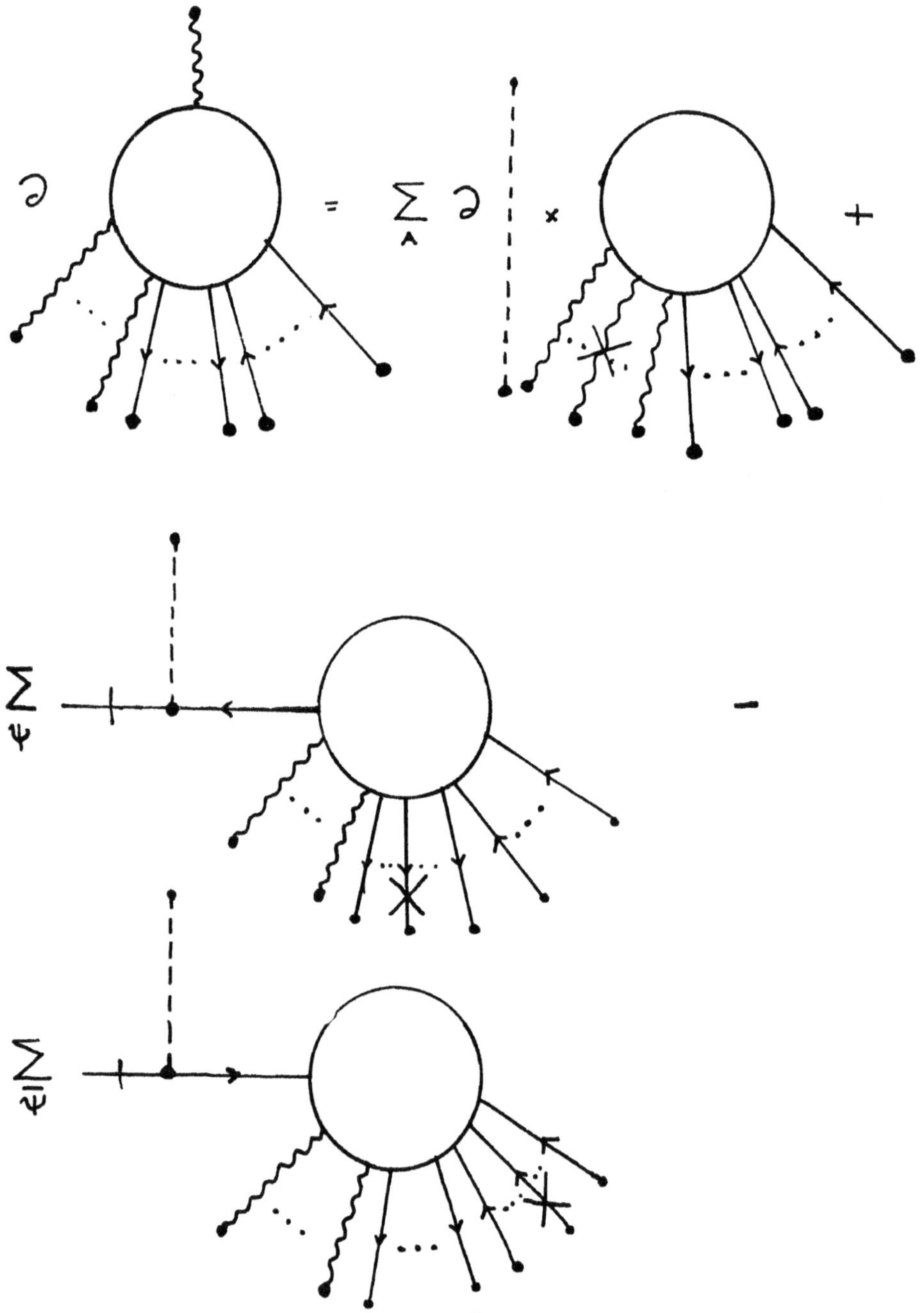

Fig. 5.1
Graphical structure of Eq. (5.12) for the case where all Q_j are elementary fields. A dashed line has been used to denote the scalar propagator αD_F .

Then (5.10) may be written

$$\text{(5.15)}\quad \frac{1}{\alpha}\partial^2\partial_\mu \frac{1}{i}\frac{\delta G}{\delta J^\mu(x)}$$

$$= -i\left(\frac{1}{i}\partial^\mu J_\mu(x) - ie\bar{\xi}(x)\frac{1}{i}\frac{\delta G}{\delta\bar{\xi}(x)} + ie\xi(x)\frac{1}{i}\frac{\delta G}{\delta\xi(x)}\right)$$

which, by (5.14), becomes

$$\text{(5.16)}\quad \frac{i}{\alpha}\partial^2\partial^\mu K_\mu(x)$$

$$= \partial^\mu\frac{\delta\Gamma}{\delta K^\mu(x)} - e\eta(x)\frac{\delta\Gamma}{\delta\eta(x)} + e\bar{\eta}(x)\frac{\delta\Gamma}{\delta\bar{\eta}(x)} .$$

Taking various functional derivatives of (5.17) yields the following facts about the vertex functions:

a) The longitudinal part of $\Gamma^{(2,0)}_{\mu\nu}(k,-k)$ is simply

$$-\frac{i}{\alpha}k_\mu k_\nu$$

with no non contribution from non-trivial graphs.

b) $\Gamma^{(B,0)}_{\mu_1\ldots\mu_B}(k_1\ldots k_B)$ is purely transverse with respect to any of its arguments for $B > 2$:

$$k_i^{\mu_i}\,\Gamma^{(B,0)}_{\mu_1\cdots\mu_i\cdots\mu_B}(k_1\ldots k_i\ldots k_B) = 0 \qquad (B>2)$$

c) The original Ward-Takahashi identity [24]

$$(p+q)^\mu\,\Gamma^{(2,1)}_\mu(-p-q,p,q) = e\Gamma^{(0,2)}(p,-p) - e\Gamma^{(0,2)}(-q,q) .$$

6. ABSOLUTE CONVERGENCE PROOF FOR SUBTRACTED INTEGRALS

That the perturbative Green functions of theories with all masses nonzero, as well as those of QED as defined in Section 5, are in fact Lorentz covariant tempered distributions is a consequence of the following theorem (generalization of [4]) combined with the ε-limit theorem of Ref. [25]:

Theorem. Let Γ be a connected diagram with at least two external vertices and no tadpole subgraphs. The integral

$$\int d^{4n}p\, d^{4m}k\, f(p)\, R_{\Gamma\varepsilon}(p,k) \tag{6.1}$$

converges absolutely for arbitrary $\varepsilon > 0$ and $f \in S(R^{4n})$ provided that (notation: $r(\gamma) = \mathbf{I}R$ superficial divergence (2.14), $\delta(\gamma)$ = subtraction degree used to define $R_{\Gamma\varepsilon}$)

i) for every non-trivial 1PI $\gamma \subseteq \hat{\Gamma}$, $r(\gamma) \geq \delta(\gamma) + 1$;

ii) if γ is a non-trivial 1PI subgraph of $\hat{\Gamma}$ and $\lambda_1 \ldots \lambda_c$ are mutually disjoint, non-trivial 1PI subgraphs properly contained in γ , then $r(\gamma/\{\lambda_i\} \geq \delta(\gamma/\{\lambda_i\})$;

iii) for every set $\{\gamma_i\}$ of mutually disjoint, non-trivial 1PI subgraphs of $\hat{\Gamma}$,

$$r\left(\hat{\Gamma}/\{\gamma_i\}\right) + \sum_i \max\{0,\ \delta(\gamma_i)+1\} > 0 \ . \tag{6.2}$$

Recall that $\hat{\Gamma}$ was constructed from Γ by assigning to each external vertex V_{Ei} , at which external momentum q_i enters Γ , an auxiliary line (q-line) with propagator

$$\left(q_i^2 - \mu^2 + i\varepsilon(\vec{q}_i^{\,2} + \mu^2)\right)^{-\nu} \qquad \mu^2 > 0 \ ,$$

with the q-line meeting at a new internal vertex, V_∞ . The exponent ν is taken sufficiently large (it is easy to see that this is always possible) that $\delta(\gamma)$ may be chosen to satisfy

$$\delta(\gamma) < 0$$

$$\delta(\gamma) \leq r(\gamma) - 1$$

for all non-trivial 1PI subgraphs and reduced subgraphs γ containing q-lines.

Introducing the subtracted integrand for $\hat{\Gamma}$,

(6.3) $$R_{\hat{\Gamma}\epsilon}(p,k) = \prod_{i=1}^{n+1} \left(q_i(p)^2 - \mu^2 + i\epsilon(\vec{q}_i(p)^2 + \mu^2)\right)^{-\nu} R_{\Gamma\epsilon}(p,k)$$

$$= S_{\hat{\Gamma}} \sum_{U \in \mathcal{F}_{\hat{\Gamma}}} \prod_{\gamma \in U} \left(-t_{p^\gamma}^{\delta(\gamma)} S_\gamma\right) \prod_i \left(q_i^2 - \mu^2 + i\epsilon(\vec{q}_i^{\,2} + \mu^2)\right)^{-\nu} I_{\Gamma\epsilon}(U)$$

we obtain a convenient sufficient condition for the absolute convergence of (6.1). Since $f \in S(R^{4n})$, there are positive constants K, K' such that

(6.4) $$|f(p)| \leq K' \prod_i \left(q_i^0(p)^2 + \vec{q}_i(p)^2 + \mu^2\right)^{-\nu}$$

$$\leq K \prod_i \left| q(p)^2 - \mu^2 + i\epsilon(\vec{q}(p)^2 + \mu^2)\right|^{-\nu} ,$$

and hence the absolute convergence of (6.1) is implied by that of

(6.4) $$\int d^{4n}p \, d^{4m}k \; R_{\hat{\Gamma}\epsilon}(p,k) \; .$$

We are now in a position to apply the power-counting theorem [21] quoted in Section 2. Since Γ is a 1PI vacuum bubble (no external momenta), we have a simplified statement of the convergence criteria: the integral (6.4) converges absolutely if for every choice of basis $u_1 \ldots u_a \; v_1 \ldots v_b$ for $\mathbb{L}(\hat{\Gamma})$, the space of linear forms in k and p,

(6.5) $$\deg_{v|u} R_{\hat{\Gamma}\epsilon}(p(u,v), k(u,v)) + 4b < 0$$

(6.6) $$\deg_{u|v} R_{\hat{\Gamma}\epsilon}(p(u,v), k(u,v)) + 4a > 0$$

Notice the UV-IR symmetry of the convergence conditions. A remarkable fact, which will be emphasized in this section, is that (6.5) and (6.6) can be verified, under the assumptions of the theorem stated above, in a completely symmetric way. Before doing so, we introduce the concept of complete forest, which is crucial to efficient UV and IR power counting. The definition and expansion theorem stated below are due to Zimmermann [4].

Suppose that S is a linear subspace of $L(\hat{\Gamma})$. If $\bar{\gamma}$ is a reduced 1PI subgraph of $\hat{\Gamma}$, we shall say that $\bar{\gamma}$ lies along S

$(\bar{\gamma} \,||\, S)$ if for all lines $L \in L(\bar{\gamma})$, $k_L^{\gamma} \in S$ (see (3.17-20)), and that $\bar{\gamma}$ is <u>oblique to S</u> $(\bar{\gamma} \not| S)$ if for all $L \in \mathbb{L}(\bar{\gamma})$, $k_L^{\gamma} \notin S$. A $\hat{\Gamma}$-forest C is said to be <u>complete</u> with respect to S ($\mathcal{C}_S(\hat{\Gamma})$ = set of all such forests) if

i) $\hat{\Gamma} \in C$;

ii) for all $\gamma \in C$, either $\bar{\gamma}(C) \,||\, S$ or $\bar{\gamma}(C) \not| S$,

where $\bar{\gamma}(C) = \gamma/\gamma_1 \ldots \gamma_r$, γ_i = maximal elements of C properly contained in γ. The usefulness of complete forests is based on the following result of Zimmermann [4]:

<u>Lemma 6.1.</u> Let $\hat{\Gamma}$ be a 1PI diagram with $\delta(\hat{\Gamma}) < 0$. Then $R_{\hat{\Gamma}\varepsilon}$ has the expansion

$$R_{\hat{\Gamma}\varepsilon} = \sum_{C \in \mathcal{C}_S(\hat{\Gamma})} Y_{\hat{\Gamma}}(C) \quad , \tag{6.7}$$

where $Y_\gamma(C)$, $\gamma \in C$ is defined recursively by

$$Y_\gamma(C) = I_{\bar{\gamma}(C)} S_\gamma \prod_{\alpha=1}^{r} f_{\gamma_\alpha} Y_{\gamma_\alpha}(C) \tag{6.8}$$

with

$$f_{\gamma_\alpha} = \begin{cases} 1 - \tau_{\gamma_\alpha} & \text{if } \bar{\gamma}_\alpha(C) \not| S \,,\ \bar{\gamma}(C) \,||\, S \\ -\tau_{\gamma_\alpha} & \text{otherwise} \end{cases} ,$$

$$\tau_\gamma \equiv t_{p^\gamma}^{\delta(\gamma)} .$$

We now present some power-counting lemmas (actually the heart of the proof of the Theorem), the first of which relates the UV and IR degrees of $\tau_\gamma Y_\gamma$, $\gamma \in C$, to those of Y_γ :

<u>Lemma 2.</u> Let τ_γ and Y_γ be defined as above with respect to an arbitrary basis $u_1 \ldots u_a \; v_1 \ldots v_b$ of $\mathbb{L}(\hat{\Gamma})$. Let S be the subspace of $\mathbb{L}(\hat{\Gamma})$ spanned by $u_1 \ldots u_a$. Then the following inequalities are valid (abbreviations: $Y_\gamma \equiv Y_\gamma(C)$, $\bar{\gamma} \equiv \bar{\gamma}(C)$):

(6.9) if $\tau_\gamma Y_\gamma \neq 0$,

$$\overline{\deg}_{vp^\gamma|u}\, \tau_\gamma Y_\gamma \le \begin{cases} \overline{\deg}_{v|up^\gamma} Y_\gamma + \delta(\gamma) & \text{for } \bar{\gamma} \parallel S \\ \overline{\deg}_{vp^\gamma|u} Y_\gamma & \text{for } \bar{\gamma} \nparallel S\,; \end{cases}$$

(6.10) if $\tau_\gamma Y_\gamma \neq 0$,

$$\overline{\deg}_{v|p^\gamma u}\, \tau_\gamma Y_\gamma \le \overline{\deg}_{v|p^\gamma u} Y_\gamma \qquad \text{for } \bar{\gamma} \parallel S\,;$$

(6.11)

$$\overline{\deg}_{v|p^\gamma u}\, (1-\tau_\gamma) Y_\gamma \le \overline{\deg}_{vp^\gamma|u} Y_\gamma - \max\{\delta(\gamma)+1, 0\} \qquad \text{for } \bar{\gamma} \nparallel S;$$

(6.12) if $\tau_\gamma Y_\gamma \neq 0$,

$$\underline{\deg}_{up^\gamma|v}\, \tau_\gamma Y_\gamma \ge \underline{\deg}_{up^\gamma|v} Y_\gamma \qquad \text{for } \bar{\gamma} \parallel S\,;$$

(6.13) if $\tau_\gamma Y_\gamma \neq 0$,

$$\underline{\deg}_{u|p^\gamma v}\, \tau_\gamma Y_\gamma \ge \begin{cases} \underline{\deg}_{up^\gamma|v} Y_\gamma - \delta(\gamma) & \text{for } \bar{\gamma} \parallel S \\ \underline{\deg}_{u|p^\gamma v} Y_\gamma & \text{for } \bar{\gamma} \nparallel S\,; \end{cases}$$

(6.14)

$$\underline{\deg}_{up^\gamma|v}\, (1-\tau_\gamma) Y_\gamma \ge \underline{\deg}_{u|p^\gamma v} Y_\gamma + \max\{\delta(\gamma)+1, 0\} \qquad \text{for } \bar{\gamma} \nparallel S.$$

Proof. Let us write

$$(6.15)\qquad Y_\gamma = D(p^\gamma, u, v)^{-1} \sum_k f_k(p^\gamma)\, g_k(u,v) \,,$$

where f_k are independent monomials of degree $\nu(k)$ in p^γ,

g_k are polynomials in u and v,

D is a product of factors

$$(6.16)\qquad F_{Li} = \ell_L^{\gamma 2} - m_{Li}^2 + i\epsilon\left(\vec{\ell}_L^{\gamma 2} + m_{Li}^2\right)$$

with $$\ell_L = P_L(p^\gamma) + U_L(u) + V_L(v) \,.$$

We observe that $P_L(p^\gamma)$ can only be nonvanishing for $L \in \bar{\gamma}$ or $L \in \bar{\lambda}$, where λ is a maximal element of C contained in γ, with $\bar{\lambda} \nparallel S$, $\gamma \parallel S$. Moreover, $V_L(v) = 0$ for $L \in \bar{\gamma}(C) \parallel S$ and $V_L(v) \neq 0$ for $L \in \bar{\gamma} \nparallel S$. It is then straightforward to verify the following relations (abbreviations: $\underline{\deg}_u \equiv \underline{\deg}_{u|vp^\gamma}$, $\underline{\deg}_{up^\gamma} \equiv \underline{\deg}_{up^\gamma|v}$):

$$(6.17)\qquad \underline{\deg}_{up^\gamma} Y_\gamma = \min_k \{\nu(k) + \underline{\deg}_u g_k\} + \underline{\deg}_{up^\gamma} D^{-1} ,$$

$$(6.18)\qquad \underline{\deg}_u Y_\gamma = \min_k \{\underline{\deg}_u g_k\} + \underline{\deg}_u D^{-1} ,$$

$$(6.19)\qquad \underline{\deg}_{up^\gamma} \tau_\gamma Y_\gamma = \min_{\nu(k) \leq \delta(k)} \{\nu(k) + \underline{\deg}_u g_k + \underline{\deg}_{up^\gamma} t_{p^\gamma}^{\delta(\gamma)-\nu(k)} D^{-1}\}$$

$$(6.20)\qquad \underline{\deg}_u \tau_\gamma Y_\gamma = \min_{\nu(k) \leq \delta(k)} \{\underline{\deg}_u g_k + \underline{\deg}_u t_{p^\gamma}^{\delta(\gamma)-\nu(k)} D^{-1}\}$$

(6.21) $$\underline{\deg}_{up^\gamma}(1-\bar{\tau}_\gamma)Y_\gamma \geq \min_k \left\{\nu(k) + \underline{\deg}_u g_k + \underline{\deg}_{up^\gamma}(1-t^{\delta(\gamma)-\nu(k)})D^{-1}\right\},$$

(6.22) $$\underline{\deg}_{up^\gamma} t_{p^\gamma}^{\delta(\gamma)-\nu(k)} D^{-1} \geq \underline{\deg}_{up^\gamma} D^{-1},$$

(6.23) $$\underline{\deg}_u t_{p^\gamma}^{\delta(\gamma)-\nu(k)} D^{-1} \geq \begin{cases} \underline{\deg}_{up^\gamma} D^{-1} - \delta(\gamma) + \nu(k) & \text{if } \bar{\gamma} \parallel S \\ \underline{\deg}_u D^{-1} & \text{if } \bar{\gamma} \not\parallel S, \end{cases}$$

(6.24) $$\underline{\deg}_{up^\gamma}(1-t^{\delta(\gamma)-\nu(k)})D^{-1} \geq \delta(\gamma)+1-\nu(k)+\underline{\deg}_u D^{-1}.$$

Equation (6.12) follows from (6.19) and (6.22) ;
(6.13) follows from (6.20) and (6.23) ;
(6.14) follows from (6.21) and (6.24) .

This establishes the IR estimates of Lemma 6.2. The UV estimates (6.9-11) are derived similarly [26].

With the aid of Lemma 6.2, one obtaines the following power-counting inequalities for Y_γ :

Lemma 6.3 For $\hat{\Gamma}$, S , C and 1PI $\gamma \in C$ as in Lemma 6.2,

(6.25) $$\overline{\deg}_{vp^\gamma|u} Y_\gamma \leq \delta(\gamma) - \overline{M}(\gamma) \qquad \text{if } \bar{\gamma} \not\parallel S,$$

(6.26) $$\overline{\deg}_{v|up^\gamma} Y_\gamma \overset{*}{<} -\overline{M}(\gamma) \qquad \text{if } \bar{\gamma} \parallel S,$$

(6.27) $$\underline{\deg}_{up^\gamma|v} Y_\gamma \geq \delta(\gamma)+1-\underline{M}(\gamma) \qquad \text{if } \bar{\gamma} \parallel S,$$

(6.28) $$\underline{\deg}_{u|p^\gamma v} Y_\gamma \overset{*}{>} -\underline{M}(\gamma) \qquad \text{if } \bar{\gamma} \not\parallel S,$$

where $$\underline{M}(\gamma) = 4 \times \sum_{\substack{\lambda\varepsilon C \\ \lambda\subseteq\gamma \\ \bar\lambda(C)||S}} (\text{no. of independent loops of } \bar\lambda(C))$$

$$\bar{M}(\gamma) = 4 \times \sum_{\substack{\lambda\varepsilon C \\ \lambda\subseteq\gamma \\ \bar\lambda(C)\nparallel S}} (\text{no. of independent loops of } \bar\lambda(C))$$

and $\overset{*}{<}$ means $\begin{cases} < & \text{if r.h.s.} \neq 0 \\ = & \text{if r.h.s.} = 0 \end{cases}$

$\overset{*}{>}$ means $\begin{cases} > & \text{if r.h.s.} \neq 0 \\ = & \text{if r.h.s.} = 0 \end{cases}$

Lemma 6.4 Let $\hat\Gamma$, S and C be as in Lemma 6.2. Let λ be a maximal element of C properly contained in $\gamma \varepsilon C$. Then, assuming hypothesis (ii) and (iii) of the Theorem,

(6.29) $\text{if } \tau_\lambda Y_\lambda \neq 0, \quad \overline{\deg}_{v p^\gamma|u} S_\gamma \tau_\lambda Y_\lambda \leq -\bar{M}(\lambda) + \delta(\lambda)$,

(6.30) $\text{if } \tau_\lambda Y_\lambda \neq 0, \quad \overline{\deg}_{v|u p^\gamma} S_\gamma \tau_\lambda Y_\lambda \overset{*}{<} -\bar{M}(\lambda) \quad \text{for } \bar\lambda \parallel S$,

(6.31) $\overline{\deg}_{v|u p^\gamma} S_\gamma (1-\tau_\lambda) Y_\lambda < -\bar{M}(\lambda) \quad \text{for } \bar\lambda \nparallel S, \bar\gamma \parallel S$,

and

(6.32) $\text{if } \tau_\lambda Y_\lambda \neq 0, \quad \underline{\deg}_{u|v p^\gamma} S_\gamma \tau_\lambda Y_\lambda \overset{*}{>} -\underline{M}(\gamma)$

(6.33) $\text{if } \tau_\lambda Y_\lambda \neq 0, \quad \underline{\deg}_{u p^\gamma|v} S_\gamma \tau_\lambda Y_\lambda \geq \delta(\lambda) + 1 - \underline{M}(\gamma) \quad \text{for } \bar\lambda \parallel S$,

(6.34) $\underline{\deg}_{u p^\gamma|v} S_\gamma (1-\tau_\lambda) Y_\lambda \geq \max\{\delta(\lambda)+1, 0\} - \underline{M}(\gamma)$

$\text{for } \bar\lambda \nparallel S, \ \bar\gamma \parallel S$.

Lemmas 6.3 and 6.4 are proved by mathematical induction, starting from

$$(6.35)\qquad \overline{\deg}_{vp^{\gamma}|u}\, Y_\gamma \le \delta(\gamma) - \overline{M}(\gamma) \qquad \text{for} \quad \bar{\gamma} \not\parallel S ,$$

$$(6.36)\qquad \overline{\deg}_{v|up^{\gamma}}\, Y_\gamma = 0 \overset{*}{<} -\overline{M}(\gamma) \qquad \text{for} \quad \bar{\gamma} \parallel S ,$$

$$(6.37)\qquad \underline{\deg}_{up^{\gamma}|v}\, Y_\gamma \ge \delta(\gamma) + 1 - \underline{M}(\gamma) \qquad \text{for} \quad \bar{\gamma} \parallel S ,$$

$$(6.38)\qquad \underline{\deg}_{u|vp^{\gamma}}\, Y_\gamma \ge 0 \overset{*}{>} -\underline{M}(\gamma) \qquad \text{for} \quad \bar{\gamma} \not\parallel S$$

for minimal $\gamma \in C$ and assuming, as induction hypothesis, that Lemma 6.3 holds for all maximal elements of C contained in a given $\gamma \in C$. The proofs, given in [26] and [27] (in the latter case, the IR subtraction degree $\rho(\gamma)$ must be replaced by $\delta(\gamma) + 1$ and hypothesis (ii) and (iii) of the Theorem must be applied) will not be given here.

To complete the proof of the Theorem, let us suppose that C is any forest of $\hat{\Gamma}$ which is complete with respect to S, the subspace of $\mathbb{L}(\hat{\Gamma})$ spanned by $u_1 \dots u_a$ (with $u_1 \dots u_a\ v_1 \dots v_b$ an arbitrary basis of $\mathbb{L}(\hat{\Gamma})$). We wish to show

$$(6.39)\qquad \overline{\deg}_{v|u}\, R_{\hat{\Gamma}_C}(C) + 4b < 0$$

$$(6.40)\qquad \underline{\deg}_{u|v}\, R_{\hat{\Gamma}_C}(C) + 4a > 0$$

First of all, we have

$$(6.41)\qquad \overline{\deg}_{v|u}\, I_{\bar{\hat{\Gamma}}(C)} \le \begin{cases} 0 & \text{for} \quad \bar{\hat{\Gamma}}(C) \parallel S \\ \delta(\bar{\hat{\Gamma}}(C)) - 4m(\bar{\hat{\Gamma}}(C)) & \text{for} \quad \bar{\hat{\Gamma}}(C) \not\parallel S \end{cases}$$

$$(6.42)\qquad \underline{\deg}_{u|v}\, I_{\bar{\hat{\Gamma}}(C)} \ge \begin{cases} r(\bar{\hat{\Gamma}}(C)) - 4m(\bar{\hat{\Gamma}}(C)) & \text{for} \quad \bar{\hat{\Gamma}}(C) \parallel S \\ 0 & \text{for} \quad \bar{\hat{\Gamma}}(C) \not\parallel S , \end{cases}$$

where $m(\bar{\hat{\Gamma}})$ = no. of independent loops of $\bar{\hat{\Gamma}}$. Moreover, if $V(\gamma)$

is a reduced vertex of $\Gamma(C)$ (formed by contracting γ to a point), we have, as a consequence of Lemma 6.4,

$$\overline{\deg}_{v|u} S_\Gamma f_\gamma Y_\gamma(C) \leq -\overline{M}(\gamma) + \delta(\gamma) \quad \text{for } \bar{\hat{\Gamma}} \times S \tag{6.43}$$

$$\overline{\deg}_{v|u} S_\Gamma f_\gamma Y_\gamma(C) \overset{*}{<} -\overline{M}(\gamma) \quad \text{for } \bar{\hat{\Gamma}} \parallel S \tag{6.44}$$

$$\underline{\deg}_{u|v} S_\Gamma f_\gamma Y_\gamma(C) \geq \max\{0, \delta(\gamma)+1\} - \underline{M}(\gamma) \quad \text{for } \bar{\hat{\Gamma}} \parallel S \tag{6.45}$$

$$\underline{\deg}_{u|v} S_\Gamma f_\gamma Y_\gamma(C) \overset{*}{>} -\underline{M}(\gamma) \quad \text{for } \bar{\hat{\Gamma}} \times S . \tag{6.46}$$

Combining (6.41), (6.43) and (6.45), as well as (6.42), (6.44) and (6.46), and using

$$4a \geq \begin{cases} 4m(\bar{\hat{\Gamma}}) + \sum_{\substack{\max \gamma \\ \in C}} \underline{M}(\gamma) & \text{for } \bar{\hat{\Gamma}} \parallel S \\ \sum_{\substack{\max \gamma \\ \in C}} \underline{M}(\gamma) & \text{for } \bar{\hat{\Gamma}} \times S \end{cases} \tag{6.47}$$

$$4b \leq \begin{cases} 4m(\bar{\hat{\Gamma}}) + \sum_{\substack{\max \gamma \\ \in C}} \overline{M}(\gamma) & \text{for } \bar{\hat{\Gamma}} \times S \\ \sum_{\max \gamma \in C} \overline{M}(\gamma) & \text{for } \bar{\hat{\Gamma}} \parallel S \end{cases} \tag{6.48}$$

with equality in (6.47) or (6.48) only if the right-hand side does not vanish, we obtain

$$\overline{\deg}_{v|u} R_{\hat{\Gamma}_C}(C) + 4b \begin{cases} < 0 & \text{if } \bar{\hat{\Gamma}}(C) \parallel S \\ \leq \delta(\hat{\Gamma}) & \text{if } \bar{\hat{\Gamma}}(C) \times S \end{cases} \tag{6.49}$$

(6.50)
$$\underline{\deg}_{u1v} R_{\hat{\Gamma}_C}(C) + 4a \begin{cases} \geq r(\bar{\hat{\Gamma}}(C)) + \sum_{\substack{\max \gamma \\ \in C}} \max\{0, \delta(\gamma)+1\} & \text{if } \bar{\hat{\Gamma}}(C) \| S \\ > 0 & \text{if } \bar{\hat{\Gamma}}(C) \nparallel S. \end{cases}$$

Since by hypothesis $r(\bar{\hat{\Gamma}}) + \sum_{\gamma} \max\{0,\delta(\gamma)+1\}$ is positive, and since moreover $\delta(\hat{\Gamma})$ is negative, inequalities (6.39-40), and hence (6.5-6) are established.

* * * * *

To complete this section we show that the diagrams of QED, as specified in the preceding section, satisfy the hypothesis of the Theorem. First of all, for 1PI $\gamma \subseteq \Gamma$,

(6.51) $$r(\gamma) - \delta(\gamma) = n_f(\gamma) - n_c(\gamma)$$

where $n_f(\gamma)$ = no. of fermion lines of γ

$n_c(\gamma)$ = no. of fermion mass-counterterm vertices of γ

But the right-hand side of (6.52) is just the number of fermion lines with distinct momenta, which is strictly positive for 1PI subgraphs (here the absence of a four-photon interaction term in L_{eff} is crucial), and non-negative for 1PI reduced subgraphs (since all the lines of a reduced graph may be photon lines). Thus hypothesis (i) and (ii) are satisfied (the case $\gamma \subseteq \hat{\Gamma}$ but $\gamma \nsubseteq \Gamma$ was treated earlier).

To check hypothesis (iii), we note that for $\Lambda = \hat{\Gamma}/\{\gamma_i\}$,

(6.52) $$r(\Lambda) + \sum_i \max\{0, \delta(\gamma_i)+1\} = 4 + \sum_{V \in \mathcal{V}(\Lambda)} (r_V' - 4)$$

where

$$r_V' = \begin{cases} \nu_b(V) + 2\nu_f(V) + 2\nu_g(V) \\ \qquad \text{if } V \in \mathcal{V}(\Gamma) \text{ or } V = V(\gamma),\ \delta(\gamma)+1 \leq 0 \\ 1 + \frac{1}{2}\nu_f(V) + 2\nu_g(V) \\ \qquad \text{if } V = V(\gamma),\ \delta(\gamma)+1 > 0 \end{cases},$$

$$\left.\begin{matrix}\nu_b(V)\\ \nu_f(V)\\ \nu_q(V)\end{matrix}\right\} = \text{no. of} \left\{\begin{matrix}\text{photon}\\ \text{fermion}\\ \text{q-}\end{matrix}\right\} \text{lines meeting at } V$$

$V(\gamma)$ = reduced vertex with γ contracted to a point.

Hence

$$r(\Lambda) + \sum_i \max\{0, \delta(\gamma_i)+1\}$$

(6.53)

$$= \sum'_{\substack{V\in\mathcal{V}(\Gamma)\\ \text{or } \left\{\begin{smallmatrix}V=V(\gamma_i)\\ \delta(\gamma_i)+1\leq 0\end{smallmatrix}\right.}} \left[\nu_b(V) + 2\nu_f(V) + 4(\nu_q(V)-1)\right]$$

$$+ \sum_{\substack{V=V(\gamma_i)\\ \delta(\gamma_i)+1>0}} \left[1 + \tfrac{1}{2}\nu_f(V) + 4\sum_{V'\in\mathcal{V}(\gamma_i)} \nu_q(V')\right]$$

$$+ \sum_{\substack{V=V_\infty\\ \text{or } V=V(\gamma_\infty),\ V_\infty\in\gamma_\infty}} \left[\nu_b(V) + 2\nu_f(V)\right]$$

where Σ' in the first sum indicates that terms for $V = V(\gamma_\infty)$, $V_\infty \in V(\gamma_\infty)$ are omitted (these are treated separately in the third summation). It is an easy exercise to verify that each summand on the right-hand side of (6.53) is strictly positive, and that the set of summands is non-empty. Thus (iii) is also fulfilled in QED.

7. FORMULATION OF THE PURE YANG-MILLS MODEL

In this final section, I shall apply BPHZ normal product techniques to the formulation of the pure, massless Yang-Mills model, with classical Lagrangian

(7.1)
$$\mathcal{L} = -\tfrac{1}{4} G_{\mu\nu}\cdot G^{\mu\nu}$$

$$G^i_{\mu\nu} = \partial_\mu A^i_\nu - \partial_\nu A^i_\mu + g\,\epsilon_{ijk} A^j_\mu A^k_\nu$$

$$i = 1,2,3$$

My reasons for studying this particular model are twofold: first, it is an excellent testing ground for the subtraction procedures tailored to zero-mass theories which Zimmermann and I have been developing over the past year or two [21,27-29]; and second, it provides a relatively simple and easily grasped illustration of the powerful techniques developed by Becchi, Rouet and Stora to treat general non-Abelian gauge theories [30,31].

To quantize the Yang-Mills model in a covariant (indefinite-metric) gauge, one adds, just as in QED, a term $-1/2\alpha\,(\partial_\mu A^\mu)^2$ to the Lagrangian (7.1). In QED, the breaking of gauge invariance and introduction of indefinite metric associated with this new Lagrangian are not harmful, due to the fact that $\partial_\mu A^\mu$ is a free field whose Green functions satisfy the identity (5.12). The latter, and related identities, allow one to show the decoupling of unphysical polarization modes of the photon, as well as the α independence of scattering amplitudes (where definable) and Green functions of appropriately chosen gauge-invariant composite fields. In the Yang-Mills model, however, $\partial_\mu A^\mu$ is no longer free and in order to derive an identity analogous to (5.12) it is necessary to introduce additional fields, the scalar-fermion ghosts of Faddeev and Popov. The formal Lagrangian then becomes

$$\hat{\mathcal{L}} = -\frac{1}{4}G_{\mu\nu}\cdot G^{\mu\nu} - \frac{1}{\alpha}\left(\frac{1}{2}(\partial_\mu A^\mu)^2 + D_\mu\bar{C}\cdot\partial^\mu C\right) \tag{7.2}$$

$$\hat{D}_\mu\bar{C} = \partial_\mu\bar{C} + g\,A_\mu\times\bar{C}\ .$$

As observed by Becchi et. al. [30], the formal derivation of the desired identity (Slavnov identity) for $\partial_\mu A^\mu$ is very simple, once one has observed that the action integral $\int\hat{L}(x)dx$ is invariant under the infinitesimal (Slavnov) transformation

$$\hat{s} = \int dy\left[\hat{D}^\nu\bar{C}(y)\cdot\frac{\delta}{\delta A^\nu(y)} - \partial^\mu A_\mu(y)\cdot\frac{\delta}{\delta C(y)} - \frac{g}{2}\,\bar{C}\times\bar{C}(y)\cdot\frac{\delta}{\delta\bar{C}(y)}\right] \tag{7.3}$$

with anti-commuting character attributed to C, $\bar{C}$, $\delta/\delta C$, $\delta/\delta\bar{C}$. Schwinger's action principle (based on the application of naive field equations and equal-time commutators or anti-commutators) then yields formally

$$\begin{aligned} &i\langle T\,\hat{s}\,\big(C(x)\prod_i A_{\mu_i}(z_i)\big)\rangle \\ &= \int dw\,\langle T\,(\hat{s}\,\hat{\mathcal{L}}(w))\,C(x)\prod_i A_{\mu_i}(z_i)\rangle = 0\ , \end{aligned} \tag{7.4}$$

that is [32,33]

$$\sum_j \langle T\, \hat{D}^{\mu_j}\, \bar{C}(z_j)\, C(x) \prod_{i\neq j} A_{\mu_i}(z_i) \rangle \tag{7.5}$$

$$= \langle T\, \partial_\mu A^\mu(x) \prod_i A_{\mu_i}(z_i) \rangle \ .$$

The graphical representation of (7.5), shown in Fig. 7.1, should be compared with Fig. 5.1 (the QED analogue).

Within the context of dimensional renormalization, the action principle (7.4) can be shown [5,6] to hold for fully renormalized Green functions. In the BPHZ framework, life is not so simple: there are correction terms on the right-hand side of (7.4). That under quite general circumstances such correction terms can be eliminated by appropriate choices of counterterms in the Lagrangian (as well as in the definition of the Slavnov transformation) is the main result of Becchi et. al., [30,31]. I shall now present a highly simplified version [33] of that analysis, made possible by the relative simplicity of the model under consideration. For the more general arguments, the reader is referred to Refs. [30] and [31].

We define the normal-product Green functions of the pure Yang-Mills model by the Gell-Mann-Low formula (1.4) with effective Lagrangian

$$\mathcal{L}_{eff} = \mathcal{L}_{eff\,1} + \mathcal{L}_{eff\,2}$$

$$\begin{aligned} \mathcal{L}_{eff\,1} = \; & -\tfrac{1}{4}(\partial_\mu A_\nu - \partial_\nu A_\mu)^2 - (1+f_1)g\, \partial_\mu A_\nu \cdot A^\mu \times A^\nu \\ & -\tfrac{g^2}{4}(1+f_2)(A_\mu \cdot A^\mu)^2 - \tfrac{g^2}{4}(1+f_3)(A_\mu \cdot A^\nu)^2 \\ & -\tfrac{1}{2\alpha}(1+f_4)(\partial_\mu A^\mu)^2 - \tfrac{1}{\alpha}\partial_\mu \bar{C} \cdot \partial^\mu C \\ & -\tfrac{g}{\alpha}(1+a)\, A_\mu \times \bar{C} \cdot \partial^\mu C \end{aligned} \tag{7.6}$$

$$\mathcal{L}_{eff\,2} = (1-s)^2 M^2 \left(\tfrac{1}{2} A_\mu A^\mu + \bar{C} C\right) \ . \tag{7.7}$$

The free propagators are taken to be

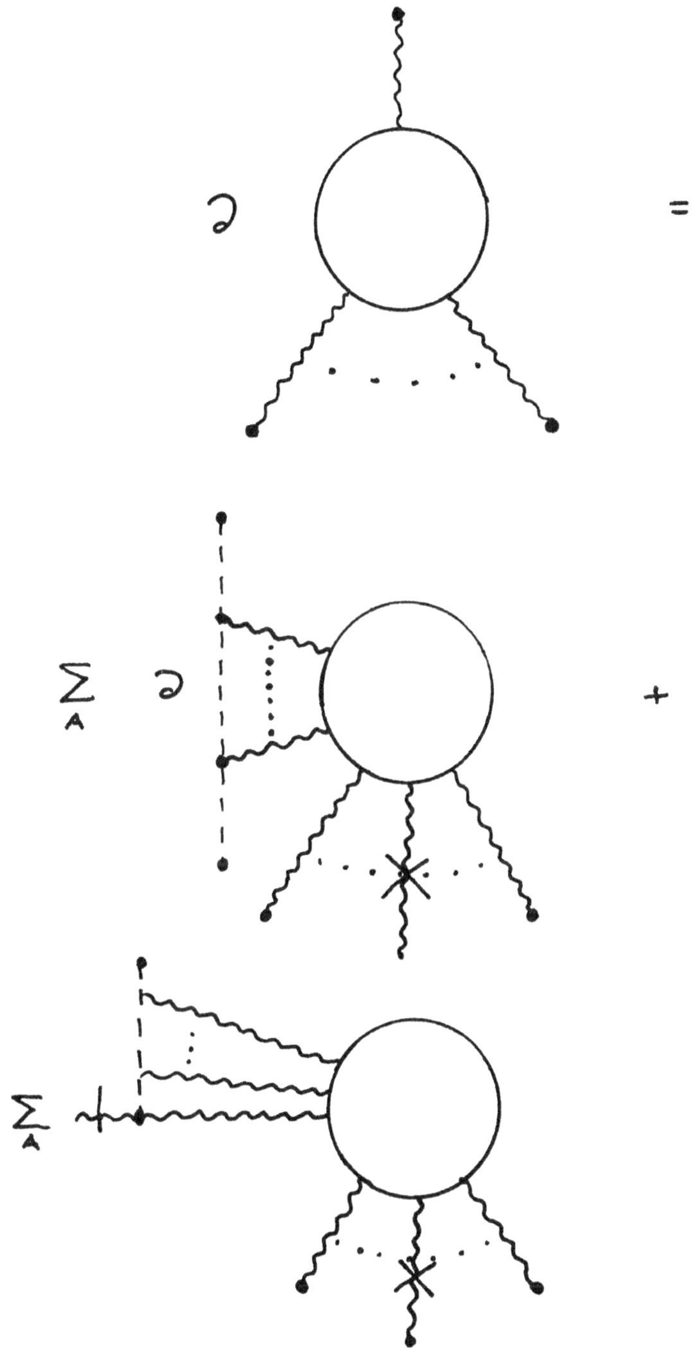

Fig. 7.1

Graphical structure of Eq. (7.5). A dashed line has been used to denote the $C\bar{C}$ propagator.

$$(7.8)\quad \langle T A_\mu^{k(0)}(0)\,\tilde{A}_\nu^{\ell(0)}(-k)\rangle_0$$

$$= \delta_{k\ell}\left[\frac{-i}{k^2-(1-s)^2M^2+i\epsilon(\vec{k}^2+(1-s)^2M^2)}\right]$$

$$\times\left[g_{\mu\nu}-\frac{(1-\alpha)k_\mu k_\nu}{k^2-(1-s)^2\alpha M^2+i\epsilon(\vec{k}^2+(1-s)^2\alpha M^2)}\right]$$

$$\langle T C^{k(0)}(0)\,\tilde{\tilde{C}}^{\ell(0)}(-k)\rangle_0$$

$$= \delta_{k\ell}\left[\frac{-i}{k^2-(1-s)^2\alpha M^2+i\epsilon(\vec{k}^2+(1-s)^2\alpha M^2)}\right]$$

and the right-hand side of (1.4) is made precise, to every order in g, by the Wick expansion for free-field time-ordered functions and the forest formula

$$(7.9)\quad R_{\Gamma\epsilon} = S_\Gamma \sum_{U\in\mathcal{F}_\Gamma}\prod_{\gamma\in U}(-\tau_\gamma S_\gamma)\,I_{\Gamma\epsilon}(U)$$

The integrand $I_{\Gamma\varepsilon}(U)$ corresponding to the forest U is parametrized as in Section 3, except that now an s variable associated with line L or vertex V is replaced by s^γ, where γ is the smallest element of U containing L or V. Moreover, τ_γ is no longer the simple Taylor operator with respect to external momenta $t^{\delta(\gamma)}_{p^\gamma}$, but rather

$$(7.10)\quad \tau_\gamma = t^{\delta(\gamma)}_{p^\gamma s^\gamma} + t^{\delta(\gamma)-1}_{p^\gamma(s^\gamma-1)} - t^{\delta(\gamma)-1}_{p^\gamma(s^\gamma-1)}\,t^{\delta(\gamma)}_{p^\gamma s^\gamma}$$

with

$$(7.11)\quad \delta(\gamma) = 4 - N(\gamma) + \sum_{V_i\in\mathcal{V}(\gamma)}(\delta_i-4)$$

$$N(\gamma) = \text{no. of external lines of } \gamma\ .$$

The subtraction scheme just described is a special case of a more general one appropriate to a wide class of theories with vanishing masses [27]. The effect of the subtractions (7.10) is to establish a single set of normalization conditions for the one-parametric family of models with s between 0 and 1. For vertex functions of the elementary fields, these are tabulated below, with vector propagators denoted by wavy lines, Faddeev-Popov propagators by dashed lines and $D^{(a)}_{xy}$ standing for a derivative of order a with respect to x and y evaluated at x=y=0.

VERTEX FUNCTION	NORMALIZATION CONDITIONS	
	AT s = 1	AT s = 0
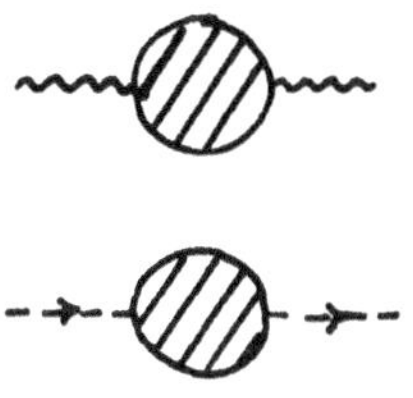	$D^{(0)}_{p(s-1)}$, $D^{(1)}_{p(s-1)}$	$D^{(2)}_{ps}$
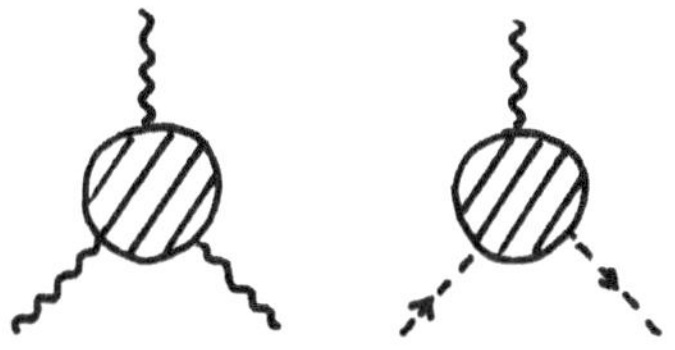	$D^{(0)}_{p(s-1)}$	$D^{(1)}_{ps}$
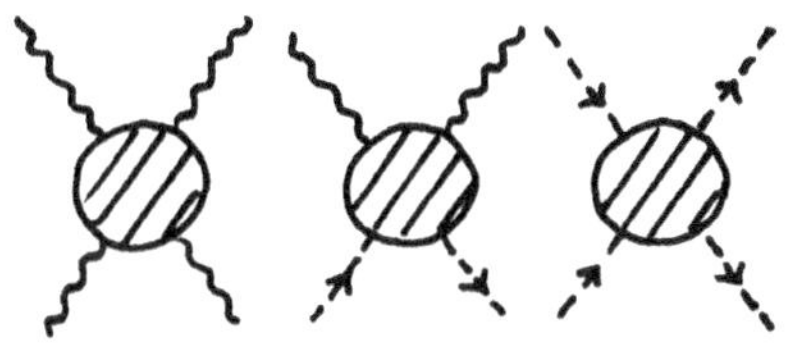	—	$D^{(0)}_{ps}$

Note that the subtractions at s = 1 (zero mass) are just as crucial to infrared convergence as those at s = 0 (non-zero mass). The latter obviously serve to avoid logarithmic IR divergences in the highest order Taylor terms needed for UV convergence (this was discussed in Section 2). The s = 1 subtractions, on the other hand, provide just enough momentum factors in the numerator to make self-energy and 3-vertex subgraphs harmless from the standpoint of IR power counting. Because of the possible overlaps of 3-vertex parts, it is helpful to use Zimmermann's complete forest decomposition, as discussed in Section 6, to verify the IR power counting criterion. This has been done for quite general normal-product Green functions in Ref. [27].

The normal-product Green functions (1.4), with $Q_i(x_i)$ generalized to include a factor $(s-1)^a$ as well as a formal field monomial, have essentially the same properties as those of purely massive theories, namely, derivative rules, normalization conditions (discussed briefly above), field equations and action principles and Zimmermann identities. The last of these include identities of the type

$$\langle T N_\delta[(s-1)^a P(x)] X\rangle = (s-1)^a \langle T N_{\delta-a}[P(x)] X\rangle + \sum_{M}^{\dim M = \delta} C(M) \langle T N_\delta[M(x)] X\rangle \tag{7.12}$$

X = product of elementary fields

(Note: each factor of (s-1) in M(x) contributes 1 to dim M).

Applied recursively, (7.12) may be used to express Green functions of normal products involving (s-1) factors entirely in terms of those without.

Let us now turn to derivation of the Slavnov identity, using the general strategy of Becchi et. al. For the time being the parameters f_1, f_2, f_3, f_4, a in L_{eff} are left free. In addition, the Slavnov transformation itself is defined with undetermined normalization factors for bilinear products:

$$s = \int dy \Big[D^\nu \bar{C}(y) \cdot \frac{\delta}{\delta A^\nu(y)} - \partial_\mu A^\mu(y) \cdot \frac{\delta}{\delta C(y)} - \frac{g}{2}(1+b)\,\bar{C}\times\bar{C}(y) \frac{\delta}{\delta \bar{C}(y)}\Big] \tag{7.13}$$

$$D^\nu \bar{C} = \partial^\nu \bar{C} + g(1+a)\, A^\nu \times \bar{C} .$$

The action principle for this transformation then takes the form,

$$i\langle T\, s X\rangle = \int dy \langle T N_5[s\mathcal{L}_{eff}(y) + B(y)] X\rangle \tag{7.14}$$

$$X = \prod_i A_{\mu_i}(z_i) \prod_j C(u_j) \prod_k \bar{C}(v_k)$$

$$s\, C(u) = -\partial_\mu A^\mu(u)$$

$$s\, \bar{C}(v) = -\frac{g}{2}(1+b)\, N_2[\bar{C}\times\bar{C}(v)]$$

$$s\, A_\mu(z) = N_2[D_\mu \bar{C}(z)] ,$$

where B(y) is a linear combination of monomials of dimension 5 (including 1 for each factor (s-1)). Using (7.12), the right-hand side of (7.14) may be rewritten as

$$(7.15)\quad (s-1)^2 \int dy \langle T N_3[E(y)] X \rangle + \int dy \langle T N_5[Q(y)] X \rangle$$

where E(y) and Q(y) are field products (no (s-1) factors) of dimension 3 and 5, respectively, and the first term vanishes identically for s=1 . If we can show that the parameters a, b and f_i can be so chosen that Q vanishes to all orders in g , we shall have the desired Slavnov identity ("softly" broken for s≠1). At first sight the possibility of doing this seems rather unlikely, since there are many more than 6 scalar, iso-scalar monomials with the same Faddeev-Popov charge as $\bar{C}$. We now come to the crucial step in the BRS analysis: a second application of the action principle to obtain the necessary constraints on Q . In our case, applying the action principle to both sides of (7.14) (with (7.15)), we have, for s=1 ,

(7.16)

$$i \sum_i \langle T N_3 \Big[\Big(\tfrac{g}{2}(a-b) - \lambda \Big) \partial_{\mu_i} (\bar{C} \times \bar{C})(z_i) + \Big(\tfrac{g^2}{2}(a-b)(1+a) - \tau \Big) A_{\mu_i} \times (\bar{C} \times \bar{C})(z_i) \Big] X_{\hat{z}_i} \rangle$$

$$- i \sum_j \langle T N_3 [\partial^\mu D_\mu \bar{C}(u_j)] X_{\hat{u}_j} \rangle$$

$$= \int dy \langle T N_6 [s\, Q(y)] X \rangle + O(gQ)$$

$$\Big(X_{\hat{u}_\ell} \equiv \prod_i A_{\mu_i}(z_i) \prod_{j \neq \ell} C(u_j) \prod_k \bar{C}(v_k) \text{, etc.} \Big).$$

The symbol O(gQ) indicates that if the coefficients of Q vanish in (n-1)st order in g , the O(gQ) coefficients vanish in order n . Choosing a and b appropriately and applying $\bar{C}$ field equation (and anti-commuting property of $\delta/\delta C(u)$)

$$(7.17)\quad \langle T N_3 [\partial^\mu D_\mu \bar{C}(u)] \frac{\delta}{\delta C(u)} X \rangle = 0 \qquad (s=1)$$

Equation (7.16) becomes

$$\int dy \langle T N_6[s Q(y)] X \rangle = O(gQ) . \tag{7.18}$$

Using the linear independence of N_6 integrated normal products, modulo formal integrations by parts, we have

$$\int dx\, Q^{(n-1)}(x) = 0 \Rightarrow \int dx [s Q(x)]^{(n)} = 0 \Rightarrow \int dx\, s_0 Q^{(n)} = 0 \tag{7.19}$$

where

$$s_0 = \int dx \left(\partial^\nu \bar{C}(x) \cdot \frac{\delta}{\delta A^\nu(x)} - \partial^\mu A_\mu(x) \frac{\delta}{\delta C(x)} \right) .$$

As inductive hypothesis, let us assume $\int Q^{(n-1)}(x) dx$ vanishes as a formal integrated polynomial. Then, by (7.19), $\int dx\, s_0 Q^{(n)} = 0$, and hence $\int dx\, s_0^2 Q^{(n)} = 0$. The latter condition implies that $\int dx Q^{(n)}$ has the general form

$$\int dx\, Q^{(n)}(x) = \int dx\, R(x) \cdot \bar{C}(x) \tag{7.20}$$

$$\begin{aligned} R(x) = \partial^\mu \big[& r_1 A_\nu (A_\mu \cdot A^\nu) + r_2 A_\mu (A_\nu \cdot A^\nu) + r_3 \partial^\nu A_\mu \times A_\nu \\ & + r_4 A^\nu \times \partial_\mu A_\nu + r_5 A_\mu \times \partial^\nu A_\nu + r_6 \partial^2 A_\mu \big] \\ & + r_7 A_\mu (A^\mu \cdot \partial_\rho A^\rho) + r_8 A_\mu (\partial^\rho A^\mu \cdot A_\rho) \\ & + r_9 A_\mu (A^\rho \cdot \partial_\mu A_\rho) . \end{aligned}$$

Applying

$$\int dx\, s_0 Q^{(n)}(x) = 0$$

to this expression, one concludes that

$$r_3 - r_4 = 0 = r_5 = r_7 = r_8 = r_9 \quad .$$

The remaining four coefficients can be annihilated by appropriate choice of f_i, i 1,...,4. Thus the coefficients of $\mathcal{L}_{eff}$ and s can be so chosen that to every order in g, $\int dx N_6[Q(x)] = 0$ and hence the Slavnov identity is established. For further details, the reader is referred to Ref. [34].

ACKNOWLEDGMENT

I would like to thank the Max-Planck-Institut für Physik in Munich for making possible my visit there in July and August of 1975. A preliminary version of these lecture notes was prepared during that visit.

REFERENCES

1. W. Zimmermann, 1970 Brandeis lectures, Lectures on Elementary Particles and Quantum Field Theory, eds. S. Deser, M. Grisaru and H. Pendleton, MIT Press, Cambridge, Mass., 1971.

2. N. N. Bogoliubov and D. W. Shirkov, Introduction to the Theory of Quantized Fields, Interscience, New York, 1959.

3. K. Hepp, Commun. Math. Phys. 2, 301 (1966).

4. W. Zimmermann, Commun. Math. Phys. 15, 208 (1969).

5. G. 't Hooft and M. Veltman, Nucl. Phys. B44, 189 (1972).

6. P. Breitenlohner and D. Maison, Max-Planck-Institut preprints MPI-PAE/PTh 15/75 (1975) and 25/74 (1974).

7. J. H. Lowenstein, 1972 Maryland lectures, Univ. of Md. Technical Report 73-068.

8. J. H. Lowenstein, Phys. Rev. D4, 2281 (1971).

9. Y. M. Lam, Phys. Rev. D6, 2154 (1972).

10. M. Gomes and J. H. Lowenstein, Phys. Rev. D7, 550 (1973).

11. J. H. Lowenstein, Commun. Math. Phys. 24, 1 (1971).

12. J. H. Lowenstein and B. Schroer, Phys. Rev. D6, 1553 (1972).

13. J. H. Lowenstein and B. Schroer, Phys. Rev. D7, 1929 (1973).

14. J. Schwinger, Phys. REv. 82, 664 (1951).

15. J. S. Bell and R. Jackiw, Nuovo Cimento 60A, 47 (1969).

16. S. L. Adler, Phys. Rev. 177, 2426 (1969).

17. C. G. Callan, Jr., Phys. Rev. D2, 1541 (1970).

18. K. Symanzik, Commun. Math. Phys. 18, 227 (1970).

19. G. Wick, Phys. Rev. 80, 268 (1950).

20. W. Zimmermann, Commun. Math. Phys. 11, 1 (1968).

21. J. H. Lowenstein and W. Zimmermann, Commun. Math. Phys. 44, 73 (1975).

22. F. Dyson, Phys. Rev. 75, 1736 (1949).

23. P. Blanchard and R. Seneor, CERN preprint TH 1420 (1971).

24. J. C. Ward, Phys. Rev. 78, 1824 (1950); Y. Takahashi, Nuovo Cimento 6, 370 (1957).

25. J. H. Lowenstein and E. Speer, "Distributional Limits of Renormalized Feynman Integrals with Zero-Mass Denominators", to be published in Commun. Math. Phys.

26. M. Gomes, J. H. Lowenstein and W. Zimmermann, Commun. Math. Phys. 39, 81 (1974).

27. J. H. Lowenstein, "Convergence Theorems for Renormalized Feynman Integrals with Zero-Mass Propagators", to be published in Commun. Math. Phys.

28. J. H. Lowenstein and W. Zimmermann, Nucl. Phys. B86, 77 (1975).

29. J. H. Lowenstein and W. Zimmermann,"Infrared Convergence of Feynman Integrals for the Massless A^4-Model", to be published in Commun. Math. Phys.

30. C. Becchi, A. Rouet and R. Stora, "Renormalization of Gauge Field Models", Centre de Physique Theorique, C.N.R.S., Marseille report, 1974 and references cited therein.

31. C. Becchi, A. Rouet and R. Stora, 1975 Erice lectures, this volume.

32. A. A. Slavnov, Teor. i Mat. Fiz. 10, 153 (1972) [Theor. Math. Phys. 10, 99 (1972)].

33. J. C. Taylor, Nucl. Phys. B33, 436 (1971).

34. J. H. Lowenstein, "Auxiliary Mass Formulation of the Pure Yang-Mills Model", to be published in Nucl. Phys. B.

REMARK ON EQUIVALENT FORMULATIONS FOR BOGOLIUBOV'S METHOD OF RENORMALIZATION (*)

Wolfhart Zimmermann

Max-Planck-Institut für Physik und Astrophysik,
München, Fed. Rep. of Germany

In this note the equivalence between Bogoliubov's original work [1] and the method of renormalization developed in Lowenstein's lecture series [2] will be sketched. In Bogoliubov's approach the propagators are regularized according to Pauli-Villars so that the unrenormalized Feynman integrals become finite. The renormalized Feynman integrals are then formed by subtracting counter terms which are constructed according to certain combinatorial rules. As has been shown by Hepp the renormalized Feynman integrals thus defined approach distributions when the regularization is removed and the limit $\varepsilon \to +0$ is taken.

In the formulation of Ref. [2] no regularization is used, instead the renormalized Feynman integrals are defined for $\varepsilon > 0$ by subtracting terms from the integrand in momentum space before the integration over the loop momenta is carried out.

In order to compare both methods we write the unrenormalihed Feynman integral in the form

$$(1) \quad J_\Gamma^{unren} = \int dk \, I_\Gamma, \quad dk = dk_1 \cdots dk_m$$

$$(2) \quad I_\Gamma = \prod_{ab\sigma} \Delta_F^{ab\sigma}(l_{ab\sigma})$$

G. Velo and A. S. Wightman (eds.), Renormalization Theory, 161–170.

$$\Delta_F^{ab\sigma} = \frac{i\, P_{ab\sigma}(\ell_{ab\sigma})}{\ell_{ab\sigma}^2 - \mu_{ab\sigma}^2 + i\varepsilon(\vec{\ell}_{ab\sigma}^{\,2} + \mu_{ab\sigma}^2)} \tag{3}$$

Γ is a one-particle irreducible diagram with vertices V_c and lines $L_{ab\sigma}$ connecting the vertices V_a and V_b. $\ell_{ab\sigma}$ denotes the internal momentum assigned to the line $L_{ab\sigma}$. It is a linear combination of the external momenta and integration momenta. The integral (1) is regularized by substituting the Pauli-Villars propagators

$$\Delta_{Fab\sigma}^{reg} = i\, P_{ab\sigma}(\ell_{ab\sigma})\left\{\frac{1}{\ell_{ab\sigma}^2 - \mu_{ab\sigma}^2 + i\varepsilon(\vec{\ell}_{ab\sigma}^{\,2} + \mu_{ab\sigma}^2)} + \sum_r \frac{c_r^{ab\sigma}(M)}{\ell_{ab\sigma}^2 - M^2 m_{ab\sigma}^{(r)2} + i\varepsilon(\vec{\ell}_{ab\sigma}^{\,2} + M^2 m_{ab\sigma}^{(r)2})}\right\} \tag{4}$$

for $\Delta_{Fab\sigma}$ in (1-2). The $m_{ab\sigma}^{(r)}$ are constants, the $c_r^{ab\sigma}(M)$ are bounded functions of M which are arranged such that

$$\overline{\deg}\, \Delta_{Fab\sigma}^{reg} < -4$$

For the unrenormalized integral in regularized form we thus obtain

$$J_{unren}^{reg} = \int dk\, I_\Gamma^{reg} \tag{5}$$

$$I_\Gamma^{reg} = \prod_{ab\sigma} \Delta_{Fab\sigma}^{reg}(\ell_{ab\sigma}) \tag{6}$$

Multiplying out all terms of $\Delta_{Fab\sigma}^{reg}$ in the definition of I_Γ^{reg} we obtain

$$I_\Gamma^{reg} = I_\Gamma + \sum_\tau I_\Gamma^\tau \tag{7}$$

where

(8)
$$I_\Gamma^\tau = \prod_{ab\sigma} \Delta_{F\tau}^{ab\sigma}$$

$$\Delta_{F\tau}^{ab\sigma} = i\, P_{ab\sigma}\left\{\frac{d^{ab\sigma\tau}}{\ell_{ab\sigma}^2 - m_{ab\sigma\tau}^2 + i\varepsilon(\cdots)}\right\}$$

The masses $m_{ab\sigma\tau}$ are

$$(9)\quad m_{ab\sigma\tau}\begin{cases}\text{either} = \mu_{ab\sigma}\\ \text{or} \quad = M m_{ab\sigma}^{(r)}\end{cases}$$

where the second value involving the auxiliary mass M is attained for at least one line $\ell_{ab\sigma}$ of the diagram.

For the coefficients $d_{ab\sigma\tau}$ we have

$$(10)\quad d_{ab\sigma\tau}\begin{cases}\text{either} = 1\\ \text{or} \quad = c_\tau^{ab\sigma}(M)\end{cases}$$

Appling Bogoliubov's R-operation R as described in Ref. [2] to I^{reg} we obtain the renormalized regularized integral

$$(11)\quad J_{ren}^{reg} = \int dk\, R\, I_\Gamma^{reg} \quad \text{for } \varepsilon>0.$$

According to Hepp's theorem the limit

$$(12)\quad \lim_{\varepsilon\to+0}\lim_{M\to\infty} J_{ren}^{reg} = \lim_{\varepsilon\to+0}\lim_{M\to\infty}\int dk\, R\, I_\Gamma^{reg}$$

exists [1]. Apart from minor differences which will be discussed below (12) represents the renormalized Feynman integral of Bogoliubov's original approach.

Since the R-operation is linear we find

$$J^{reg}_{ren} = \int dk \left(RI_\Gamma + \sum_\tau RI_{\Gamma\tau} \right) \tag{13}$$

The convergence theorems of [2] imply that the integrals

$$\int dk\, RI_\Gamma \tag{14}$$

and

$$\int dk\, RI_{\Gamma\tau} \tag{15}$$

are separately convergent. Thus

$$J^{reg}_{ren} = \int dk\, RI_\Gamma + \sum_\tau \int dk\, RI_{\Gamma\tau} \tag{16}$$

The integral (14) does not depend on M and its limit for $\varepsilon \to +0$ represents the renormalized Feynman integral of [2]. Hence both methods are equivalent if

$$\lim_{\varepsilon\to+0} \lim_{M\to\infty} \int dk\, R I^{reg}_\Gamma = \lim_{\varepsilon\to+0} \int dk\, R I_\Gamma \tag{17}$$

or

$$\lim_{M\to\infty} \int dk\, RI_{\Gamma\tau} = 0 \quad \text{for each } \tau. \tag{18}$$

For this we need only prove that

$$\lim_{M\to\infty} \int dk\, |R_\tau| = 0 \tag{19}$$

where the integrand R_τ is defined by

$$\mathcal{R}\, I_{\Gamma\tau} = FR_\tau \tag{20}$$

with F denoting the product of the bounded functions $d_{ab\sigma\tau}$ of M. R_τ is obtained from I_Γ by substituting M $\mu_{ab\sigma}$ for one or several masses.

Let H be a hyperplane in the integration space described by a set T of Z independent parameters. By a slight extension of the power counting argument of Ref. [2] it can be shown that

$$\overline{\text{degr}}_{T,M} R_\tau < -Z \tag{21}$$

where the degree is also taken with respect to M. We write R_τ in the form

$$R_\tau = \frac{P}{D} \tag{22}$$

with the common denominator D. The polynomial P we expand with respect to M^2

$$P = \sum P_\nu M^{2\nu}$$

(21) then implies

$$\overline{\text{degr}}_T \frac{P_\nu}{D} < -Z - 2\nu \tag{23}$$

With this information we will be able to estimate the behavior of

$$A = \int dk \frac{P_\nu}{D} \tag{24}$$

for large M. Renaming variables, the denominator D can be written as

$$D = \prod_{a=1}^{N} \left(\ell_a^2 - M^2\rho_a^2 + i\varepsilon(\vec{\ell}_a^{\,2} + M^2\rho_a^2)\right) \prod_{b=1}^{N'} \left(\ell_b'^2 - m_b^2 + i\varepsilon(\vec{\ell}_b'^{\,2} + m_b^2)\right) \tag{25}$$

$$N \geq 1$$

The integral A is majorized by the corresponding Euclidean integral

(26) $$B = \int \frac{|P_\nu|\, dk}{\prod_{a=1}^{N}(\ell_a^2 + M^2\rho_a^2)\prod_{b=1}^{N'}(\ell_b^2 + m_b^2)}$$

$$|A| \leqq f(\varepsilon)B$$

Let s of the internal momenta ℓ_i be variable on the hyperplane H, the remaining ℓ_i be constant. Then

(27) $$\overline{\deg}_{r_T} \frac{P_\nu}{\prod_a(\ell_a^2 + M^2\rho_a^2)^{(1-\frac{\nu}{N})}\prod_b(\ell_b'^2 + m_b^2)} < -Z - 2\nu(1 - \frac{s}{N})$$

In any case

(28) $$\overline{\deg}_{r_T} \frac{P_\nu}{\prod(\ell_a^2 + M^2\rho_a^2)^{(1-\frac{\nu}{N})}\prod(\ell_b'^2 + m_b^2)} < -Z$$

since $\nu > 0$ and $0 \leqq s \leqq N$. Hence

(29) $$G = \int \frac{|P_\nu|\, dk}{\prod(\ell_a^2 + M^2\rho_a^2)^{(1-\frac{\nu}{N}-\alpha)}\prod(\ell_b'^2 + m_b^2)}$$

satisfies the requirements of the power counting theorem provided α is small enough. Each M-dependent denominator we factorize into

$$\ell_a^2 + M^2\rho_a^2 = (\ell_a^2 + M^2\rho_a^2)^{(1-\frac{\nu}{N}-\alpha)}(\ell_a^2 + M^2\rho_a^2)^{\frac{\nu}{N}+\alpha}$$

With the estimates

$$\frac{1}{\ell_a^2 + M^2\rho_a^2} \leqq \frac{1}{\ell_a^2 + \rho_a^2} \text{ for } M > 1, \quad \frac{1}{\ell_a^2 + M^2\rho_a^2} \leqq \frac{1}{M^2\rho_a^2}$$

we find

168.

$$\mathfrak{Z} \leqq \frac{1}{M^{2\nu+2\alpha N}\prod_a \rho_a^{2(\frac{\nu}{N}+\alpha)}}$$

(30) $$\int \frac{|P_\nu|\,dk}{\prod_a (\ell_a'^2+\rho_a^2)^{(1-\frac{\nu}{N}-\alpha)}\prod_b(\ell_b'^2+m_b^2)}$$

It follows

(31) $$|M^{2\nu}A| \leqq \frac{C}{M^{2\alpha N}}\int \frac{|P_\nu|\,dk}{\prod_a (\ell_a^2+\rho_a^2)^{(1-\frac{\nu}{N}-\alpha)}\prod_b(\ell_b'^2+m_b^2)}$$
$$\text{for } M \geqq 1.$$

Hence

(32) $$\lim_{M\to\infty}\int \frac{P_\nu M^{2\nu}dk}{D} = 0 \text{ since } N \geqq 1,\ \alpha > 0.$$

This completes the equivalence proof. We finally discuss some minor differences. In order to obtain Bogoliubov's original formulation in corrdinate space we modify (1) and (5) by introducing one external momentum q_a for each vertex of the diagram and take the Fourier transform with respect to all q_a. Up to a factor the Fourier transform of the unrenormalized integral becomes just the product of the propagators

(33) $$G_\Gamma^{\text{unren}} = \prod_{ab\sigma} \tilde{\Delta}^{\text{reg}}_{Fab\sigma}(x_a - x_b)$$

Here $\tilde{\Delta}_F^{\text{reg}}$ is the Fourier transform of the Pauli-Villars propagator (3). For the Fourier transform of (5) we find

(34) $$G_\Gamma^{\text{ren}} = \sum_{F\in\mathcal{F}}\prod_{\gamma\in F}(-M_\gamma)\,G_\Gamma^{\text{unren}}$$

with the sum extending over all forests $\mathcal{F}$ of Γ. The action

$$g = M_\gamma f$$

of M_γ on a function f of the coordinates $x^{i_1}, \ldots, x^{i_a}$ associated with the subdiagram γ is defined by

$$\delta(q_{i_1}+\cdots+q_{i_a})\tilde{g} = \delta(q_{i_1}+\cdots+q_{i_a})\, t^{\delta(\gamma)}_{q_{i_1}\cdots q_{i_a}}\tilde{f} \tag{35}$$

for the Fourier transforms of f and g. $\delta(\gamma)$ is the degree of the diagram γ.

There are still two differences between (34) and the expressions considered by Bogoliubov, Parasiuk and Hepp. In their work

(i) the ε-term is treated in a covariant manner even before the limits $M\to\infty$ and $\varepsilon\to+0$ are taken,

(ii) fewer subtractions are used, namely only those which coorespond to generalized vertices.

The first point has been clarified by Hepp and Speer who showed that the difference between the two expressions is of order ε in the limit $\varepsilon\to+0$ [3]. In order to check (ii) we must prove that all terms may be omitted for which any $\bar{\gamma}$ with $\gamma\varepsilon F$ contains some lines connecting a vertex to itself. $\bar{\gamma}$ is the reduced diagram

$$\bar{\gamma} = \gamma/\gamma_{j_1}\cdots\gamma_{j_c} \tag{36}$$

with $\gamma_{j_1}, \ldots, \gamma_{j_c}$ being the maximal elements of F contained in γ.

For the proof we start with some combinatorial definitions. A full vertex part v of Γ is defined by a set of vertices $\mathcal{V}$ of Γ and the set of all lines of Γ connecting any two vertices of $\mathcal{V}$. Let γ be a subdiagram of Γ. We define the full vertex part v_γ by adding to γ all lines connecting any two vertices of γ. Let A be a forest. We define a completion $C = \bar{A}$ by adding all full vertex parts v_γ with $\gamma\varepsilon A$. A forest C is called complete if $\gamma\varepsilon C$ always implies $v_\gamma\varepsilon C$. Hence a complete forest C is identical with its completion. Two forests are called equivalent if they have the same completion. Let K be an equivalence class of forests with the completion C. Then we have

$$\sum_{A\in K}\prod_{\gamma\in A}(-M_\gamma)G_\Gamma^{\text{unren}} = \prod_{\gamma\in C}F(\gamma)G_\Gamma^{\text{unren}} \tag{37}$$

(i) $F(\gamma) = - M_\gamma$ if γ is a full vertex part and if there is no other $\mu\varepsilon C$ with $v_\mu = \gamma$,

(ii) $F(\gamma) = 1 - M_\gamma$ if γ is a full vertex part and if there is at least one other $\mu\varepsilon C$ with $v_\mu = \gamma$.

(iii) $F(\gamma) = - M_\gamma$ if γ is no full vertex part.

(37) is proved by multiplying out all factors $1 - M_\gamma$. If all $\gamma\varepsilon C$ belong to class (i) the forest C contains only full vertex parts. We show that

$$\prod_{\gamma\in C} F(\gamma)\, G_\Gamma^{\text{unren}} = 0 \tag{38}$$

if C contains at least one element ρ which is not a full vertex part. In that case $\gamma = v_\rho$ is also an element of C. Among all elements μ of C with the property

$$\rho \subseteq \mu \subset \gamma$$

we choose the maximal one. With this the product (38) contains two factors

$$(1 - M_\gamma)(-M_\mu) \tag{39}$$

which yield an expression of the form

$$(1 - M_\gamma)\prod \Delta_F^{a_i b_i \sigma_i}(x_{a_i} - x_{b_i})(-M_\mu)\Phi \tag{40}$$

The product Π extends over all lines which belong to γ but not to μ. The Fourier transform of

$$\prod \Delta_F^{a_i b_i \sigma_i}(-M_\mu)\Phi$$

is a polynomial in the momenta associated with γ with a degree not greater than $\delta(\mu)$. On the other hand M_γ acts on the Fourier transform as a Taylor operator in the same variables up to the order $\delta(\gamma) \geq \delta(\mu) + 2$. Hence (38) vanishes, and there remains the formula

$$G_\Gamma^{\text{ren}} = \sum_{A\in\mathcal{V}} \prod_{\gamma\in A} (-M_\gamma)\, G_\Gamma^{\text{unren}} \tag{41}$$

where $\mathcal{V}$ is the class of all forests consisting of full vertex parts only. This completes the proof that only those subtractions survive which correspond to generalized vertices.

(*) This note is based on work by K. Hepp and the author which has not been published except for a short account in 'Lectures on Elementary Particles and Quantum Field Theory', Brandeis University, Vol. I, MIT Press, Cambridge, Mass. (1970).

[1] N.N. Bogoliubov and D.W. Shirkov, Uspechi fiz. Nauk 55, 149 (1955), Introduction to the Theory of Quantized Fields, Interscience Publ., New York (1959),

N.N. Bogoliubov and O. Parasiuk, Acta Math. 97, 227 (1957) K. Hepp, Comm. Math. Phys. 2, 301 (1966);

[2] J.H. Lowenstein, Lectures in this volume. This report contains further references;

[3] K. Hepp, unpublished and E.R. Speer, 'Seminars on Renormalization Theory', Technical Report, University of Maryland (1972).

THE POWER COUNTING THEOREM FOR FEYNMAN INTEGRALS WITH MASSLESS PROPAGATORS

W. Zimmermann

Max-Planck-Institut für Physik und Astrophysik,
München, Fed. Rep. of Germany

Abstract. Dyson's power counting theorem is extended to the case where some of the mass parameters vanish. Weinberg's ultraviolet convergence conditions are supplemented by infrared convergence conditions which combined are necessary and sufficient for the absolute convergence of Euclidean Feynman integrals.

In any renormalization scheme based on Feynman integrals in momentum space the power counting theorem plays a decisive part. The power counting technique was developed by Dyson in his work on quantum electrodynamics for checking the convergence of renormalized Feynman integrals [1]. A precise formulation of the theorem was stated and proved rigorously by Weinberg for the case of non-vanishing propagator masses and Euclidean metric [2]. In this lecture an extension of the power counting theorem to arbitrary masses will be discussed which was proposed in Ref.[3]. Only the case of Euclidean metric will be treated, for the passage to Minkowski metric and the distributional limit the reader is refered to the lecture series by Lowenstein. The results are consistent with recent work by Bergère and Lam, as well as by Trute and Pohlmeyer, on the asymptotic behavior of parametrized Feynman integrals for small mass values [4-6].

The Euclidean contribution of a one-particle-irreducible Feynman diagram to a Green's function has the form

(1) $$J = \int dk\, R(k,p)$$

G. Velo and A. S. Wightman (eds.), Renormalization Theory, 171–184.

(2) $$R = \frac{P}{\prod_{j=1}^{n} (\ell_j^2 + m_j^2)^{n_j}}$$

where

$$k = (k_1, \cdots, k_m), \quad p = (p_1, \cdots, p_N)$$
$$dk = dk_1 \cdots dk_m$$
$$m_j \geqq 0, \qquad n_j > 0.$$

k_j and p_j are Euclidean four vectors. For Euclidean four vectors w we use the notation

$$w = (w_0, w_1, w_2, w_3), \; dw = dw_0 dw_1 dw_2 dw_3$$
$$w^2 = \sum_{j=0}^{3} w_j^2$$

The vectors ℓ_j are linear combinations

$$\ell_j = K_j(k) + P_j(p)$$

of the vectors $k_1, \ldots, k_m$ and $p_1, \ldots, p_N$ with $K_j \neq 0$. P is a polynomial in the components of k and p.

Feynman integrals folded with suitable test functions of the external momenta can be majorized by integrals which are again of the type (1-2) with p = 0. This information was used by Lowenstein and Speer in proving the existence of the distributional limit for Feynman integrals in Minkowski space [7]. Convergence criteria for integrals (1-2) are thus relevant for the existence of Feynman integrals as distributions as well as the point-wise existence at given external momenta.

For the formulation of the power counting theorem we will need some general definitions. Let L denote the space of the linear forms

(3) $$\ell = \sum_{j=1}^{n} a_j k_j + \sum_{j=1}^{N} b_j p_j$$

which will be interpreted as inhomogeneous linear forms in the integration variables $k_1, \ldots, k_m$. Elements of L are called linearly (in)dependent if their homogeneous parts (in k) are linearly (in)dependent. A set of elements in L is called a basis of L if their homogeneous parts form a basis for the space of the homogeneous forms in k.

Without loss of generality we may assume that a basis

$$\ell_{j_1}, \ldots, \ell_{j_m}$$

of L exists consisting of linear forms which occur in the denominators of (2). If this should not be the case numerator and denominator of the integrand R could be multiplied by suitable factors $k_i^2 + m^2$.

We will further need certain subintegrals which we set up as follows. Let

$$(4) \quad u_1 = \ell_{i_1}, \ldots, u_a = \ell_{i_a}, \; v_1 = \ell_{j_1}, \ldots, v_b = \ell_{j_b}$$

be a basis of L with Jacobian one (relative to $k_1, \ldots, k_m$). Using (4) as new integration variables for (1) we obtain

$$J = \int du\, dv\, R$$

$$u = (u_1, \ldots, u_a), \quad v = (v_1, \ldots, v_b)$$
$$du = du_1 \cdots du_a, \quad dv = dv_1 \cdots dv_b$$

where P and ℓ_j are expressed in terms of u, v and p through

$$k = k(u, v, p).$$

We consider a hyperplane H_u defined by the condition that the linear forms

$$\ell_{i_1} = u_1, \ldots, \ell_{i_a} = u_a$$

have constant values u. Likewise we define a hyperplane H_v by the condition that the linear forms

$$\ell_{j_1} = v_1, \ldots, \ell_{j_b} = v_b$$

have constant values v. The subintegrals along H_u and H_v are

$$J(H_u) = \int dv\, R, \quad J(H_v) = \int du\, R$$

We distinguish two different definitions for the dimension of a subintegral. The upper dimension $\overline{\dim}$ refers to the behavior for large values of the integration variables. The lower dimension $\underline{\dim}$ refers to the behavior for small values of the integration variables. We define

$$\overline{\dim}\, J(H_u) = \overline{\deg}_{v/u} R + 4b$$

$$\underline{\dim}\, J(H_v) = \underline{\deg}_{u/v} R + 4a$$

The upper degree $\overline{\deg}_{v/u}$ (or lower degree $\underline{\deg}_{u/v}$) denotes the leading power of ρ in the limit $\rho\to\infty$ (or $\rho\to 0$) if $v_j = \rho\hat{v}_j$ (or $u_j = \rho\hat{u}_j$) is substituted into R with constant values of $p_1, \dots, p_N$ and $u_1, \dots, u_a$ (or $v_1, \dots, v_b$).

Weinberg's hypothesis of the power counting theorem may now be stated as follows:

Ultraviolet Convergence Condition.

The inequality

$$\overline{\dim}\, J(H_u) > 0 \tag{5}$$

holds for any basis (4) and for any hyperplane H_u defined by constant values of $u_1, \dots, u_a$.

In particular, the upper dimension of the full integral J should be negative. Weinberg's condition (5) is sufficient for the absolute convergence of J provided all masses are different from zero

$$m_j > 0, \qquad j = 1, \dots, n. \tag{6}$$

For the general case the following condition is proposed in addition:

Infrared Convergence Condition

The inequality

(7) $$\underline{\dim}\, J(H_v) > 0$$

holds for any basis (4) satisfying

(8) $$m_{i_1} = \cdots = m_{i_a} = 0$$

and corresponding hyperplane H_v defined by constant values of $v_1, \ldots, v_b$.

It will be seen that the ultraviolet and infrared convergence conditions combined are necessary and sufficient for the absolute convergence of the integral (1-2). More elegantly, the generalized power counting theorem may be formulated as follows:

Power Counting Theorem

Necessary and sufficient for the absolute convergence of the integral (1-2) is the condition that for any subintegral

(9) $$J(H) = \int_H dk\, R$$

along a hyperplane H the upper dimension be negative and the lower dimension be positive,

(10) $$\overline{\dim}\, J(H) < 0 \quad \text{and}$$

(11) $$\underline{\dim}\, J(H) > 0.$$

The ultraviolet and infrared convergence conditions (5, 7-8) are obviously special cases of (10) and (11). Moreover, it is easy to see that such dimensional rules are necessary for the absolute convergence. As example we assume

$$\underline{d} = \underline{\dim}\, J(H) < 0$$

in violation of (7) or (11) for some hyperplane H_u parametrized by momenta $u_1, \ldots, u_a$. A scaling parameter is introduced by seting

$$u_j = \rho u'_j .$$

Then, for an appropriate region of the u'_j and sufficiently small ρ, we have

$$|R| \geq c|\rho|^{\delta} \quad \text{where} \quad \delta = \underline{\deg} R$$

Using ρ as one of the integration parameters the subintegral

$$\int_{|\rho|\leq 1} d\rho\,|R| \approx \int_{|\rho|\leq 1} |\rho|^{d-1} d\rho$$

is divergent.

More involved is the proof that the stated conditions are also sufficient for the absolute convergence. As hypothesis of the proof we will use the ultraviolet and infrared convergence conditions in their special forms (5) and (7-8).

We begin stating a lemma on the infrared convergence of certain integrals which are homogeneous in the integration variables.

Lemma

Consider integrals of the form

$$F = \int_{u_i^2 \leq 1} du_1 \ldots du_a \frac{M}{\prod_j (U_j^2)^{n_j}} \tag{12}$$

where the U_j are linear combinations of the Euclidean four-vectors $u_1, \ldots, u_a$ and M is a monomial in the components of $u_1, \ldots, u_a$. M may be factorized as

$$M = \prod_{i=1}^{a} M_i \tag{13}$$

where M_i is a monomial of u_i. For any subset

$$u_{i_1}, \ldots, u_{i_c} \tag{14}$$

of the integration variables we form the integral

$$F_{i_1 \ldots i_c} = \int du_{i_1} \ldots du_{i_c} \frac{M_{i_1} \ldots M_{i_c}}{\prod_{j\,i_1 \ldots i_c} (U_j^2)^{n_j}} \tag{15}$$

where the product $\Pi_{j i_1 \ldots i_c}$ extends over all U_j which are linear combinations of vectors (14) only. The integrals (15) are called sections of (12).

The statement is that the integral (12) is absolutely convergent if the dimension $d_{i_1 \ldots i_c}$ of each section (15) is positive:

$$\text{(16)} \qquad \dim F_{i_1 \ldots i_c} = d_{i_1 \ldots i_c} > 0 .$$

This condition includes the dimension of the full integral which we denote by d,

$$d = \dim F = d_{1 \ldots a} > 0 .$$

The proof of this Lemma is straigthforward by recursively estimating integrals of the form

$$\int_{u_{i_1}^2 \leq \cdots \leq u_{i_a}^2 \leq 1} du_1 \cdots du_a \frac{M}{\prod_j (U_j^2)^{n_j}}$$

For details we refer to Ref.[3].

We return to the discussion of the integral (1-2). The main task of the convergence proof will be to disentangle infrared regions from ultraviolet regions of integration. To the subintegrals involving ultraviolet regions the original form of Weinberg's theorem valid for non-vanishing masses will be applied. The convergence of subintegrals extending over infrared regions will be checked using the Lemma stated above.

We first divide the region of integration into domains where some momenta ℓ_j are inside a sphere of radius r while others lie outside. Let S_0 be the set of all momenta with $m_j = 0$. Let S be any subset

$$S \subseteq S_0$$

T denotes the complementary set

$$T = S_0 \setminus S$$

We require that with a momentum ℓ_j the set S should contain any ℓ_i which satisfies

$$\ell_i^2 \equiv \ell_j^2 , \quad m_i = 0$$

We decompose the integral (1) into

(17) $$J = \sum_S A_S$$

where

(18) $$A_S = \int\limits_{\substack{\ell_i^2 \leq r^2 \text{ in } S \\ \ell_i^2 \geq r^2 \text{ in } T}} dk \frac{P}{\prod_j (\ell_j^2 + m_j^2)^{n_j}}$$

For studying A_S we select momentum vectors

(19) $$u_1 = \ell_{i_1} , \dots , u_a = \ell_{i_a}$$

in S which form a basis of S. Then $\ell_j \varepsilon S$ is a linear combination of $u_1, \dots, p_1, \dots ,$

$$\ell_i = U_i + Q_i$$

$$U_i = \sum_{j=1}^{a} c_{ij} u_j , \quad Q_i = \sum_{j=1}^{N} d_{ij} p_j$$

We say that S or the integral A_S has zero external momenta if $Q_i = 0$ for all $\ell_i \varepsilon S$.

For r small enough the term A_S vanishes unless all external momenta vanish. For the proof we observe that $u_\alpha^2 \leqq r^2$ since the u_α occur among the $\ell_i \varepsilon S$. The U_j are of the form

$$U_j = \sum_{\alpha=1}^{a} \eta_\alpha u_\alpha$$

where $| \eta_\alpha | \leqq \eta$ with η being a characteristic number of the integral. Now

$$|Q_j| \leqq |\ell_j| + |U_j| \leqq (1+\eta) r$$

implies

$$r \geqq \frac{|Q_j|}{1+\eta} \quad \text{for any } Q_j ,$$

if the domain of integration is not empty. If at least one $Q_j \neq 0$ we may choose r such that

$$0 < r < \frac{|Q_j|}{1+\eta} \tag{20}$$

But then the domain of integrations is empty and $A_S = 0$. Hence for r small enough we find

$$J = \sum_S A_S \tag{21}$$

where S is restricted to those subsets for which $Q_j = 0$ for any $\ell_j \varepsilon S$.

In each integral A_S we introduce new variables of integration as follows. By adding suitable vectors

$$v_1 = \ell_{j_1}, \ldots, v_b = \ell_{j_b}, \quad a+b=m, \tag{22}$$

we extend (19) to a basis

$$u_1, \ldots, u_a, v_1, \ldots, v_b \tag{23}$$

of L with Jacobian one (relative to $k_1, \ldots, k_m$). Then each $\ell_j \varepsilon S$ is a linear combination of $u_1, \ldots, u_a$. The remaining ℓ_j are linear combinations of $u_1, \ldots, u_a, v_1, \ldots, v_b, p_1, \ldots, p_N$. We next write the numerator P as a polynomial in u

$$P = \sum_\alpha C_{S\alpha} M_\alpha$$

$$M_\alpha = \prod_{i=1}^{a} M_{i\alpha_i}, \quad \alpha = (\alpha_1, \ldots, \alpha_a), \tag{24}$$

$$M_{i\alpha_i} = u_{i0}^{\alpha_{i0}} u_{i1}^{\alpha_{i1}} u_{i2}^{\alpha_{i2}} u_{i3}^{\alpha_{i3}}, \quad \alpha_i = (\alpha_{i0}, \ldots, \alpha_{i3})$$

with the coefficients being polynomials in $v_1, \ldots, v_b, p_1, \ldots, p_N$. Then

$$A_S = \sum_\alpha A_{S\alpha}$$

$$(25)\quad A_{S\alpha} = \int\limits_{\ell_j^2 \leq r^2 \text{ in } S} du \frac{M_\alpha}{\prod_S (\ell_j^2)^{n_j}} \int\limits_{\ell_j^2 \geq r^2 \text{ in } T} dv \frac{C_{S\alpha}}{\prod_T (\ell_j^2)^{n_j} \prod_U (\ell_j^2 + m_j^2)^{n_j}}$$

$$U = (\ell_1, \cdots, \ell_n) \setminus S_0$$

The integral

$$(26)\quad \int\limits_{\ell_j^2 \geq r^2 \text{ in } T} dv \frac{C_{S\alpha}}{\prod_T (\ell_j^2)^{n_j} \prod_U (\ell_j^2 + m_j^2)^{n_j}}$$

has no infrared region, for

$$\ell_j^2 \geqq r^2 \quad \text{if} \quad \ell_j \in T$$

$$\text{and} \quad \ell_j^2 + m_j^2 \geqq m_j^2 > 0 \quad \text{if} \quad \ell_j \in U$$

As a consequence of the ultraviolet convergence conditions (5) the integral (26) is expected to be absolutely convergent and uniformly bounded in the region $u_j^2 \leq r^2$. The appropriate estimates can be found in Refs.[3, 8]. Thus

$$(27)\quad |J| \leqq \sum_{S\alpha} D_{S\alpha} \int\limits_{\ell_j^2 \leqq r^2} du \frac{|M_\alpha|}{\prod_S (\ell_j^2)^{n_j}}$$

$$\leqq \sum_{S\alpha} C_{S\alpha} \int\limits_{u_j^2 \leqq 1} du \frac{|M_\alpha|}{\prod_S (\ell_j^2)^{n_j}}$$

In the last line the domain of integration was enlarged by restricting the momenta u_j only. The integration variables were rescaled by the factor r using that the integrand is homogeneous in u. For the remaining integrals in the estimate we use the Lemma in order to verify the infrared convergence. According to this we have convergence for the integrals in (27) if the dimension of each section is positive. In order to check the dimension of the full integral

(28) $$I = \int_{u_i^2 \leq 1} du \frac{M_\alpha}{\prod_S (\ell_j^2)^{n_j}}$$

we form the subintegral

(29) $$J(H) = \int du_1 \cdots du_a \frac{P}{\prod (\ell_j^2 + m_j^2)^{n_j}}$$

of (1) along a hyperplane H defined by constant values of $v_1, \ldots, v_b$. By the hypothesis (5-6) the lower dimension δ of J(H) is positive,

(30) $$\delta = 4a + \underline{\deg}_u P - \underline{\deg}_u \Pi(\ell_j^2 + m_j^2)^{n_j} > 0.$$

This implies

$$\begin{aligned} 0 < \delta &= 4a + \underline{\deg}_u P - \deg \Pi_S (\ell_j^2)^{n_j} - \deg_u \Pi_T (\ell_j^2)^{n_j} \\ &\leqq 4a + \underline{\deg}_u P - \deg \Pi_S (\ell_j^2)^{n_j} \\ &\leqq 4a + \deg M_\alpha - \deg \Pi_S (\ell_j^2)^{n_j} = d. \end{aligned}$$

Hence the dimension of (29) is positive.

We further have to verify that the dimension $d_{i_1 \ldots i_c}$ of each section

(31) $$I_{i_1 \cdots i_c} = \int du_{i_1} \cdots du_{i_c} \frac{M_{\alpha\, i_1 \cdots i_c}}{\prod_{i_1 \cdots i_c} (\ell_j^2)^{n_j}}$$

is positive. Here $M_{\alpha\, i_1 \ldots i_c}$ is the restriction of the product (24) to factors depending on u_{i_α}. $\Pi_{i_1 \ldots i_c}$ denotes the product over all factors for which $\ell_j \varepsilon S$ is a linear combination of the vectors $u_{i_1} \ldots u_{i_c}$ only. Useful information is obtained by comparing the expansion

(32) $$P = \sum C_\alpha M_\alpha$$

of P with respect to monomials M_α in $u_1, \ldots, u_a$ with the expansion

(33) $$P = \sum C'_{\alpha'} M'_{\alpha'}$$

with respect to independent monomials $M'_{\alpha'}$ in $u_{i_1}, \ldots$ only. We know that $M_{\alpha i_1 \ldots i_c}$ occurs as a factor of at least one monomial M_α with $C_\alpha \neq 0$. Since the monomials M_α are linearly independent the factor $C'_{\alpha'}$ of $M'_{\alpha'} = M_{\alpha i_1 \ldots i_c}$ in (33) must also be different from zero. This implies the inequality

$$\underline{\deg}_{u'} P \leq \deg M_{\alpha i_1 \ldots i_c} \tag{34}$$

which will be crucial for the proof of the theorem. $\underline{\deg}_{u'}$ denotes the lower degree with respect to the variables $u' = (u_{i_1}, \ldots, u_{i_c})$. We now form the subintegral

$$J(H') = \int du_{i_1} \cdots du_{i_c} \frac{P}{\prod_j (\ell_j^2 + m_j^2)^{n_j}} \tag{35}$$

along a hyperplane H' defined by constant values of $v_1, \ldots, v_b$ and the momenta u_j which do not belong to u'. The lower dimension δ' of (35) is positive by hypothesis (5-6),

$$\begin{aligned} 0 < \delta' &= 4c + \underline{\deg}_{u'} P - \underline{\deg}_{u'} \prod_j (\ell_j^2 + m_j^2)^n \\ &\leq 4c + \underline{\deg}_{u'} P - \deg \Pi_{i_1 \ldots i_c} (\ell_j^2)^{n_j} \end{aligned}$$

With (34)

$$\begin{aligned} d_{i_1 \ldots i_c} &= 4c + \deg M_{\alpha i_1 \ldots i_c} - \deg \prod_{i_1 \ldots i_c} (\ell_j^2)^{n_j} \\ &\geq 4c + \underline{\deg}_{u'} P - \deg \prod_{i_1 \ldots i_c} (\ell_j^2)^{n_j} \end{aligned}$$

follows. Hence the dimension $d_{i_1 \ldots i_c}$ of (35) is positive. According to the Lemma each integral on the right hand side of (27) converges. This completes the proof that the ultraviolet and infrared convergence conditions (5, 7-8) are sufficient for the absolute convergence of the integral (1-2). Since the dimensional rules (10-11) are necessary for the absolute convergence and moreover imply (5, 7-8) the power counting theorem follows.

The theorem was stated for integrals of the general type (1-2) without reference to diagrams. This considerably simplifies convergence proofs to all orders, particularly if the absence of overlapping ultraviolet and infrared divergencies has to be shown. In special situations, however, it can be helpful to rephrase the convergence conditions with the terminology of Feynman diagrams.

As is well-known the ultraviolet convergence condition has a simple diagrammatic interpretation for Feynman integrals of a diagram Γ without counter terms or spin polynomials. In this case it is equivalent to the requirement that the upper dimension be negative for each one-particle-irreducible subdiagram of Γ.

Similarly, a diagrammatic criterion was found by Mack for the infrared convergence of an unrenormalized Feynman integral without spin [9]. Mack's rule will now be derived from the infrared convergence condition (7-8). Up to numerical factors the unrenormalized Feynman integral of a diagram Γ involving scalar particles only is of the form

$$J = \int \frac{dk}{\prod(\ell_j^2 + m_j^2)} \tag{36}$$

Let H_v denote a hyperplane defined by a basis (4) which satisfies (5). The infrared convergence condition states that the lower dimension of

$$J(H_v) = \int \frac{du}{\prod(\ell_j^2 + m_j^2)} \tag{37}$$

be positive. Along with (37) we introduce the integral

$$J_S = \int \frac{du}{\prod_S \ell_j^2} \tag{38}$$

where the product Π_S extends over the set S of all ℓ_j with $m_j = 0$ which are linear combinations of $u_1, \ldots, u_a$ only. J_S is the Feynman integral of the reduced diagram Γ' obtained from Γ by contracting all lines which do not belong to S. Each line of the reduced diagram Γ' is massless. Since none of the internal momenta of Γ' depends on v_i or p_j the sum of external momenta vanishes at each vertex of Γ. The integrals (37) and (38) have the same lower dimension. For

$$\underline{\mathrm{degr}}(\ell_j^2 + m_j^2) = 0 \text{ if } \ell_j \notin S.$$

Therefore

$$\overline{\dim}\, J_S = \underline{\dim}\, J_S = \underline{\dim}\, J(H_v) > 0.$$

We thus find Mack's infrared convergence condition for unrenormalized Feynman integrals without spin: For each reduced diagram of Γ with vanishing external momenta and masses the dimension should be positive [9].

REFERENCES

1. F.L. Dyson, Phys. Rev. 75, 1736 (1949)

2. S. Weinberg, Phys. Rev. 118, 838 (1960)

3. J. Lowenstein and W. Zimmermann, Comm. Math. Phys. 44, 73 (1975)

4. M. Bergère and P. Lam, Comm. Math. Phys. 39, 1 (1974) and to be published

5. H. Trute, to be published

6. K. Pohlmeyer, to be published

7. J. Lowenstein and E. Speer, to be published

8. Y. Hahn and W. Zimmermann, Comm. Math. Phys. 10, 330 (1968)

9. G. Mack in: Scale and Conformal Symmetry in Hadron Physics, ed. R. Gatto, J. Wiley, New York (1973)

SOME RESULTS ON DIMENSIONAL RENORMALIZATION

P. Breitenlohner* and D. Maison**

Max-Planck-Institute fur Physik und Astrophysik,
München, Germany

I. DEFINITION OF DIMENSIONALLY REGULARIZED AMPLITUDES

In this lecture [1] we will give a definition of dimensionally regularized Feynman amplitudes and rules for the treatment of covariants (e.g. spin polynomials) which are quite different in spirit[2] from the ones given by Speer, but give equivalent results. Given a connected Feynman graph G with vertices $\mathcal{V}=\{V_i, i=1,\cdots,M\}$ lines $\mathcal{L}=\{\ell_j, j=1,\ldots,L\}$, $h=L-M+1$ loops and incidence matrix $e=(e_{ji})$, its amplitude is formally given by

$$\mathcal{T}(x)=\lim_{\varepsilon\to 0} \hbar^{h-1} \prod_{\mathcal{V}} X_i \prod_{\mathcal{L}} \Delta_j^{\varepsilon}(e_{ji}x_i)$$

$$\Delta_j^{\varepsilon} = Z_j\left(-i\frac{\partial}{\partial u_j}\right) \int_0^{\infty} d\alpha_j (-2i\alpha_j)^{-n/2} \exp\left\{-\frac{i}{4\alpha_j}(x+u_j)^2 - i\alpha_j(m_j^2-i\varepsilon)\right\}\Big|_{u_j=0}$$

for any (positive integer) dimension n. Taking the Fourier transform and interchanging integrations we obtain

$$\mathcal{T}(\underline{p}) = h^{h-1}(2\pi)^{n/2}\delta(\Sigma p_i)i^h(2i)^{-hn/2} \lim_{\varepsilon\to 0} \mathcal{T}_{\varepsilon}(p_1,\cdots,p_{M-1})$$

$$\mathcal{T}_{\varepsilon}(\underline{p}) = \int_0^{\infty} d\underline{\alpha}\ I_{\varepsilon}(\underline{p},\underline{u},\underline{\alpha})\Big|_{\underline{u}=0} \ ; \ M = \begin{pmatrix} 0 & -2e^{+} \\ 2e & -4\alpha \end{pmatrix} \qquad (1)$$

$$I_{\varepsilon}=\Pi X_i \Pi Z_j\left(-i\frac{\partial}{\partial u_j}\right)\left(\frac{\det M}{(-4)^L}\right)^{-n/2} \exp i\{(p^{+}u^{+})M^{-1}\binom{p}{u} - \Sigma\alpha_j(m_J^2-i\varepsilon)\}$$

*Present address: ITP SUNY at Stony Brook, Stony Brook, NY 11794

**Present address: New York University New York, NY 10003

G. Velo and A. S. Wightman (eds.), Renormalization Theory, 185–192.

where p and e are now restricted to M-1 vertices and $\alpha = \mathrm{diag}(\alpha_j)$. For simplicity we assume the vertex factors X_i to be independent of external momenta p_i which is no loss of generality due to the identity (2) given below. Furthermore their dependence on internal momenta $(-i\partial/\partial u_j)$ can be absorbed into the spin polynomials Z_j.

We take Eq. (1) as a starting point to define the regularized amplitude. First we replace n by a complex variable ν in $\mathcal{I}_\varepsilon$ and second all covariants contained in X_i or Z_j as well as $\underline{u}$ and $\partial/\partial\underline{u}$ are treated as abstract objects, satisfying some algebraic relations. Consequently (at least the $\underline{u}$-dependent part of) the exponential in I_ε has to be treated as formal power series until all derivatives with repect to $\underline{u}$ are performed and $\underline{u}$ is set to zero. Due to our choice of the ε-dependence the α-integrals for the coefficients of the covariants converge absolutely for $\varepsilon > 0$ and $\mathrm{Re}(\nu)$ sufficiently negative; they have a meromorphic continuation to the entire ν-plane.

At this point we can prove two usefull identities

$$(p - ie^{+}\,\partial/\partial u^{+})\, I_\varepsilon = 0 \tag{2}$$

$$i\int_0^\infty d\alpha_j\,(\Box_{u_j} + m_j^2 - i\varepsilon)\, I_\varepsilon = -\int_0^\infty d\alpha_j\;\partial I_\varepsilon/\partial\alpha_j = I_\varepsilon\Big|_{\alpha_j=0} \tag{3}$$

which express momentum conservation at each vertex and the fact that we may use equations of motion inside regularized Feynman amplitudes. Both of these identities follow from the simple structure of I_ε: apart from mass terms all the $\underline{\alpha}$-dependence comes through M which itself is linear in $\underline{\alpha}$. For the second one we need, however, $g_{\mu\mu} = \nu$, and this forces us to treat the covariants as abstract objects instead of just using 4-dimensional covariants.

The two identities (2,3), together with combinatorial considerations, allow us to derive the action principle which is most easily expressed in terms of the generating functional $Z(\underline{a},\underline{\lambda})$. The total Lagrangian $\mathcal{L}$ is split into a free part $\mathcal{L}_0$ and an interaction part $\mathcal{L}_{int}$ as usual and depends on quantized fields $\underline{\phi}$, external field (sources) $\underline{a}$ and space-time independent parameters $\underline{\lambda}$ (e.g. masses and coupling or renormalization constants). Z is given by

$$Z(\underline{a},\underline{\lambda}) + \langle T\{\exp \frac{i}{\hbar}\int \mathcal{L}_{int}(\underline{\phi}(x),\underline{a}(x),\underline{\lambda})dx\}\rangle^{o}_{conn.} \tag{4}$$

and the action principle says

$$\frac{\hbar}{i}\,\frac{\delta Z}{\delta\underline{a}(x)} = \langle T\{\frac{\delta\mathcal{L}}{\delta\underline{a}(x)}\,\exp\frac{i}{\hbar}\mathcal{L}_{int}\}\rangle^{o}_{conn} \tag{5}$$

$$\frac{\hbar}{i}\frac{\partial Z}{\partial \underline{\lambda}} = \langle T\{\frac{\partial \mathcal{L}}{\partial \underline{\lambda}} \exp \frac{i}{\hbar} \mathcal{L}_{int}\}\rangle^{o}_{conn.} \tag{6}$$

$$0 = \langle T\{[\underline{P}(\underline{\phi}(x))(\frac{\delta \mathcal{L}}{\delta \underline{\phi}} - \partial_{\mu}\frac{\delta \mathcal{L}}{\delta \partial_{\mu}\underline{\phi}})](x) \exp \frac{i}{\hbar}\mathcal{L}_{int}\}\rangle^{o}_{conn.} \tag{7}$$

for arbitrary local polynomials $\underline{P}(\underline{\phi}(x))$. All these relations are valid for dimensionally regularized Feynman amplitudes in the sense of formal power series in $\hbar$.

II. LORENTZ COVARIANTS; RENORMALIZED AMPLITUDES

Treating Lorentz covariants as abstracts objects means first that any covariant (say p_{μ}) is treated as one object and not as a collection of its (four) components and second that we have to specify some rules how to manipulate them. Starting from any expression meaningful when interpreted in 4 dimensions, we may perform the usual manipulations with the following three exceptions: i) $g_{\mu\mu}=\nu$ and not 4; ii) $\{\gamma_{\mu},\gamma_5\}\neq 0$ and iii) we cannot use Fierz transformations (at least not in the form valid in 4 dimensions).

Such transformations will in general introduce some ν-dependence in addition to the ν-dependence of the original coefficients. In order to define a unique Laurent expansion around $\nu=4$ for (the coefficients of) such abstract covariants it is therefore necessary to reduce all covariants to a normal form (NF). Such NF's need not be unique, and in fact are not if γ_5 is involved, but all remaining relations must have ν-independent coefficients. This can be achieved by antisymmetrizing all products of γ-matrices (using $\{\gamma_{\mu},\gamma_{\nu}\}=2g_{\mu\nu}\mathbb{1}$) and contracting as many indices as possible (using $g_{\mu\nu}g_{\nu\lambda}=g_{\mu\lambda}$, $g_{\mu\nu}p_{\nu}=p_{\mu}$, $p_{\mu}q_{\mu}=(pq)$ and $g_{\mu\mu}=\nu$).

Once all covariants are reduced to their NF's renormalization is performed as usual. Recursively (in the number of loops or powers of $\hbar$) we define counterterms in the Lagrangian which cancel precisely the singular part of the Laurent expansion for all Feynman amplitudes up to a certain order in $\hbar$. It should be clear from Speer's lectures that these singular parts are in fact polynomials (of minimal degree) in the external momenta and can be absorbed into counterterms. This can, however, as well, be proven entirely within the framework of dimensional regularization without using analytic renormalization as an intermediate step.

Finally we interpret all covariants as 4-dimensional ones

(some expressions vanish at this point, $\{\gamma_\mu,\gamma_5\}$ is typical example) still keeping ν complex in the coefficients. If all masses are different from zero the limit $\varepsilon\to o$ exists (in $\mathcal{S}'$) even if some momenta p_i are set to zero (weak symptotic limit). This limit is analytic in a neighborhood of $\nu=4$. Its value for $\nu=4$ defines dimensionally renormalized Green's functions to any order in $\hbar$.

The validity of the action principle for the renormalized Green's functional can be proven in most cases as a consequence of the uniqueness of the Laurent expansion. Basically the action principle states the equality of two regularized amplitudes. Reducing both sides to NF may or may not involve an additional ν-dependence. If no such additional ν-dependence is involved, the action principle holds for the renormalized functional in its naive form, otherwise we expect anomalies similar in structure to the "normal product corrections" in the BPHZ scheme. Clearly whenever the derivation of the action principle in its naive form depends neither on $g_{\mu\mu}=4$ nor on $\{\gamma_\mu,\gamma_5\}=o$ nor on Fierz transformations there will be no anomaly. Otherwise the presence or absence of anomalies has to be studied in more detail; they are known to be present in the case of axial vector Ward identities and trace identities. Expressions like $\bar\psi\{\overleftrightarrow{\partial},\gamma_5\}\psi$ behave very much the same as $N_4(\bar\psi\gamma_5\psi)-N_3(\bar\psi\gamma_5\psi)$ in the BPHZ scheme and relations analogous to Zimmermann identities can be derived [3].

III. GREEN'S FUNCTIONS FOR PURELY MASSLESS THEORIES

It was supposed for a long time that dimensional renormalization does provide a means to define Green's functions for theories with massless particles. Recently we could prove that this is indeed the case for a wide class of purely massless theories[4]. In the more general case of theories with massless as well as massive particles this can be true only if certain normalization conditions are satisfied by choosing suitable finite counterterms[5], and it is not yet clear to what extent these counterterms modify the action principle. In the purely massless case, however, those normalization conditions are satisfied automatically and there are no such counterterms. As a consequence the action principle holds as in the massive case. As a possible application we would like to mention a nonabelian gauge field coupled to a massless spinor (quark) field. The validity of the Slavnov-Taylor identities in such theories is an immediate consequence of the action principle[6].

Theorem: The contributions of arbitrary Feynman graphs to dimensionally renormalized Green's functions for purely massless theories exist as tempered distributions, provided all internal vertices (i.e. terms in the interaction Lagrangian or integrated composite operators) have power counting degree not less than four.

Contributions of graphs containing detachable subgraphs vanish. All renormalization parts with positive superficial degree of divergence $\omega \geq 1$ vanish at the origin in momentum space up to order $\omega - 1$.

Remark: The theorem may even be extended to graphs containing one internal vertex of dimension three, if this vertex is not part of any tadpole.

In the following we will try to sketch the idea of the proof, refering to ref.[4] for any details. Clearly all the infrared problems show up only in the limit $\varepsilon \to o$, quite contrary to the treatment of Lowenstein and Speer. This is due to our choice of ε-dependence in Eq.(1). As a consequence we can prove the vanishing of contributions from detachable subgraphs, whereas Speer has to define them to be zero and to prove the consistency of this definition.

It might be worth mentioning that all our infrared problems refer to off-shell Green's functions. We do not attempt to define the S-matrix for massless particles (strong asymptotic limit).

The main difficulty of the limit $\varepsilon \to o$ arises from the fact that $(p^2+io)^\omega = \lim_{\varepsilon \to o}(p^2+i\varepsilon)^\omega$ (considered as distribuiton over $\mathbb{R}^n$) is meromorphic in ω with simple poles at $\omega = -n/2, -n/2-1, \cdots$ whereas $(p^2+i\varepsilon)^\omega$ or $(p^2-m^2+io)^\omega$ are entire functions of ω[6].

As in the massive case we subdivide the domain of $\underline{\alpha}$-integration in Eq.(1) and introduce scaling variables in each sector in order to display the behavior of (det M) if some or all α's vanish (UV divergencies). In addition we have to display simultaneously the possible degeneracy of the quadratic form emerging from M^{-1} by setting all p's attached to internal vertices and all u's to zero. This can be done using an idea of Pohlmeyer[8].

In order to simplify the argument let us assume a connected graph G without tadpoles, such that all external vertices have dimension 1 or 3/2 (ordinary Green's functions) and all internal vertices have dimension 4 (strictly renormalizable theories). The argument can be generalized to the case of arbitrary Green's functions (including those for any number of composite operators) and to nonrenormalizable interactions.

Let G_∞ be the graph obtained from G by contracting all its external vertices to a new one V_∞. For any maximal set $\mathcal{C}_\infty$ of nonoverlapping 1PI subgraphs of G_∞ and any mapping $\sigma : \mathcal{C}_\infty \to \mathcal{L}$ such that for any $H \in \mathcal{C}_\infty$ $\sigma(H)$ is a line of H but not a line of any $H' \in \mathcal{C}_\infty$ properly contained in H, we can define a sector $\mathcal{D}(\mathcal{C}_\infty, \sigma)$ of $\underline{\alpha}$-space. All these sectors are a partition of the whole domain of integration. In each sector we introduce scaling variables $(\underline{t},\underline{\beta})$ and ($\underline{\alpha}$-independent) linear combinations $\underline{q}$ of the K-1 linearly independent external momenta $\underline{p}$ suitable for that sector.

Performing an overall scaling integral the contribution of each sector to the Feynman amplitude is a sum of terms of the form

$$\lim_{\varepsilon\to 0} \Gamma\left(\frac{4-\nu}{2} h_G - d_G\right) \int_0^1 \prod \frac{dt_H}{t_H} t_H^{(4-\nu)h_H/2-d_H} \int_0^1 \prod d\beta \times$$
$$\times\, h(\underline{t},\underline{\beta},\nu) Q(\sqrt{\eta_H} q_H)(-id(\sqrt{\eta_H} q_H)+\varepsilon)^{d_G-(4-\nu)h_G/2} \tag{8}$$

$Q(\underline{q})$ is a homogeneous polynomial of degree r, d(q) a quadratic form over K-1 copies of Minkowski space derived from a positive definite (K-1)×(K-1) matrix, both with coefficients which are C^∞ in $(\underline{t},\underline{\beta})$. h is an analytic function, η_H are certain product of t's, $d_G=(\omega_G-r)/2$, $d_H=\omega_H/2$ and ω is the superficial degree of divergence.

The whole degeneracy of the quadratic form is exhibited by the η's. The possible singularities resulting from such a degeneracy of the quadratic form can, however, be controlled by the following

Lemma: Let η_j,ρ_j, $j=1,\ldots,J$ and $\gamma_\ell,k_\ell,\kappa_\ell,\ell=1,\ldots,L$ be real numbers in the domain $0\leq\eta_J\leq\ldots\leq\eta_1$, $0<\rho_j<2$ for all j, $\gamma_\ell\geq 0$ and $k_\ell,\kappa_\ell>0$ for all ℓ. In addition take $\rho\in\mathbb{C}$, d_{ij} a positive definite matrix defining the quadratic form $d(\underline{q})=\Sigma\, d^{ij}_{ij}\, q_i q_j$ over J copies of 4 dimensional Minkowski space and $Q(\underline{q})$ a homogeneous polynomial of degree r. Then for $\mathrm{Re}(\rho)<\Sigma\rho_j+\Sigma k_\ell+r/2$

$$\frac{Q(\sqrt{\eta_j}q_j)\prod\eta_j^{\rho_j}\prod\gamma_\ell^{k_\ell}}{(-id(\sqrt{\eta_j}q_j)+i\Sigma\gamma_\ell\kappa_\ell+0)^\rho} = \lim_{\varepsilon\to 0}\frac{Q(\sqrt{\eta_j}q_j)\prod\eta_j^{\rho_j}\prod\gamma_\ell^{k_\ell}}{(-id(\sqrt{\eta_j}q_j)+i\Sigma\gamma_\ell\kappa_\ell+\varepsilon)^\rho} \tag{9}$$

exists in $\mathcal{S}'(\mathbb{R}^{4J})$, is continuous in $(\underline{\eta},\underline{\gamma})$ and C^∞ in the coefficients of Q and d.

This lemma together with some careful estimates of the various degrees ω suffices to prove the existence of the expression (8). We use the fact that either t_H does not occur in any η_H or $d_H<-\sigma_H$ with $\sigma_H>0$ large enough to provide the powers $\eta_H^{\rho_H}$ necessary to apply the lemma. Consequently there exists some $\delta>0$ such that the limit (8) exists as a meromorphic function for all $|\nu-4|<\delta$.

In other words the limit exists for all ν satisfying $0<|\nu-4|<\delta$.

We may choose to define ε-dependent counterterms such that the poles at $\nu=4$ (UV divergencies) are removed from the amplitude as long as ε is still finite. The renormalized amplitude for $\varepsilon>0$ is analytic at $\nu=4$ and can be expressed as a sum of terms like (8), each of them having poles at $\nu=4$. Clearly the limit $\varepsilon\to 0$ of the renormalized amplitude is again analytic at $\nu=4$.

Finally another application of the lemma leads to the conclusion that all contributions explicitely proportional to ε vanish in the limit. They may arise from i) the ε-dependent part of counterterms, ii) the ε in Eq. (3) or iii) detachable subgraphs.

Let us finally sketch the proof of the lemma. Its formulation is more general than used so far (L=o); the mass terms κ_ℓ are needed in case of nonrenormalizable interactions and theories including massive as well as massless lines. Once the existence of the limit is proven, differentiability follows easily; any derivative reproduces the form of (9). Due to the ordering of η's and the 'positive definiteness' of $d(\underline{q})$ there exists a transformation $S:\underline{q}\to\underline{q}'$, C^∞ in $\underline{\eta}$, such that $d(\sqrt{\eta_j}q_j)=\Sigma\eta_j q_j'^2$. Decomposing $Q(\sqrt{\eta_j}q_j)$ into monomials in $\underline{q}'$ we get terms

$$\lim_{\varepsilon\to o}\Pi[\eta_j^{\rho_j} Q_j(\sqrt{\eta_j}q_j')]\Pi\gamma_\ell^{k_\ell}(-i\Sigma\eta_j q_j'^2+i\Sigma\gamma_\ell\kappa_\ell+\varepsilon)^{-\rho}$$

and may now repeatedly use the Mellin integral representation

$$(A+B)^{-\alpha}=\int_{b-i\infty}^{b+i\infty}\frac{d\beta}{2\pi i}\,\frac{\Gamma(\alpha-\beta)\Gamma(\beta)}{\Gamma(\alpha)}\,A^{-(\alpha-\beta)}B^{-\beta}$$

valid for $\mathrm{Re}(\alpha)>b>0$ and A,B such that the integral converges absolutely. At this point it is absolutely essential to interprete $(q'^2+i0)^\omega$ is distribution. Shifting some contours but always keeping the integral absolutely convergent we pick up an ε-independent residue whereas the remaining integral vanishes in the limit $\varepsilon\to 0$. Continuity in $(\underline{\eta},\underline{\gamma})$ is easily verified from this integral representation.

REFERENCES

1. This lecture is closely related to those of C. Becchi, J. H. Lowenstein, E. R. Speer and R. Stora. Most of their references are useful for this lecture too. We are using the graph-theoretical definitions given in the Appendix.

2. P.Breitenlohner, D.Maison; Dimensional Renormalization and the Action Principle, Preprint MPI-PAE/PTh 25/74, May 1975 (revised version); submitted to Comm. Math. Physics.

3. ------; Anomalies in Dimensional Renormalization, in preparation.

4. ------; Dimensionally Renormalized Green's Functions for Theories with Massless Particles I, Preprint MPI-PAE/PTh 15/75, July 1975.

5. ------; Dimensionally Renormalized Green's Functions for Theories with Massless Particles II, in preparation.

6. ------; Dimensional Renormalization of Massless Yang-Mills Theories, Preprint MPI-PAE/PTh 26/75, August 1975.

7. I. M. Gelfand, G. E. Shilov; Generalized Functions, Vol I, Academic, New York 1964.

8. K. Pohlmeyer; Large Momentum Behavior of the Feynman Amplitudes in the ϕ_4^4 Theory, Preprint DESY 74/36, August 1974.

ADIABATIC LIMIT IN PERTURBATION THEORY

H. Epstein and V. Glaser

C.E.R.N., Geneva

ABSTRACT

We show that, with correct mass and wave function renormalization, the time-ordered products for Wick polynomials $T(\mathcal{L}(y_1)\ldots\mathcal{L}(y_n))$ constructed by a method outlined in a previous paper[1] are such that the vectors of the form

$$\int T(\mathcal{L}(y_1)\ldots\mathcal{L}(y_n))\; g(y_1)\ldots g(y_n)\; \Psi dy_1\ldots dy_n$$

have limits when g tends to a constant, provided Ψ is chosen in a suitable dense domain. It follows that the S matrix has unitary adiabatic limit as an operator-valued formal power series in Fock space.

INTRODUCTION

In references [1,2] we have described an inductive method of constructing the nth order of renormalized perturbation theory. The outcome is an operator valued formal power series denoted $S(\underline{g})$. Here $\underline{g} = \{g_j^r(x)\}$ is a multicomponent smooth function over $\mathbb{R}^4$, with fast decrease at infinity,

$$S(\underline{g}) = \sum_{n=1}^{\infty} \frac{i^n}{n!} \int T^{r_1,\ldots,r_n}_{j_1,\ldots,j_n}(x_1,\ldots,x_n) g^{r_1}_{j_1}(x_1)\ldots g^{r_n}_{j_n}(x_n) d^4x_1\ldots d^4x_n$$

Editor's Footnote: Instead of a summary of the lectures of H. Epstein, we are here printing a long unpublished account of the joint work by Epstein and Glaser on the subject of the lectures. It was presented to the Meeting on Renormalization Theory C.N.R.S. Marseille June 14-18, 1971 and was circulated as a CERN preprint: Ref. TH.1344-CERN.

G. Velo and A. S. Wightman (eds.), Renormalization Theory, 193–254.

The summation over n is formal. The $T(x_1,\dots,x_n)$ are distribution-valued operators acting in the Fock space of (e.g.) one neutral scalar free field with mass $m > 0$, denoted $A(x)$. The lowest order terms of $S(\underline{g})$ are:

$$S(g) = 1 + i \int \sum_{j,r} \mathcal{L}_j^{(n)}(x) g_j^r(x) dx + \text{higher orders in } \underline{g} \qquad \text{(i)}$$

$$\mathcal{L}_j^{(r)} \text{ is for example } \frac{\nu_j!}{(\nu_j - r)!} : A^{\nu_j - r}(x): \ , \ 0 \leq r \leq \nu_j$$

More complicated interactions are easily treated by the same method. We consider here, for simplicity, the case when j takes only one value, i.e.

$$\mathcal{L}^{(r)}(x) = \frac{\nu!}{(\nu-r)!} : A(x)^{\nu-r} : \ , \text{ with } \ \nu \geq 3 \ .$$

$S(g)$ is then the generating functional of the time-ordered products of Wick polynomials in the free field:

$$T^{r_1 \dots r_n}(x_1,\dots,x_n) = T(\mathcal{L}^{(r_1)}(x_1) \dots \mathcal{L}^{(r_n)}(x_n))$$

and is also the S-matrix for the theory with interaction Lagrangian density indicated by (i), i.e. with the space-time switching-off factors $\underline{g}$, the incoming asymptotic field being $A(x)$. In the same (switched off) theory the generating function for the time-ordered products of the interacting "fields" $\hat{\mathcal{L}}^{(r)}(x)$ is :

$$\mathcal{V}(\underline{g};\underline{h}) = S(\underline{g})^{-1} S(\underline{g}+\underline{h})$$

i.e.

$$\hat{T}(x_1,\dots,x_n;\underline{g}) = \frac{\delta}{i\delta \underline{h}(x_1)\dots i\delta \underline{h}(x_n)} \mathcal{V}(g;h)\Big|_{\underline{h}=0} ,$$

$\hat{T}(X;g) \equiv T(\hat{\mathcal{L}}(x_1;\underline{g})\dots\hat{\mathcal{L}}(x_n;\underline{g}))$, and the corresponding Green functions are obtained by taking matrix elements in Ω , the free ($\equiv$ incoming) vacuum. It was shown in [1,2] that these Green functions have limits when $g^{(r)}(x) \to$ constants: this is the easy part of the adiabatic limit, which is based solely on the clustering properties due to $m > 0$.

However, this is insufficient to prove the unitarity of the S-matrix. To investigate this question, we assume that the inductive construction has been done so that the two-point Green function (in the adiabatic limit), in momentum space, has a pole at $p^2 = m^2$ with residue 1. This gives the possibility for A(x) to remain the incoming asymptotic field of the theory after the adiabatic limit has been taken: the latter is e.g. defined as the limit when

$$g^{(0)}(x) \to \lambda , \quad g^{(r)}(x) \to 0 \quad \text{for} \quad r \geq 1 .$$

(The method also works if $g^{(r)}(x) \to \lambda_r$ for all $r \leq \nu-3$, $\lambda_r \to 0$ for $r \geq \nu-2$). Our purpose is to show that

$$S_n(\underline{g}) = \int T^{0,0,\ldots 0}(x_1,\ldots,x_n) g^0(x_1)\ldots g^0(x_n) dx_1 \ldots dx_n \tag{ii}$$

tends to a limit as an operator in Fock space in a strong sense: more precisely, if we replace $\underline{g}(x)$ by $\underline{g}_\varepsilon(x) = \{g^0(\varepsilon x), 0, \ldots 0\}$, $g^0 \in \mathcal{S}$, and if $\Psi \in D_2$, then

$$\operatorname*{strong\,lim}_{\varepsilon \to 0} S_n(\underline{g}_\varepsilon)\Psi$$

exists. Here D_2 is the (dense) subspace of Fock space formed by all vectors with only a finite number of non-vanishing N-particle wave functions, which must be Hölder continuous, with compact support (or rapid decrease at infinity). The choice of the domain D_2 is suggested by the simplest example: a vector with $\mathcal{C}^\infty$ wave functions is transformed by $\int :A(x)^\nu: g(x) dx$ into a vector with wave functions no longer $\mathcal{C}^\infty$ but only Hölder continuous.

This fact, and the unitarity of $S(\underline{g}_\varepsilon)$ for $\varepsilon > 0$, imply that, in the adiabatic limit, the S-matrix will remain unitary, in the sense of formal power series in the coupling constant $g^0(0)$.

In Section 2, we have assembled the mathematical information about distributions having the "adiabatic property". In Section 3 the induction hypothesis is described (note in particular the conditions we impose on the "self-energy kernels"). Section 4 contains the rather straightforward verifications which show that the induction works.

This is a slightly revised version of our communication to the Marseille Meeting on Renormalization Theory, CNRS 1971.

1. THE PROBLEM

We wish to show that, when $g \in \mathcal{S}(R^4)$ and g tends to a constant λ (in some sense to be precised) the operator

$$S_n(g) = \int T^{(0,\ldots,0)}(x_1,\ldots,x_n) g(x_1)\ldots g(x_n) dx \ldots dx_n \tag{1}$$

tends to an operator in Fock space, in some sense, to be precised, strong enough to yield the unitarity of the limiting S matrix considered as a formal power series in .

Since the operator (1) is unbounded, the convergence may only be expected to hold when this operator is applied to a vector chosen in a suitable domain D_2. A natural restriction to impose on such vectors is that they contain a finite number of particles. Moreover, since the kernels of the operator (1) are well known to have singularities such as poles, etc., one must expect that the wave functions of the admissible vectors should be rather regular. But we wish to be able to multiply two of the limiting operators, hence, they should map D_2 into itself. Now, the simplest example [take $T^{(0)}(x) = \mathcal{L}^{(0)}(x) = :A(x)^{\nu}:$] shows that a vector with $\mathcal{C}^{\infty}$ wave functions (in momentum space) is transformed by $\int \mathcal{L}^{(0)}(x)dx$ into a vector whose wave functions are no longer $\mathcal{C}^{\infty}$ but only Hölder-continuous. Thus we are led to conjecture that D_2 might be formed of vectors with a finite number of non-zero Hölder-continuous wave functions. This turns out to be correct.

As to the sense in which $S_n(g)$ tends to a limit, the simplest conjecture would be that, if $\Psi \in D_2$, the vector-valued tempered distribution

$$\int e^{i\sum_{j=1}^{n} q_j x_j} T(x_1,\ldots,x_n)\Psi \, dx_1\ldots dx_n \qquad (2)$$

might actually be a (vector-valued) continuous function of $q_1,\ldots,q_n$ in a neighbourhood of zero. However, the simplest examples dispel this illusion, so that a more subtle way has to be found for defining the value of (2) at zero. The next section describes a class of distributions for which it is possible to define a value at a certain point. It turns out that the method of evaluation given there can be used to define the "value at 0" of (2) and to find the sense in which $S_n(g)\Psi$ converges to that value as $g(x) \to 1$.

<u>Outline of the Method</u> Suppose T is a distribution on $\mathbb{R}^N$ and that there exist constants $\delta > 0$, $M \geq 0$, $C \geq 0$, such that for every $\phi \in \mathcal{S}(\mathbb{R}^N)$,

$$|\langle T,\phi\rangle| \leq C\{|\int \phi(q)dq| + \sum_{|\alpha|\leq M} \sup_q |q|^{N+\delta+|\alpha|}|D^{\alpha}\phi(q)|\}$$

"adiabatic property"

Then it will be seen in Section 2 that the limit

$$\lim_{\substack{\varepsilon\to 0\\ \varepsilon>0}} \varepsilon^{-N} \int T(q)\phi(q/\varepsilon)dq$$

exists and is of the form const. $\int \phi(q)dq$; we then take this constant to be the "value" or "adiabatic limit" of the distribution T at 0.

The idea of our method is to prove by induction on n that (2), as a vector valued distribution in $q_1,\ldots,q_n$, satisfies such a condition. It will be recalled that the inductive construction of the $T(X)$ operators, or, equivalently of the advanced, retarded operators etc., proceeds in two steps.

a) One constructs the "discontinuity", or difference between a totally advanced and totally retarded operator of n "Lagrangians" $\mathcal{L}^{(r_1)}(x_1),\ldots,\mathcal{L}^{(r_n)}(x_n)$. This difference is expressed in terms of products of chronological operators with fewer arguments. Thus, one must verify that the product of two operator-valued distributions having the above property also has it. This is essentially due to the fact that the "adiabatic property" for distributions is stable under tensor products. It follows that the "discontinuity" possesses the adiabatic property.

b) The totally advanced operator of n Lagrangians is then obtained essentially by a dispersion relation in many variables, i.e., by convoluting the discontinuity with a generalized Cauchy kernel; this operator also preserves the adiabatic property.

However, there are technical difficulties:

i) The convolution is performed over n-1 of the n four-vector variables, say, $q_1,\ldots,q_{n-1}$; as a consequence the variable $Q\sum_{j=1}^{n} q_j$ has to be treated separately; in this variable, one must prove the Holder continuity of the matrix elements of (2), except those corresponding to self-energy kernels.

ii) The self-energy kernels must have been properly renormalized i.e., correct mass and wave function renormalization must have been performed (see I, Section VIII). This insures that they possess the adiabatic property, and that their adiabatic limit vanishes twice on the mass shell, and is necessary in order to prove the correct properties of the product of a self-energy kernel with another kernel.

In Section 2 we have assembled the mathematical information regarding distributions having the"adiabatic property". In

Section 3, the induction hypothesis is described. Section 4 contains the rather straightforward verifications needed to show that the induction works. Except for the treatment of self-energy kernels, this can be omitted by the reader without great loss of information.

2. SOME MATHEMATICAL PRELIMINARIES

2.1 Notation for Holder Continuous Functions

(We are interested in functions defined over momentum space, hence, the natural unit of length is the mass m.)

Let f be a function defined on $\mathbb{R}^N$ and taking its values in a Banach space whose norm is denoted B. We denote, for $\delta > 0$:

$$H^{\delta}_{x;B}(f(x)) = \max\{\sup_{x} B(f(x)), \sup_{\substack{x,h \\ |h|\leq m}} |h/m|^{-\delta} B(f(x+h) - f(x))\} \tag{3}$$

In particular, if f is a numerical function, we omit B and denote

$$H^{\delta}_{x}(f(x)) = \max\{\sup_{x}|f(x)|, \sup_{\substack{x,h \\ |h|\leq m}} |h/m|^{-\delta}|f(x+h) - f(x)|\} \tag{4}$$

Let f be a function of two variables $x \in \mathbb{R}^N$ and $y \in \mathbb{R}^M$. It is easy to verify that, if $\rho > 0$, $\sigma > 0$, $\delta > 0$:

$$H^{\rho}_{x;H^{\sigma}_{y;B}}(f(x,y)) = H^{\sigma}_{y;H^{\rho}_{x;B}}(f(x,y)) \tag{5}$$

$$H^{\rho}_{x;H^{\sigma}_{y;B}}(f(x,y)) \leq 2\, H^{\rho+\sigma}_{x,y;B}(f(x,y)) \tag{6}$$

$$H^{\delta}_{x,y;B}(f(x,y)) \leq 2\, H^{\delta}_{x;H^{\delta}_{y;B}}(f(x,y)) \tag{7}$$

Note that, for a fixed f, $H^{\delta}_{x}(fx))$ is an increasing function of δ.

2.2 "Adiabatic limit" of a Distribution at a Point

Definition 1: We say that a distribution $T \in \mathcal{D}'(\mathbb{R}^N)$ with compact support satisfies an adiabatic norm of order δ at 0 if

$\delta > 0$ and if there exist constants $C > 0$ and $P > 0$ such that, for every $(\mathbb{R}^N)$,

$$|\langle T,\phi\rangle| < C\{|I(\phi)| + ||\phi||_{\delta,P}\} \tag{8}$$

Here, and in the rest of this section, we use the notation:

$$||\phi||_{\delta,P} = \sum_{0\leq|\alpha|\leq P} \sup_{x \in R^N} |x|^{N+\delta+|\alpha|} \, D^{\alpha}\phi(x)|$$

$|x|$ being the Euclidean norm in $\mathbb{R}^N$.

$I(\phi)$ denotes $\int_{\mathbb{R}^N} \phi(x)dx$.

Clearly this notion is invariant against linear (non-singular) changes of co-ordinates. Moreover, if T has this property, and if h is a function in $(\mathbb{R}^N)$, we have

$$|\langle hT,\phi\rangle| = |\langle T,h\phi\rangle| \leq C\{|\int h(x)\phi(x)dx| + \sum_{|\alpha|<P} \sum_{\gamma<\alpha} \frac{\alpha!}{\gamma!(\alpha-\gamma)!} \sup_x |x|^{N+\delta+|\alpha|} |D^{\alpha-\gamma}h(x)| \cdot |D^{\gamma}\phi(x)\} \leq$$

$$\leq C'\{|\int h(x)\phi(x)dx| + ||\phi||_{\delta,P}\}$$

However,

$$|\int h(x)\phi(x)dx| = |\int [h(x) - h(0)]\phi(x)dx + h(0)\int \phi(x)dx| \leq$$

$$\leq |h(0)| \; |\int \phi(x)dx| + \sup_x |x|^{N+\delta} |\phi(x)| \cdot \int \frac{|h(y) - h(0)|dy}{|y|^{N+\delta}}$$

If $0 < \delta < 1$ the last integral exists (since h is continuously differentiable and bounded) and is bounded by const. $\delta^{-1}(1-\delta)^{-1}$. Hence

$$|\langle hT,\phi\rangle| \leq C''_{\delta} \{|I(\phi)| + ||\phi||_{\delta,P}\} \quad \text{if} \quad 0 < \delta < 1$$

The same calculation shows that, under the same condition, and if $h(0) \neq 0$,

$$I(\phi)| \leq C'''[|\int \phi(x)h(x)dx| + ||\phi||_{\delta,0}]$$

Hence, in the preceding definition one can replace $I(\phi)$ by

$\int h(x)\phi(x)dx$ with any $h \in \mathcal{S}(\mathbb{R}^N)$ such that $h(0) \neq 0$, provided $0 < \delta < 1$.

This gives a meaning to the following extension of Definition 1.

Definition 2: We say that a tempered distribution $T \in \mathcal{S}'(\mathbb{R}^N)$ satisfies an adiabatic norm of order δ at 0 if there is a function $h \in \mathcal{D}(\mathbb{R}^N)$ with $h(0) = 1$ such that

$$|\langle hT,\phi\rangle| \leq C\{|I(\phi)| + ||\phi||_{\delta,P}\}$$

for some $C \geq 0$, some $P \geq 0$, and all $\phi \in \mathcal{S}(\mathbb{R}^N)$.

While we have given this definition for completeness, we shall always restrict our attention to the local behaviour of distributions in this paper.

Lemma 1 If a distribution $T \in \mathcal{S}'(\mathbb{R}^N)$ satisfies the adiabatic norm (8) then the "adiabatic limit"

$$\lim_{\substack{\varepsilon\to 0\\ \varepsilon>0}} \langle T(\varepsilon x),\phi(x)\rangle = \lim_{\substack{\varepsilon\to 0\\ \varepsilon>0}} \langle T,\phi_\varepsilon\rangle = LI(\phi) \tag{9}$$

exists for every $\phi(x) \in \mathcal{S}(\mathbb{R}^N)$ and is a distribution in ϕ of the form indicated in (9), where L is a constant independent of ϕ . Here

$$\phi_\varepsilon(x) = \varepsilon^{-N}\phi(\frac{x}{\varepsilon}) \quad , \quad \varepsilon > 0 \tag{10}$$

Proof

In view of the formula:

$$\langle T,\phi_\varepsilon\rangle = \langle T,\phi\rangle - \int_\varepsilon^1 \langle T,\frac{\partial}{\partial\eta}\phi_\eta\rangle\, d\eta \tag{11}$$

it is enough to prove that the last integral converges absolutely at $\varepsilon = 0$ in order to establish the existence of the adiabatic limit. Nowe we have

$$\frac{\partial}{\partial\varepsilon}\phi_\varepsilon(x) = -\frac{1}{\varepsilon}\chi_\varepsilon(x) \quad \text{with} \quad \chi(x) = \sum_{i=1}^N \frac{\partial}{\partial x_i}\{x_i\phi(x)\} \equiv \partial(x\phi(x)),$$

$$\chi_\varepsilon(x) = \varepsilon^{-N}\chi(\frac{x}{\varepsilon}) \tag{12}$$

and, therefore, by inserting in (8)

$$|\langle T,\frac{\partial}{\partial\varepsilon}\phi_\varepsilon\rangle| \leq \varepsilon^{\delta-1} C||\partial(x\phi)||_{\delta,P} \leq \text{const.}\ \varepsilon^{\delta-1} \tag{13}$$

Here we have used the fact that

$$I(-\frac{1}{\varepsilon}\chi_\varepsilon) = -\varepsilon^{-1}I(\chi) = 0$$

(change of integration variables $y = x/\varepsilon$ and Gauss theorem since χ is a divergence) and the formula

$$\sup_{x\in\mathbb{R}^N}|F(x)| = \sup_{y\in\mathbb{R}^N} F(\varepsilon y) \quad \text{for any} \quad \varepsilon > 0 \tag{14}$$

in evaluating the sup appearing in (8). Since $\delta > 0$ the integral (11) is absolutely convergent at $\varepsilon = 0$. Hence, the adiabatic limit (9) exists and is by a well-known theorem again a distribution in $\mathcal{S}'$, say S:

$$\lim_{\varepsilon\to+0} \langle T,\phi_\varepsilon\rangle = \langle S,\phi\rangle$$

Now, by the same argument which led us to (13) we get for any $\phi \in \mathcal{S}$ and $\varepsilon > 0$:

$$|\langle T,\phi_\varepsilon\rangle| \le C\{|I(\phi)| + \varepsilon^\delta||\phi||_{\delta,P}\} \tag{15}$$

and upon taking the limit $\varepsilon \to +0$:

$$|\langle S,\phi\rangle| \le C|I(\phi)|$$

By applying this inequality to $\psi(x) = \phi(x) - I(\phi)\cdot\phi_0(x) \in \mathcal{S}$, where ϕ is any element $\in \mathcal{S}$ and ϕ_0 a fixed element such that $I(\phi_0) = 1$, we get $\langle S,\psi\rangle = \langle S,\phi\rangle - \langle S,\phi_0\rangle\cdot I(\phi) = 0$ in view of $I(\psi) = 0$. This proves the lemma if we set $\langle S,\phi_0\rangle = L$.

In the next Lemma we shall need the inequality:

$$|\langle T,\phi\rangle - LI(\phi)\rangle| = |\int_0^1 \langle T, \frac{\partial}{\partial\varepsilon}\phi_\varepsilon\rangle d\varepsilon| \le \delta^{-1} C||\partial(x\phi)||_{\delta,P} \tag{16}$$

which is an immediate consequence of the equality (11) with $\varepsilon = 0$ and the inequality (13).

<u>Lemma 2</u> Suppose for a $T \in \mathcal{S}'(\mathbb{R}^N)$ the adiabatic norm

$$|\int T(x+y)\phi(x)dx| \le C\{|I(\phi)| + ||\phi||_{\delta,P}\} \tag{17}$$

holds uniformly for all y in a closed convex set $\mathcal{K} \subset \mathbb{R}^N$ (i.e., C is a constant <u>independent</u> of $y \in \mathcal{K}$). Then (by the preceding lemma) the adiabatic limit

$$\lim_{\substack{\varepsilon\to 0\\ \varepsilon>0}} \int T(\varepsilon x+y)\phi(x)dx = \lim_{\varepsilon\to+0} \int T(x+y)\phi_\varepsilon(x)dx = L(y)I(\phi) \tag{18}$$

exists for all $y \in \mathcal{K}$ and $L(y)$ is a uniformly bounded and uniformly Hölder continuous function on $\mathcal{K}$ with the index of continuity δ. More precisely

$$\text{a)}\ |L(y)| \leq C\,, \qquad \text{b)}\ |L(y+h) - L(y)| \leq M|h|^{\delta} \tag{19}$$

$$\text{for all } y \text{ and } y+h \in \mathcal{K}\,.$$

With C and M, two constants independent of y.

<u>Proof</u> The inequality (19a) follows immediately from the inequality (15) applied to this case upon passage to the limit $\varepsilon \to +0$. In order to obtain (19b) we take any fixed $\phi \in \mathcal{S}$ such that $I(\phi) = 1$ and write for an $\varepsilon > 0$ to be chosen later:

$$\begin{aligned} |L(y+h) - L(y)| &\leq \left|\int T(x+y+h)\phi_\varepsilon(x)dx - L(y+h)\right| + \\ &+ \left|\int T(x+y)\phi_\varepsilon(x)dx - L(y)\right| + \left|\int \{T(x+y+h) - T(x+y)\}\phi_\varepsilon(x)dx\right| \\ &\leq 2\delta^{-1}C\,||\partial(x\phi_\varepsilon)||_{\delta,P} + \left|\int T(x+y)\{\phi_\varepsilon(x-h) - \phi_\varepsilon(x)\}dx\right| \end{aligned} \tag{20}$$

The last inequality follows from the inequality (16) in view of $I(\phi_\varepsilon) = I(\phi) = 1$. Now, it is easily seen that:

$$||\partial(x\phi_\varepsilon)||_{\delta,P} = \varepsilon^{\delta}\,||\partial(x\phi)||_{\delta,P} \tag{21}$$

On the other hand we write:

$$\phi_\varepsilon(x-h) - \phi_\varepsilon(x) = \int_0^1 dt\,\frac{d}{dt}\,\phi_\varepsilon(x-th) = -\int_0^1 h\partial\phi_\varepsilon(x-th)dt \tag{22}$$

with
$$h\partial = \sum_{i=1}^{n} h_i \frac{\partial}{\partial x_i}$$

so that the last term in the inequality (20) becomes

$$\begin{aligned} \left|\int T(x+y)\{\phi_\varepsilon(x-h) - \phi_\varepsilon(x)\}dx\right| &= \left|\int_0^1 dt \int T(x+y+th)h\partial\phi_\varepsilon(x)dx\right| \\ &\leq C\,||h\partial\phi_\varepsilon(x)||_{\delta,P} = C\varepsilon^{\delta-1}||h\partial\phi||_{\delta,P} \end{aligned} \tag{23}$$

The inequality in (23) is obtained by application of the Ineq. (17) to the test function $h\partial\phi_\varepsilon$ in view of the fact that by convexity $y + th \in \mathcal{K}$ for all $0 \leq t \leq 1$. By the theorem of Gauss the integral term does not contribute. As to the last equality, we obtain it by using the trick (14). Combining (21) and (23) we finally arrive at:

$$|L(y+h) - L(y)| \leq C[2\delta^{-1}||\partial(x\phi)||_{\delta,P} + ||\frac{h}{\varepsilon}\partial\phi||_{\delta,P}]\varepsilon^{\delta} \tag{24}$$

By choosing $\varepsilon = |h|$ we see that the expression in square brackets remains bounded for all h and hence (19b) is also proven.

Lemma 3 Under the condition of Lemma 2 we have, for any y and $y + h \in \mathcal{H}$ the inequality:

$$|\int T(x+y+h) - T(x+y)\}\phi(x)dx| \leq$$

$$\leq M\{|h|^{\delta}|I(\phi)| + |h|^{\theta\delta}||\phi||_{(1-\theta)\delta,P+1}\} \tag{25}$$

Here $0 < \delta \leq 1$, $0 \leq \theta < 1$ and M is a constant independent of y, h and ϕ.

Proof Choose a $u \in \mathcal{D}(\mathbb{R}^N)$ such that $u(x) = 1$ for $0 \leq |x| \leq 1/2$, $u(x) = 0$, for $|x| \geq 1$ and $0 \leq u(x) \leq 1$ everywhere and write:

$$\phi(x) = u(\frac{x}{|h|})\phi(x) + v(\frac{x}{|h|})\phi(x) \equiv \phi_1(x) + \phi_2(x) \tag{26}$$

where $v = 1 - u$.

Note the following support properties of u and v:

$$\text{supp. } u(\frac{x}{|h|}) \subset \{|x| \leq |h|\}, \quad \text{supp. } v(\frac{x}{|h|}) \subset \{|x| \geq \frac{|h|}{2}\}$$
$$\text{supp. } D^{\alpha}u(\frac{x}{|h|}) = \text{supp. } D^{\alpha}v(\frac{x}{|h|}) \subset \{\frac{|h|}{2} \leq |x| \leq |h|\} \tag{27}$$

for any $\alpha \neq 0$.

With the shorthand notation $\Delta = T(x+y+h) - T(x+y)$ the left-hand side of (25) is upon insertion of (26) bounded by

$$|\langle\Delta,\phi\rangle| \leq |\langle\Delta,\phi_1\rangle| + |\langle\Delta,\phi_2\rangle| \tag{28}$$

The first term we treat as follows:

$$|\langle\Delta,\phi_1\rangle| \leq |\langle\Delta - L(y+h) + L(y),\phi_1\rangle| + |L(y+h) - L(y)|\,|I(\phi_1)| \leq$$

$$\leq \text{const. } \{||\partial(x\phi_1)||_{\delta,P} + |h|^{\delta}\,|I(\phi_1)|\} \tag{29}$$

Here L is the adiabatic limit of the distribution T, and we have used the inequality (19b) of Lemma 2 and the inequality (16). Now we evidently have

$$\|\partial(x\phi_1)\|_{\delta P} \le \text{const.}\ \|\phi_1\|_{\delta,P+1} =$$

$$= \text{const.} \sum_{0<|\alpha|<P+1} \sup_{|x|\le|h|} |x|^{N+\delta+|\alpha|}|D^\alpha\phi_1(x)| \tag{30}$$

We differentiate ϕ_1 using Leibniz's formula and obtain:

$$|x|^{N+|\alpha|+\delta}|D^\alpha\phi_1(x)| \le \sum_{\beta+\gamma=\alpha} \frac{\alpha!}{\beta!\gamma!} \frac{|x|^{N+\delta+|\beta|+|\gamma|}}{|h|^{|\gamma|}} |(D^\gamma u)(\tfrac{x}{|h|})|\ |D^\beta\phi(x)|$$

Since $|D^\gamma u(x)| \le$ const. for all $|\gamma| \le P+1$ and $|x|^\omega \le |h|^\omega$ for all $\omega \ge 0$ in the sup region, we get:

$$\|\partial(x\phi_1)\|_{\delta,P} \le \text{const.}|h|^{\theta\delta} \sum_{|\alpha|\le P+1} \sup_{|x|\le|h|} |x|^{N+(1-\theta)\delta+|\alpha|}|D^\alpha\phi(x)|$$

$$\le \text{const.}\ |h|^{\delta\theta}\|\phi\|_{(1-\theta)\delta,P+1} \tag{31}$$

for all $0 \le \theta \le 1$. As for the integral term in (29) we can bound it as follows:

$$|I(\phi_1)| \le |I(\phi)| + |I(\phi_2)| \le |I(\phi)| + \int_{|x|\ge\frac{|h|}{2}} |\phi(x)|d^Nx \le$$

$$\le |I(\phi)| + \sup_{x\in R^N} |x|^{N+\eta}|\phi(x)| \int_{|y|\ge\frac{|h|}{2}} \frac{d^Ny}{|y|^{N+\eta}} \le \tag{32}$$

$$\le |I(\phi)| + \frac{\text{const.}}{\eta}|h|^{-\eta} \sup_x |x|^{N+\eta}|\phi(x)|$$

for any $\eta > 0$. By choosing $\eta = (1-\theta)\delta$ with $0 \le \theta < 1$, and inserting the inequalities (32) and (31) into (29) we see that the contribution $|\langle\Delta,\phi_1\rangle|$ satisfies the inequality (25).

The last term in (28) we treat as in formula (23):

$$|\langle\Delta,\phi_2\rangle| = \left|\int_0^1 dt \int T(x+y+th)h\partial\phi_2(x)dx\right| \le$$

$$\leq \text{const.}\,|h| \sum_{1\leq|\alpha|\leq P+1} \sup_{|x|\geq\frac{|h|}{2}} |x|^{N-1+\delta+|\alpha|}\,|D^\alpha\phi_2(x)| \leq$$

$$\leq \text{const.}\,|h|^{\theta\delta} \sum_{1\leq|\alpha|\leq P+1} \sup_{|x|\geq\frac{|h|}{2}} |x|^{N+(1-\theta)\delta+|\alpha|}\,|D^\alpha\phi_2(x)| \tag{33}$$

In the last step we have written $|h| = |h|^{\theta\delta}\,|h|^{1-\theta\delta}$, $0 \leq \theta \leq 1$, and used the fact that $|h|^\omega \leq 2^\omega|x|^\omega$ for all $\omega \geq 0$ in the sup region. Again, by Leibniz's formula and with the abbreviation $\delta' = (1-\theta)\delta$,

$$\sup_{|x|\geq\frac{|h|}{2}} |x|^{N+\delta'+|\alpha|}\,|D^\alpha\phi_2(x)| \leq \sup |x|^{N+\delta'+|\alpha|}\,|v(\tfrac{x}{|h|})D^\alpha\phi(x)| \; +$$

$$+ \text{const.} \sum_{\substack{\gamma+\beta=\alpha\\ \gamma\neq 0}} \sup_{\frac{|h|}{2}\leq|x|\leq|h|} |x|^{N+\delta'+|\gamma|+|\beta|}\,\frac{1}{|h|^{|\gamma|}}\,|(D^\gamma v)(\tfrac{x}{|h|})|\,|D^\beta\phi(x)| \tag{34}$$

[compare (27)]. The remark that $|x|^{|\gamma|} \leq |h|^{|\gamma|}$ in the sup region $|h|/2 \leq |x| \leq |h|$ leads to the conclusion that

$$|\langle\Delta,\phi_2\rangle| \leq \text{const.}\,|h|^{\theta\delta}\,||\phi||_{(1-\theta)\delta,P+1}$$

which completes the proof.

We shall need the following special case.

Lemma 4 Suppose that, for a certain $T \in \mathcal{S}'(\mathbb{R}^N)$ the inequality

$$|\int T(x+y)\phi(x)dx| \leq C||\phi||_{\delta,P}$$

holds uniformly for all y in a closed convex set $\mathcal{K} \subset \mathbb{R}^N$ (i.e., C is a constant independent of $y \in \mathcal{K}$). Then, we have, for any y and $y + h \in \mathcal{K}$:

$$|\int\{T(x+y+h) - T(x+y)\}\phi(x)dx| \leq$$
$$\leq M|h|^{\theta\delta}\,||\phi||_{(1-\theta)\delta,P+1}$$

Here $0 < \delta \leq 1$, $0 \leq \theta < 1$, and M is a constant independent of y, h, and ϕ.

Proof This has already been proved in the course of proving Lemma 3.

Lemma 5. Let T be a distribution with compact support which, in an open neighbourhood of 0, coincides with a Hölder continuous function. Then T satisfies an adiabatic norm at 0.

Proof First suppose that T vanishes in a neighbourhood of 0. Then, it is clear, by a partition of the unit, that there are constants $C \geq 0$, $a > 0$, and $P \geq 0$, such that

$$|\langle T,\phi\rangle| \leq C \sum_{|\alpha|\leq P} \sup_{\substack{x \\ |x|\geq a}} |D^\alpha \phi(x)| \leq$$

$$\leq C'_\omega \sum_{|\alpha|\leq P} \sup_{\substack{x \\ |x|\geq a}} a^{\omega+|\alpha|} |D^\alpha \phi(x)| \leq$$

$$\leq C'_\omega \sum_{|\alpha|\leq P} \sup_x |x|^{\omega+|\alpha|} |D^\alpha \phi(x)|$$

for any $\omega \geq 0$.

It suffices now to prove the lemma in case T is a Hölder continuous function with compact support. Then T satisfies

$$|T(x) - T(0)| < C|x|^\delta \quad \text{with} \quad 0 \leq \delta < 1$$

and

$$\int T(x)\phi(x)dx = T(0)\int \phi(x)dx + \int [T(x) - T(0)]\phi(x)dx ,$$

$$\left|\int [T(x) - T(0)]\phi(x)dx\right| \leq C\{ \int_{|x|<R} |x|^\delta |\phi(x)|dx + |T(0)| \int_{|x|\geq R} |\phi(x)|dx\}$$

$$\leq C'\{\sup_x |x|^{N+\delta-\varepsilon}|\phi(x)|\frac{R^\varepsilon}{\varepsilon} + |T(0)|\sup_x |x|^{N+\rho}|\phi(x)|\frac{1}{\rho R^\rho}\}$$

for any $\varepsilon > 0$ and $\rho > 0$.

In particular, choosing $\varepsilon = (1 - \theta)\delta$, with $0 < \theta < 1$ and $\rho = \theta\delta$ we have

$$\left|\int T(x)\phi(x)dx\right| \leq C_{\theta\delta}\{\sup_x |x|^{N+\theta\delta}|\phi(x)| + |I(\phi)|\}$$

Thus, we see that T satisfies an adiabatic norm of order $\theta\delta$ where $0 < \theta < 1$.

Tensor Product Rule

Lemma 6 Let $T \in \mathcal{S}'(\mathbb{R}^N)$ and $S \in \mathcal{S}'(\mathbb{R}^M)$ be two distributions with compact support. Assume that there exist numbers $\omega > 0$,

$P \geq 0$, $C \geq 0$, $\lambda > 0$, $Q \geq 0$, $D > 0$ such that, for any $\phi \in \mathcal{S}(\mathbb{R}^N)$ and $\psi \in \mathcal{S}(\mathbb{R}^M)$,

$$|\langle T,\phi\rangle| \leq C\{|\int_{\mathbb{R}^N}\phi(x)dx| + \sum_{|\alpha|\leq P} \sup_{x\in\mathbb{R}^N}|x|^{\omega+|\alpha|}|D_x^\alpha\phi(x)|\}$$

$$|\langle S,\psi\rangle| \leq D\{|\int_{\mathbb{R}^M}\psi(y)dy| + \sum_{|\beta|\leq Q} \sup_{y\in\mathbb{R}^M}|y|^{\lambda+|\beta|}|D_y^\beta\psi(y)|\}$$

Then, if $\lambda = M + \delta_2$, $\omega = N + \delta_1$, $0 < \delta_2 < 1$, $0 < \delta_1 < 1$, $\delta = \min(\delta_1,\delta_2)$, there exists for each $\varepsilon > 0$ with $\varepsilon \leq N$, $\varepsilon \leq M$, a constant K_ε such that, for any $\chi \in \mathcal{E}(\mathbb{R}^N\times\mathbb{R}^M)$

$$|\langle T\otimes S,\chi\rangle| \leq K_\varepsilon\{|\int\chi(x,y)dxdy| +$$

$$+ \sum_{|\alpha|\leq P+Q} \sup_{x,y} |x,y|^{N+M+\delta+|\alpha|-\varepsilon}|D^\alpha_{x,y}\chi(x,y)|\}$$

<u>Proof</u> Let $\chi \in \mathcal{D}(\mathbb{R}^N \times \mathbb{R}^M)$ with support in an open-bounded ball containing supp. $T \times$ supp. S.

$$|\int T(x)S(y)\chi(x,y)dxdy| = |\int T(x)dx\int S(y)\phi(x,y)dy|$$

$$\leq C\{|\int S(y)dy\int\chi(x,y)dx| + \sum_{|\alpha|\leq P} \sup_{x\in\mathbb{R}^N}|x|^{\omega+|\alpha|}|\int S(y)D_x^\alpha\chi(x,y)dy|\}$$

$$\leq CD\{|\int\chi(x,y)dxdy| + \sum_{|\beta|\leq Q} \sup_{y\in\mathbb{R}^M}|y|^{\lambda+|\beta|}|\int D_y^\beta\,\chi(x,y)dx|$$

$$+ \sum_{|\alpha|\leq P} \sup_{x\in R^N} |x|^{\omega+|\alpha|}|\int D_x^\alpha\,\chi(x,y)dy| +$$

$$+ \sum_{\substack{|\alpha|\leq P\\|\beta|\leq Q}} \sup_{\substack{x\in\mathbb{R}^N\\y\in\mathbb{R}^M}}|x|^{\omega+|\alpha|}|y|^{\lambda+|\beta|}|D_x^\alpha D_y^\beta\,\chi(x,y)|\}\ ;$$

$$\sup_{y\in\mathbb{R}^M}|y|^{\lambda+|\beta|}|\int D_y^\beta\,\chi(x,y)dx| \leq \text{const.}\,\varepsilon^{-1}\sup_{\substack{y\in\mathbb{R}^M\\x\in\mathbb{R}^N}}|y|^{\lambda+|\beta|}|x|^{N-\varepsilon}|D_y^\beta\,\chi(x,y)|$$

$$\leq \text{const.}\ \varepsilon^{-1}\sup_{\substack{y\in\mathbb{R}^M\\x\in\mathbb{R}^N}} |x,y|^{\lambda+N-\varepsilon+|\beta|}|D^\beta\chi(x,y)|$$

for any ε such that $0 < \varepsilon \leq N$. The terms

$$\sup_{x \in \mathbb{R}^N} |x|^{\omega+|\alpha|} \left| \int D_x^\alpha \chi(x,y)dy \right|$$

can be treated similarly, and the result follows.

The result can be extended to the case $\chi \in \mathcal{S}(\mathbb{R}^N \times \mathbb{R}^M)$ by noticing that $\langle T \otimes S, \chi\rangle = \langle T \otimes S, h\chi\rangle$ for a fixed $h \in \mathcal{D}(\mathbb{R}^{N+M})$ and using the remarks at the beginning of Section 2.2.

2.3 Study of $D^\alpha \tilde{\omega} * \phi$

In this subsection, $\tilde{\omega}$ is a tempered distribution over $\mathbb{R}^N$, $N \geq 2$ with the following property:

$$\tilde{\omega} = \sum_{\substack{\beta \\ |\beta|=1}} D^\beta \chi_\beta$$

where, for each β (with $|\beta| = 1$), χ_β is the distribution defined by a function over $\mathbb{R}^N$ (also denoted by χ_β) ; this function is assumed to be $\mathcal{C}^\infty$ everywhere except at 0, and homogeneous of degree $-N+1$, i.e.,

$$\chi_\beta(\rho q) = \rho^{-N+1} \chi_\beta(q) \quad \text{for all } q \neq 0 \quad \text{and } \rho > 0$$

As a consequence, in the complement of the origin

$$|D^\alpha \chi_\beta(q)| < \frac{C_\alpha}{|q|^{N+|\alpha|-1}} \quad ,$$

$$|D^\alpha \tilde{\omega}(q) < \frac{B_\alpha}{q^{N+|\alpha|}}$$

This implies , in particular, that χ_β is locally absolutely integrable.

We wish to investigate, for $\phi \in \mathcal{S}(\mathbb{R}^N)$, the ($\mathcal{C}^\infty$) function obtained by the convolution $D^\alpha \tilde{\omega} * \phi(q) = \tilde{\omega} * D^\alpha \phi(q)$

$$\begin{aligned} &= \sum_{|\beta|=1} \chi_\beta * D^{\alpha+\beta} \phi(q) = \\ &= \int \sum_{|\beta|=1} \chi_\beta(q-q')D^{\alpha+\beta} \phi(q')dq' \end{aligned} \tag{35}$$

and we assume, in the following, $q \neq 0$.

The integral (35) can be broken into two terms by dividing the range of integration into $\{q':|q'| \leq 1/2|q|\}$ and $\{q':|q'| \geq 1/2|q|\}$. In the first of these regions we have:

$$|q'| \leq 1/2|q| \ , \quad |q-q'| \geq 1/2 \ |q| > 0$$

Hence,

$$\int_{|q'|\leq 1/2|q|} \sum_{|\beta|=1} \chi_\beta(q-q')D^{\alpha+\beta}\phi(q')dq' =$$

$$= \int_{|q'|\leq 1/2|q|} D^\alpha \tilde{\omega}(q-q')\phi(q')dq' \quad +$$

$$+ \int_{|q'|=1/2|q|} d\sigma(q') \sum_{|\beta|=1} \frac{q'^\beta}{|q'|} \chi_\beta(q-q')D^\alpha\phi(q') +$$

$$+ \sum_{\substack{\rho+\sigma+\gamma=\alpha \\ |\rho|=1}}' \int_{|q'|=1/2|q|} d\sigma(q') \frac{q'^\rho}{|q'|} D^\sigma \tilde{\omega}(q-q')D^\gamma\phi(q')$$

[The symbol Σ' means that not all values of ρ,σ,γ such that $\rho + \sigma + \gamma = \alpha$, $|\rho| = 1$ are actually present in the sum; $d\sigma(q')$ is the surface element of the sphere $\{q':|q'| = 1/2|q|\}$.]
We shall first deal with the surface terms:

$$\left| \int_{|q'|=1/2|q|} d\sigma(q') \frac{q'^\rho}{|q'|} D^\sigma \tilde{\omega}(q-q')D^\gamma\phi(q') \right| \leq$$

$$\leq \int_{|q'|=1/2|q|} d\sigma(q')|q-q'|^{-N-|\sigma|}|D^\gamma\phi(q')| \leq$$

$$\leq \text{const.}|q|^{-|\sigma|-r-1}(1+|q|)^{-s} \sup_{q'}|q'|^r(1+|q'|)^s|D^\gamma\phi(q')| =$$

$$= \text{const.}|q|^{-r-|\alpha-\gamma|}(1+|q|)^{-s} \times \sup_{q'} |q'|^r(1+|q'|)^s|D^\gamma\phi(q')|$$

for all $s \geq 0$ and any real r ; and, similarly

$$\left| \int_{|q'|=1/2|q|} d\sigma(q') \sum_{|\beta|=1} \frac{q'^\beta}{|q'|} \chi_\beta(q-q')D^\alpha\phi(q') \right| \leq$$

$$\leq \text{const.}|q|^{-r}(1+|q|)^{-s} \sup_{q'} |q'|^r(1+|q'|)^s|D^\alpha\phi(q')|$$

for all $s \geq 0$ and any real r . The term

$$\int_{|q'|\le 1/2|q|} D^\alpha \tilde{\omega}(q-q')\phi(q')dq'$$

may be written

$$D^\alpha\tilde{\omega}(q) \int_{|q'|\le 1/2|q|} \phi(q')dq' \; +$$

$$+ \int_{|q'|\le 1/2|q|} [D^\alpha\tilde{\omega}(q-q') - D^\alpha\tilde{\omega}(q)]\phi(q')dq'$$

Here, the first term is bounded in modulus by

$$|q|^{-N-|\alpha|} \left| \int_{|q'|\le 1/2|q|} \phi(q')dq' \right|$$

This, in turn, is majorized by

$$|q|^{-N-|\alpha|} \left| \int \phi(q')dq' \right| + |q|^{-N-|\alpha|} \int_{|q'|\ge 1/2|q|} |\phi(q')|dq' \le$$

$$\le |q|^{-N-|\alpha|} \left| \int \phi(q')dq' \right| + |q|^{-N-|\alpha|} [\sup_{q'} |q'|^r |\phi(q')|] \times$$

$$\times \int_{|q'|\ge 1/2|q|} \frac{dq'}{|q'|^r} \le \frac{C}{r-N} |q|^{-r-|\alpha|} \sup_{q'} |q'|^r |\phi(q')| \; +$$

$$+ |q|^{-N-|\alpha|} \left| \int \phi(q')dq' \right|$$

for any $r > N$.

The modulus of the second term is bounded by

$$\left| \int_0^1 dt \int_{|q'|\le 1/2|q|} \sum_{|\beta|=1} (-q')^\beta D^{\alpha+\beta}\tilde{\omega}(q-tq')\phi(q')dq' \right| \le$$

$$\le C \int_0^1 dt \int_{|q'|\le 1/2|q|} |q'|\; |q-tq'|^{-N-|\alpha|-1} |\phi(q')|dq' \le$$

$$\le C \int_{|q'|\le 1/2|q|} |q|^{-N-|\alpha|-1} |q'|\; |\phi(q')|dq' \le$$

$$\le C|q|^{-N-|\alpha|-1} \left[\int_{|q'|\le 1/2|q|} dq' |q'|^{\varepsilon-N} \right] \sup_{q''} |q''|^{N+1-\varepsilon} |\phi(q'')|$$

for any $\varepsilon > 0$,

$$\leq \frac{C}{1-\delta} |q|^{-N-|\alpha|-\delta} \sup_{q'} |q'|^{N+\delta} |\phi(q')|$$

for any $\delta < 1$.

On the other hand, if $r < N$,

$$\left| \int_{|q'|\leq 1/2|q|} D^{\alpha}\widetilde{\omega}(q-q')\phi(q')dq' \right| < C \int_{|q'|\leq 1/2|q|} |q|^{-N-|\alpha|} \frac{|q'|^r}{|q'|^r} |\phi(q')|dq' \leq$$

$$\leq C|q|^{-N-|\alpha|} [\sup_{q''}|q''|^r|\phi(q'')|] \int_{|q'|\leq 1/2|q|} |q'|^{-r}dq'$$

$$\leq \frac{\text{Const}}{N-r} |q|^{-r-|\alpha|} \sup_{q'}|q'|^r|\phi(q')|$$

We now turn to the contributions of the region $\{q':|q'| > 1/2|q|\}$:

$$\left| \int_{|q'|\geq 1/2|q|} \chi_{\beta}(q-q')D^{\alpha+\beta} \phi(q')dq' \right| \leq$$

$$\leq C \int_{|q'|\geq 1/2|q|} |q-q'|^{-N+1}|D^{\alpha+\beta} \phi(q')|dq' \leq$$

$$\leq C \int_{|q'|\geq 1/2|q|} |q-q'|^{-N+1}|q'|^{-r-1}|q'|^{r+1}|D^{\alpha+\beta} \phi(q')|dq' \leq$$

$$\leq C \sup_{q''}|q''|^{r+1}|D^{\alpha+\beta}\phi(q'')| \int_{|q'|\geq 1/2|q|} dq'|q-q'|^{-N+1}|q'|^{-r-1},$$

for any $r > 0$.

Note that since $|q'| > 1/2|q| > 0$ in the range of integration of the last integral, this integral is well defined; by homogeniety it is equal to const. $|q|^{-r}$.

A more precise evaluation shows that this integral is bounded by $C/r|q|^{-r}$ where $C > 0$ is independent of r. Thus, the quantity we consider is bounded by

$$C/r|q|^{-r} \sup_{q'}|q'|^{r+1}|D^{\alpha+\beta}\phi(q')| \quad \text{for any } r > 0 .$$

We can now reassemble all the pieces of information we have gathered as follows:

1) if $N + 1 > r \geq N$, we have

$$|q|^{r+|\alpha|}|D^\alpha \widetilde{\text{Ш}}*\phi(q)| \leq$$

$$\leq \frac{C}{N+1-r} \sum_{|\gamma|\leq|\alpha|+1} \sup_{q'} |q'|^{r+|\gamma|}|D^\gamma\phi(q')| \tag{36}$$

$$+ C|q|^{r-N}|\int_{|q'|\leq 1/2|q|} \phi(q')dq'|$$

2) if $0 < r < N$, we have

$$|q|^{r+|\alpha|}|D^\alpha \widetilde{\text{Ш}}*\phi(q)| \leq$$

$$\leq \frac{C}{r(N-r)} \sum_{|\gamma|\leq|\alpha|+1} \sup_{q'} |q'|^{r+|\gamma|}|D^\gamma\phi(q')| \tag{37}$$

3) if $r > N$, say $r = N + \delta$ with $0 < \delta < 1$, (36) yields:

$$|q|^{N+\delta+|\alpha|}|D^\alpha \widetilde{\text{Ш}}*\phi(q)| \leq C|\int\phi(q')dq'| +$$

$$+ \frac{C}{\delta(1-\delta)} \sum_{|\gamma|\leq|\alpha|+1} \sup_{q'} |q'|^{N+\delta+|\gamma|}|D^\gamma\phi(q')| .$$

Finally, we note that if $h \in \mathcal{S}(\mathbb{R}^N)$ we have

$$\int h(q)\widetilde{\text{Ш}}*\phi(q)dq = \int \phi(q')\psi(q')dq'$$

where $\psi(q') = \int\widetilde{\text{Ш}}(q-q')h(q)dq$ is a $\mathcal{E}^\infty$ bounded function over $\mathbb{R}^N$.

Let now F be a tempered distribution over $\mathbb{R}^N$, which, for simplicity, we shall take with compact support; let F have the adiabatic property at 0 , i.e., let us assume that, for some $P \geq 0$, $1 > \delta > 0$, $C > 0$, and for all $\phi \in \mathcal{S}$:

$$|\langle F,\phi\rangle| \leq C\{|\int\phi(q)dq| + ||\phi||_{\delta,P}\} \tag{38}$$

Then $\widetilde{\text{Ш}}*F$ also has the adiabatic property at 0 . Indeed, first note that $\langle F,\phi\rangle = \langle F,h\phi\rangle$ where $h \in \mathcal{D}(\mathbb{R}^N)$ is equal to 1 on the support of F and, hence

$$|\langle F,\phi\rangle| \leq C'\{|\int\phi(q)h(q)dq| + ||\phi||_{\delta,P}\} \tag{39}$$

Applying the preceding inequalities, in the case $r = N + \delta$, [$\widetilde{\sqcup\!\sqcup}(q)$ being replaced there by $\check{\sqcup\!\sqcup}(q) = \widetilde{\sqcup\!\sqcup}(-q)$, which has the same properties] we obtain:

Lemma 7 Let F be a distribution over $\mathbb{R}^N$ satisfying the bound (39) with $0 < \delta < 1$ [for some $h \in \mathcal{S}(\mathbb{R}^N)$]. Then, there exists a constant K_δ (depending on δ) such that

$$|<\widetilde{\sqcup\!\sqcup}*F,\phi>| = |<F,\check{\sqcup\!\sqcup}*\phi>| \leq K_\delta[|\int \phi(q)dq| + ||\phi||_{\delta,P+1}]$$

2.4 Adiabatic Limits of Derivatives

We first consider a distribution F with compact support over $\mathbb{R}^N$ such that

$$|<F,\phi>| \leq C \sum_{|\alpha|\leq M} \sup_q |q|^{N+1+\delta+|\alpha|} |D^\alpha\phi(q)| , \quad \delta > 0 . \tag{40}$$

Clearly, this implies that the adiabatic limit of F at 0 is 0. But it also implies that, for $|\beta| = 1$

$$|<D^\beta F,\phi>| = |<F,D^\beta\phi>| \leq$$

$$\leq C \sum_{|\alpha|\leq M} \sup_q |q|^{N+1+\delta+|\alpha|} |D^{\alpha+\beta}\phi(q)|$$

$$< C \sum_{|\alpha|<M+1} \sup_q |q|^{N+\delta+|\alpha|} |D^\alpha\phi(q)|$$

Thus, $D^\beta F$ also has a zero adiabatic limit at 0. Conversely, suppose $\{T^\beta : |\beta| = 1\}$ is a family of distributions having adiabatic limits at 0 (which we may assume to be 0) and satisfying

$$D^\gamma T^\beta - D^\beta T^\gamma = 0 \quad \text{for any } \beta \text{ and } \gamma \text{ with } |\beta| = |\gamma| = 1,$$

$$|<T^\beta,\phi>| \leq C \sum_{|\alpha|\leq M} \sup_q |q|^{N+\delta+|\alpha|} |D^\alpha\phi(q)| \tag{41}$$

We wish to show that

$$T = \int_0^1 dt \sum_{|\beta|=1} q^\beta T^\beta(tq)$$

makes sense as a distribution, i.e., that, for any $\phi \in \mathcal{S}$,

$$\psi(q) = \int_0^1 \frac{dt}{t^{N+1}} q^\beta \phi(\tfrac{q}{t}) \quad (|\beta| = 1)$$

is an admissible test function for T^β .

Indeed, we have, for $q \neq 0$

$$|q|^{N+\delta+|\alpha|}|D^\alpha\psi(q)| = |\int_0^1 \frac{dt}{t^{N+1}}[\alpha_j(D^{\alpha-\beta}\phi)(\tfrac{q}{t})t^{-|\alpha|+1} +$$

$$+ q^\beta(D^\alpha\phi)(\tfrac{q}{t})t^{-|\alpha|}]|\ |q|^{N+\delta+|\alpha|} \leq$$

$$\leq \int_0^1 dt[|\tfrac{q}{t}|^{N+|\alpha|+\delta}|\alpha|t^\delta|(D^{\alpha-\beta}\phi)(\tfrac{q}{t})| +$$

$$+|\tfrac{q}{t}|^{N+|\alpha|+\delta+1}t^\delta|(D^\alpha\phi)(\tfrac{q}{t})|] \leq$$

$$\leq |\alpha| \sup_q |q|^{N+|\alpha-\beta|+1+\delta}|D^{\alpha-\beta}\phi(q)| +$$

$$+ \sup_q |q|^{N+|\alpha|+\delta+1}|D^\alpha\phi(q)|$$

As a consequence, T exists and satisfies:

$$|<T,\phi>| \leq C' \sum_{|\alpha|\leq M} \sup_q |q|^{N+1+\delta+|\alpha|}|D^\alpha\phi(q)|$$

Now let T' be any solution of

$$D^\beta T' = T^\beta \qquad \text{for all } \beta \text{ with } |\beta| = 1$$

then $T' - T$ is a constant. Hence T' also has an adiabatic limit $T'(0)$ at 0, which is precisely equal to $T' - T$. Finally, the result of this discussion can be expressed as follows:

Let T' be a distribution whose first derivatives $D^\beta T' = T^\beta$ have a zero adiabatic limit at 0, i.e., satisfy (41). Then 1) T' has an adiabatic limit $T'(0)$ at 0; 2) $T'-T'(0)$ satisfies:

$$|<T'-T'(0),\phi>| \leq C' \sum_{|\alpha|\leq M} \sup_q |q|^{N+1+\delta+|\alpha|}|D^\alpha\phi(q)|$$

Of course, if instead of assuming that the T^β have a zero adiabatic limit at 0 we assume that they have non-vanishing adiabatic limits, denoted $T^\beta(0)$, we find:

$$|<T'-T'(0) - \sum_{|\beta|=1} q^\beta T^\beta(0),\phi>| \leq C' \sum_{|\alpha|\leq M} \sup_q |q|^{N+1+\delta+|\alpha|}|D^\alpha\phi(q)|$$

Conversely, suppose

$$|\langle T,\phi\rangle| \leq C[|\int \phi(q)dq| + \sum_{|\beta|=1} |\int q^{\beta}\phi(q)dq| + $$
$$+ \sum_{|\alpha|\leq M} \sup_q |q|^{N+1+\delta+|\alpha|} |D^{\alpha}\phi(q)|]$$

Then, for $|\gamma| = 1$

$$|\langle D^{\gamma}T,\phi\rangle| = |\langle T,D^{\gamma}\phi\rangle| \leq C[\sum_{|\beta|=1} |\int q^{\beta}D^{\gamma}\phi(q)dq| +$$
$$+ \sum_{|\alpha|\leq M} \sup_q |q|^{N+1+\delta+|\alpha|} |D^{\alpha+\gamma}\phi(q)|] \leq$$
$$\leq C'[|\int \phi(q)dq| + \sum_{|\alpha|\leq M+1} \sup_q |q|^{N+\delta+|\alpha|} |D^{\alpha}\phi(q)|]$$

3. INDUCTION HYPOTHESIS

We shall recall that the "chronological products" $T^{r_1 \ldots r_n}(x_1,\ldots,x_n)$ of $\mathcal{L}^{(r_1)}(x_1),\ldots,\mathcal{L}^{(r_n)}(x_n)$ have been constructed in I by induction on n . The first step is to obtain the difference

$$A(Y;j) - R(Y;j) = D(Y;j)$$

(where $j \in \{1,\ldots,n\}$ and $Y = \{1,\ldots,j-1,j+1,\ldots,n\}$) in terms of already constructed quantities. Here we shall need the fact that D(Y;j) can be expressed in terms of already known "connected" operators, namely the Steinmann monomials for a smaller number of fields $\mathcal{L}^{(r)}(x)$. Indeed, from the Steinmann arrow calculus (see I, Section VIII) we have, for example if $Y = \{1,\ldots,n-1\}$, j=n,

$$D(Y;n) = AY;n) - R(Y;n) \equiv Y\uparrow n - Y\downarrow n =$$
$$= 1\uparrow 2\uparrow \ldots \uparrow n - 1\downarrow 2\downarrow \ldots \downarrow n = 1\uparrow 2\uparrow 3\uparrow \ldots \uparrow n - 1\downarrow 2\uparrow 3\uparrow \ldots \uparrow n + 1\downarrow 2\uparrow 3\uparrow \ldots \uparrow n$$
$$- 1\downarrow 2\downarrow 3\downarrow \ldots \downarrow n = [1,2\uparrow \ldots \uparrow n] + 1\downarrow [2\uparrow \ldots \uparrow n - 2\downarrow \ldots \downarrow n] =$$
$$= \sum_{p=1}^{n-1} \sum_{\substack{I\cup J = \{1,\ldots,p-1\} \\ I\cap J = \emptyset}} [I\downarrow p, J\downarrow (p+1)\uparrow \ldots \uparrow n] \tag{42}$$

If we insert intermediate states in these products and take vacuum expectation values, we see that the vacuum state contribution vanishes and we only have to deal with "connected" products of "connected" operators.

For any GRP, or chronological, or anti-chronological product of the fields $\mathcal{L}^{(r_1)}(x_1),\dots,\mathcal{L}^{(r_n)}(x_n)$, a formula of the type [I, (35)] holds. For example, if we denote (as in I, Section VIII) $\mathcal{A}^S(\mathcal{L}^{(r_1)}(x_1),\dots,\mathcal{L}^{(r_n)}(x_n))$ the GRP corresponding to the cell S we have

$$\mathcal{A}^S(\mathcal{L}^{(0)}(x_1)\dots\mathcal{L}^{(0)}(x_n)) = \\ = \sum_r (\Omega,\mathcal{A}^S(\mathcal{L}^{(r_1)}(x_1),\dots,\mathcal{L}^{(r_n)}(x_n))\Omega) \;:\frac{A(X)^r}{r!}: \tag{43}$$

Thus, the operator on the left-hand side is given by the various kernels appearing on the right-hand side. Denote

$$K(q_1,\dots,q_n) = \\ = \int e^{i\sum_{j=1}^{n} q_j x_j}(\Omega,\mathcal{A}^S(\mathcal{L}^{(r_1)}(x_1),\dots,\mathcal{L}^{(r_n)}(x_n))\Omega)dx_1\dots dx_n \tag{44}$$

The expression $K(q_1+p_{I_1}-p'_{J_1},\dots,q_n+p_{I_n}-p'_{J_n})$ will be called a (connected) kernel with n vertices, ℓ' incoming external lines ℓ outgoing external lines: here

$$\begin{aligned} &I_1,\dots,I_n \quad \text{is a partition of } \{1,\dots,\ell\} \\ &J_1,\dots,J_n \quad \text{is a partition of } \{1,\dots,\ell'\} \\ &|I_k| + |J_k| = r_k \quad \text{for } k=1,\dots,n, \\ &p_I = \sum_{j\in I} p_j, \quad p'_J = \sum_{j\in J} p'_j, \\ &\ell+\ell' = \sum_{k=1}^{n} r_k \end{aligned} \tag{45}$$

[Of course $K(q_1,\dots,q_n)$ is of the form $\delta(q_1+\dots+q_n)F(q_1,\dots,q_{n-1})$.]

We shall say that the kernel

$$\delta(q_1+\dots+q_n+p_1+\dots+p_\ell-p'_1-\dots-p'_{\ell'})F(q_1+p_{I_1}-p'_{J_1},\dots,q_{n-1}+p_{I_{n-1}}-p'_{J_{n-1}})$$

possesses the <u>property</u> A <u>with omitted variable</u> q_n if the following conditions are satisfied.

There exist: an open neighbourhood V_1 of the set

$$\{q_1,\ldots,q_{n-1},p'_1,\ldots,p'_{\ell'}: q_1 = q_2 = \ldots = q_{n-1} = 0,$$
$$p'^2_j = m^2,\ p'^0_j > 0,\ 1 \le j \le \ell'\};$$

an open neighbourhood V_2 of

$$\{Q \in \mathbb{R}^4, p_1,\ldots,p_\ell : Q = 0,\ p_j^2 = m^2, p_j^0 > 0,\ 1 \le j \le \ell\};$$

positive numbers $\lambda > 0$, $M \geq 0$ and, for each $R > 0$, and sufficiently small $\delta > 0$, a constant $K_\delta > 0$ such that:

if ψ is a C^∞ function of $q_1,\ldots q_{n-1}, p'_1,\ldots,p'_{\ell'}$, with support in $V_1 \cap \{q,p' : |q_j| < R,\ 1 \le j \le n-1,\ |p'_k| < R,$ $1 \le k \le \ell'\}$ then

$$E(Q,p) = \int \delta(Q + \sum_{j=1}^{\ell} p_j - \sum_{k=1}^{\ell'} p'_k) F(q+\hat{p}-\hat{p}') \prod_{j=1}^{\ell'} d\sigma(p'_j) \cdot$$
$$\cdot\ \psi(q_1,\ldots,q_{n-1}; p'_1,\ldots,p'_{\ell'}) dq_1 \ldots dq_{n-1}$$

is Hölder continuous in Q and $p = (p_1,\ldots,p_\ell)$ in V_2 [here we have used the abbreviations:

$$q = q_1,\ldots,q_{n-1}$$
$$\hat{p} = \hat{p}_1,\ldots,\hat{p}_{n-1}, \qquad \hat{p}_j = p_{I_j}$$
$$\hat{p}' = \hat{p}'_1,\ldots,\hat{p}'_{n-1}, \qquad \hat{p}'_j = p_{J_j}$$
$$d\sigma(k) = \delta(k^2-m^2)\theta(k^0)d^4k].$$

Moreover, (for suffiencently small $\delta > 0$) and $|Q| < R$, $|p_j| < R$

$$H^{\lambda\delta}_{Q,p}(E(Q,p)) \le K_\delta[H^\delta_{p'}(\int\psi(q;p')dq) +$$

$$+ \sum_{\substack{\alpha \\ |\alpha|\le M}} \sup_q |q|^{4(n-1)+|\alpha|+\lambda\delta} H^\delta_{p'}(D^\alpha_q \psi(q;p'))] \tag{46}$$

Note that here q stands for $(q_1,\ldots,q_{n-1})$ and $|q|^2 = \sum_{j=1}^{n-1} |q_j|^2$.

We shall say that a kernel $K(q_1 + p_{I_1} - p'_{J_1},\dots,q_n + p_{I_n} - p'_{J_n})$ [where $K(q_1,\dots,q_n) = \delta(q_1+\dots+q_n)F(q_1,\dots,q_{n-1})$] has the property B if there exist: an open neighbourhood V_1 of

$$\{q_1,\dots,q_n,p'_1,\dots,p'_{\ell'}:q_1=\dots=q_n = 0,\ p'^2_j = m^2,\ p'^0_j > 0, 1\le j\le \ell'\}$$

a neighbourhood V_2 of

$$\{p_1,\dots,p_\ell:\ p_j^2 = m^2,\ p_j^0 > 0\ ,\ 1 \le j \le \ell\}$$

numbers $\lambda > 0$ and $M \ge 0$, and for each $R > 0$ and sufficiently small $\delta > 0$, a constant C_δ such that if $\psi(q_1,\dots,q_n;p'_1,\dots,p'_{\ell'})$ is a $\mathcal{C}^\infty$ function with support in

$$V_1 \cap \{q,p':\ |q_j| < R,\ 1 \le j \le n,\ |p'_k| < R,\ 1 \le k \le \ell'\}$$

then

$$E'(p) = \int K(q_1+p_{I_1}-p'_{J_1}\ ,\dots,q_n + p_{I_n} - p'_{J_n}) \prod_{j=1}^{\ell'} d\sigma(p'_j)$$
$$\psi(q_1,\dots,q_n;p'_1,\dots,p'_{\ell'})dq_1\dots dq_n$$

is a Hölder continuous function of p in V_2 , satisfying

$$H_p^{\lambda\delta}(E'(p)) \le C_\delta[H_{p'}^{\delta}(\int \psi(q;p')dq_1\dots dq_n) + \\ + \sum_{|\alpha|\le M} \sup_q |q|^{4n+\lambda\delta+|\alpha|}\ H_{p'}^{\delta}(D_q^\alpha\psi(q;p'))] \tag{47}$$

Note that here q stands for $(q_1,\dots,q_n)$ and $|q|^2 = \sum_{j=1}^{n} |q_j|^2$.

We intend to prove (by induction on the number n of vertices) that the kernels defined by (43) all have the property B; the kernels having exactly one incoming and one outgoing external line ($\ell = \ell' = 1$) will be called self-energy kernels; we shall also prove by induction that all the kernels defined by (43), except self-energy kernels possess the property A . As for the self-energy kernels $K_2(p,p',q_1,\dots,q_n)$ they may be either of the form:

$$K(q_1+p,\ q_2,\dots,q_{n-1},q_n-p') = \delta(\sum_{j=1}^{n} q_j+p-p')S(q_1+p,q_2,\dots,q_{n-1}) \tag{48}$$

(calling q_1 the vertex carrying the outgoing line p and q_n the vertex carrying the ingoing line p') or of the form

$$K(q_1,\ldots,q_{n-1},q_n+p-p') = \delta(\sum_{j=1}^{n} q_j+p-p')S(q_1,\ldots,q_{n-1}) \quad (49)$$

(in case the two external lines are attached to the same vertex q_n). For such kernels we shall prove (by induction on n) that, provided correct mass and wave function renormalizations have been performed, the distributions S have the following property (C).

1) Property C_1 In the first case, there exist a neighbourhood V_1 of $\{q_1,\ldots,q_{n-1} = 0\}$, a positive number $M \geq 0$, and given any $\delta > 0$ sufficiently small, any $R > 0$ and any $p \in \mathbb{R}^4$ with $p^2 < 4m^2$, positive numbers $A(p) \geq 0$, $B(p) \geq 0$, $K_\delta(p) \geq 0$ such that if ψ is a $\mathcal{C}^\infty$ function of $q_1,\ldots,q_{n-1}$ with support in $V_1 \cap \{q: |q_j| \leq R, \ 1 \leq j \leq n-1\}$, then

$$E(p) = \int S(q_1+p,q_2,\ldots,q_{n-1})\psi(q_1,\ldots,q_{n-1})dq_1\ldots dq_{n-1}$$

satisfies

$$|E(p)| \leq A(p)\ |\int\phi(q)dq| + B(p) \sum_{\substack{\beta\\|\beta|=1}} |\int q^\beta\psi(q)dq| +$$

$$+ K_\delta(p) \sum_{|\alpha|\leq M} \sup_q |q|^{4(n-1)+\delta+1+|\alpha|}|D^\alpha\psi(q)| . \quad (50)$$

$$(q = (q_1,\ldots,q_{n-1}))$$

2) Property D In the same case, $A(p)$ and $B(p)$ vanish on the mass hyperboloid, i.e., for $p^2 = m^2$, $p^0 > 0$: in other words, if $p^2 = m^2$, $p^0 > 0$ then

$$|E(p)| \leq K_\delta(p) \sum_{|\alpha|\leq M} \sup_q |q|^{4(n-1)+\delta+1+|\alpha|}|D^\alpha\psi(q)|$$

Note that, according to Lemma 3 (50) implies the Hölder continuity of $E(p)$ in $\{p:p^2 < m^2\}$.

3) Property C_2 In the second case, there exist similar objects V_1, M, K_δ such that, if ψ is $\mathcal{C}^\infty$ with support in

$$V_1 \cap \{q, |q_j| < R, \ \text{all } j \leq n-1\},$$

$$|\int S(q_1,\ldots,q_{n-1})dq_1\ldots dq_{n-1}\psi(q_1,\ldots,q_{n-1})| \leq$$

$$\leq K_\delta \sum_{|\alpha|\leq M} \sup_q |q|^{4(n-1)+1+\delta+|\alpha|}|D^\alpha\psi(q)| \quad (51)$$

(Actually, the proof shows, and needs, the fact that all these inequalities can be extended, in the obvious way, to Banach-space-valued test functions. We neglect this point here for notational reasons.)

Properties D and C_2 clearly express the correct mass and wave function renormalizations necessary for the existence of the adiabatic limit.

We shall frequently appeal in the following to Lemma 8.

Lemma 8 Let $S(q_1,\dots,q_{n-1})$ be a tempered distribution on $\mathbb{R}^{4(n-1)}$. Assume that there exist a neighbourhood V_1 of the origin in $\mathbb{R}^{4(n-1)}$, positive numbers $R > 0$, $M \geq 0$, $\delta > 0$, $(0 < \delta < 1)$, and K_δ such that, for every $p \in R^4$ with $p^2 = m^2$, $p^0 > 0$, $|p| < 2R$ and for every $\psi \in \mathcal{S}(\mathbb{R}^{4(n-1)})$ with support in $V_1 \cap \{q = q_1,\dots, q_{n-1} : |q| < R\}$, denoting

$$E(p;\psi) = \int S(p+q_1,q_2,\dots,q_{n-1})\psi(q_1,\dots,q_{n-1})dq_1\dots dq_{n-1} ,$$

$$|E(p;\psi)| \leq K_\delta \sum_{|\alpha|\leq M} \sup_q |q|^{4(n-1)+1+\delta+|\alpha|} |D^\alpha\psi(q)|$$

Let

$$F(p;\chi) = \int \delta((p-Q)^2-m^2)\theta(p^0-Q^0)S(p-Q+q_1,q_2,\dots,q_{n-1}) \chi(q,Q)dq\, d^4Q$$

[where $q_1 = (q_1,\dots,q_{n-1})$]. Suppose χ is a $\mathcal{C}^\infty$ function of q and Q with support in

$$\{q,Q:\ q \in V_1, |q| < R, |Q| < R\}$$

Then $F(p;\chi)$ is Hölder continuous in p in the open set

$$\{p : |p| < R\}$$

where it satisfies

$$H_p^{\tau\delta}(F(p,\chi)) \leq C_\delta \sum_{|\alpha|\leq M+1} \sup_{Q,q} |Q,q|^{4n+\lambda\delta+|\alpha|} |D^\alpha\chi(q,Q)|$$

(Here C_δ is a positive constant depending on δ, τ and λ; τ and λ are any two numbers such that $\tau + \lambda < 1$, $\tau \geq 0$, $\lambda > 0$.)

Proof

$$F(p;\chi) = \int \delta((p-Q)^2-m^2)\theta(p^0-Q^0)dQ\ V(p-Q,Q),$$

$$V(r,Q) = \int S(r+q_1,\dots,q_{n-1})\chi(q,Q)dq$$

$$|F(p;\chi)| \le \sup_{\substack{Q,r \\ |Q|<R \\ |r|<2R \\ r^2=m^2,\, r^0>0}} |Q|^{3-\varepsilon\delta}\, V(r,Q)| \times$$

$$\times \int_{|Q|<R} \frac{d^3\vec{Q}}{2m|\vec{Q}|^{3-\varepsilon\delta}}$$

The last integral is $(2\pi/\varepsilon m)R^{\varepsilon\delta}$. But, according to our hypotheses

$$|V(r,Q)| \le K_\delta \sum_{|\alpha|\le M} \sup_q |q|^{4(n-1)+1+\delta+|\alpha|}|D_q^\alpha\ \chi(q,Q)|\ ,$$

$$|F(p;\chi)| < \frac{K'_\delta}{\varepsilon\delta} \sum_{|\alpha|\le M} \sup_{\substack{Q,q \\ |Q|<R \\ |q|<R}} |q|^{4(n-1)+1+\delta+|\alpha|}|Q|^{3-\varepsilon\delta}|D_q^\alpha\chi(q,Q)|$$

$$|F(p;\chi)| < \frac{K'_\delta}{\varepsilon\delta} \sum_{|\alpha|\le M} \sup_{Q,q} |q,Q|^{4n-\varepsilon\delta+\delta+|\alpha|}|D^\alpha\chi(q,Q)|$$

$$(0 < \varepsilon < 1)$$

Since this inequality holds in the convex set $\{p:|p| < R\}$, and since

$$F(p+h;\chi) = F(p;\chi_h)\ ,\ \chi_h(q,Q) = \chi(q,Q-h)$$

we can apply Lemma 4 and obtain

$$H_p^{\tau\delta}(F(p;\chi)) \le C_\delta\{\sum_{|\alpha|\le M+1} \sup_{Q,q} |q,Q|^{4n+\lambda\delta+|\alpha|}|D^\alpha\chi(q,Q)|\}$$

where $\tau + \lambda = 1-2\varepsilon$, $\lambda > 0$.

Lemma 9 Let $S(q_1,\dots,q_{n-1})$ be a tempered distribution on $\mathbb{R}^{4(n-1)}$. Assume that there exists a neighbourhood V_1 of 0 in $\mathbb{R}^{4(n-1)}$, positive numbers $R>0$, $M\ge 0$, $\delta > 0$, $(0 < \delta < 1)$, and K_δ such that for every $\psi \in \mathcal{S}(\mathbb{R}^{4(n-1)})$ with support in

$V_1 \cap \{q: |q| < R\}$,

$$\left| \int S(q_1,\ldots,q_{n-1})\psi(q_1,\ldots,q_{n-1})dq_1\ldots dq_{n-1}\right| \le$$

$$\le K_\delta \sum_{|\alpha|\le M} \sup_q |q|^{4(n-1)+1+\delta+|\alpha|} |D^\alpha\psi(q)|$$

Let
$$F(p;\chi) = \int \delta((p-Q)^2-m^2)\theta(p^0-Q^0)S(q_1,\ldots,q_{n-1}) \chi(q_1,\ldots,q_{n-1},Q)dq_1\ldots dq_{n-1}\, d^4Q$$

Suppose χ is a $\mathcal{C}^\infty$ function of q and Q with support in

$$\{q,Q: q\in V_1,\ |q| < R\ ,\ |Q| < R\}$$

Then $F(p;\chi)$ is Hölder continuous in p in $\{p:|p| < R\}$, where it satisfies

$$H_p^{\tau\delta}(F(p;\chi)) \le C_\delta \sum_{|\alpha|\le M+1} \sup_{Q,q} |Q,q|^{4n+\lambda\delta+|\alpha|} |D^\alpha\chi(q,Q)|$$

Here $C_\delta > 0$ depends on δ, τ, λ; τ and λ are any two numbers such that $\tau+\lambda < 1$, $\tau \ge 0$, $\lambda > 0$.

Proof The proof is a simpler version of the proof of the preceding lemma.

4. PASSAGE FROM n-1 TO n

The inductive proof involves several steps. The first is

Step 1

Lemma 10 (Phase space lemma) Let ϕ be a function defined on $\mathbb{R}^{4\ell}$ in the neighbourhood of

$$\{p = \{p_1,\ldots,p_\ell\} : p_j^2 = m^2\ ,\ p_j^0 > 0\ ,\ 1 \le j \le \ell\}$$

with $\ell \ge 2$, and taking its value in a Banach space. The norm of a vector u in this Banach space will be denoted $B(u)$.
Let

$$F_\ell(Q;\phi) = \int \delta(Q-\sum_{j=1}^{\ell} p_j)\{\prod_{j=1}^{\ell} \delta(p_j^2-m^2)\theta(p_j^0)d^4p_j\}\phi(p_1,\ldots,p_\ell)$$

Then, for all $Q \in \mathbb{R}^4$ and every $h \in R^4$ with $|h| \leq m$, and for $0 < \delta \leq 1/2$,

$$B(F_\ell(Q+h;\phi) - F_\ell(Q;\phi)) \leq C_\ell |Q|^{2(\ell-1)} |\frac{h}{m}|^\delta H^\delta_{p;B}(\phi(p)) \quad (52)$$

The proof is straightforward and tedious and will not be given here.

This lemma shows our assertion is true for $n = 1$. Indeed in that case $(\Omega, T^r(x)\Omega) = (\Omega, \mathcal{L}^{(r)}(x)\Omega)$ is only different from 0 if $r = \nu$; in this case it is equal to $\nu!$ Thus, for ℓ outgoing and ℓ' incoming external lines, $E(Q,p)$ simply becomes:

$$E(Q,p) = \nu! \int \delta(Q + \sum_{j=1}^{\ell} p_j - \sum_{k=1}^{\ell'} p'_k)\psi(p') \prod_{k=1}^{\ell'} d\sigma(p'_k) =$$

$$= F_{\ell'}(Q+P;\psi) \ , \ P = \sum_{j=1}^{\ell} p_j \ .$$

We always assume $\nu = \ell + \ell' \geq 3$; if $\ell' < 2$, $E(Q,p)$ is then zero in a neighbourhood of $\{Q,p : Q = 0,\ p_j^2 = m^2,\ p_j^0 > 0\}$ by energy momentum conservation. For $\ell' \geq 2$, Lemma 10 shows that property A holds.

Step 2

As a second step we shall verify that if two kernels have the property A, then a "connected product" of these two kernels also possesses this property. The q variables attached to the first kernel (resp. the second kernel) are denoted $(q_1,\ldots,q_n)$ [resp. $(q_{n+1},\ldots q_{n+m})$]. If a vertex of the product kernel has an external line, then this vertex also has an external line in one of the original kernels and can be an "omitted variable" for this kernel. We can always assume that this variable, the omitted variable for the product kernel is either q_n or q_{n+m}. (See Fig. 1.) The proof will only be outlined; in particular, we omit the (straightforward) considerations concerning the supports and the neighbourhoods V_1, V_2, etc., which occur in this proof.

First case: In this first case (omitted variable = q_n) we have to study

$$E(Q,p) = \int \delta(Q+P-P')F(q_1+p_{I_1}-p'_{J_1}-k_{L_1},\ldots,q_{n-1}+p_{I_{n-1}}-p'_{J_{n-1}}-k_{L_{n-1}})$$
$$\prod_{j=1}^{s} d\sigma(k_j)\delta(q_{n+1}+\ldots+q_{n+m} + P_2 - P'_2 + \sum_{j=1}^{s} k_j)$$

$$G(q_{n+1}+p_{I_{n+1}}+k_{L'_1}-p'_{J_{n+1}},\dots,q_{n+m-1}+p_{I_{n+m-1}}+k_{L'_{m-1}}-p'_{J_{n+m-1}})$$

$$\prod_{j=1}^{\ell'} d\sigma(p'_j)\psi(q_1,\dots,q_{n-1},q_{n+1},\dots,q_{n+m};p'_1,\dots,p'_{\ell'})$$

$$dq_1\dots dq_{n-1}\, dq_{n+1}\dots dq_{n+m}$$

Here $I_1,\dots,I_n$, $I_{n+1},\dots,I_{n+m}$ are disjoint sets of indices whose union is $\{1,\dots,\ell\}$; $J_1,\dots,J_n$, $J_{n+1},\dots,J_{n+m}$ are disjoint sets of indices whose union is $\{1,\dots,\ell'\}$. $L_1,\dots,L_n$ (resp. $L'_1,\dots,L'_m$) are disjoint sets of indices whose union is $\{1,\dots,s\}$. The abbreviations

$$p_I = \sum_{j\in I} p_j\ ,\ p'_J = \sum_{j\in J} p'_j\ ,\ k_L = \sum_{j\in L} k_j\ ,\ P = \sum_{j=1}^{\ell} p_j\ ,$$

$$P' = \sum_{j=1}^{\ell'} p'_j$$

have been used;

$$P_2 = \sum_{j=n+1}^{n+m} p_{I_j}\ ;\ P'_2 = \sum_{j=n+1}^{n+m} p'_{J_j}\ ;\ P_1 = P-P_2\ ,\ P'_1 = P'-P'_2\ .$$

To simplify the notation we shall now denote

$$q = (q_1,\dots,q_{n-1}) \quad q' = (q_{n+1},\dots,q_{n+m-1}) \quad p = (p_1,\dots,p_\ell)$$

$$\hat{p} = (p_{I_1},\dots,p_{I_{n-1}})\ ,\ \check{p} = (p_{I_{n+1}},\dots,p_{I_{n+m-1}})$$

$$\hat{p}' = (p'_{J_1},\dots,p'_{J_{n-1}})\ ,\ \check{p}' = (p'_{J_{n+1}},\dots,p'_{J_{n+m-1}})$$

$$\hat{k} = (k_{L_1},\dots,k_{L_{n-1}}),\ \check{k} = (k_{L'_1},\dots,k_{L'_{m-1}})$$

(Note that $\hat{k}$ and $\check{k}$ are not independent!). Furthermore, let

$$I_1\cup\dots\cup I_n = \{1,\dots,t\}\ ,\ I_{n+1}\cup\dots\cup I_{n+m} = \{t+1,\dots,\ell'\}$$

we denote $p'' = \{p'_1,\dots,p'_t\}$, $p''' = \{p'_{t+1},\dots,p'_{\ell'}\}$

We now make a change of variables by replacing the variable q_{n+m} by

$$Q_2 - \sum_{j=n+1}^{n+m-1} q_j$$

and denote

$$\chi(q,q',Q_2;p') = \psi(q,q',Q_2 - \sum_{j=n+1}^{n+m-1} q_j;p')$$

Then,

$$E(Q,p) = \int \delta(Q+P-P')F(q+\hat{p}-\hat{p}'-\hat{k}) \prod_{j=1}^{s} d\sigma(k_j)$$

$$\delta(Q_2+P_2-P_2' + \sum_{j=1}^{s} k_j)G(q'+\check{p}-\check{p}'+\check{k}) \prod_{j=1}^{\ell'} d\sigma(p_j')$$

$$\chi(q,q',Q_2;p')dq\ dq'\ dQ_2$$

$$E(Q,p) = \int \delta(Q-Q_2+P_1-P_1' - \sum_{j=1}^{s} k_j)F(q+\hat{p}-\hat{p}'-\hat{k})$$

$$\left[\prod_{j=1}^{s} d\sigma(k_j)\right] \left[\prod_{j=1}^{t} d\sigma(p_j')\right] dq_1 \ldots dq_{n-1}\ dQ_2$$

$$U(Q_2;q,Q_2;p'',k;p)$$

where

$$U(Q_2;q,Q_3;p'',k;p) =$$

$$= \int \delta(Q_2+P_2+ \sum_{j=1}^{s} k_j-P_2')G(q'+\check{p}-\check{p}'-\check{k}) \prod_{j=t+1}^{\ell'} d\sigma(p_j') \tag{53}$$

$$\chi(q,q',Q_3;p')dq'$$

$U(Q_2;q,Q_3;p'',k;p)$ is Hölder continuous in Q_2, p, k, p'', and since the second kernel has the property A (with q_{n+m} omitted), for all sufficiently small $\delta > 0$,

$$H^{\kappa\delta}_{Q_2,p'',p,k} (D_q^{\beta}\ U(Q_2;q,Q_3;p'',k;p)) \leq$$

$$\leq K_\delta[H^{\delta}_{p'}(\int D_q^{\beta}\ \chi(q,q',Q_3;p')dq') \ +$$

$$+ \sum_{|\alpha| \leq M_2} \sup_{q'} |q'|^{4(m-1)+\kappa\delta+|\alpha|} H^{\delta}_{p'}(D^{\alpha}_{q'} D^{\beta}_{q}\chi(q,q',Q_3;p'))]$$

where κ is a certain constant, $0 < \kappa < 1$. When χ is $\mathfrak{C}^\infty$, U is $\mathfrak{C}^\infty$ in q, Q_3. Note also that

$$\int D_q^\beta U(Q_2;q,Q_3;p'',k;p)dQ_3$$

is obtained by replacing in (53), $\chi(q,q',Q_3;p')$ by $\int D^\beta \chi(q,q',Q_3;p')dQ_3$. As a consequence, for all sufficiently small

$$H^{\kappa\delta}_{Q_2,p'',p,k}\left(\int D_q^\beta U(Q_2;q,Q_3;p'',k;p)dQ_3\right) \leq$$

$$\leq K_\delta[H^\delta_{p'}\left(\int D_q^\beta \chi(q,q',Q_3;p')dq'\, dQ_3\right) +$$

$$+ \sum_{|\alpha|\leq M_2} \sup_{q'} |q'|^{4(m-1)+\kappa\delta+|\alpha|} H^\delta_{p'}\left(\int D^\alpha_{q'} D^\beta_q \chi(q,q',Q_3;p')dQ_3\right)]$$

Let

$$V(p,Q;Q_2;Q_3) = \int \delta(Q-Q_2+P_1-P_1' - \sum_{j=1}^{s} k_j)$$

$$F(q+\hat{p}-\hat{p}'-\hat{k}) \left[\prod_{j=1}^{t} d\sigma(p'_j)\right] \left[\prod_{j=1}^{s} d\sigma(k_j)\right]$$

$$U(Q_2;q,Q_3;p'',k;p)$$

This is a Hölder continuous function of p, Q, Q_2, and there is a constant τ $(0 < \tau < \kappa)$ and, for each sufficiently small $\delta > 0$, a constant K'_δ such that

$$H^{\tau\delta}_{Q_2,p,Q}(V(p,Q;Q_2;Q_3)) \leq$$

$$\leq K'_\delta\ [H^{\kappa\delta}_{Q_2,P'',p,k}\left(\int U(Q_2;q,Q_3;p'',k;p)dq\right) +$$

$$+ \sum_{|\beta|\leq M} \sup_q |q|^{4(n-1)+\tau\delta+|\beta|} H^{\kappa\delta}_{Q_2,p''p,k}(D^\beta_q U(Q_2;q,Q_3;p'',k;p))]$$

Note again that $\int U(Q_2;q,Q_3;p'',k,p)dq$ is obtained by replacing in (53), $\chi(q,q',Q_3;p')$ by $\int \chi(q,q',Q_3;p')dq$. From the preceding estimates it follows that

$$H^{\tau\delta}_{Q_2,p,Q}(V(p,Q;Q_2;Q_3)) \leq K''_\delta[H^\delta_{p'}(\int \chi(q,q',Q_3;p')dqdq') +$$

$$+ \sum_{|\alpha|\leq M_2} \sup_{q'} |q'|^{4(m-1)+\kappa\delta+|\alpha|} H^\delta_{p'}(\int D^\alpha_{q'}\chi(q,q',Q_3;p')dq)$$

$$+ \sum_{|\beta|\leq M_1} \sup_{q} |q|^{4(n-1)\tau\delta+|\beta|} H^\delta_{p'}(\int D^\beta_{q}\,\chi(q,q',Q_3;p')dq')$$

$$+ \sum_{\substack{|\alpha|\leq M_2\\ |\beta|\leq M_1}} \sup_{q,q'} |q'|^{4(m-1)+\kappa\delta+|\alpha|}|q|^{4(n-1)+\tau\delta+|\beta|}$$

$$H^\delta_{p'}(D^\beta_q D^\alpha_{q'}\,\chi(q,q',Q_3;p'))] = B(Q_3) \qquad (54)$$

A similar calculation shows that

$$H^{\tau\delta}_{p,Q}(\int V(p,Q;0;Q_3)dQ_3)$$

is bounded by the quantity obtained by replacing, in (54) $\chi(q,q',Q_3;p')$ by $\int \chi(q,q',Q_3;p')dQ$. We have

$$E(Q,p) = \int V(p,Q;Q_2;Q_2)dQ_2 =$$

$$= \int [V(p,Q;Q_2;Q_2) - V(p,Q;0;Q_2)]dQ_2 + \int V(p,Q;0;Q_3)dQ_3$$

A bound has already been found for the last integral. Denoting

$$W(p,Q;Q_2,Q_3) = V(p,Q;Q_2;Q_3) - V(p,Q;0;Q_3)$$

the Hölder continuity of $V(p,Q:Q_2;Q_3)$ in p, Q, Q_2 yields:

$$H^{\frac{\tau}{2}\delta}_{p,Q}(W(p,Q;Q_2;Q_3)) \leq |Q_2|^{\frac{\tau}{2}\delta}B(Q_3)$$

[B being defined by (54). Hence

$$H^{\frac{\tau}{2}\delta}_{p,Q}(\int W(p,Q;Q_2;Q_2)dQ_2) \leq$$

$$\leq \int |Q_2|^{\frac{\tau}{2}\delta}B(Q_2)dQ_2 \leq \frac{C}{\varepsilon} \sup_{Q_2} |Q_2|^{4+\frac{\tau}{2}\delta-\varepsilon} B(Q_2)$$

We choose $\varepsilon = (\tau/4)\delta$. Finally, putting together all the bounds we obtain:

$$H^{\frac{\tau}{2}\delta}_{p,Q}(E(Q,p)) \leq K'''_{\delta}\{\sup_{Q_2} |Q_2|^{4+\frac{\tau}{4}\delta} B(Q_2) +$$

$$+ H^{\delta}_{p'}(\int \chi(q,q',Q_2;p')dqdq'dQ_2)$$

$$+ \sum_{|\alpha|\leq M_2} \sup_{q'} |q'|^{4(m-1)+\tau\delta+|\alpha|}$$

$$H^{\delta}_{p'}(\int D^{\alpha}_{q'}\chi(q,q',Q_2;p')dQ_2 dq) +$$

$$+ \sum_{|\beta|\leq M_1} \sup_{q} |q|^{4(n-1)+\tau\delta+|\beta|} H^{\delta}_{p'}(\int D^{\beta}_{q}\chi(q,q',Q_2;p')dQ_2 dq')$$

$$+ \sum_{\substack{|\alpha|\leq M_2\\ |\beta|\leq M_1}} \sup_{q,q'} |q'|^{4(m-1)+\tau\delta+|\alpha|}|q|^{4(n-1)+\tau\delta+|\beta|}$$

$$H^{\delta}_{p'}(\int D^{\alpha}_{q'}D^{\beta}_{q}\chi(q,q',Q_2;p')dQ_2)\}$$

We recall, from the proof of the tensor product rule, that if f is a function of $x \in \mathbb{R}^N$ and $y \in \mathbb{R}^M$, with compact support, we have

$$\sup_{x} |x|^{N+|\alpha|+\delta} |\int D^{\alpha}_{x} f(x,y)dy| \leq$$

$$\leq \text{const}\ \frac{1}{\varepsilon} \sup_{x,y} |x,y|^{N+M+\delta-\varepsilon+|\alpha|} |D^{\alpha}_{x} f(x,y)|$$

A repeated use of this device shows that there exists a constant λ with $0 < \lambda < 1$ such that for sufficiently small $\delta > 0$,

$$H^{\lambda\delta}_{p,Q}(E(Q,p)) \leq A_{\delta}\{H^{\delta}_{p'}(\int \chi(q,q',Q_2;p')dqdq'dQ_2)$$

$$+ \sum_{|\alpha|\leq M_1+M_2} \sup_{q,q',Q_2} |q,q',Q_2|^{4(m+n-1)+\lambda\delta+|\alpha|}$$

$$H^{\delta}_{p'}(D_{q,q'}\chi(q,q',Q_2;p'))\}$$

This, expressed in terms of ψ, is just the result required.

Second case: the omitted variable is q_{n+m} With the same notations as in the previous case, we have to study

$$E(Q,p) = \int \delta(Q+P+P')F(q+\hat{p}-\hat{p}-\hat{k})\delta(q_1+\ldots+q_n+P_1-P_1' - \sum_{j=1}^{s} k_j)$$

$$G(q'+\check{p}-\check{p}'+\check{k}) \prod_{j=1}^{s} d\sigma(k_j) \prod_{k=1}^{\ell'} d\sigma(p'_k)$$

$$\psi(q,q_n,q';p')dq\, dq_n dq'$$

We make the change variable

$$q_n = Q_1 - \sum_{j=1}^{n-1} q_j$$

and denote

$$\chi(q,q',Q_1;p') = \psi(q,Q_1 - \sum_{j=1}^{n-1} q_j,q';p')$$

Then,

$$E(Q,p) = \int \delta(Q_1+P_1-P_1'- \sum_{j=1}^{s} k_j)F(q+\hat{p}-\hat{p}'-\hat{k}) \prod_{j=1}^{s} d\sigma(k_j) \prod_{k=1}^{t} d\sigma(p_k')$$

$$dq\, dQ_1 \delta(Q-Q_1+P_2+ \sum_{j=1}^{s} k_j-P_2')G(q'+\check{p}-\check{p}'+\check{k})$$

$$\prod_{k=t+1}^{\ell'} d\sigma(p_k') \chi(q,q',Q_1;p')dq'$$

The principle is the same: we define

$$U(Q-Q_1,p;q,Q_3;p'',k) = \tag{55}$$

$$= \int \delta(Q-Q_1+P_2+ \sum_{j=1}^{s} k_j - P_2')G(q'+\check{p}-\check{p}'+\check{k}) \chi(q,q',Q_3;p')dq' \prod_{j=t+1}^{\ell'} d\sigma(p_j'),$$

$$V(p,Q;Q_1,Q_3) = \int \delta(Q_1+P_1-P_1'- \sum_{j=1}^{s} k_j)F(q+\hat{p}-\hat{p}'-\hat{k}) \prod_{j=1}^{s} d\sigma(k_j)$$

$$\prod_{k=1}^{t} d\sigma(p_k') U(Q-Q_1,p;q,Q_3;p'',k)dq, \tag{56}$$

$$W(p,Q;Q_1,Q_3) = V(p,Q;Q_1,Q_3) - V(p,Q;0,Q_3)$$

so that

$$E(Q,p) = \int W(p,Q;Q_1,Q_1)dQ_1 + \int V(p,Q;0,Q_3)dQ_3$$

We note that $V(p,Q;Q_1,Q_3)$ is Hölder continuous in p, Q <u>and</u> Q_1. The rest of the proof is quite similar to that of the preceding case.

<u>Step 3</u>

Now, let K_1 be a kernel (not of the self-energy type) having the property A (with respect to the omission of any q_j carrying an external line). Let $K_2(k;p';q_{n+1},\ldots,q_{n+m})$ be a self-energy kernel [with ingoing (resp. outgoing) external line p' (resp. k)], of the form either:

$$\delta(k-p' + \sum_{j=n+1}^{n+m} q_j)\, S\,(q_{n+1}+k, q_{n+2},\ldots,q_{n+m-1}) \tag{57}$$

or

$$\delta(k-p' + \sum_{j=n+1}^{n+m} q_j)\, S\,(q_{n+1},\ldots,q_{n+m-1}) \tag{58}$$

Assume that, in either case, K_2 possesses the property C. We wish to show that a product kernel obtained by connecting K_1 and K_2 by one line again possesses the property A. We have to consider two cases, according to which kernel comes first.

<u>First case: K_1 comes first</u> Note that (Fig. 2) the connecting line between the two kernels is an external line for the first kernel; it is attached to the vertex q_1, for example, which is thus an omissible variable for K_1. If q_1 carries another external line (which then remains an external line for the product kernel), it may be the omitted variable for the product kernel. Thus we have to distinguish three cases.

a) <u>The omitted variable for the product is not q_1 but belongs to K_1</u>

Suppose q_n is the omitted variable

$$E(Q,p) = \int \delta(Q+P-P')\, F\,(q_1+p_{I_1}-p'_{J_1}-k,\ldots,q_{n-1}+p_{I_{n-1}}-p'_{J_{n-1}})$$

$$d\sigma(k) \prod_{j=1}^{\ell'} d\sigma(p'_j) K_2(k;p'_{\ell'};q_{n+1},\ldots,q_{n+m})$$

$$\psi(q_1,\ldots,q_{n-1},q_{n+1},\ldots,q_{n+m};p')dq_1\ldots dq_{n-1}dq_{n+1}\ldots dq_{n+m}$$

Because K_2 contains

$$\delta(k - p + \sum_{j=n+1}^{n+m} q_j)$$

this can be rewritten, using the abbreviation

$$Q_2 = \sum_{j=n+1}^{n+m} q_j$$

as

$$E(Q,p) = \int \delta(Q+P-P')\, F\, (q_1+Q_2+p_{I_1}-p'_{J_1}-p'_{\ell'},\dots,q_{n-1}+p_{I_{n-1}} - p'_{J_{n-1}})$$

$$\prod_{j=1}^{\ell'} d\sigma(p'_j) K_2(k;p'_{\ell'};q_{n+1},\dots,q_{n+m}) d\sigma(k)$$

$$\psi(q_1,\dots,q_{n-1},q_{n+1},\dots,q_{n+m};p')dq_1\dots dq_{n-1}dq_{n+1}\dots dq_{n+m}$$

Denote

$$\chi(q_1,\dots,q_{n-1},q_{n+1},\dots,q_{n+m};p') =$$

$$= \psi(q_1-Q_2,\dots,q_{n-1},q_{n+1},\dots,q_{n+m};p')$$

we see that

$$E(Q,p) = \int \delta(Q-P-P')\, F\, (q_1+p_{I_1}-p'_{J_1}-p'_{\ell'},\dots,q_{n-1}+p_{I_{n-1}}-p'_{J_{n-1}})$$

$$\prod_{j=1}^{\ell'} d\sigma(p'_j)dq_1\dots dq_{n-1} \int K_2(k;p'_{\ell'};q_{n+1},\dots,q_{n+m})d\sigma(k)$$

$$\chi(q_1,\dots,q_{n-1},q_{n+1},\dots,q_{n+m};p')dq_{n+1}\dots dq_{n+m}$$

If K_2 is of the form (57), the second integral can be written

$$W(q;p') = V(q;p'_{\ell'};p')$$

where for $r \in \mathbb{R}^4$, we define

$$V(q;r;p') = \int dQ_2 \delta((r-Q_2)^2-m^2)\, \theta\, (r^0-Q_2^0) \quad \times$$

$$\times \int S(r-Q_2+q_{n+1},q_{n+2},\dots,q_{n+m-1})\, \chi_1(q,q',Q_2;p')dq'$$

Here

$$q = (q_1,\dots,q_{n-1}),\ q' = (q_{n+1},\dots,q_{n+m-1}), p' = (p'_1,\dots,p'_{\ell'}),$$

$$\chi_1(q,q',Q_2;p') = \chi(q_1,\dots,q_{n-1},q_{n+1},\dots,q_{n+m-1},Q_2 - \sum_{j=n+1}^{n+m-1} q_j;p')$$

By applying Lemma 8, we find that $V(q;r;p')$ is actually Hölder continuous in r in a neighbourhood of $\{r:r^2= m^2,\ r^0 > 0\}$, where it satisfies:

$$H^{\kappa\delta}_{p',r}(V(q;r;p')) \le C_\delta \sum_{|\alpha|\le M+1} \sup_{q',Q_2} |q',Q_2|^{4m+\kappa\delta+|\alpha|}$$

$$H^{\delta}_{p'}(D^{\alpha}_{q',Q_2}\chi_1(q,q',Q_2;p'))$$

As a consequence the same inequality holds for $H^{\kappa\delta}_{p'}(W(q;p'))$, since $W(q;p') = V(q;p'_{\ell'};p')$. Now

$$E(Q,p) = \int \delta(Q+P-P')\, F\,(q+\hat{p}-\hat{p}') \prod_{j=1}^{\ell'} d\sigma(p'_j)dq\ W(q;p')$$

and the result follows from property A of K_1.

If K_2 is of the form (58), the second integral can be written:

$$W(q;p') = \int dQ_2\, \delta((p'_{\ell'}-Q_2)^2-m^2)\, \theta\,(p'^0_{\ell'}-Q^0_2)\ \times$$

$$\times \int S(q_{n+1},\dots,q_{n+m-1})\chi_1(q,q',Q_2;p')dq'$$

where χ_1 is defined as in the preceding case, and the conclusion is the same (using Lemma 9).

b) <u>The omitted variable for the product is</u> q_1

We can then represent $K_1(q_1,\dots,q_n)$ as

$$\delta(\sum_{j=1}^{n} q_j)\, F\,(q_2,\dots,q_n)$$

Then,

$$E(Q,p) = \int \delta(Q+P-P')\, F\,(q_2+p_{I_2}-p'_{J_2},\dots,q_n+p_{I_n}-p'_{J_n}) \prod_{j=1}^{\ell'} d\sigma(p'_j)$$

$$dq_2\dots dq_n \int d\sigma(k)K_2(k;\, p'_{\ell'};q_{n+1},\dots,q_{n+m})$$

$$\psi(q_2,\dots,q_n,q_{n+1},\dots,q_{n+m};p')dq_{n+1}\dots dq_{n+m}$$

From this point on, the proof is identical to that of case a) (with χ replaced by ψ).

c) <u>The omitted variable for the product is</u> q_{n+m}

We represent K_1 as in case b) so that, if K_2 is of the form (57)

$$E(Q,p) = \int \delta(Q+P-P')\, F\,(q_2+p_{I_2}-p'_{J_2},\dots,q_n+p_{I_n}-p'_{J_n})$$

$$\delta(P-P'_1-k+\sum_{j=1}^{n} q_j)\, S\,(k+q_{n+1},\dots,q_{n+m-1}) \sum_{j=1}^{\ell'} d\sigma(p'_j)d\sigma(k)$$

$$\psi(q_1,\dots,q_n,q_{n+1},\dots,q_{n+m-1};p')dq_1\dots dq_{n+m-1}$$

Let us perform the change of variable

$$q_1 = Q_1 - \sum_{j=2}^{n} q_j$$

and denote

$$q = (q_2,\dots,q_n),\ \hat{p} = (p_{I_2},\dots,p_{I_n}),\ \hat{p}' = (p'_{J_2},\dots,p'_{J_n}),$$

$$q' = (q_{n+1},\dots,q_{n+m-1})$$

and

$$\chi(q,q',Q_1;p') = \psi(Q_1 - \sum_{j=2}^{n} q_j,q_2,\dots,q_n,q_{n+1},\dots,q_{n+m-1};p')$$

Then

$$E(Q,p) = \int \delta(Q+P-P')\, F\,(q+\hat{p}-\hat{p}') \prod_{j=1}^{\ell'} d\sigma(p')dq \quad \times$$

$$\times \int dQ_1\ \delta((p'_{\ell'}-Q+Q_1)^2-m^2)\ \theta\ (p'^0_{\ell'}-Q^0+Q_1^0) \quad \times$$

$$\times \int S(p'_{\ell'}-Q+Q_1+q_{n+1},\dots,q_{n+m-1})\ \chi\ (q,q',Q_1;p')dq'$$

Defining

$$V(q;r;p') = \int dQ_1\ \delta((r+Q_1)^2-m^2)\ \theta\ (r^0+Q_1^0)$$

$$\int S(r+Q_1+q_{n+1},\dots,q_{n+m-1})\ \chi\ (q,q',Q_1;q')dq'$$

we see, as in case a), that $V(q;r;p')$ is Hölder continuous in r. As a consequence (by property A of K_1)

$$E(Q,Q',p) = \int \delta(Q+P-P')\, F(q+\hat{p}-\hat{p}') \sum_{j=1}^{\ell'} d\sigma(p_j') \times$$

$$\times\; V(q;p_\ell',-Q';p')$$

is Hölder continuous in p,Q and Q'. Setting $Q' = Q$ we finally obtain the required bound for

$$H^{\lambda\delta}_{p,Q}(E(Q,p)).$$

The proof is similar if K_2 is of the form (58).

Second case: K_2 comes first The omitted variable may be (see Fig. 3) q_1, q_{n+1}, or a variable belonging to K_1 and different from q_1, say q_n.

a) The omitted variable for the product is $q_n \neq q_1$

$$E(Q,p) = \int S(p_1+q_{n+1},\dots,q_{n+m-1})\, \delta\Big(\big(p_1+\sum_{j=n+1}^{n+m} q_j\big)^2-m^2\Big)$$

$$\theta\big(p_1^0 + \sum_{j=n+1}^{n+m} q_j^0\big)\, \delta(Q+P-P')\, F\big(q_1+p_1+\sum_{j=n+1}^{n+m} q_j+p_{I_1}-p'_{J_1},\dots,$$

$$q_{n-1}+p_{I_{n-1}}-p'_{J_{n-1}}\big) \prod_{j=1}^{\ell'} d\sigma(p_j')dq_1\dots dq_{n-1}dq_{n+1}\dots dq_{n+m}$$

$$\psi(q_1,q_2,\dots,q_{n-1},q_{n+1},\dots,q_{n+m};p') \quad =$$

$$= \int S(p_1+Q_2+q_{n+1},q_{n+2},\dots,q_{n+m-1})\, \delta((p_1+Q_2)^2-m^2)$$

$$\theta(p_1^0+Q_2^0)dQ_2 dq' \int \delta(Q+P-P')$$

$$F(q_1+p_1+p_{I_1}-p'_{J_1},\dots,q_{n-1}+p_{I_{n-1}}-p'_{J_{n-1}}) \prod_{j=1}^{\ell'} d\sigma(p_j')$$

$$\psi(q_1-Q_2,q_2,\dots,q_{n-1},q_{n+1}+Q_2,q_{n+2},\dots, -\sum_{j=n+1}^{n+m-1} q_j;p')$$

$$dq\dots dq_{n-1}$$

Calling $\chi(q,q',Q_2;p') =$

$$= \psi(q_1-Q_2,\dots,q_{n-1},q_{n+1}+Q_2,\dots,q_{n+m-1}, -\sum_{j=n+1}^{n+m-1} q_j;p')$$

where
$$q = (q_1,\dots,q_{n-1}),\ q' = (q_{n+1},\dots,q_{n+m-1})$$

and
$$L(Q,p;q',Q_2) = \int \delta(Q+P-P')$$

$$F(q_1+p_1+p_{I_1}-p'_{J_1},\dots,q_{n-1}+p_{I_{n-1}}-p'_{J_{n-1}})$$

$$\prod_{j=1}^{\ell'} d\sigma(p'_j)dq_1\dots dq_{n-1}\ \chi(q,q',Q_2;p')$$

we can estimate the Hölder continuity of $L(Q,p;q',Q_2)$ by using the property A of K_1. Then, applying Lemma 8 to the expression

$$\int \delta((p_1+Q_2)^2-m^2)\ \theta\ (p_1^0+Q_2^0)\ S\ (q_{n+1}+p_1+Q_2,q_{n+2},\dots,q_{n+m-1})$$
$$L(Q,p;q',Q_2)dq'dQ_2$$

we easily find that the product kernel does have the property A with q_n omitted. The case when K_2 is of the form (58) is treated similarly, using Lemma 9.

b) <u>The omitted variable for the product is q_1</u>

In this case, we find [if K_2 is of the form (57)]

$$E(Q,p) = \int \delta((p_1+Q_2)^2-m^2)\ \theta\ (p_1^0+Q_2^0)dQ_2dq'$$
$$S(p_1+Q_2+q_{n+1},q_{n+2},\dots,q_{n+m-1}) \int \delta(Q+P-P')$$
$$F(q_2+p_{I_2}-p'_{J_2},\dots,q_n+p_{I_n}-p'_{J_n}) \prod_{j=1}^{\ell'} d\sigma(p'_j)$$
$$\chi(q,q',Q_2;p')dq$$

where
$$q_2= (q_2,\dots,q_n)\ ,\ q' = (q_{n+1},\dots,q_{n+m-1})$$

and
$$\chi(q,q',Q_2;p') =$$
$$= \psi(q_2,\dots,q_n,q_{n+1} + Q_2,q_{n+2},\dots, - \sum_{j=n+1}^{n+m-1} q_j;p')$$

and the end of the argument is similar to the preceding one.

Again the treatment is similar for the case when K_2 is of the form (58).

c) The omitted variable for the product is q_{n+1}

If K_2 is of the form (57)

$$E(Q,p) = \int S(p_1+Q-Q_1 - \sum_{j=n+2}^{n+m} q_j, q_{n+2},\ldots,q_{n+m-1})$$

$$\delta((p_1+Q-Q_1)^2-m^2)\theta(p_1^0+Q^0-Q_1^0)dQ_1\,dq_{n+2}\ldots dq_{n+m}$$

$$\int \delta(Q+P-P')\,F(q_2+p_{I_2}-p'_{J_2},\ldots,q_n+p_{I_n}-p'_{J_n}) \prod_{j=1}^{\ell'} d\sigma(p'_j)$$

$$\psi(Q_1 - \sum_{j=2}^{n} q_j, q_2,\ldots,q_n,q_{n+2},\ldots,q_{n+m})dq_2\ldots dq_n =$$

$$= \int S(p_1+Q-Q_1+q_{n+1},q_{n+2},\ldots,q_{n+m-1})$$

$$\delta((p_1+Q-Q_1)^2-m^2)\,\theta\,(p_1^0+Q^0-Q_1^0)dQ_1\,dq_{n+1}\ldots dq_{n+m-1}$$

$$\int F(q_2+p_{I_2}-p'_{J_2},\ldots,q_n+p_{I_n}-p'_{J_n}) \prod_{j=1}^{\ell'} d\sigma(p'_j)$$

$$dq_2\ldots dq_n\;\chi(q,q',Q_1;p')$$

where $q = q_2,\ldots,q_n$, $q' = q_{n+1},\ldots,q_{n+m-1}$, and

$$\chi(q,q',Q_1;p') = \psi(Q_1 - \sum_{j=2}^{n} q_j, q_2,\ldots,q_n,q_{n+2},\ldots,q_{n+m-1}, -\sum_{j=n+1}^{n+m-1} q_j;p')$$

The rest of the proof continues as in case b).

If K_2 is of the form

$$K_2(p;p';q_{n+1},\ldots,q_{n+m}) = \delta(p-p' + \sum_{j=n+1}^{n+m} q_j)\,S\,(q_{n+2},\ldots,q_{n+m})$$

[Note that we have changed the names of the variables from (58) in order to call q_{n+1} the variable carrying the external outgoing line], then

$$E(Q,p) = \int S(q_{n+2},\dots,q_{n+m})\delta((p_1+Q-Q_1)^2-m^2)$$

$$\theta(p_1^0+Q^0-Q_1^0)dq_{n+2}\dots dq_{n+m}dQ_1 \int \delta(Q+P-P')$$

$$F(q_2+p_{I_2}-p'_{J_2},\dots,q_n+p_{I_n}-p'_{J_n}) \prod_{j=1}^{\ell'} d\sigma(p'_j)$$

$$\psi(Q_1-\sum_{j=2}^{n} q_j,q_2,\dots,q_n,q_{n+2},\dots,q_{n+m};p')$$

$$dq_2\dots dq_n$$

Owing to the property A of K_1 and to Lemma 9 we see that, calling $p_1+Q = r$, $E(Q,p)$ is Hölder continuous in $Q,r,p_2,\dots,p_\ell$ with norms as required by the property A .

Step 4

We now verify that a connected product of two self-energy kernels possessing the properties C and D again possesses these properties.

First case: The two kernels are of type (48). (See Fig. 4.) Then, the product kernel is also of type (48), i.e.,

$$\delta(\sum_{j=1}^{n+m} q_j+p-p') \quad S(q_1+p,q_2,\dots,q_n,q_{n+1},\dots,q_{n+m-1})$$

with

$$S(q_1+p,q_2,\dots,q_{n+m-1}) =$$

$$= S_1(q_1+p,q_2,\dots,q_{n-1})\,\delta((p+\sum_{j=1}^{n} q_j)^2-m^2)\,\theta(p^0+\sum_{j=1}^{n} q_j^0)$$

$$S_2(p+\sum_{j=1}^{n} q_j+q_{n+1},q_{n+2},\dots,q_{n+m-1})$$

Thus

$$E(p) = \int S(q_1+p,q_2,\dots,q_{n+m-1})\,\psi(q_1,\dots,q_{n+m-1})dq_1\dots$$

$$\dots dq_{n+m-1}$$

$$= \int \delta((p+Q)^2-m^2)\,\theta(p^0+Q^0)\,S_1((p+Q)+q_1,q_2,\dots,q_{n-1})$$

$$S_2((p+Q+q_{n+1},q_{n+2},\dots,q_{n+m-1})\,\chi(q,q',Q)dq\,dq'\,dQ$$

where $q = (q_1,\dots,q_{n-1})$, $q' = (q_{n+1},\dots,q_{n+m-1})$ and

$$\chi(q,q',Q) = \psi(q_1+Q,q_2,\dots,q_{n-1}, -\sum_{j=1}^{n-1} q_j, q_{n+1},\dots,q_{n+m-1})$$

From here on, the proof is analogous to that of Lemma 8: denote

$$W(k,Q) = \int S_1(k+q_1,\dots,q_{n-1})S_2(k+q_{n+1},\dots,q_{n+m-1})$$

$$\chi(q,q',Q)dq\,dq'$$

Then

$$|E(p)| = |\int \delta((p+Q)^2-m^2)\,\theta(p^0+Q^0)\,W(p+Q,Q)dQ| \leq$$

$$\leq \int \frac{d^3\vec{Q}}{2m} \sup_{\substack{k \\ k^2=m^2 \\ k^0>0}} |W(k,Q)|$$

$$\leq \frac{C}{\varepsilon\delta} \sup_{\substack{k,Q \\ k^2-m^2,k^0>0}} |Q|^{3-\varepsilon\delta}|W(k,Q)|$$

But, by property D of S_1 and S_2, we have, for a certain $\delta > 0$

$$|E(p)| \leq \frac{C}{\varepsilon\delta} K_\delta \sum_{\substack{|\alpha|\leq M \\ |\alpha'|\leq M'}} \sup_{q,q',Q} |Q|^{3-\varepsilon\delta}|q|^{4(n-1)+\delta+1+|\alpha|}$$

$$|q'|^{4(m-1)+\delta+1+|\beta|}|D_q^\alpha D_{q'}^\beta \chi(q,q',Q)|$$

$$\leq K'_\delta \sum_{|\alpha|\leq M+M'} \sup_{q,q',Q} |q,q',Q|^{4(n+m-1)+(2-\varepsilon)\delta+1+|\alpha|}|D^\alpha\chi(q,q',Q)|$$

This shows that S has the property C_1 with $A(p) = B(p)=0$, and, in particular, property D .

<u>Second case</u>: The first kernel is of type (57), the second type (58). (See Fig. 5.) In this case

$$S(p+q_1,q_2,\dots,q_n,q_{n+1},\dots,q_{n+m-1}) =$$
$$= S_1(q_1+p,q_2,\dots,q_{n-1})\delta((p+\sum_{j=1}^{n} q_j)^2-m^2)\theta(p^0+\sum_{j=1}^{n} q_j^0) \times$$

$$\times S_2(q_{n+1},\ldots,q_{n+m-1}),$$

$$E(p) = \int S(p+q_1,\ldots,q_{n+m-1})\,\psi(q_1,\ldots,q_{n+m-1})dq_1\ldots dq_{n+m-1}$$

$$= \int \delta((p+Q)^2-m^2)\,\theta(p^0+Q^0)\,W((p+Q),Q)dQ\ ,$$

with

$$W(k,Q) = \int S_1(k+q_1,q_2,\ldots,q_{n-1})$$
$$S_2(q_{n+1},\ldots,q_{n+m-1})\,\chi(q,q',Q)dq\,dq'\ ,$$

$$q = \{q_1,\ldots,q_{n-1}\}\ ,\qquad q' = \{q_{n+1},\ldots,q_{n+m-1}\}\ ,$$

$$\chi(q,q',Q) = \psi(q_1+Q,q_2,\ldots,q_{n-1},-\sum_{j=1}^{n-1} q_j,q_{n+1},\ldots,q_{n+m-1})$$

and the rest of the proof follows the same line as in the first case, with the same conclusion [$A(p) = B(p) = 0$].

<u>Third case</u>: K_1 is of type (49), K_2 of type (48). (See Fig. 6.)

$$S(p+q_n,q_1,\ldots,q_{n-1},q_{n+1},q_{n+2},\ldots,q_{n+m-1}) =$$

$$= S_1(q_1,\ldots,q_{n-1})\,\delta((p+\sum_{j=1}^{n} q_j)^2-m^2)\,\theta(p^0+\sum_{j=1}^{n} q_j^0)\ \times$$

$$\times\ S_2(p+\sum_{j=1}^{n} q_j + q_{n+1},\ldots,q_{n+m-1})\ ,$$

$$E(p+) = \int \delta((p+Q)^2-m^2)\,\theta(p^0+Q^0)dQW((p+Q),Q)$$

$$W(k,Q) = \int S_1(q_1,\ldots,q_{n-1})S_2(k+q_{n+1},q_{n+2},\ldots,q_{n+m-1})$$

$$\chi(q,q',Q)dq\,dq',$$

$$q = (q_1,\ldots,q_{n-1})\ ,\quad q' = (q_{n+1},\ldots,q_{n+m-1})$$

$$\chi(q,q',Q) = \psi(q_1,\ldots,q_{n-1},Q-\sum_{j=1}^{n-1} q_j,q_{n+1},\ldots,q_{n+m-1})$$

and the rest of the proof is similar to that of the first case, with the same conclusion [A(p) = B(p) = 0] .

Fourth case: K_1 and K_2 are both type (49). (See Fig. 7.)

$$E(p) = \int \delta((p+Q)^2-m^2)\,\theta\,(p^0+Q^0)\,dQ\,W(Q) \ ,$$

$$W(Q) = \int S_1(q_1,\ldots,q_{n-1})S_2(q_{n+1},\ldots,q_{n+m-1})$$

$$\psi(q_1,\ldots,q_{n-1},Q-\sum_{j=1}^{n-1} q_j,q_{n+1},\ldots,q_{n+m-1})$$

$$dq_1\ldots dq_{n-1}\,dq_{n+1}\ldots dq_{n+m-1}$$

The conclusion is the same as in the first three cases [A(p) = B(p) = 0].

Step 5

The result of steps 2,3,4 is that if the kernels defined by (43) satisfy the induction hypothesis for a number of vertices $\leq n-1$, then the kernels corresponding to D(Y;n) [see (42)] possess the same property. It has been seen in I that A(Y;n) can be obtained from D(Y;n) as follows. Denote

$$(\Omega,A^r(x_1,\ldots,x_{n-1};x_n)\Omega) =$$

$$= \int e^{-i\sum_{j=1}^{n} x_j q_j}\,\delta(\sum_{j=1}^{n} q_j)\,a_n^r(q_1,\ldots,q_{n-1})\,dq_1\ldots dq_n$$

$$(\Omega,D^r(x_1,\ldots,x_{n-1};x_n)\Omega) = \tag{59}$$

$$= \int e^{-i\sum_{j=1}^{n} x_j q_j}\,\delta(\sum_{j=1}^{n} q_j)\,d_n^r(q_1,\ldots,q_{n-1})\,dq_1\ldots dq_n$$

Then, $a_n^r(q_1,\ldots,q_{n-1})$ is given, up to a polynomial in $q_1,\ldots,q_{n-1}$ of degree ω ,by

$$a_n^r(q) = \int dt(1-t)^{\omega} \sum_{\substack{\alpha\\|\alpha|=\omega+1}} \frac{\omega+1}{\alpha!}\,q^{\alpha}D^{\alpha}\widetilde{\amalg}(tq-q')\,d_n^r(q')\,dq' \tag{60}$$

where $\tilde{\omega}(q)$ is a certain distribution over $\mathbb{R}^{4(n-1)}$, homogeneous of degree $-4(n-1)$, and $\mathcal{C}^\infty$ except at the origin. Moreover,

$$\tilde{\omega}(q) = \sum_{\substack{\beta \\ |\beta|=1}} D^\beta \chi_\beta(q)$$

where χ_β is, for each β (with $|\beta| = 1$) a homogeneous distribution of degree $-4(n-1)+1$, $\mathcal{C}^\infty$ except at 0. The formula (60) makes sense because $d_n^r(q)$ is $\mathcal{C}^\infty$ (in fact vanishes) in a neighbourhood of 0.

Step 5 consists of examining how this operation preserves the properties postulated in the inductive hypothesis. Let F be a tempered distribution over $\mathbb{R}^N$ vanishing in

$$\{q \in \mathbb{R}^N, |q| < R\}, \quad (R > 0).$$

We wish to study the tempered distribution T symbolically defined by

$$T(q) = \int_0^1 dt(1-t)^\omega \int \sum_{|\alpha|=\omega+1} \frac{\omega+1}{\alpha!} q^\alpha D^\alpha \tilde{\omega}(tq-q')F(q')dq' \tag{61}$$

It can be shown that this formula makes sense if F vanishes in a neighbourhood of 0 and has a degree of growth at infinity $\leq \omega$ in the sense of Definition 2 of I (Section VI). Moreover

$$D^\alpha T(q) = \int D^\alpha \tilde{\omega}(q-q')F(q')dq'$$

for all α with $|\alpha| \geq \omega + 1$. (Note that $|D^\alpha \tilde{\omega}(q)| < C|q|^{-N-|\alpha|}$ for $q \neq 0$.) Let w be a $\mathcal{C}^\infty$ function such that:

$$0 \leq w \leq 1 \;;\; w(q) = 1 \quad \text{for} \quad |q| > R-\varepsilon R; w(q) = 0 \quad \text{for} \quad |q| < R-2\varepsilon R$$

(here $0 < 2\varepsilon < 1$). Then

$$T(q) = \int_0^1 dt(1-t)^\omega \sum_{|\alpha|=\omega+1} \int \frac{\omega+1}{\alpha!} q^\alpha D^\alpha \tilde{\omega}(tq-q')w(q')F(q')dq'$$

If $|q| < R - 2\varepsilon R$, $D^\alpha \tilde{\omega}(tq-q')w(q')$ is a $\mathcal{C}^\infty$ function of q' and $T(q)$ is the result of applying the distribution F to the test function

$$\psi(q') = \int_0^1 dt(1-t)^\omega \sum_{|\alpha|=\omega+1} \frac{\omega+1}{\alpha!} q^\alpha D^\alpha \widetilde{\sqcup\!\sqcup}(tq-q')w(q')$$

$D_q^\gamma D_{q'}^\beta \psi(q')$ is given by a sum of terms of the form:

$$\text{const.} \int_0^1 dt(1-t)^\omega t^{|\sigma|} q^{\alpha-\rho} D^{\alpha+\sigma+\tau} \widetilde{\sqcup\!\sqcup}(tq-q') D^\lambda w(q')$$

where $\rho + \sigma = \gamma$ and $\rho \leq \alpha, \tau + \lambda = \beta$.

If $|q| < R(1-2\varepsilon)$ this expression is bounded in modulus by

$$\text{const.} \int_0^1 dt \frac{|q|^{|\alpha|-|\rho|} |D^\lambda w(q')|}{|tq-q'|^{N+|\alpha|+|\sigma|+|\tau|}} \leq$$

$$\leq \text{const.} \frac{|q|^{|\alpha|-|\rho|} |D^\lambda w(q')|}{[|q'|-|q|]^{N+|\alpha|+|\sigma|+|\tau|}}$$

Moreover, if $|\lambda| \geq 1$, $D^\lambda w(q')$ vanishes for $|q'| > R(1-\varepsilon)$; if $|q| < R(1-3\varepsilon)$, we have

$$|q'|-|q| > \varepsilon R, \; |q'|-|q| \geq (1 - \frac{1-3\varepsilon}{1-2\varepsilon})|q'| = \frac{\varepsilon}{1-2\varepsilon} |q'|$$

Hence, $|q'|-|q| >$ const. $(1+|q'|)$, and for all values of λ , the above expression is bounded by:

$$\frac{C_\beta |q|^{|\alpha|-|\beta|}}{(1+|q'|)^{N+|\alpha|+|\beta|+|\sigma|}} \leq \frac{C'_\beta |q|^{(|\alpha|-|\gamma|)^+}}{(1+|q'|)^{N+|\alpha|+|\beta|}} =$$

$$= \frac{C'_\beta |q|^{(\omega+1-|\gamma|)^+}}{(1+|q'|)^{N+\omega+1+|\beta|}}$$

(here a^+ means $\max\{a,0\}$).

Now for all $\phi \in \mathcal{S}$, and for every sufficiently small $\varepsilon > 0$

$$|\langle F,\phi\rangle| \leq C(\varepsilon) \sum_{|\beta|\leq P} \sup_{q'} (1+|q'|)^{N+\omega+|\beta|+\varepsilon} |D^\beta \phi(q')|$$

Applying this to $\phi = \psi$, we find, for $|q| < R(1-3\varepsilon)$:

$$|D^{\gamma}T(q)| \leq \text{const.}\, |q|^{(\omega+1-|\gamma|)^{+}}$$

Suppose now that F vanishes in an open ball $\{q: |q-a| < R\}$ centered at some point $a \in \mathbb{R}^N$. Then

$$\int_0^1 dt(1-t)^{\omega} \int \sum_{|\alpha|=\omega+1} (q-a)^{\alpha} \frac{(\omega+1)}{\alpha!} D^{\alpha}\tilde{\omega}(t(q-a)+q'-a)F(q')dq'$$

is (by the previous argument) $\mathcal{C}^{\infty}$ in the open ball $\{q: |q-a| < R$ and has the same derivatives of order $\omega + 1$ as $T(q)$. Therefore, it differs from $T(q)$ by a polynomial of degree ω ; hence $T(q)$ is $\mathcal{C}^{\infty}$ in the ball $\{q: |q-a| < R\}$. We conclude that: if F vanishes in an open set $\mathcal{O}$, then T is $\mathcal{C}^{\infty}$ in $\mathcal{O}$. (This is a natural generalization of an obvious property of the Cauchy kernel.) Since it is easy to see that if F is $\mathcal{C}^{\infty}$ with compact support, T is $\mathcal{C}^{\infty}$ everywhere, it follows that, in general T is $\mathcal{C}^{\infty}$ in any open set where F is $\mathcal{C}^{\infty}$.

Now let h be a $\mathcal{C}^{\infty}$ function with compact support equal to one in a (bounded) open set $\mathcal{O}$. If we write F as $(1-h)F + hF$, the contribution of $(1-h)F$ to (61) is $\mathcal{C}^{\infty}$ in $\mathcal{O}$ by the preceding argument. As to the contribution of hF it can be written:

$$G(q) = \int [\tilde{\omega}(q-q') - \sum_{|\alpha|\leq\omega} \frac{q^{\alpha}}{\alpha!} D^{\alpha}\tilde{\omega}(-q')]\, F(q')h(q')dq'$$

(Indeed all the terms of the above expression make sense because hF has compact support and vanishes in a neighbourhood of 0 ; moreover it has the same derivatives of order $\omega + 1$ as G, as well as zero derivatives at 0 of order $\leq \omega$; hence, it coincides with G.) We note that

$$\begin{aligned} &\frac{q^{\alpha}}{\alpha!} \int D^{\alpha}\tilde{\omega}(-q')\, F(q')\, h(q')dq' = \\ &= \frac{q^{\alpha}}{\alpha!} \int D^{\alpha}\tilde{\omega}(-q')\, F(q')\, u(q')\, h(q')dq' \end{aligned} \tag{62}$$

where u is $\mathcal{C}^{\infty}$ and vinishes in a neighbourhood of 0. Since $u(q')h(q')D^{\alpha}\tilde{\omega}(-q')$ is a $\mathcal{C}^{\infty}$ function with compact support (62) is of the form Cq^{α}.

In conclusion we see that

$$T(q) = S(q) + \int \widetilde{\sqcup\!\sqcup}(q-q')\, h(q')\, F(q')dq' \tag{63}$$

where $S(q)$ is a tempered distribution which is $\mathcal{C}^\infty$ in the bounded open set $\mathcal{O}$ where $h(q') = 1$.

We now assume that the kernel

$$\delta\Big(\sum_{j=1}^{n} q_j + \sum_{j=1}^{\ell} p_j - \sum_{j=1}^{\ell'} p'_j\Big) F(q_1+p_{I_1}-p'_{J_1},\ldots,q_{n-1}+p_{I_{n-1}}-p'_{J_{n-1}})$$

has the property A (the omitted variable being q_n); furthermore we suppose that $F(q) = 0$ for $|q| < m$ $[q = (q_1,\ldots,q_{n-1})]$, and define T by (61), so that (63) holds for suitable h . The expression

$$\int \delta(Q+P-P')\, T(q+\hat{p}-\hat{p}') \prod_{j=1}^{\ell'} d\sigma(p'_j)\, \psi(q;p')dq \tag{64}$$

has two terms corresponding to the decomposition (63). The most interesting is

$$\int \delta(Q+P-P')\, F(q'+\hat{p}-\hat{p}')\, h(q'+\hat{p}-\hat{p}')\, \chi(q';p') \prod_{j=1}^{\ell'} d\sigma(p'_j)dq' \quad , \tag{65}$$

$$\chi(q';p') = \int \widetilde{\sqcup\!\sqcup}(q-q')\, \psi(q;p')dq$$

Suppose, for example, that the support of χ is contained in $\{p' : |\hat{p}'-\hat{p}'_0| < r\}$ and that p is chosen such that $|\hat{p}-\hat{p}_0| < r$; let h have support in $\{q : |q-\hat{p}_0+\hat{p}'_0| < 3r\}$ and let h take the value 1 in $\{q : |q-\hat{p}_0+\hat{p}'_0 < 2r\}$. Then, the support of $h(q'+\hat{p}-\hat{p}')\,\chi(q';p')$ is contained in $\{q' : |q'| < 5r\}$; moreover, if supp. $\psi \subset \{q,p' : |q| < r/2,\ |\hat{p}'_0-\hat{p}'| < r/2\}$ it is contained in the region $\mathcal{O} = \{q : |q+\hat{p}-\hat{p}'-\hat{p}_0+\hat{p}'_0| < 2r\}$ where $h(q+\hat{p}-\hat{p}') = 1$ and where $S(q+\hat{p}-\hat{p}')$ is $\mathcal{C}^\infty$.

Considering χ as a function of q' with values in the Hölder continuous functions of p', $h(q'+\hat{p}-\hat{p}')$ as a $\mathcal{C}^\infty$ function of q' with values in the Hölder continuous functions of p and p' and using the fact that (see Section 2.3) convolution with $\widetilde{\sqcup\!\sqcup}$ preserves the adiabatic property, we find (with evident conditions on the support of ψ):

$$H^{\lambda\delta}_{p,p'}(\int h(q'+\hat{p}-\hat{p}')\,\chi\,(q';p')dq') \;+$$

$$+ \sum_{|\alpha|\leq M} \sup_{q'} |q'|^{4(n-1)+\lambda\delta+|\alpha|} H^{\lambda\delta}_{p,p'}(D^{\alpha}_{q'}[h(q'+\hat{p}-\hat{p}')\,\chi\,(q';p')])$$

$$\leq C_\delta\,\{H^{\delta}_{p'}(\int \psi(q;p')dq) \;+$$

$$+ \sum_{|\alpha|\leq M+1} \sup_{q} |q|^{4(N-1)+\delta+|\alpha|}\, H^{\delta}_{p'}(D^{\alpha}_{q}\,\psi(q;p'))\}$$

which proves that this part of the kernel defined by T has the property A with the omitted variable q_n .

Let us consider now the contribution of $S(q)$ to (64):

$$\int \delta(Q+P-P')\,S\,(q+\hat{p}-\hat{p}')\,\psi\,(q;p')dq \prod_{j=1}^{\ell'} d\sigma(p'_j)$$

If, as we suppose, the support of ψ is contained in $\{q:|q| < R\}$, we may consider $\psi(q;p')$ as a $\mathcal{C}^\infty$ function of q with values in the Hölder continuous functions of p', and $S(q+\hat{p}-\hat{p}')$ as a $\mathcal{C}^\infty$ function of q with values in the Hölder continuous functions of p and p' . We have

$$H^{\delta}_{p,p'}(\int S(q+\hat{p}-\hat{p}')\,\psi\,(q;p')dq) \leq$$

$$\leq C_\delta\,\{H^{\delta}_{p'}(\int \psi(q;p')dq) + \sup_{q} |q|^{4(n-1)+\delta}\, H^{\delta}_{p'}(\psi(q;p'))\}$$

(same calculation as in 2.2).

Applying now Lemma 10 (phase space Lemma) we find that the kernel defined by T has the property A with omitted variable q_n provided the number ℓ' of its incoming external lines is at least 2. However, if $\ell' \leq 1$ and if T (and hence F) is not a self-energy kernel, it automatically has the property A (because it vanishes in a neighbourhood of the mass shell and $q = 0$). We now reach the following conclusion.

Let K be a kernel, not of the self-energy type, arising from $a^r_n(1,\ldots,n-1;n)$. The induction hypothesis for kernels with at most n-1 vertices implies that K has the property A with omitted variable q_n , provided q_n carries an external line. But, if q_j is another variable carrying an external line, K can be written as

$$K' + \sum_{k=1}^{s} R_k$$

where K' is a kernel (with the same vertices and external lines) arising from $a_n^r(i_1,\ldots,i_{n-1};j)$ (hence having the property A with q_j as the omitted variable, and, for each k, R_k is a connected product of two kernels with at most n-1 vertices, having the property A by Steps 2 and 3. Hence, K has the property A with the omitted variable being any variable q_j carrying at least one external line. The same argument yields the same conclusion for the vacuum expectation values of all the G.R.P. with n vertices.

The case of self-energy kernels

We have seen in Step 4 that if a self-energy type kernel K is the connected product of two self-energy kernels [with at most (n-1) vertices each], it follows from the induction hypothesis that it satisfies property C_1 [with $A(p) = B(p) = 0$] and D. If, on the other hand, K is the connected product of any two kernels not of the self-energy type, it vanishes in a neighbourhood of $p^2 < 4m^2$, $q = 0$, hence, it again satisfies C_1 and D $[A(p) = B(p) = 0]$ or C_2 . Hence, the kernels of the self-energy type arising from $(\Omega, D^r(Y;j)\Omega)$ either possess property C_1 , with $A(p) = B(p) = 0$, and, hence, property D, or property C_2 (here $|Y| + 1 = n$).

Now let

$$T(p+q_1,\ldots,q_{n-1})\,\delta\,(p + \sum_{j=1}^{n} q_j - p')$$

be a kernel of type (48) arising from $(\Omega, A^r(Y;n)\Omega)$, (where $Y = \{1,\ldots,n-1\}$). According to the above remarks

$$T(q_1,\ldots,q_{n-1}) \equiv T(q) =$$

$$= \int_0^1 dt\,(1-t)^{\omega} \int \sum_{|\alpha|=\omega+1} \frac{\omega+1}{\alpha!}\, D^{\alpha}\widetilde{\Box}\,(tq-q')\, F\,(q')dq' \; + \qquad (66)$$

$$+ \text{ polynomial of degree } \omega$$

where $F(p+q_1,\ldots,q_{n-1})$ satisfies the property C_1 with $A(p) = B(p) = 0.$ The same is, of course true for $F(q)h(q)$ if $h \in \mathcal{D}(\mathbb{R}^{4(n-1)})$.

Hence, if we perform the decomposition (53)

$$T(q) = S(q) + \int \widetilde{\sqcup\!\sqcup}(q-q')\,h(q')\,F(q')dq' \ ,$$

$$h \in \mathcal{S},\ h(q) = 0 \ \text{ for } \ |q| > 2R \ , \ h(q) = 1 \ \text{ for } \ |q| < \frac{3}{2} R \ ,$$

[possible because $F(q)$ vanishes for $|q| < M$], we see that:

1) $S(p+q_1,\ldots,q_{n-1})$ is infinitely differentiable for $|q| < R/2$, $|p| < R$. Hence, S and its derivatives of all orders have adiabatic limits in this region.

2) For $|\beta| = 1$, $D_q^\beta\,[h(p+q_1,q_2,\ldots,q_{n-1})\,F(p+q_1,q_2,\ldots,q_{n-1})]$ has an adiabatic limit at $q = 0$ for every p such that $p^2 < 4m^2$, uniformly in p in every compact there. It follows from our study of the convolution by $\widetilde{\sqcup\!\sqcup}$, that

$$D^\beta \widetilde{\sqcup\!\sqcup} * (hF)(p+q_1,\ldots,q_{n-1})$$

also has an adiabatic limit at $q = 0$, uniformly in p in every compact in $\{p:p^2 < 4m^2\}$.

Hence, the same is true for $D^\beta T(p+q_1,q_2,\ldots,q_{n-1})(|\beta| \leq 1)$. On the other hand, we know that, if $\phi \in \mathcal{S}(\mathbb{R}^4)$, β arbitrary

$$\int D^\beta T(p+q_1,q_2,\ldots,q_{n-1})\,\phi(p)d^4p \tag{67}$$

is an infinitely differentiable function of $q_1,\ldots,q_{n-1}$ in a neighbourhood of zero. The value at $q = 0$ of this expression is of the form:

$$\int G_+^\beta(p)\,\phi(p)\,dp$$

Here $G_+^\beta(p)$ is a tempered distribtuion, the boundary value from the forward tube of a function $L^\beta(k)$, holomorphic in

$$\{k \in \mathbb{C}^4 : k^2 \neq m^2 \quad \text{and} \quad k^2 \notin 4m^2+\mathbb{R}^+\}$$

Clearly, for $|\beta| \leq 1$, since $\int G_+^\beta(p)dp\phi(p)$ can be obtained by taking the adiabatic limit of (67) at $q = 0$, we have $G_+^\beta(p) =$ adiabatic limit of $D^\beta T(p+q_1,q_2,\ldots,q_{n-1})$ for $p^2 < 4m^2$ (hence G_+^β is Hölder continuous there).

The same argument holds for the corresponding kernel associated with $(\Omega, R(Y;n)\Omega)$ and its first derivatives; their adiabatic limits coincide with the boundary values $G_-^{\beta}(p)$ of $L^{\beta}(k)$ from the backward tube in the region $p^2 < 4m^2$. However, the two kernels we have considered differ by a sum of connected products of kernels of fewer vertices; hence, their adiabatic limits and those of their first derivatives coincide in the region $p^2 < 4m^2$. By the edge-of-the-wedge theorem, $L^{\beta}(k)$ is actually holomorphic in

$$\{k \in \mathbb{C}^4 : k^2 \notin 4m^2 + \mathbb{R}^+\}$$

and, in view of its polynomial boundedness

$$L^{\beta}(k) = \sum_{|\alpha| \leq N} c_{\alpha}^{\beta}(k^2) k^{\alpha}$$

where the $c_{\alpha}^{\beta}(z)$ are holomorphic in $\{z \in \mathbb{C} : z \notin 4m^2 + \mathbb{R}^+\}$ (we have applied the Bogoliubov-Vladmimirov theorem [3]; see also [4]).

Let $P(p+q_1, q_2, \ldots, q_{n-1})$ be the polynomial:

$$\sum_{|\alpha| \leq N} [c_{\alpha}^{0}(m^2)(p+q_1)^{\alpha} + c_{\alpha}'(m^2)[(p+q_1)^2 - m^2](p+q_1)^{\alpha}]$$

$$+ \sum_{\substack{|\beta|=1 \\ \beta_{1\mu}=0\,\forall\mu}} c_{\alpha}^{\beta}(m^2)(p+q_1)^{\alpha} q^{\beta}$$

where $c_{\alpha}'(z) = (d/dz)c_{\alpha}^{0}(z)$.

At $q = 0$, this expression takes the value

$$\sum_{|\alpha| \leq N} c_{\alpha}^{0}(m^2) + 2\, c_{\alpha}'(m^2)(p^2 - m^2) p^{\alpha}$$

and, for $p^2 = m^2$, this coincides with $L(p)$.

If γ is a multi-index with $|\gamma| = 1$ and $\gamma_{1\mu} = 0$ for all μ, (i.e., γ refers only to $q_2, \ldots, q_{n-1}$)

$$D^{\gamma} P(p+q_1, q_2, \ldots, q_{n-1})\Big|_{q=0} = \sum_{\alpha} c_{\alpha}^{\gamma}(m^2) p^{\alpha}$$

and this coincides with $L^{\gamma}(p)$ if $p^2 = m^2$.

If $|\gamma| = 1$ and γ refers to q_1 only,

$$D^{\gamma}P(p+q_1,q_2,\ldots,q_{n-1})\Big|_{q_2=\ \ldots\ =q_{n=1}=0} =$$

$$= \sum_{\alpha} [c^0_{\alpha}(m^2)D^{\gamma}(p+q_1)^{\alpha} + 2c'_{\alpha}(m^2)(p+q_1)^{\gamma}(p+q_1)^{\alpha}$$

$$+ c'_{\alpha}(m^2)[(p+q_1)^2-m^2]D^{\gamma}(p+q_1)^{\alpha}]$$

At $q_1 = 0$, $p^2= m^2$, this becomes:

$$\sum_{\alpha} c^0_{\alpha}(m^2)D^{\gamma}_p\, p^{\alpha} + 2c'_{\alpha}(m^2)p^{\gamma}p^{\alpha} =$$

$$= \sum_{\alpha} D^{\gamma}_{q_1}[c^0_{\alpha}((p+q_1)^2)(p+q_1)^{\alpha}] = D^{\gamma}_{q_1}L(p+q_1)\Big|_{q_1=0}$$

Hence, we have found a polynomial $P(q_1,\ldots,q_{n-1})$ such that $D^{\beta}T(p+q_1,q_2,\ldots,q_{n-1}) - D^{\beta}P(p+q_1,q_2,\ldots,q_{n-1})$ has adiabatic limit zero at $q = 0$ for all p such that $p^2=m^2$ and all β such that $|\beta| \leq 1$. We can thus redefine T, by subtracting P from it, in such a way that the new kernel has the properties C_1 and D. It will follow that all the kernels with the same vertices and external lines attached to them, arising from the other G.R.P. have the property D: indeed, such a kernel differs from T by a sum of connected products of kernels with fewer vertices; such products have, as we have seen, the property C_1 with A p) = B(p) = 0, and hence the property D.

An unsatisfactory feature of the above argument is that the polynomial P as it is defined here could have degree > 2, and this, in general clashes with the minimality of renormalization. To avoid this trouble we may first assume that renormalization is being performed so as to preserve Lorentz invariance. In this case we have

$$L^0(k) = c(k^2)$$

$$L^{\beta}(k) \equiv L_{j\mu}(k) = c_j(k^2)k_{\mu} \quad \text{if } |\beta| = \beta_{j\mu} = 1$$

The polynomial P then becomes

$$c_0(m^2) + c'_0(m^2)(q_1^2-m^2) + \sum_{j=2}^{n-1} c_j(m^2)(q_1,q_j)$$

and has degree ≤ 2.

More generally, we see that, if the two-point function $H(p+q_1,q_2,\dots,q_{n-1})$ has been defined so that it vanishes at $q=0$ for any p such that $p^2=m^2$, together with all its first derivatives in $q_1,\dots,q_{n-1}$, the corresponding kernels possess the properties C_1 and D.

Suppose now that $S(q_n,q_2,\dots,q_{n-1})\,\delta\,(p-p'+\sum_{j=1}^{n} q_j)$ is a self-energy kernel of type (15) arising from some G.R.P. Then, as it is well known, it coincides in a real neighbourhood of zero with a function analytic in a complex neighbourhood of zero. By subtracting from S a polynomial of degree 1, we can make it vanish twice at 0. The kernel thus obtained will clearly possess the property C_2.

Finally, we have proved that all connected kernels possessing external lines and n vertices satisfy the properties expressed in the induction hypothesis. As to the kernels with no external lines (which play no role in the inductive construction of connected kernels with external lines), we also wish them to have adiabatic limits such that the condition

$$(\Omega,S(g)\Omega)\xrightarrow[g\to\lambda]{}1$$

be satisfied.

Let $\delta(q_1+\dots+q_n)\,F\,(q_1,\dots,q_{n-1})$ be such a connected vacuum kernel. $F(q)$ is analytic in some complex open neighborhood of 0.

$$\int \delta(q_1+\dots+q_n)\,F\,(q_1,\dots,q_{n-1})\,\psi\,(q_1,\dots,q_n)dq_1\dots dq_n =$$

$$= \int\delta(Q)\,F\,(q)\,\chi\,(q,Q)dq\,dQ = \int F(q)\,\chi\,(q,0)dq\ ,$$

where

$$q=(q_1,\dots,q_{n-1})\quad\text{and}\quad \chi(q,Q)=\psi(q_1,\dots,q_{n-1},Q-\sum_{j=1}^{n-1}q_j)$$

If we now replace χ by χ_ε given by

$$\chi_\varepsilon(q,Q)=\varepsilon^{-4n}\,\chi(\frac{q}{\varepsilon},\frac{Q}{\varepsilon})$$

we find : $\varepsilon^{-4}\int F(\varepsilon q)\,\chi\,(q,0)dq$. This, in general diverges like ε^{-4} unless $F(q)$ vanishes sufficiently strongly at 0. If we redefine F, by subtracting a polynomial of the 4th degree, so that it takes the form

$$F(q) = \sum_{|\alpha|=5} q^{\alpha} F_{\alpha}(q)$$

(the F_{α} being analytic at 0) we find:

$$\left| \int F(q)\, \chi(q,0)\, dq \right| \leq \frac{C}{\varepsilon} \sup_{q} |q|^{4(n-1)+5-\varepsilon} |\chi(q,0)|$$

$$\leq \frac{C}{\varepsilon} \sup_{q,Q} |q,Q|^{4n+1-\varepsilon} |\chi(q,0)|$$

Thus, the kernel does have a zero adiabatic limit at 0.

ACKNOWLEDGEMENTS

The authors are much indebted to Professor K. Hepp, whose question this paper tries to answer, and to Drs. J. Bros, and R. Stora for advice. They wish to thank Drs. P. Blanchard and R. Seneor for useful discussions.

REFERENCES

[1] H. Epstein and V. Glaser - CERN Preprint TH. 1156 (1970), reprinted in Prépublications de la R.C.P. No 25, Vol. 11, Strasbourg (1970) , and Proceedings of the 1970 Summer School of Les Houches.

[2] H. Epstein and V. Glaser - Ann Inst. H. Poincaré 19,211(1973)

[3] N.N. Bogoliubov and V.S. Vladimirov, Nauchnye Dokl. Vysshei Shkoly, N. 3, 26 (1958) and N. 2, 179 (1959).

[4] J. Bros. H. Epstein, and V. Glaser, Commun. Math. Phys. 6, 77 (1967).

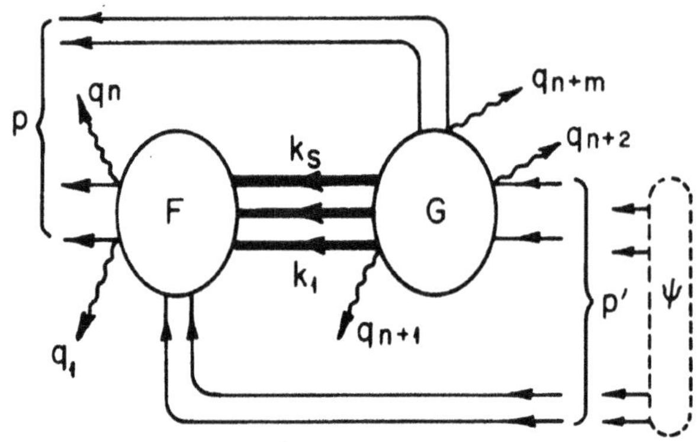

Figure 1: Multiplication of two operators not of self-energy type.

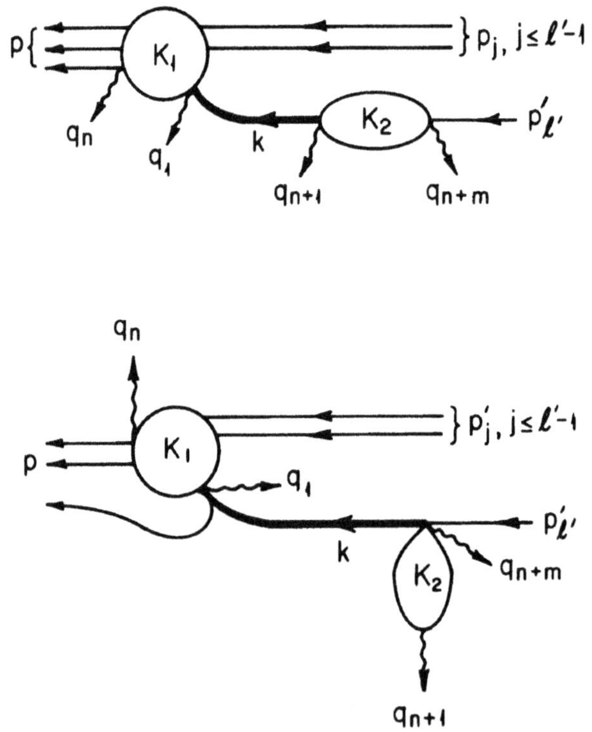

Figure 2: Multiplication of one operator not of self-energy type by an operator of self-energy type (1st kind top, 2nd kind bottom).

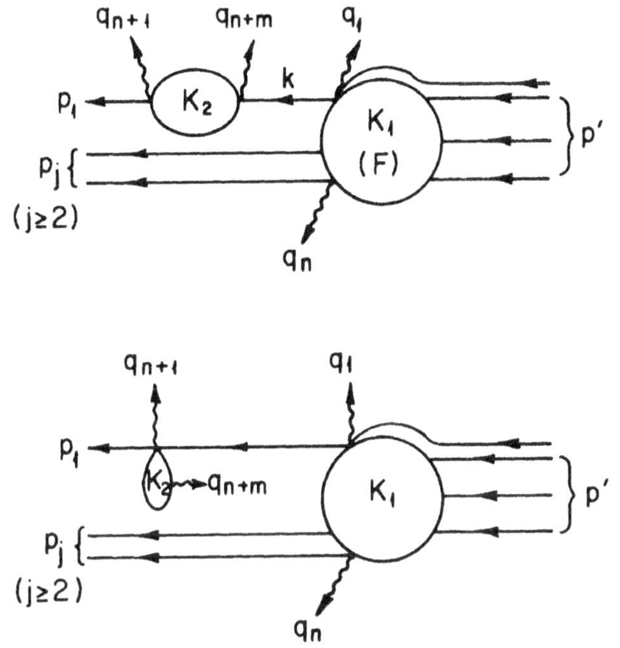

Figure 3: Multiplication of a self-energy-type operator (1st kind top, 2nd kind bottom) by an operator not of self-energy type.

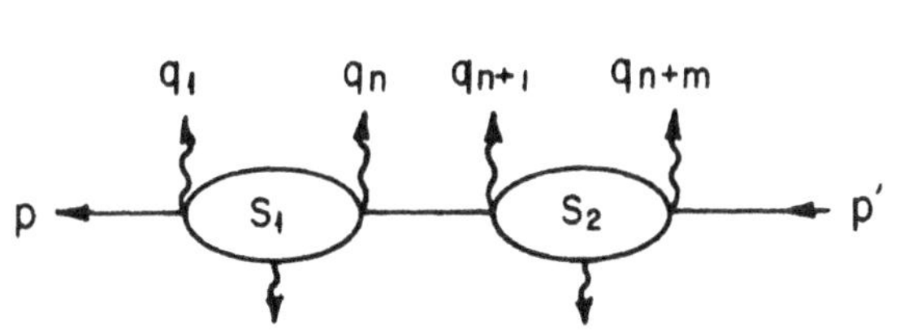

Figure 4: Multiplication of two self-energy-type operators of 1st kind.

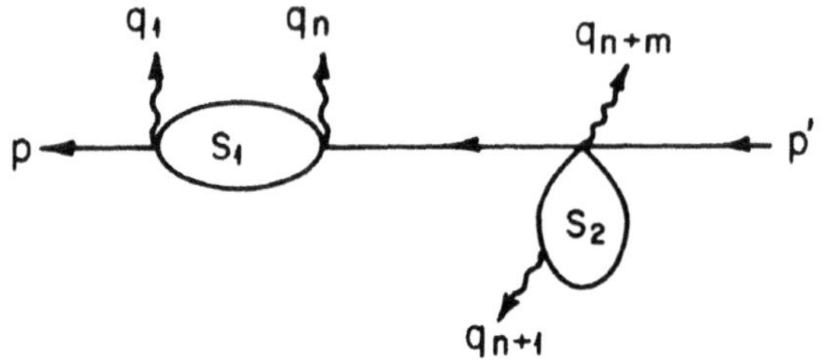

Figure 5: Multiplication of a self-energy-type operator of 1st kind by one of 2nd kind.

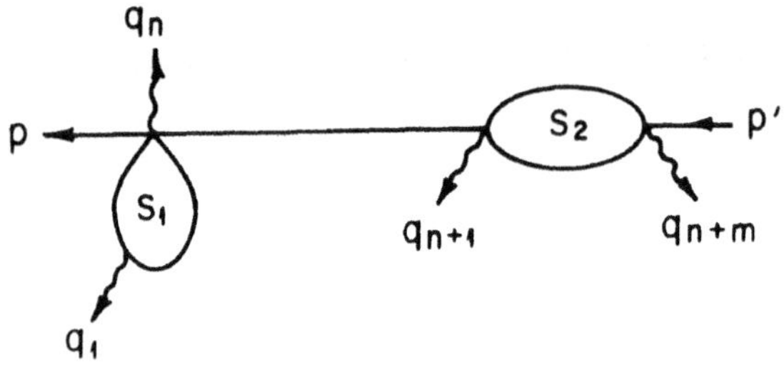

Figure 6: Multiplication of a self-energy-type operator of 2nd kind by one of 1st kind.

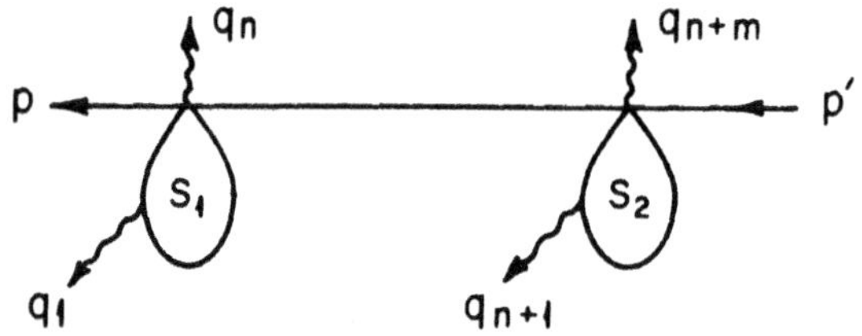

Figure 7: Multiplication of two self-energy-type operators of 2nd kind.

EXISTENCE OF GREEN'S FUNCTIONS IN PERTURBATIVE Q. E. D.

R. SENEOR

Centre de Physique Théorique, Ecole Polytechnique
91120 Palaiseau (France)

0. Introduction

The purpose of this lecture is to report on some work, done in collaboration with P. Blanchard [1], which shows how, in the framework developped by H. Epstein and V. Glaser [2] one can prove the existence of Green's functions in quantum electrodynamics (Q. E. D.).

The starting point is a family $\{T_r(X)\}$ of T-products satisfying all the required properties (see H. Epstein's lectures). The theory (Q. E. D.) is specified by the lowest order term

$$T(x) = \mathcal{L}(x) = \sum_{\mu=1}^{4} \bar{\psi}(x)\gamma_\mu \psi(x) A_\mu(x)$$

The notations are the following ones. For $X = (x_1,..,x_n) \in \mathbf{R}^{4n}$

$$T_r(X) = T_{r_1,\ldots,r_n}(x_1,\ldots,x_n) = T(\mathcal{L}_{r_1}(x_1),\ldots,\mathcal{L}_{r_n}(x_n))$$

with

$$r_i = (r_i^1, r_i^2, r_i^3) \qquad i = 1,\ldots,n$$

and $r_i^j = (a_i^j, \alpha_i^j)$. The subscript j in r_i^j indicates the type of field : j = 1 is for $\bar{\psi}$, j = 2 is for ψ and j = 3 is for A_μ. The indices a_i^j takes the values 1 or 0. The

G. Velo and A. S. Wightman (eds.), Renormalization Theory, 255–267.

indices α_i^j takes the values 1, 2, 3 or 4 and correspond to the Spinor or to the tensor indices. The derivated lagrangian $\mathcal{L}_{r_i}(x_j)$ is defined by

$$\mathcal{L}_{r_i}(x_i) = \prod_{j:a_i^j=1} \frac{\delta}{\delta\phi^j_{\alpha_i^j}} \mathcal{L}(x_j)$$

where ϕ^j stands for $\bar{\psi}$ when $j = 1$, ψ when $j = 2$, and A when $j = 3$.

However from now on we omit any references to Spinor or tensor indices. To clear what will follow we introduce also a diagramatic representation. Let $J(X)$ be the set of indices which numbers the x-variables, then the vacuum expectation $<T_r(X)>$ will be represented as a diagram with $|X|$ vertices and $\sum_{i\in J(X)} \sum_{j=1}^{3} a_i^j$ external lines ; more precisely with $\sum_{i\in J(X)} (a_i^1 + a_i^2)$ external fermions and $\sum_{i\in J(X)} a_i^3$ external photons.

According to the symmetry of the theory, there is no diagrams with no fermions and an odd number of photons, and the number of fermions is always even.

Finally let us point out that the degree of singularity at the origin of the distribution $<T_r(X)>$ is given by

$$\omega = 4 - \frac{3}{2} \sum_{i\in J(X)} (a_i^1 + a_i^2) - \sum_{i\in J(X)} a_i^3 \tag{0.1}$$

and the only divergent diagrams are

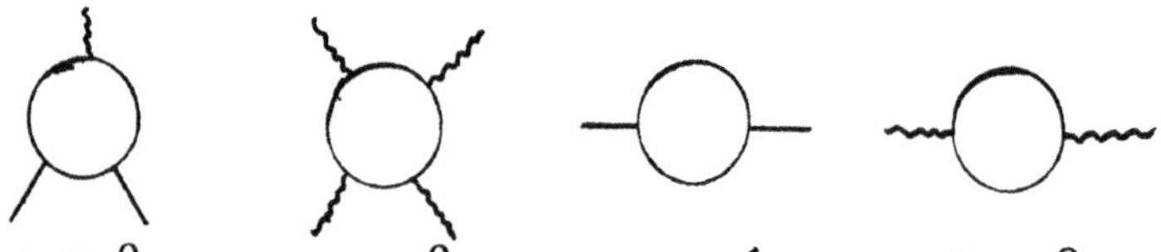

$\omega = 0$ $\omega = 0$ $\omega = 1$ $\omega = 2$

where the straight lines are fermions and the wriggled ones are photons lines.

Now let us go back to our purpose which is to study the nth order term of the perturbative expansion of $\hat{T}_r(X;g)$ (for $X = (x_1,\ldots,x_m)$) :

$$[\hat{T}_r(X;g)]_{(n)} = \frac{i^n}{n!}\int\ldots\int R(y_1,\ldots,y_n;x_1,\ldots,x_m)g(y_1)$$

$$\ldots g(y_n)dy_1\ldots dy_n = \frac{i^n}{n!}\int\ldots\int \mathcal{L}(y_1)\downarrow\ldots\mathcal{L}(y_n)\downarrow$$

$$T(\mathcal{L}_{r_1}(x_1)\ldots\mathcal{L}_{r_m}(x_m)) = \frac{i^n}{n!}\int Y\downarrow\ T_r(X)g(Y)dY$$

and to prove the existence of an "adiabatic limit" for $[\hat{T}_r(X;g)]_{(n)}$ when $g(y)$ tends to the constant function λ.

More precisely we want to show that $\lim_{g\to\lambda} \langle[\hat{T}_r(X;g)]_{(n)}\rangle$ exists as a tempered distribution, the limit being taken in a suitable way.

However, more generally, in order to recover all the properties of the T-products, we will also study

$$[\hat{T}(X_1)\ldots\hat{T}(X_p)]_{(n)} = \frac{i^n}{n!}\int Y\downarrow\{T(X_1)\ldots T(X_p)\}g(Y)dY$$

$$= \frac{i^n}{n!}\sum_{\substack{Y_1\cup\ldots\cup Y_p = Y\\ Y_i\cap Y_j=\Phi,\ i\neq j}} \left(\int Y_1\downarrow T(X_1)g(Y_1)dY_1\right)\ldots\left(\int Y_p\downarrow T(X_p)g(Y_p)dY_p\right)$$

for disjoint sets $X_1,\ldots,X_p$.

To shorten, from now on, $O(X)$ will denote indifferently a T-product or a product of T-products. The existence of the weak adiabatic limit (i.e. in term of vacuum expectation values) is proved recursively. The induction is based on the following remark. By definition (see H. Epstein's lecture)

$$Y\downarrow T(X) = \sum_{I\subset Y} (-1)^{|I|}\bar{T}(I)T(Y\backslash I,X) \tag{0.2}$$

and the support is in $\{y\}_Y \subset \{x\}_X + \bar{V}^-$. Rewritting (0.2) as

$$Y\downarrow T(X) - T(Y,X) = \sum_{\substack{I\subset Y\\ I\neq\Phi}} (-1)^{|I|}\bar{T}(I)T(Y\backslash I,X) \tag{0.3}$$

one sees that the right hand side depends only on T-products of "length" less than $|Y| + |X|$. On the other hand

$$Y\uparrow T(X) - Y\downarrow T(X) = \sum_{I\subset Y} (-1)^{|I|}[T(Y\backslash I,X),\bar{T}(I)] \tag{0.4}$$

has also a right hand side depending only on T-products of "length" less than $|Y| + |X|$, and because of the support properties of $Y{\uparrow}T(X)$ or $Y{\downarrow}T(X)$ one can, by a cutting procedure (see H. Epstein's lecture), recover $Y{\downarrow}T(X)$ from the difference (and also $T(Y,X)$ from (0.3)). The principle of the proof is therefore

1) to introduce some norm such that if $\langle Y{\downarrow}O(X)\rangle$ is bounded by such a norm then $\langle Y{\downarrow}O(X)\rangle$ has an adiabatic limit (in the sense of this norm)

2) to prove by induction on $|Y| + |X|$, that the difference defined by (0.4) obeys such a norm i.e., roughly speaking, that this norm is compatible with tensor products.

3) to prove this norm for $\langle Y{\downarrow}O(X)\rangle$ knowing it for the difference (0.4), i.e. that this norm is compatible with the cutting procedure.

Let us now describe these steps with more details.

1. The adiabatic norms

We introduce adiabatic norm in position and momentum space.

Definition : A distribution $T \in \mathcal{S}'(\mathbb{R}^N)$ satisfies an adiabatic norm of degree δ at ∞, $\delta > 0$, if $\exists$ $C \geq 0$ and $M \geq 0$ such that $\forall$ $\varphi \in \mathcal{S}(\mathbb{R}^N)$ one has

$$|\langle T,\varphi\rangle| \leq C \sum_{|\alpha|\leq M} \mathrm{Sup}(1 + |x|)^{|\alpha|-\delta}|D^\alpha\varphi(x)| \ .$$

From this definition follows

Lemma 1 : If $T \in \mathcal{S}'(\mathbb{R}^N)$ has an adiabatic norm at ∞, then $\forall\varphi \in \mathcal{S}(\mathbb{R}^N)$

$$\lim_{\substack{\varepsilon\to 0\\ \varepsilon>0}} \langle T(x),\varphi(\varepsilon x)\rangle = L\varphi(0)$$

with L independent of φ.

Remark that formally $L = \int T(x)dx$.

Proof : One writes $\langle T(x),\varphi(\varepsilon x)\rangle = I(\varepsilon)$ as

$$I(\varepsilon) = I(1) - \int_\varepsilon^1 \frac{d}{d\tau} I(\tau)$$

and one gets that $\left|\frac{d}{d\tau} I(\tau)\right| \leq \frac{C^t}{\tau^{1-\delta}}$ which is therefore integrable, thus the integral exists for $\varepsilon = 0$ from which follows the existence of $\lim_{\varepsilon \to 0} I(\varepsilon) = \langle S,\varphi\rangle$.

Now to compute the limit (which is a temperate distribution as weak limit of temperate distributions) one remarks that

$$\sum_{|\alpha|\leq M} \operatorname{Sup}(1+|x|)^{|\alpha|-\delta}|D^{\alpha}\varphi(\varepsilon x)|$$

is bounded by $C(|\varphi(0)| + \varepsilon^{\delta} \sum_{1\leq|\alpha|\leq M} \operatorname{Sup}(1+\|x\|)^{|\alpha|-\delta}|D^{\alpha}\varphi(x)|)$

Thus $\lim_{\varepsilon\to 0} I(\varepsilon) \leq C|\varphi(0)|$. Now taking $\varphi(x) = \psi(x) - \psi(0)h(x)$ with $h(x) \in \mathcal{S}(\mathbb{R}^N)$ and $h(0) = 1$ fixed, $\forall\ \psi(x) \in \mathcal{S}(\mathbb{R}^N)$ one gets that

$$\langle S,\psi\rangle = \psi(0)\langle S,h\rangle$$

and we set $\langle S,h\rangle = L$.

Equivalently one can define the notion of adiabatic limit in momentum space with

Definition : A distribution $\tilde{T} \in \mathcal{S}'(\mathbb{R}^N)$ satisfies an adiabatic norm of order δ at 0, $\delta > 0$ if $\exists\ C \geq 0$ and $P \geq 0$ such that $\forall\ \tilde{\varphi} \in \mathcal{S}(\mathbb{R}^N)$ one has

$$|\langle\tilde{T},\tilde{\varphi}\rangle| \leq C\{|\textstyle\int\tilde{\varphi}(p)\,dp| + \sum_{0\leq|\alpha|\leq P} \operatorname{Sup}|p|^{N+\delta+|\delta|}|D^{\alpha}\tilde{\varphi}(p)|\}$$

As a consequence

Lemma 2 : If $\tilde{T} \in \mathcal{S}'(\mathbb{R}^N)$ has an adiabatic norm at 0, then $\forall\ \tilde{\varphi} \in \mathcal{S}(\mathbb{R}^N)$

$$\lim_{\substack{\varepsilon\to 0\\ \varepsilon>0}} \langle\tilde{T}(\varepsilon p)\tilde{\varphi}(p)\rangle = \lim_{\substack{\varepsilon\to 0\\ \varepsilon>0}} \langle\tilde{T},\tilde{\varphi}_{\varepsilon}\rangle = \tilde{L}\int\tilde{\varphi}(p)\,dp$$

exists, with $\tilde{\varphi}_{\varepsilon} = \frac{1}{\varepsilon^N}\tilde{\varphi}(\frac{p}{\varepsilon})$.

Formally this means that $\tilde{L} = \lim_{\varepsilon\to 0} \tilde{T}(\varepsilon p) = \tilde{T}(0)$. One can show that these two definitions are equivalent. In fact :

If T has an adiabatic limit at ∞ of order $\delta < 1$, then $\tilde{T}$ has an adiabatic limit at 0 of order δ', $\delta' < \delta$. Reciprocally, if $\tilde{T}$ has an adiabatic limit at 0 of order $\delta < 1$, then T has an adiabatic limit at ∞ of order δ', $\delta' < \delta$.

Finally we need for the sequel some new norms corresponding to infrared singular behavior.

Definition : Let D be some integer, $\delta > 0$, then $T \in \mathcal{S}'(\mathbb{R}^N)$ has an adiabatic norm at ∞ with defect index max(0,1-D) if $\exists$ $C \geq 0$ and $P \geq 0$ such that

$$|\langle T,\varphi\rangle| \leq C \sum_{|\alpha|\leq P} \operatorname{Sup}(1+|x|)^{|\alpha|-(D-1)-\delta}|D^\alpha\varphi(x)|$$

The equivalent norm in momentum space if given by

$$|\langle\tilde{T},\tilde{\varphi}\rangle| \leq \tilde{C} \sum_{|\beta|\leq\tilde{P}} \operatorname{Sup}|p|^{N+|\beta|+D-1+\delta}(1+|p|)^{\tilde{L}}|D^\beta\tilde{\varphi}(p)|$$

for some $\tilde{C}$, $\tilde{P}$ and $\tilde{L}$.

The defect index max(0,1-D) measures the lack of convergence towards an adiabatic limit.

Exercices : Let $\delta^+(p;m) = \theta(p^0)\delta(p^2-m^2)$. Show that for $m \neq 0$, $\delta^+(p;m)$ has defect index 0 and indeed that the adiabatic limit is 0. Show that $\delta^+(p;0)$ has defect index 3. (Hint: use the fact that δ^+ is a positive measure).

To simplify the presentation we will not take into account the power counting behaviour as it has been done in [1]. This would have led us to introduce norms such that

$$\sum_{|\alpha|\leq P} \operatorname{Sup}|x|^{\mu_\alpha^+}(1+|x|)^{-\mu_\alpha^+ + |\alpha|-(D-1)-\delta}|D^\alpha\varphi(x)|$$

for a distribution singular at the origin of order ω, with $\mu_\alpha^+ = (-\omega+|\alpha|-\varepsilon)^+$ for some ε.

2. The induction

We develop here step 2 of the proof as explained in the introduction.

We first derive an algebraic formula which allows to do the induction with respect to $|Y|+|X|$.

Let Y and X be such that $|Y| = n$ and $|X| = s$ then

we claim that

$$Y\uparrow O(X) - Y\downarrow O(X) = \sum_{\substack{Y_1\cup\ldots\cup Y_p=Y \\ Y_i\cap Y_j=\emptyset\ i\neq j}}' [Y_1\updownarrow Y_2\updownarrow\ldots\updownarrow Y_k, Y_{k+1}\updownarrow\ldots\updownarrow Y_p\updownarrow O(X)] \tag{2.1}$$

where Σ' means that the sum extends on some choice of $\updownarrow$ to be $\uparrow$ or $\downarrow$ and on some partitions of Y.

Exercice : Show this formula using the arrow calculus [Hint : Do it recursively starting from $y_1\uparrow O(X) - y_1\uparrow O(X) = [y_1, O(X)]$ and the fact that $y_j\updownarrow$ acts as a derivation]. ■

One sees that the difference is a sum of commutators which all have O(X) on the right and since $Y_1\cup\ldots\cup Y_k$ cannot be empty this means that one has $|Y_{k+1}\cup\ldots\cup Y_p| < n$.

On the other hand since O(X) is $T(X_1)\ldots T(X_q)$, from

$$Y\uparrow T(X_1)\ldots T(X_q) = \sum_{\substack{Y_1\cup\ldots\cup Y_q \\ Y_i\cap Y_j=\emptyset\ i\neq j}}' (Y_1\uparrow T(X_1))\ldots(Y_q\uparrow T(X_q))$$

(and the same for $Y\downarrow\ldots$) and from the structure of the arrow calculus one has that the right hand side can be expressed in term of commutators of products of $Y\updownarrow T(X)$ with $|Y| < n$ and $|X| \leq s$.

We can now formulate the induction hypothesis.

Suppose that s and n being fixed positive integers we have shown that for

$$|X| \leq s \quad \text{and} \quad |Y| < n$$

$Y\uparrow O(X)$ and $Y\downarrow O(X)$ have an adiabatic limit, then we will prove that $Y\uparrow O(X) - Y\downarrow O(X)$ has, for $|X| = s$ and $|Y| = n$ an adiabatic limit.

To see where are the difficulties in this proof, let us consider a term in the commutator (2.1). Its vacuum expectation value is of the form

$$\langle T_{r'}(Y').T_{r''}(Y'',X)\rangle$$

where we have noted $Y_1\updownarrow\ldots\updownarrow Y_p\updownarrow O(X)$ as T(Y,X) (remember that such a term can be expand in term of products of T-products). Here $Y' \cup Y'' = Y$ and $Y' \cap Y'' = \phi$.

According to the Wick theorem it is a sum of terms like

$$<T_{r'+s'}(Y')><T_{r''+s''}(Y'',X)> \prod_{j=1}^{\ell} P_j(\partial)\,\Delta_j^+$$

with $P_j(\partial) = g_{\mu\nu}$ or $\frac{1}{i} \sum_{\mu=1}^{4} \gamma_\mu \frac{\partial}{\partial x_j^\mu} + m_j$ and Δ_j^+ is the Fourier transform of $\delta^+(p;m_j)$. For fermions $m_j = m$ and for photon $m_j = 0$. The Fourier transform of such a term is

$$\tilde{t}_r(q,p) = \int\cdots\int \tilde{t}_{r'+s'}(q_1'+k_1,\ldots,q_\ell'+k_\ell,q_{\ell+1}',\ldots,q_{\nu-1}')$$

$$\tilde{t}_{r''+s''}(q_1''-k_{i(\mu)},\ldots,q_\mu''-k_{i(\mu)},p_1-k_{i(\mu+1)},\ldots,p_{s-1}-k_{i(\mu+s-1)})$$

$$\delta^{(4)}(\sum_1^{\ell} k_i - \sum_1^{\nu} q_i')\prod_{j=1}^{\ell} P_j(k_j)\,\delta^+(k_j;m_j)\,dk_j \qquad (2.2)$$

where i is some mapping of $(1,\ldots,\mu+s-1)$ on $(1,\ldots,\ell)$. One sees on formula (2.2) that formally due to the $\delta^{(4)}$ function, $t_r(q,p)$ "vanishes" for $\|q\|$, the Euclidean norm of $(q_1',\ldots,q_\nu',q_1'',\ldots,q_\mu'')$, closed to zero if at least one of the intermediate particles (represented by the δ^+-function) is a fermion. This argument can be made rigorous and therefore the only case to consider is when in $\tilde{t}_r(q,p)$, the absorptive part, all intermediate particles are photons.

Finally remark also that because of the commutator form we can deal only with connected terms with respect to the Y's and in particular $\ell \geq 1$ in (2.2).

As a consequence, in order to prove an adiabatic norm for the absorptive part one has first to control the infrared behaviour of $<T_r(Y)>$ with only photon external lines and then the one of $<T_r(Y,X)>$ with photon external lines issued from the Y's, whatever are the other lines.

Remark that since we are interested in the behaviour of $Y{\downarrow}O(X)$ as a distribution, the x's variables are always smeared with test functions and therefore if there are photons lines issued from x-points they are always smeared.

3. The case of $T_r(Y)$ with only photon external lines .

Our purpose is to define a norm which allows us to control the infrared behaviour. Let $\tilde{t}_r(q)$ be the Fourier transform of $\langle T_r(Y)\rangle$. According to the consideration of the previous section, diagramatically $\tilde{t}_r(q)$ can be represented as a sum of diagrams :

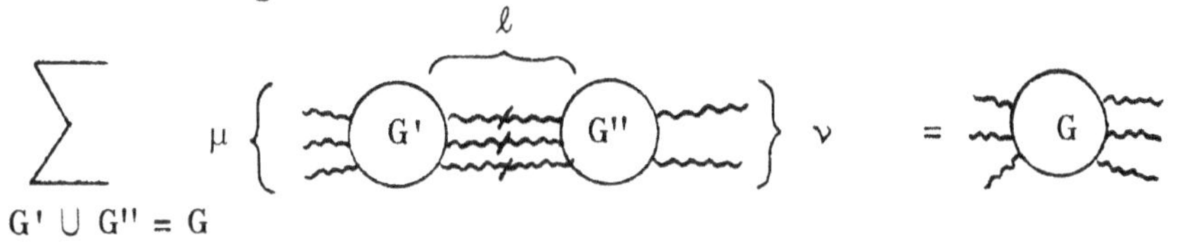

where the intermediate lines  are on mass shell.

Let us first look at some simple homogeneous considerations, i.e. replace q by λq and look at the behaviour in λ. Suppose one has defined a number $D(G)$ depending only on the number of external lines of G and suppose $\tilde{t}_r(\lambda q) \sim \lambda^{D(G)}$ then roughly one should have

$$\lambda^{D(G)} \sim \sum_{G'\cup G''} \lambda^{D(G')}\lambda^{2\ell-4}\lambda^{D(G'')}$$

where $\lambda^{2\ell-4}$ came from the fact that

$$\delta^{(4)}(\lambda(\Sigma q-\Sigma k))\prod_{i=1}^{\ell}\delta^{+}(\lambda k_i)\,d^4\lambda k_i = \lambda^{2\ell-4}\delta^{(4)}(\Sigma q-\Sigma k)\prod_{i=1}^{\ell}\delta^{+}(k_i)\,d^4k_i$$

If $D(G) \leq 0$ there is no adiabatic limit (for $D = 0$, it can still behaves like a logarithm).

Since the behaviour for $D(G)$ is dominated by the worst behaviour in the sum one should have

$$D(G) \leq \inf_{G'\cup G''=G} (D(G')+D(G'')+2\ell-4) \qquad (3.1)$$

Setting $D(G) = D(\ell) = a\ell+b$, ℓ being the number of external lines of G, this last condition gives $D(\ell) \geq a(\ell-2)+2$ with $a+1 \geq 0$.

One sees that the self energy should behave like λ^2 and if we ask that the four photons diagrams behave like λ one gets that one can take $a = -\frac{1}{2}$ and thus

$$D(2\ell) = -\ell + 3 \qquad (3.2)$$

Applying steps 2 and 3 of the inductive procedure one can show recursively :

<u>Lemma 1</u> : Let $\sum_j a_j^i = 0$ $i = 1,2$ than $\exists\, C, P, \delta$ such that

$$|\langle t_r(\xi), \varphi(\xi)\rangle| \quad C \sum_{P \geq |\alpha| \geq \max(0,D)} \sup_{\xi} (1+|\xi|)^{|\alpha|-D+1-\delta} |D^\alpha \varphi|$$

for $\xi = (\xi_1, \ldots, \xi_{|Y|-1}), \xi_j = y_j - y_{|Y|}$ and $\varphi \in \mathcal{S}(\mathbb{R}^{4(|Y|-1)})$

$$D = -\frac{1}{2} \sum a_j^3 + 3.$$

For step 2, this lemma follows from the fact that the tensor product rule as defined by H. Epstein can be extended to such norms. For step 3 there is some problem which will be explained later.

4. The case of $\int T_r(Y;X) f(X)\, dX$

We again follow the lines of 3 and try to define an index $D'(G)$. The condition we add this time is $D'(G) = 1$ when there is no external photon lines issued from Y's. One can take

$$D' = - \sum_{j \in J(Y)} a_j^3 + 1$$

In fact consider

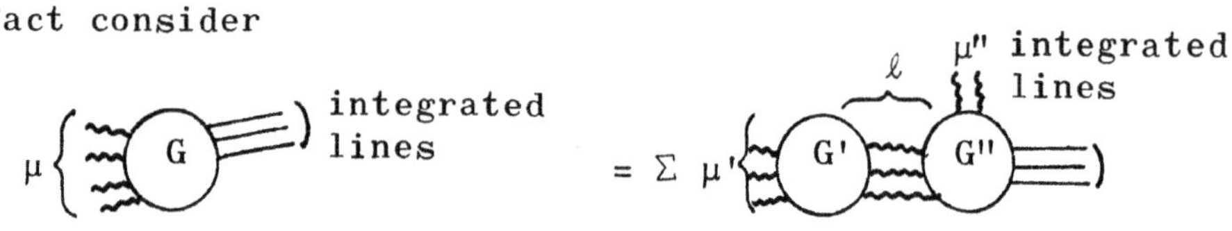

with G' and G'' any two diagrams which union is G. One gets

$$D(G') + D(G'') + 2\ell - 4 = -\frac{1}{2}(\mu' + \ell) + 3 - (\mu'' + \ell'') + 1 + 2\ell - 4$$

$$= -(\mu' + \mu'') + 1 + \left[\frac{1}{2}(\mu' + \ell) - 1\right] + \ell - \ell'' \geq D(G)$$

since $\mu' + \ell$ should be even and different from zero and since $\ell \geq \ell''$ where ℓ'' is the number of photon lines issued from y's variables.

Then one can show

<u>Lemma 2</u> : Let $\sum_{i \in J(Y)} a_i^j = 0$ $j = 1,2$, then $\exists\, C, M, N, P, \delta$

and ε such that

$$|\langle t_r(\xi,\zeta),\varphi(\xi,\zeta)\rangle| \leq C \sum_{M\geq|\alpha|\geq 0} \sum_{N\geq|\beta|\geq 0} \mathrm{Sup}(1+\|\xi\|)^{|\alpha|-D'+1-\delta}$$

$$\|\zeta\|^{\mu^+_\beta}(1+\|\zeta\|)^P |D^\alpha_\xi D^\beta_\zeta \varphi(\xi,\zeta)|$$

with $\xi_j = y_j - x_{|X|}$ $j \in J(Y)$ $\zeta_j = x_j - x_{|X|}$ $j \in J(X)$ and

$$\mu^+_\beta = (-\omega + |\beta| - \varepsilon)^+ \qquad \omega = 4 - \frac{3}{2} \sum_{j\in J(X)} (a^1_j + a^2_j) - \sum_{j\in J(X)} a^3_j$$

$$D' = - \sum_{j\in J(Y)} a^3_j + 1.$$

The norm in this lemma is very complicated since it includes beside the infrared behaviour, the correct power counting behaviour in the x-variables. Remark also that when $D' = 1$, the norm in the ξ-variables is an adiabatic norm at ∞, which is the required result.

Again the difficult part is related to step 3.

5. The cutting procedure

One has to consider two cases according to whether $X = \phi$ or not.

When $X = \phi$, i.e. as in section 3, one proceeds exactly as explained in Epstein's lecture or [2], the only trouble is the fact that when one replaces φ by $\omega\varphi$ the lower bound on the derivatives in lemma 1 disappears. But the only diagrams for which this lower bound is different from zero are the self energies and the four photon diagrams for which fortunately $\omega = 2$ and 0 respectively.

One uses then the ambiguities resulting from this singular behaviour during the cutting procedure to get the correct result. Let us see in more details what happens in the case of a self energy.

Let $t(X)$ be a distribution obtained from the cutting procedure and singular at the origin of order $\omega = 1$ and satisfying $\forall\ \varphi \in \mathcal{S}(\mathbb{R}^N)$

$$\langle t(x)\varphi(x)\rangle \leq C^t \sum_{|\alpha|\geq 0} \mathrm{Sup}\,(1+|x|)^{|\alpha|-1-\delta} |D^\alpha \varphi|$$

Then $\exists$ c and c_μ $\mu = 1,\dots,N$ such that

$$\bar{\bar{t}}(x) = t(x) - c\,\delta^N(x) + \sum_1^N c_\mu \frac{\partial}{\partial x_\mu} \delta^N(x)$$

satisfies

$$<\bar{\bar{t}},\varphi> \leq C^t \sum_{|\alpha|\geq 2} \operatorname{Sup}\,(1+|x|)^{|\alpha|-1-\delta} |D^\alpha \varphi(x)|$$

and one has :

$$\lim_{\varepsilon\to 0} <\bar{\bar{t}}(x),\varphi(\varepsilon x)> = \lim_{\varepsilon\to 0} <x_\mu\, \bar{\bar{t}}(x),\ \varphi(\varepsilon x)> = 0$$

This last property means that $\tilde{\bar{\bar{t}}}(q)$ "vanishes" with all its first derivatives at the origin.

The case of the four photons is treated in the same way.

Now when $X \neq \phi$, the trouble comes from the fact that one has only support properties in the ξ-variables (see lemma 2). So in applying the usual cutting procedure one has to introduce some modification. One replace the standard $\sqcup\!\sqcup$ by another one related to a cone V^+_ρ defined by $V^+_\rho = \{y \in \mathbb{R}^4 \,|\, y_0 > \rho|\vec{y}| \;;\; \rho < 1\}$, thus $V^+_\rho \supset V^+$. Define $a(\xi.\zeta) = <Y \uparrow O(X)>$, $r(\xi,\zeta) = <Y\downarrow O(X)>$ and $d(\xi,\zeta) = a(\xi,\zeta) - r(\xi.\zeta)$ then $a(\xi.\zeta) = \sqcup\!\sqcup d(\xi,\zeta) + b(\xi.\zeta)$ with $b(\xi,\zeta) = a(\xi,\zeta) - \sqcup\!\sqcup a(\xi.\zeta) + \sqcup\!\sqcup r(\xi,\zeta)$.

The support of $a(\xi,\zeta)$ is contained in $C_+ = \{\xi,\zeta \,|\, \xi_i - \zeta_{u(i)} \in V^+$ for at least one mapping of $(1,\dots,|Y|)$ on $(1,\dots,|X|-1)\}$ and that of r in C_-. Then one can see that on the support of b one of this two conditions is satisfied :

$$\xi - \zeta \in V^+,\ \xi \notin V^+_\rho \quad \text{or} \quad \xi - \zeta \in V^-,\ \xi \notin V^-_\rho$$

and we claim that if 2 vectors satisfy one of these conditions, then there exist $C(\rho)$ such that

$$\|\xi\| \leq C(\rho)\, \|\zeta\|$$

in the Euclidean norm.

This shows that for $b(\xi,\zeta)$ the support in the ζ-variables (the x's) allow us to control the support in the ξ-variables and thus to get the required result. In particular to be convinced, notice that if Supp f(X) is a compact set, then, the Fourier transform of b is analytic and therefore possess an adiabatic limit.

6. Conclusion

This lecture gives an idea of the construction from which follows the existence of Green's functions in Q. E. D.. As a by product, this is given by lemma 2, one gets that the power counting behaviour of the adiabatic limit is the same as for the free theory. Remark also that the vanishing at 0 of the photon self energies does not mean one can perform the photon mass normalization (i.e. the vanishing on the light cone).

Finally, this proof of the existence of Green's functions extends to ϕ^{2n}, $n \geq 2$, zero mass theories, and even can be extended, in principle, to any theory involving massive and non massive particles.

References

[1] P. Blanchard and R. Sénéor, Green's functions for theories with massless particles (in perturbation theory), Ann. Inst. Henri Poincaré, vol.XXIII, n°2, 1975, p.147-209 ; CERN. preprint TH 1420 and Contribution to the meeting on Renormalization Theory, C. N. R. S. Marseille, June 1971.

[2] H. Epstein and V. Glaser, The role of locality in perturbation theory, Ann. Inst. Henri Poincaré, vol. XIX, n° 3, 1973, p.211-295.

GAUGE FIELD MODELS

C. BECCHI
University of Genova
Italy

A. ROUET
Max-Planck Institut für
Physik und Astrophysik
München

R. STORA
Centre de Physique Théorique
C.N.R.S. - Marseille

Lectures given by C. BECCHI

INTRODUCTION

The renormalized versions of the Schwinger Quantum Action Principle [1] (Q.A.P.) have been among the most important subjects of the first week of this School [2,3]. It turns out that the naive identities, which can be deduced on a very formal ground, expressing in terms of the Green functions the quantum equivalent of the Noether theorem are affected with quantum correction depending on the particular renormalization rules. In some "clever" renormalization scheme [3], (e.g. in those which are based on the dimensional regularization of the Feynman integrals) these quantum corrections seem to satisfy a not yet well defined criterion of minimality. In a more general framework they have known power counting properties [2,4]. This weak form of the Q.A.P. proved in the framework of the Bogoliubov-Parasiuk-Hepp-Zimmermann [5] (B.P.H.Z.) renormalization scheme has been stated in Stora's lectures [6]. Exploiting the B.P.H.Z. version of the Q.A.P. Stora has been able to prove rather general results on the renormalization of lagrangian theories with broken symmetries [7].

The purpose of these lectures is to continue Stora's

G. Velo and A. S. Wightman (eds.), Renormalization Theory, 269–297.

analysis discussing with the same technique a more sophisticated class of lagrangian models, that is the nonabelian (Yang-Mills) gauge field models (G.F.M.). This is the only known class of renormalizable models involving a self interacting system of vector mesons. Hence it is of particular interest in the construction of renormalizable models based on vector intermediate bosons for the weak and electromagnetic interactions [8].

In the G.F.M. in which the gauge group is semisimple the gauge photons must be massless except if the gauge symmetry is spontaneously broken according to the Higgs-Kibble (H.K.) mechanism [9]. In order to avoid infrared difficulties we shall discard models in which massless fields are involved, limiting ourselves to the analysis of theories in which the gauge symmetry is completely broken according to the H.K. mechanism.

It turns out clearly from the study of quantum electrodynamics, (Q.E.D.), the oldest and best known gauge field model, that a gauge invariant lagrangian model cannot be directly quantized apparently preserving the local structure of the theory and leading to a renormalizable perturbation theory. (Here by renormalizable we mean satisfying the necessary power counting prescriptions.) The classical way out of this difficulty implies the introduction of a certain number of degrees of freedom which are not physically interpretable (e.g. the longitudinal, scalar, photons and the Faddeev-Popov [10] (Φ.Π.) ghosts). This procedure completely changes the renormalization problem. We have a model depending on a finite number of parameters and we have to show that the original physical properties of the theory are not affected by the introduction of the new "unphysical" degrees of freedom. More precisely we have to prove that the restriction of the scattering operator of the theory to a certain "physical" subspace of the Fock space is unitary in the perturbative sense and independent from the properties of the unphysical particles.

In much the same way as in Q.E.D. this is the consequence of a set of Ward-Takahashi identities, the Slavnov identities [11], which, owing to the Q.A.P., express the invariance of the Faddeev-Popov lagrangian of the G.F.M. under a set of nonlinear field transformations (Slavnov transformations) [12] explicitely involving the Φ.Π. ghost fields. Hence our first goal will be to prove that one can fulfill the Slavnov identities to all orders of the renormalized perturbation theory.

Then we shall discuss in some details the gauge independence of the physical scattering operator and we shall outline the connection between its unitarity and the Slavnov symmetry. The complete combinatorial proof of unitarity will not be examined here since it is discussed in details in the referred literature [12,13]. We shall restrict ourselves to models involving semisimple gauge groups. This will greatly simplify the analysis of the possible quantum corrections to the Q.A.P. which in technical terms will be reduced to the study of the cohomology group of the Lie algebra characterizing the gauge theory [7,13]. For what concerns the basic tools and definitions (in particular the functional method) that we shall extensively use in the following we refer to the first section of Stora's lectures.

In section two we discuss at the classical level the algebraic properties of the SU(2) Higgs-Kibble-Englert-Brout-Faddeev-Popov lagrangian and we exhibit its invariance under Slavnov transformations.

In section three we study the renormalization of the Slavnov identity in the G.F.M. involving semisimple gauge groups.

Section four deals with unitarity and gauge independence of the physical S operator in the SU(2) H.K. model.

2. THE CLASSICAL SU(2) MODEL

In this section we shall examine in some details the classical structure of the SU(2) Higgs-Kibble-Englert-Brout model and the problems which are encountered in its quantization.

Let $\{\varphi_a\} = \{\pi_\alpha, \sigma\}$ with $a = 1, \dots 4$, and $\alpha = 1, 2, 3$ be a multiplet of scalar fields and $a_{\mu\alpha}$, $\alpha = 1, 2, 3$, a multiplet of Yang-Mills photons.
An infinitesimal gauge transformation of parameter $\delta\omega_\alpha(x)$ is defined by

$$\delta\pi_\alpha(x) = \frac{e}{2}\left[\varepsilon_\alpha{}^{\beta\gamma}\,\delta\omega_\beta\,\pi_\gamma + (\sigma + F)\,\delta\omega_\alpha\right](x) \equiv \int dy\, \frac{\delta\pi_\alpha(x)}{\delta\omega_\beta(y)}\,\delta\omega_\beta(y)$$

$$\delta\sigma(x) = -\frac{e}{2}\left[\pi^\alpha\,\delta\omega_\alpha\right](x) \equiv \int dy\, \frac{\delta\sigma(x)}{\delta\omega_\alpha(y)}\,\delta\omega_\alpha(y) \tag{1}$$

$$\delta a^\mu_\alpha(x) = \left[\partial^\mu\delta\omega_\alpha + \varepsilon_\alpha{}^{\beta\gamma}\,\delta\omega_\beta\, a^\mu_\gamma\right](x) \equiv \int dy\, \frac{\delta a^\mu_\alpha(x)}{\delta\omega_\beta(y)}\,\delta\omega_\beta(y)$$

The parameter F plays the role of the σ field vacuum expectation value. A lagrangian invariant under such transformations is constructed in terms of the antisymmetric covariant field tensor :

$$G^{\mu\nu}_{\alpha} = \partial^{\mu} a^{\nu}_{\alpha} - \partial^{\nu} a^{\mu}_{\alpha} - e\, \varepsilon_{\alpha}^{\ \beta\gamma} a^{\mu}_{\beta} a^{\nu}_{\gamma} \tag{2}$$

and of the covariant derivatives :

$$(D_{\mu}\varphi)_{a} = \left\{ \partial_{\mu}\pi_{\alpha} - \frac{e}{2}\left[\varepsilon_{\alpha}^{\ \beta\gamma} a^{\mu}_{\beta}\pi_{\gamma} - (\sigma+F) a^{\mu}_{\alpha} \right], \partial^{\mu}\sigma + \frac{e}{2}\pi^{\alpha} a^{\mu}_{\alpha} \right\} \tag{3}$$

it has the form :

$$\mathcal{L}_{inv}(\varphi_{a}, a^{\mu}_{\alpha}) = -\frac{1}{4}\left(G^{\alpha}_{\mu\nu}\right)^{2} + \frac{1}{2}\left(D_{\mu}\varphi_{a}\right)^{2} - H\left([\varphi+F]^{2}\right) \tag{4}$$

Here H is a polynomial in $\pi^{\alpha}\pi_{\alpha} + (\sigma+F)^{2} \equiv [\varphi+F]^{2}$ invariant under the transformation (1) and satisfying the condition :

$$\partial_{\varphi_{a}} H\left([\varphi+F]^{2}\right)\Big|_{\varphi=0} = 0 \tag{5}$$

expressing that the classical vacuum corresponds to the configuration $\varphi_{a} = 0$. From the invariance property of H and from Eq.(5) it turns out that the mass matrix of the scalar fields :

$$\partial_{\varphi_{a}} \partial_{\varphi_{b}} H\left([\varphi+F]^{2}\right)\Big|_{\varphi=0} \tag{6}$$

has three null eigenvalues corresponding to the π_{α} modes and a positive eigenvalue corresponding to the σ field. This result expresses the content of the Goldstone theorem [14].

To discuss the quantization of this model we have to examine the wave operator corresponding to the free part (that bilinear in the fields) of the lagrangian. In particular the free lagrangian of the photon – π channel is :

$$\mathcal{L}^{\circ}_{inv} = -\frac{1}{4}\left(\partial^{\mu} a^{\nu}_{\alpha} - \partial^{\nu} a^{\mu}_{\alpha}\right)^2 + \frac{e^2}{8} F^2 (a^{\mu}_{\alpha})^2 + \frac{1}{2}(\partial_{\mu}\pi_{\alpha})^2 + \frac{eF}{2}\partial_{\mu}\pi^{\alpha} a^{\mu}_{\alpha} \qquad (7)$$

The Fourier transformed wave operator factorizes in two submatrices. The first associated with the transverse modes of the photon:

$$K^{T\,\alpha\beta}_{\mu\nu}(p) = -\delta^{\alpha\beta}\left(g^{\mu\nu} - \frac{p^{\mu}p^{\nu}}{p^2}\right)\left(p^2 - \frac{e^2F^2}{4}\right) \qquad (8)$$

exhibits the Higgs phenomenon [9] (the photons are massive). The second, corresponding to the coupled longitudinal photon - π channel, is :

$$\left\| K^{\alpha\beta}_{a_L,\pi}(p) \right\| = \delta^{\alpha\beta} \begin{vmatrix} e^2F^2 & -ip\frac{eF}{2} \\ ip\frac{eF}{2} & p^2 \end{vmatrix} \qquad (9)$$

The determinant of this matrix is identically zero. It follows that our invariant theory is not directly quantizable.

In the cases in which the theory is spontaneously broken one can eliminate the degrees of freedom with degenerate wave operator by means of a field dependent gauge transformations [15]. This reduces the degrees of freedom of the model to a minimal number, the transverse modes of the photons and the σ field, to which "physical" particles are associated. However the resulting lagrangian is not renormalizable by power counting.

Another solution is based on the Faddeev-Popov lagrangian [10]. This is the natural extension to the case of non abelian G.F.M. of the usual Q.E.D. lagrangian. It involves, beyond the scalar and vector fields, two conjugate multiplets of scalar fields $\{C_{\alpha}\}$, $\{\bar{C}_{\alpha}\}$ quantized according to the Fermi statistics in a Fock space carrying indefinite metric . The lagrangian is :

$$\mathcal{L}(\varphi_a, a^{\mu}_{\alpha}, c_{\alpha}, \bar{c}_{\alpha}) = \mathcal{L}_{inv}(\varphi_a, a^{\mu}_{\alpha}) - \frac{1}{k}\left[\frac{(g^{\alpha})^2}{2} - c_{\alpha}(m\bar{c})^{\alpha}\right] \qquad (10)$$

with

$$g_\alpha = \partial_\mu a^\mu_\alpha + \rho \pi_\alpha \tag{11}$$

and

$$(\mathfrak{M}\bar{c})_\alpha(x) = \int dy \, \frac{\delta g_\alpha(x)}{\delta\omega_\beta(y)} \, \bar{c}_\beta(y) \equiv$$
$$\equiv \left\{ \Box \bar{c}_\alpha + e\,\varepsilon_\alpha{}^{\beta\gamma} \partial_\mu (\bar{c}_\beta a^\mu_\gamma) + \frac{\rho e}{2}\left[\varepsilon_\alpha{}^{\beta\gamma} \bar{c}_\beta \pi_\gamma + (\sigma+F)\bar{c}_\alpha\right] \right\}(x \tag{12}$$

This new lagrangian can be straightforwardly quantized in a Fock space with indefinite metric ; the corresponding perturbation theory is renormalizable by power counting. Indeed now :

$$\det \left\| K^{\alpha\beta}_{a_{L},\pi}(p^2) \right\| \propto \left(p^2 - \frac{\rho e F}{2}\right)^2 \text{ and } K_{c\bar{c}}(p^2) = \frac{1}{K}\left(p^2 - \frac{\rho e F}{2}\right) \tag{13}$$

However choosing the $\Phi.\Pi.$ lagrangian we have increased the number of the unphysical degrees of freedom of our model. We are then left with the problem of proving that the restriction of the S operator to the above defined physical subspace of the free field Fock space is unitary in the perturbative sense and independent from the parameters specifying the $\Phi.\Pi.$ term of the lagrangian. In the tree approximation this is true if the matrix elements of the gauge operator $g_\alpha(x)$ within physical states vanish. This generalizes the well known Q.E.D. supplementary condition. We are now going to show that this follows from the Slavnov identity, which at the classical level expresses the invariance of the lagrangian under the following set of infinitesimal transformations (Slavnov transformations)

$$\begin{aligned}
\delta\pi_\alpha(x) &= \delta\lambda \int dy \, \frac{\delta\pi_\alpha(x)}{\delta\omega_\beta(y)} \, \bar{c}_\beta(y) \equiv \delta\lambda \, P_\alpha(x) \\
\delta\sigma(x) &= \delta\lambda \int dy \, \frac{\delta\sigma(x)}{\delta\omega_\beta(y)} \, \bar{c}_\beta(y) \equiv \delta\lambda \, P_0(x) \\
\delta a^\mu_\alpha(x) &= \delta\lambda \int dy \, \frac{\delta a^\mu_\alpha(x)}{\delta\omega_\beta(y)} \, \bar{c}_\beta(y) \equiv \delta\lambda \, P^\mu_\alpha(x) \\
\delta\bar{c}_\alpha(x) &= \delta\lambda \, \frac{e}{2} \, \varepsilon_\alpha{}^{\beta\gamma} (\bar{c}_\beta \bar{c}_\gamma)(x) \equiv \delta\lambda \, \overline{P}_\alpha(x) \\
\delta c_\alpha(x) &= \delta\lambda \, g_\alpha(x)
\end{aligned} \tag{14}$$

$\delta\lambda$ is a space-time independent infinitesimal parameter commuting with $\{\varphi_a\}$ and $\{a_{\mu\alpha}\}$ but anticommuting with the $\Phi.\Pi.$ ghost fields, and for two transformations labelled by $\delta\lambda_1$ and $\delta\lambda_2$, $\delta\lambda_1$ and $\delta\lambda_2$ anticommute.

The Slavnov transformation laws can be cast together introducing the functional differential operator :

$$\mathcal{S} = \int dx \left[P_\alpha \frac{\delta}{\delta \pi_\alpha} + P_0 \frac{\delta}{\delta \sigma} + P_\alpha^\mu \frac{\delta}{\delta a_\alpha^\mu} + \bar{P}_\alpha \frac{\delta}{\delta \bar{c}_\alpha} + g_\alpha \frac{\delta}{\delta c_\alpha} \right](x) \tag{15}$$

Then the transformation law for the field $\{\psi\} = \{\varphi_a, a_\mu^\alpha, c_\alpha, \bar{c}_\alpha\}$ is :

$$\delta\psi = \delta\lambda\, \mathcal{S}\, \psi \tag{16}$$

The invariance of $\mathcal{L}$ under the transformations (14) can be checked immediately taking into account the composition law of the gauge transformations :

$$\left[\frac{\delta}{\delta\omega_\alpha(x)}, \frac{\delta}{\delta\omega_\beta(y)} \right] = e\, \varepsilon^{\alpha\beta}_{\gamma}\, \delta(x-y) \frac{\delta}{\delta\omega_\gamma(x)} \tag{17}$$

Let us now introduce the system of external fields :

$$\{\eta\} = \{\gamma_\alpha, \gamma_0, \gamma_\alpha^\mu, \zeta_\alpha\} \tag{18}$$

whose couplings are defined by the new lagrangian ;

$$\mathcal{L}^{(\eta)}(\psi, \eta) = \mathcal{L}(\psi) + \gamma_\alpha P^\alpha + \gamma_0 P^0 + \gamma_\alpha^\mu P_\mu^\alpha + \zeta_\alpha \bar{P}^\alpha \tag{19}$$

Since, owing to the group structure underlying the Slavnov transformations, the polynomials $\{P_\alpha, P_\alpha^\mu, P_0, \bar{P}_\alpha\}$ are invariant, also $\mathcal{L}^{(\eta)}$ is invariant. This fact can be translated by means of the Q.A.P. into a system of identities relating the Green functions of the theory whenever the possible quantum corrections are absent, hence, at least in the tree approximation. These identities are easily expressed in terms of the connected Green functional Z_c [6]. Upon introducing the sources $\{\mathcal{J}\} = \{J_\alpha, J_0, j_\alpha^\mu, \xi_\alpha, \bar{\xi}_\alpha\}$ for the fields $\{\psi\} = \{\pi_\alpha, \sigma, a_\alpha^\mu, c_\alpha, \bar{c}_\alpha\}$ respectively yields:

$$0 = \mathcal{S} Z_c(\mathcal{J}, \eta) \equiv \tag{20}$$

$$\equiv \int dx \left[J_\alpha \frac{\delta}{\delta\gamma_\alpha} + J_0 \frac{\delta}{\delta\gamma_0} + j_\alpha^\mu \frac{\delta}{\delta\gamma_\alpha^\mu} - \zeta_\alpha \frac{\delta}{\delta\xi_\alpha} - \bar{\xi}_\alpha \left(\partial_\mu \frac{\delta}{\delta j_{\mu\alpha}} + \rho \frac{\delta}{\delta J_\alpha} \right) \right](x) Z_c(\mathcal{J}, \eta)$$

The corresponding identity for the vertex functional :

$$\Gamma(\psi,\eta) = Z_c(J,\eta) - \int dx\,(J,(\psi+F))(x)\Big|_{\psi+F=\frac{\delta Z_c}{\delta J}} \tag{21}$$

is

$$\int dx\Big[\frac{\delta}{\delta\pi_\alpha}\Gamma\frac{\delta}{\delta\gamma_\alpha}\Gamma + \frac{\delta}{\delta\sigma}\Gamma\frac{\delta}{\delta\gamma_0}\Gamma + \frac{\delta}{\delta a^\alpha_\mu}\Gamma\frac{\delta}{\delta\gamma^\mu_\alpha}\Gamma + \\ + \frac{i}{\delta\bar{c}_\alpha}\Gamma\frac{\delta}{\delta\zeta^\alpha}\Gamma + g_\alpha\frac{\delta}{\delta c_\alpha}\Gamma\Big](\psi,\eta)(x) = 0 \tag{22}$$

Let us now recall that we have defined physical degrees of freedom of our model the transverse components of the photons and the σ field. Comparing with Eq.(14) we see that the Slavnov transformed of the physical fields do not contain any term linear in c . In terms of Feynman diagrams this property can be translated into the following : if ψ_{phys} is any physical field, an amplitude containing $s\,\psi_{phys}$ as a vertex is one particle irreducible with respect to cuts between this vertex and the gauge function g^α, since from Eq.(10) and Eq.(14) we see that :

$$s\,g_\alpha(x) = (m\bar{c})_\alpha(x) = k\,\xi_\alpha(x) \tag{23}$$

Hence this property of the physical fields is not altered if we combine them with g_α .

From the L.S.Z. reduction formulae it turns out that we can write the restriction of the S operator to the physical space in the form [6] :

$$S_{phys} = :\exp\int dx\,dy\Big\{\sigma^{in}(x)K_\sigma(x,y)\frac{\delta}{\delta J_0(y)} + a^{T\,in}_{\mu\alpha}(x)K^{\alpha\beta}_{\mu\nu}(x,y)\frac{\delta}{\delta j^\beta_\nu(y)}\Big\}: \\ \exp\frac{i}{\hbar}Z_c(J,\eta)\Big|_{J=\eta=0} \equiv \Sigma_{phys}\,Z(J,\eta)\Big|_{J=\eta=0} \tag{24}$$

where σ^{in}, K_σ , $a^{T,in}_{\mu,\alpha}$, $K^{\mu\nu}_{\alpha\beta}$ are the canonically quantized asymptotic fields and the corresponding wave operators. Now the previously mentioned property of the physical fields implies that the operator s defined in Eq.(15) and Σ_{phys} commute, since Σ_{phys} amputates on the mass shell the physical legs. It follows that the field g^α is decoupled from the physical states :

$$\frac{\hbar}{i}\frac{\delta}{\delta\xi_\alpha(0)}\sum_{phys} s\, Z_c(J,\eta)\Big|_{J=\eta=0} = \frac{\hbar}{i}\left(\partial_\mu \frac{\delta}{\delta J^\alpha_\mu(0)} + \rho\frac{\delta}{\delta J_\alpha(0)}\right)\sum_{phys} Z_c(J,\eta)\Big|_{J=\eta=0} = 0 \tag{25}$$

This is the desired supplementary condition.

This fulfillement of the Slavnov identity can be considered as the fundamental property of the G.F.M.. However to base our renormalization program on the Slavnov identity we have to verify that our lagrangian is uniquely determined (up to the addition of a divergence and a multiplicative renormalization of the fields) by the condition of Slavnov invariance, the dimensional constraints ensuring the renormalizability of the theory being understood. This is easily verified if the gauge group is semisimple. In the non semisimple case there may arise new terms. For instance in Q.E.D. in the Stueckelberg gauge the mass term:

$$\int dx \left[\frac{a_\mu^2}{2} + c\bar{c}\right](x) \tag{26}$$

is Slavnov invariant.

If we assign dimension two to the external fields and we introduce the Φ.Π.charge $Q^{\Phi\Pi}$:

$$Q^{\Phi\Pi}\zeta_\alpha = 2\zeta_\alpha$$

$$Q^{\Phi\Pi}c_\alpha = c_\alpha\,,\quad Q^{\Phi\Pi}\gamma_\alpha = \gamma_\alpha\,,\quad Q^{\Phi\Pi}\gamma_0 = \gamma_0,\quad Q^{\Phi\Pi}\gamma^\alpha_\mu = \gamma^\alpha_\mu\,,$$

$$Q^{\Phi\Pi}\pi_\alpha = Q^{\Phi\Pi}\sigma = Q^{\Phi\Pi}a^\alpha_\mu = 0 \tag{27}$$

$$Q^{\Phi\Pi}\bar{c}_\alpha = -\bar{c}_\alpha$$

the most general dimension four lagrangian carrying null charge is :

$$\bar{\mathcal{L}}^{(\eta)}(\psi,\eta) = \bar{\mathcal{L}}(\psi) + \gamma_\alpha \mathcal{P}^\alpha + \gamma_0\mathcal{P}^0 + \gamma^\mu_\alpha \mathcal{P}^\alpha_\mu + \zeta_\alpha\bar{\mathcal{P}}^\alpha \tag{28}$$

where the polynomials $\{\mathcal{P}^\alpha, \mathcal{P}^0, \mathcal{P}^\alpha_\mu, \bar{\mathcal{P}}^\alpha\}$ have dimension non exceeding two. The Slavnov invariance of the action is expressed in the functional differential form :

$$\int dx\left[\mathcal{P}_\alpha \frac{\delta}{\delta\pi_\alpha} + \mathcal{P}_0 \frac{\delta}{\delta\sigma} + \mathcal{P}^\mu_\alpha \frac{\delta}{\delta a^\mu_\alpha} + \bar{\mathcal{P}}^\alpha \frac{\delta}{\delta\bar{c}_\alpha} + g_\alpha \frac{\delta}{\delta c_\alpha}\right](x) \int \bar{\mathcal{L}}^{(\eta)}(y)\, dy$$
$$= \bar{s} \int dy\ \bar{\mathcal{L}}^{(\eta)}(y) = 0 \tag{29}$$

Looking in particular at the coefficients of the external fields we get :

$$\bar{s}\,\mathcal{P}_\alpha = \bar{s}\,\mathcal{P}_0 = \bar{s}\,\mathcal{P}^\mu_\alpha = \bar{s}\,\bar{\mathcal{P}}_\alpha = 0 \tag{30}$$

from which the structure of the Slavnov transformations (Eq.(14)) can be reconstructed. Indeed, for example, the general form of $\bar{\mathcal{P}}_\alpha$ is :

$$\bar{\mathcal{P}}_\alpha = \frac{1}{2}\, \Gamma_\alpha^{\beta\gamma}\, \bar{c}_\beta \bar{c}_\gamma \tag{31}$$

and Eq.(30)) writes :

$$\bar{s}\,\bar{\mathcal{P}}_\alpha = \frac{1}{2}\, \Gamma_\alpha^{\beta\lambda}\,\Gamma_\lambda^{\gamma\delta}\, \bar{c}_\beta \bar{c}_\gamma \bar{c}_\delta = 0 \tag{32}$$

which is nothing but the Jacobi identity :

$$\Gamma_\alpha^{\beta\lambda}\,\Gamma_\lambda^{\gamma\delta} + \Gamma_\alpha^{\gamma\lambda}\,\Gamma_\lambda^{\delta\beta} + \Gamma_\alpha^{\delta\lambda}\,\Gamma_\lambda^{\beta\gamma} = 0 \tag{33}$$

whose solutions are stable under perturbations [13]. That is if a set of structure constants differs from another one by perturbation, to the first order in the perturbation the two sets are equivalent.

Now, concerning the external field independent part of the Lagrangian, if $\mathcal{L}(\psi)$ is Slavnov invariant a fortiori :

$$s^2 \int dx\ \bar{\mathcal{L}}(\psi)(x) \equiv \int dx\,(\partial_\mu P^\mu_\alpha + \rho P_\alpha)(x) \frac{\delta}{\delta c_\alpha(x)} \int dy\ \bar{\mathcal{L}}(\psi)(y)$$
$$= \int dx\left[(m\bar{c})_\alpha \frac{\delta}{\delta c_\alpha}\right](x) \int dy\ \bar{\mathcal{L}}(\psi)(y) = 0 \tag{34}$$

Making the dependence on the $\Phi.\Pi.$ fields explicit, $\bar{\mathcal{L}}(\psi)$ assumes the form :

$$\bar{\mathcal{L}}(\psi) = \bar{\mathcal{L}}_{INV}(\varphi_a, a^\mu_\alpha) + \Delta\mathcal{L}(\varphi_a, a^\mu_\alpha) +$$
$$+ L^{\alpha\beta\gamma\delta}\, c_\alpha c_\beta \bar{c}_\gamma \bar{c}_\delta + c_\alpha (K\bar{c})^\alpha \tag{35}$$

where $\bar{\mathcal{L}}_{\mathrm{inv}}$ is invariant under gauge transformations and the corresponding action is Slavnov invariant. By Eq.(34) it follows that :

$$L^{\alpha\beta\gamma\delta} = 0 \tag{36}$$

and that :

$$\int dx \, (m\bar{c}_\alpha)(x) \, (K\bar{c})^\alpha(x) = 0 \tag{37}$$

The general solution of Eq.(37) is, if the gauge group is semisimple :

$$(K\bar{c})_\alpha = \Gamma_{\alpha\alpha'} (m\bar{c})^{\alpha'} \tag{38}$$

where $\Gamma_{\alpha\alpha'}$ is a symmetrical matrix. Going back to Eq.(29) yields :

$$\bar{\mathcal{L}}(\psi) = \bar{\mathcal{L}}_{\mathrm{inv}}(\varphi_a, a^\mu_\alpha) + c_\alpha \Gamma^{\alpha\alpha'} (m\bar{c})_{\alpha'} - \frac{1}{2} g_\alpha \Gamma^{\alpha\alpha'} g_{\alpha'} \tag{39}$$

which is identical with Eq.(12) modulo a redefinition of g^α and c^α.

3. RENORMALIZATION OF THE SLAVNOV IDENTITY

In this section we discuss the existence of quantum extensions of G.F.M. with semisimple gauge group, satisfying the appropriate Slavnov identities.

In order to study the general case we shall use a slightly different notation labelling the quantized fields by :

$$\{\psi\} = \{\phi_i, c_\alpha, \bar{c}_\alpha\} \tag{40}$$

with $\{\phi_i\} = \{\varphi_a, a^\mu_\alpha\}$. Here $\{c_\alpha, \bar{c}_\alpha\}$ are the Φ.Π. fields, $\{a^\mu_\alpha\}$ the gauge photons, and $\{\varphi_a\}$ a system of matter fields. We also introduce the sources:

$$\{\mathcal{J}\} = \{J_i, \bar{\xi}_\alpha, \xi_\alpha\} \tag{41}$$

and the external fields :

$$\{\eta\} = \{\gamma_i, \zeta_\alpha\} \tag{42}$$

The $\Phi.\Pi.$ charge and dimension assignements are the natural extensions of those in the previous section.

The gauge function g^α is chosen linear in the fields:

$$g^\alpha(x) = g_i^\alpha \, \phi_i(x) \qquad \text{with} \qquad g^\alpha_{\beta,\mu} = g^\alpha_\beta \, \partial_\mu \tag{43}$$

A quantized extension of a G.F.M. is defined by means of an effective lagrangian :

$$\mathcal{L}_{eff}(\psi,\eta) = \mathcal{L}_{eff}(\psi) + \gamma^i P_i + \zeta^\alpha \bar{P}_\alpha \tag{44}$$

to which the Zimmermann [5] subtraction index four is assigned.

The lagrangian $\mathcal{L}_{eff}(\psi,\eta)$ is a formal power series in $\hbar$:

$$\mathcal{L}_{eff}(\psi,\eta) = \sum_0^\infty {}_n \, \hbar^n \, \overset{(n)}{\mathcal{L}}_{eff}(\psi,\eta) \tag{45}$$

Hence in particular :

$$\overset{-(0)}{P}_\alpha(\psi) = \frac{e}{2} \overset{(0)}{f}_\alpha{}^{\beta\gamma} \, \bar{c}_\beta \, \bar{c}_\gamma \tag{46}$$

and

$$\overset{(0)}{P}_i(\psi) = e\left(\overset{(0)}{t}{}^\alpha_{ij} \, \phi_j + \overset{(0)}{q}{}^\alpha\right) \qquad \text{with :} \quad \overset{(0)}{q}{}^\alpha_a = \overset{(0)}{t}{}^\alpha_{ab} \, q_b \tag{47}$$

$$\text{and} \quad \overset{(0)}{q}{}^\alpha_{\beta\mu} = \frac{1}{e} \, \delta^\alpha_\beta \, \partial_\mu$$

The coefficients $\overset{(0)}{f}_\alpha{}^{\beta\gamma}$ and $\overset{(0)}{t}{}^\alpha_{ij}$ are the structure constants and a representation of the Lie algebra associated with the gauge group.

The purpose of this section is to show that the Slavnov identity :

$$\delta Z_c(J,\eta) \equiv \int dx \left[J_i \frac{\delta}{\delta \gamma_i} - \xi_\alpha \frac{\delta}{\delta \xi_\alpha} - \bar{\xi}_\alpha g_i^\alpha \frac{\delta}{\delta J_i} \right](x) Z_c(J,\eta) = 0 \quad (48)$$

can be solved in terms of the coefficients of $\mathcal{L}_{eff}(\psi,\eta)$. These coefficients are themselves formal power series in $\hbar$.

We shall proceed in two steps. First we shall discuss the necessary condition :

$$\delta^2 Z_c(J,\eta) \equiv \int dx \left(\bar{\xi}_\alpha g_i^\alpha \frac{\delta}{\delta \gamma_i} \right)(x)\, Z_c(J,\eta) = 0 \quad (49)$$

Then we shall go back to Eq.(48).

After a Legendre transformation Eq.(49) becomes :

$$\int dx \left(\frac{\delta}{\delta C_\alpha} \Gamma\, g_\alpha^i \frac{\delta}{\delta \gamma_i} \Gamma \right)(x) = 0 \quad (50)$$

which is solved by :

$$\frac{\delta}{\delta C_\alpha(x)} \Gamma(\psi,\eta) - \kappa g_i^\alpha \frac{\delta}{\delta \gamma_i(x)} \Gamma(\psi,\eta) \equiv D_\alpha(x)\Gamma(\psi,\eta) = 0 \quad (51)$$

Now for a generic choice of the effective lagrangian the Q.A.P. yields :

$$D_\alpha(x)\,\Gamma(\psi,\eta) = \left(\Delta_\alpha(x)\cdot\Gamma\right)_3(\psi,\eta) \quad (52)$$

where the right-hand side means the insertion of the vertex $N_3(\Delta_\alpha(x))$ into the appropriate Green function. The vertex $\Delta_\alpha(x)$ is a polynomial in the fields and their derivatives whose coefficients are formal power series in $\hbar$:

$$\Delta_\alpha(x) = \sum_\nu^\infty n\, \hbar^n \Delta_\alpha^{(n)}(x) \qquad \text{with } \nu \geq 1 \quad (53)$$

To prove that Eq.(51) can be solved in terms of $\mathcal{L}_{eff}$ is equivalent to show that the equation :

$$\Delta_\alpha^{(\nu)} = 0 \quad (54)$$

can be solved in terms of $\mathcal{L}_{eff}^{(\nu)}$ for any $\nu \geq 1$. Now we know that :

$$\Delta_\alpha^{(\nu)}(x) = D_\alpha(x) \int dy\, \mathcal{L}_{eff}^{(\nu)}(\psi,\eta)(y) + \mathcal{Q}_\alpha^{(\nu)}(x, \mathcal{L}_{eff}) \qquad (55)$$

where $\mathcal{Q}_\alpha^{(\nu)}(x, \mathcal{L}_{eff})$ sums up the quantum correction to the naive version of the Q.A.P. and consequently depends on $\mathcal{L}_{eff}^{(\kappa)}$ only for κ smaller than ν. Clearly Eq.(54) can be solved if:

$$\mathcal{Q}_\alpha^{(\nu)}(x, \mathcal{L}_{eff}) = D_\alpha(x)\, \hat{\mathcal{Q}}(\mathcal{L}_{eff}) \qquad (56)$$

for some local functional $\hat{\mathcal{Q}}(\mathcal{L}_{eff})$ of dimension four. From a simple analysis it turns out that Eq.(56) is fulfilled if $\mathcal{Q}_\alpha^{(\nu)}(x, \mathcal{L}_{eff})$ and consequently $\Delta_\alpha^{(\nu)}(x)$ do not contain terms of the form :

$$\Gamma^{\widetilde{\alpha\beta}\gamma\delta}(c_\beta \bar{c}_\gamma \bar{c}_\delta)(x) \qquad (57)$$

where the symbol $\widetilde{\alpha\beta}$ means symmetry under the exchange of the two indices. Now, taking into account the anticommuting character of the operator $D_\alpha(x)$, we have

$$\left[D_\alpha(x), D_\beta(y)\right]_+ \Gamma(\psi,\eta) = \left\{D_\alpha(x)\left(\Delta_\beta(y).\Gamma\right)_3 + D_\beta(y)\left(\Delta_\alpha(x).\Gamma\right)_3\right\}(\psi,\eta) = 0 \qquad (58)$$

Since picking out of Eq.(52) the term of order $\hbar^\nu$ yields :

$$\left(\Delta_\beta(y)\,\Gamma\right)_3^{(\nu)}(\psi,\eta) = \Delta_\beta^{(\nu)}(y) \qquad (59)$$

we have to the same order the consistency condition :

$$D_\alpha(x)\, \Delta_\beta^{(\nu)}(y) + D_\beta(y)\, \Delta_\alpha^{(\nu)}(x) = 0 \qquad (60)$$

from which the absence from $\Delta_\alpha^{(\nu)}(x)$ of terms of the form (57) follows immediately.

This is a particularly simple example of our general algebraic method. The possible quantum anomalies to the Q.A.P. are constrained, in addition to power counting, by a system of consistency conditions which to the first non vanishing order in $\hbar$ assume the naive (classical) form (as in Eq.(60)). We shall apply this method to the Slavnov identity and show that if the gauge group is semisimple, the consistency conditions are sufficient to ensure that by a suitable choice of the effective lagrangian all the possible anomalies can be eliminated in the absence of the Adler-Bardeen anomaly [16]. This is the full exploitation of the idea, due to Wess and Zumino [17], of defining the Adler-Bardeen anomaly by means of a system of consistency conditions.

We now go back to the Slavnov identity. Assuming from now on that the $\Phi.\Pi.$ part of the effective lagrangian is determined in such a way as to satisfy Eq.(49) we get from the Q.A.P.:

$$\mathcal{S} Z_c (J,\eta) = (\Delta . Z_c)_S (J,\eta) \tag{61}$$

where the vertex :

$$\Delta = \int dx \, \Delta(x) \tag{62}$$

has dimension lower than six and :

$$\Delta(x) = \left[- \mathcal{S}^{(P)} \mathcal{L}_{eff}(\psi,\eta)(x) + \hbar \, \mathcal{Q}(x, \mathcal{L}_{eff}, \hbar) \right] \tag{63}$$

where :

$$\mathcal{S}^{(P)} = \int dx \left[P_i \frac{\delta}{\delta \phi_i} + \overline{P}_\alpha \frac{\delta}{\delta \overline{c}_\alpha} + g_\alpha \frac{\delta}{\delta c_\alpha} \right](x) \tag{64}$$

Furthermore there exists an integer $\nu \geq 1$ such that :

$$\Delta(x) = \sum_{\nu}^{\infty} {}_n \, \hbar^n \Delta^{(n)}(x) \tag{65}$$

and from Eq.(63):

$$\Delta^{(\nu)}(x) = - \left(\mathcal{S}^{(P^{(\nu)})} \overset{(0)}{\mathcal{L}}_{eff}(\psi,\eta) + \mathcal{S}_0 \overset{(\nu)}{\mathcal{L}}_{eff}(\psi,\eta) \right)(x) + \\ + R^{(\nu)}(\mathcal{L}_{eff}, x) \tag{66}$$

where

$$\mathcal{S}_0 = \mathcal{S}^{(P)} \Big|_{\substack{P_i = P_i^{(0)} \\ \overline{P}_\alpha = \overline{P}_\alpha^{(0)}}} \quad \text{and} \quad \mathcal{S}^{(P^{(\nu)})} = \mathcal{S}^{(P)} \Big|_{\substack{P_i = P_i^{(\nu)} \\ \overline{P}_\alpha = \overline{P}_\alpha^{(\nu)} \\ g^\alpha = 0}} \tag{67}$$

The remainder $R^{(\nu)}$ depends on the coefficients of $\mathcal{L}_{eff}(\psi,\eta)$ up to order $\nu - 1$. Making explicit the dependence on the external fields we have :

$$\Delta^{(\nu)}(x) = \Delta_0^{(\nu)}(x) + \left[\gamma^i \Delta_i^{(\nu)}\right](x) + \left[\zeta^\alpha \Delta_\alpha^{(\nu)}\right](x) \tag{68}$$

with :

$$\begin{aligned} \Delta_0^{(\nu)}(x) &= -\left[\delta^{(P^{(\nu)})} \mathcal{L}_{eff}^{(0)}(\psi) + \delta_0 \mathcal{L}_{eff}^{(\nu)}(\psi)\right] + R_0^{(\nu)}(x, \mathcal{L}_{eff}) \\ \Delta_i^{(\nu)}(x) &= \delta^{(P^{(\nu)})} P_i^{(0)}(x) + \delta_0 P_i^{(\nu)}(x) + R_i^{(\nu)}(x, \mathcal{L}_{eff}) \\ \Delta_\alpha^{(\nu)}(x) &= -\left[\delta^{(P^{(\nu)})} \bar{P}_\alpha^{(0)} + \delta_0 \bar{P}_\alpha^{(\nu)}\right](x) + R_\alpha^{(\nu)}(x, \mathcal{L}_{eff}) \end{aligned} \tag{69}$$

This is the structure of the possible anomalies to the Slavnov identity. However if Eq.(49) is satisfied, $\Delta(x)$ is constrained by a system of consistency conditions which we are now going to derive. First we introduce a new external anticommuting field $\beta(x)$ carrying $\Phi.\Pi.$ charge $Q^{\Phi\Pi} = 1$ and coupled to the vertex $N_5[\Delta(x)]$. In the theory defined by the new lagrangian :

$$\begin{aligned} \mathcal{L}_{eff}(\psi, \eta, \beta) &= \left[\mathcal{L}_{eff}(\psi, \eta) + (\beta\Delta)\right](x) \\ &= \mathcal{L}_{eff}(\psi) + \beta\Delta_0(\psi) + \gamma^i(P_i - \beta\Delta_i) + \zeta^\alpha(\bar{P}_\alpha + \beta\Delta_\alpha) \end{aligned} \tag{70}$$

the Slavnov identity writes :

$$\begin{aligned} \delta Z_c(J, \eta, \beta) &= (\Delta . Z_c)_5 (J, \eta, \beta) + (\Delta^{(\beta)} . Z_c)_5 (J, \eta, \beta) \\ &= \int dx \frac{\delta}{\delta\beta(x)} Z_c(J, \eta, \beta) + (\Delta^{(\beta)} . Z_c)_5 (J, \eta, \beta) \end{aligned} \tag{71}$$

In the right-hand side of Eq.(71) the insertion $\Delta^{(\beta)}$ lumps together the quantum corrections involving β -couplings. To order ν these new terms can only arise from the naive variation of $\Delta(x)$ since the β -dependent radiative corrections appear to an order in $\hbar$ strictly higher than ν (that of the new coupling). Hence we can write :

$$\Delta^{(\beta)} = \int dx\, \beta(x) \left\{ \left[\delta^{(\Delta)} \mathcal{L}_{eff}^{(0)}(\psi) + \gamma^i P_i^{(0)} + \zeta^\alpha \bar{P}_\alpha^{(0)} + \delta_0 \Delta \right] + \hbar^{\nu+1} Q^{(\beta)} \right\}(x) \tag{72}$$

with :

$$\delta^{(\Delta)} = \int dx \left[\Delta_i \frac{\delta}{\delta\phi_i} - \Delta_\alpha \frac{\delta}{\delta \bar{c}_\alpha} \right](x) \tag{73}$$

Now we can compute :

$$\delta^2 Z_c(J,\eta) = \delta^2 Z_c(J,\eta,\beta)\Big|_{\beta=0} = \delta\int dx \frac{\delta}{\delta\beta(x)} Z_c(J,\eta,\beta)\Big|_{\beta=0}$$
$$= -\int dx \frac{\delta}{\delta\beta(x)} \delta Z_c(J,\eta,\beta)\Big|_{\beta=0} = -\int dx\left(\left[\frac{\delta}{\delta\beta(x)}\Delta^{(\beta)}\right]\cdot Z_c\right)_{\beta}(J,\eta)=0 \tag{74}$$

since :

$$\int dx\, dy \frac{\delta^2}{\delta\beta(x)\,\delta\beta(y)} = 0 \tag{75}$$

vanishes owing to the anticommuting character of the external field .

To order ν Eq.(74) yields the desired system of consistency conditions for Δ :

$$\begin{aligned} &\delta^{(\Delta^{(\nu)})}\bar{P}_\alpha^{(0)} + \delta_0 \Delta_\alpha^{(\nu)} = 0 \\ &\delta^{(\Delta^{(\nu)})} P_i^{(0)} - \delta_0 \Delta_i^{(\nu)} = 0 \\ &\int dx\left[\delta^{(\Delta^{(\nu)})}\mathcal{L}_{eff}^{(0)}(\psi) + \delta_0 \Delta_0^{(\nu)}(\psi)\right](x) = 0 \end{aligned} \tag{76}$$

The first equation ensures that the system :

$$\Delta_\alpha^{(\nu)} = \Delta_i^{(\nu)} = 0 \tag{77}$$

can be solved in terms of $\bar{P}_\alpha^{(\nu)}$ and $P_i^{(\nu)}$. Indeed by a completely algebraic (cohomological) method which is fully exhibited in Appendix A and B of reference [13], it is possible to show that the first two consistency equations yield :

$$\begin{aligned} \Delta_\alpha^{(\nu)} &= -\left(\delta^{(\Pi^{(\nu)})}\bar{P}_\alpha^{(0)} + \delta_0 \Pi_\alpha^{(\nu)}\right) \\ \Delta_i^{(\nu)} &= \delta^{(\Pi^{(\nu)})} P_i^{(0)} + \delta_0 \Pi_i^{(\nu)} \end{aligned} \tag{78}$$

with :

$$\delta^{(\Pi^{(\nu)})} = \int dx\left(\Pi_i^{(\nu)}\frac{\delta}{\delta\phi_i} + \Pi_\alpha^{(\nu)}\frac{\delta}{\delta\bar{c}_\alpha}\right)(x) \tag{79}$$

for some $\Pi_i^{(\nu)}$ and $\Pi_\alpha^{(\nu)}$ linear in $\Delta_i^{(\nu)}$ and $\Delta_\alpha^{(\nu)}$.

Comparing Eq.(79) with Eq.(69) we get :

$$R_\alpha^{(\nu)}(x,\mathcal{L}_{eff}) = \mathcal{S}^{(P^{(\nu)})}\bar{P}_\alpha^{(0)} + \mathcal{S}_0\, \mathcal{P}_\alpha^{(\nu)}$$

$$R_i^{(\nu)}(x,\mathcal{L}_{eff}) = -\left(\mathcal{S}^{(P^{(\nu)})}P_i^{(0)} + \mathcal{S}_0\, \mathcal{P}_i^{(\nu)}\right) \qquad (80)$$

for some $\mathcal{P}_i^{(\nu)}$ and $\mathcal{P}_\alpha^{(\nu)}$ depending on the coefficients of $\mathcal{L}_{eff}(\varphi,\eta)$ up to order $\nu-1$. Then the system (77) is solved by the choice :

$$\bar{P}_\alpha^{(\nu)} = \mathcal{P}_\alpha^{(\nu)} \quad \text{and} \quad P_i^{(\nu)} = \mathcal{P}_i^{(\nu)} \qquad (81)$$

We are then left with the external field independent part of the anomaly $\Delta_0^{(\nu)}(x)$ subject to the condition :

$$\mathcal{S}_0 \int dx\, \Delta_0^{(\nu)}(x) = 0 \qquad (82)$$

and we want to show that the equation :

$$\int dx\, \Delta_0^{(\nu)}(x) = 0 \qquad (83)$$

can be solved in terms of $\mathcal{L}_{eff}^{(\nu)}(\varphi)$. Comparing with Eq.(69) we see that this is ensured if we can show that :

$$\int dx\, \Delta_0^{(\nu)}(x) = \mathcal{S}_0 \int dx\, \hat{\Delta}_0^{(\nu)}(x) \qquad (84)$$

for some $\hat{\Delta}_0^{(\nu)}$ linear in $\Delta_0^{(\nu)}$. Making explicit the dependence on the $\Phi.\Pi.$ fields we can decompose $\Delta_0^{(\nu)}$ into the form :

$$\Delta_0^{(\nu)}(x) = (\Delta^\alpha \bar{C}_\alpha)(x) + \int dy\, dz \left(C_\alpha \Delta^{\alpha,\beta\gamma}(y,z)\, \bar{C}_\beta(y)\bar{C}_\gamma(z)\right)(x) + \Delta^{\alpha\beta,\gamma\delta\eta}(C_\alpha C_\beta \bar{C}_\gamma \bar{C}_\delta \bar{C}_\eta)(x) \qquad (85)$$

To exploit the constraint (82) we first write :

$$\mathcal{S}_0^2 \int dx\, \Delta_0^{(\nu)}(x) = \int dy \left[(\mathcal{S}_0^2 C_\beta)\frac{\delta}{\delta C_\beta}\right](y) \int dx\, \Delta_0^{(\nu)}(x) = 0 \qquad (86)$$

from which

$$\Delta^{\alpha\beta,\gamma\delta\eta} = 0 \qquad (87)$$

and

$$\int dy\, dz \left(C_\alpha \Delta^{\alpha,\beta\gamma}(y,z)\, \bar{C}_\beta(y)\, \bar{C}_\gamma(z) \right)(x) = \Delta^{\widetilde{\alpha\beta}\gamma} \left(C_\alpha (s_0^2 C_\beta)\, \bar{C}_\gamma \right)(x) \tag{88}$$

follow immediately. Here $\Delta^{\widetilde{\alpha\beta}\gamma}$ is a tensor symmetrical under the permutation of the first two indices. Then applying Eq.(82) directly to $\Delta_0^{(\nu)}$ we get :

$$\Delta^{\widetilde{\alpha\beta}\gamma} = 0$$

and

$$s_0 \int dx \left(\Delta^\alpha \bar{C}_\alpha \right)(x) = 0 =$$

$$= \frac{1}{2} \int dx\, dy\, \bar{C}_\alpha(x)\, \bar{C}_\beta(y) \left\{ \frac{\delta}{\delta\omega_\alpha(x)} \Delta_\beta(y) - \frac{\delta}{\delta\omega_\beta(y)} \Delta_\alpha(x) - f^{\alpha\beta}{}_\gamma\, \delta(x-y)\, \Delta^\gamma(x) \right\} \tag{89}$$

One can show that the general solution of Eq.(89), which is nothing but the Wess-Zumino [17] consistency condition, or, in other words, the first cohomology equation for the gauge Lie algebra, is, if the gauge group is semisimple :

$$\Delta^\alpha(x) = \frac{\delta}{\delta\omega_\alpha(x)} \hat{\Delta}_0 + h^\alpha(x) \tag{90}$$

for some dimension four functional $\hat{\Delta}_0$. Here $h^\alpha(x)$ is the Adler-Bardeen [16] anomaly which has been discussed in Stora's lecture concerning the current algebra Ward identities [6]. Such an anomaly can only arise if the tree lagrangian contains $\varepsilon_{\mu\nu\rho\sigma}$ or γ_5 symbols and if the Lie algebra admits a non trivial invariant completely symmetrical $D^{\alpha\beta\gamma}$ tensor. If the Adler-Bardeen anomaly is absent we get :

$$\Delta_0^{(\nu)} = \int dx \left(\Delta^\alpha \bar{C}_\alpha \right)(x) = s_0 \hat{\Delta}_0 \tag{91}$$

which completes the proof.

4. GAUGE INVARIANCE

Given a gauge model renormalizable in such a way as to preserve the Slavnov identity to all orders of perturbation theory, there remains to show that one can interpret it in physical terms in spite of the presence of many ghost fields. First one should introduce the physical parameters (masses, coupling constants) into the theory, connecting them, through a system of normalization conditions, to the parameters which are left arbitrary in the lagrangian. Then one has to specify a physical subspace within which the S operator is unitary and independent from the properties (masses,...) of the ghost particles.

We shall discuss here in some details the gauge independence of the physical S operator of the SU(2) H.K.E.B. model whose classical limit has been discussed in the first section.

The quantized fields are

$$\{\psi\} = \{\sigma, \pi_\alpha, a^\mu_\alpha, c_\alpha, \bar{c}_\alpha\}$$

The classical lagrangian, given in Eq.(12), is invariant under a group of global (non space dependent) rotations transforming the π^α, a^α_μ, c^α, $\bar{c}^\alpha$ fields as vectors and leaving the σ field invariant. Under this restriction the Slavnov transformations, whose classical form is given in Eq.(14), depends on five parameters. The dependence on three of them, namely ρ, e, and F is explicit in Eq.(14). There are two more hidden parameters which can be introduced by the substitution :

$$\pi_\alpha \to \chi\, \pi_\alpha \qquad \rho \to \rho/\chi$$
$$\bar{c}_\alpha \to \tau\, \bar{c}_\alpha \qquad e \to e/\tau \tag{92}$$

The most general lagrangian invariant under Slavnov transformations :

$$\mathcal{L}(\psi) = -\frac{Z_a}{4}\left(G^a_{\mu\nu}\right)^2 + \frac{Z_1}{2}\left(D_\mu \varphi_a\right)^2 + \frac{\mu^2}{2}(\varphi+F)^2 - \frac{\lambda^2}{4!}\left((\varphi+F)^2\right)^2$$
$$- \left[\frac{\mu^2 F^2}{2} - \frac{\lambda^2}{4!}F^4\right] - \frac{1}{\kappa}\left[\frac{(g^\alpha)^2}{2} - c^\alpha (m\bar{c})_\alpha\right] \tag{93}$$

(compare with Eq.(4) and Eq.(12)), depends on five parameters, namely : Z_a, Z_1, μ, λ, κ. These however have to be adjusted so that the coefficient of the term linear in

σ vanishes :

$$\mu^2 + \frac{F^2\lambda^2}{3!} = 0 \tag{94}$$

The theory thus depends on ten parameters, four specifying the transformation laws, one related with the σ field vacuum expectation value, five specifying the external field independent part of the lagrangian. One can alternatively specify the position of the poles in the transverse photon, σ and c-$\bar{c}$ propagators (m, M, $m_{\phi\pi}$), the residues at these poles (Z_a^{-1}, Z_1^{-1}, κ) and a coupling constant ε, by the normalization conditions :

$$\left(g^{\mu\nu} - \frac{p^\mu p^\rho}{p^2}\right) \frac{\delta}{\delta \tilde{a}^\rho_\alpha(p)} \frac{\delta}{\delta a_{\nu\beta}(0)} \Gamma \Big|_{\psi=\eta=0} \equiv \Gamma^{\mu\nu,\alpha\beta}_{a_T a_T}(p) =$$

$$= - Z_a \, \delta^{\alpha\beta} \left(g^{\mu\nu} - \frac{p^\mu p^\nu}{p^2}\right) \left[p^2 - m^2 + O((p^2-m^2)^2)\right]$$

$$\frac{\delta}{\delta\tilde{\sigma}(p)} \frac{\delta}{\delta\sigma(0)} \Gamma \Big|_{\psi=\eta=0} = \Gamma_{\sigma\sigma}(p) = Z_1 (p^2 - M^2) + O((p^2-M^2)^2) \tag{95}$$

$$\Gamma^{\mu\nu,\alpha\beta}_{a_T a_T \sigma}(m^2, m^2, M^2) = \delta^{\alpha\beta} g^{\mu\nu} \varepsilon$$

$$\frac{\delta}{\delta\tilde{c}^\alpha(p)} \frac{\delta}{\delta\bar{c}^\beta(0)} \Gamma \Big|_{\psi=\eta=0} = \Gamma^{\alpha\beta}_{c\bar{c}}(p) = \frac{\delta_{\alpha\beta}}{\kappa} (p^2 - m^2_{\phi\pi}) + O((p^2 - m^2_{\phi\pi})^2)$$

These normalization conditions together with Eq.(94) which is equivalent to :

$$\frac{\delta}{\delta J_0(x)} Z_c(J,\eta) \Big|_{J=\eta=0} = 0 \tag{96}$$

fix the values of z_a, z_1, μ, λ, κ, ρ, e, F, leaving free the unphysical parameters τ and χ. It is easy to show that, as a consequence of the Slavnov invariance, the masses associated with the coupled longitudinal photon-π channel are pairwise degenerated with those of the c-$\bar{c}$ channel. Hence, owing to the global isotopic spin symmetry, all the ghost masses are degenerate and fixed by the normalization conditions (95).

According to the analysis of section 3 it is possible to find an effective lagrangian fulfilling the Slavnow identity (20) and the normalization conditions (95) and (96) to all orders of perturbation theory. To prove in this theory the gauge invariance of the physical S operator we have to show that:

$$\partial_\lambda \Sigma_{phys} Z(J,\eta)\Big|_{J=\eta=0} = \partial_\lambda S_{phys} = 0 \tag{97}$$

(compare with Eq.(24)), where λ is one of the parameters $\mathcal{K}$, $m_{\phi\pi}$.

Here we shall sketch out the proof of Eq.(97) referring for the details to the existing literature [13]. If "a" is a parameter of our lagrangian we define the invariant derivative with respect to a :

$$D_a = \partial_a + \frac{\partial_a \rho}{\rho}\int dx \left[J_\alpha \frac{\delta}{\delta J_\alpha} + \gamma_\alpha \frac{\delta}{\delta \gamma_\alpha}\right](x) \equiv \partial_a + \frac{\partial_a \rho}{\rho}\Lambda \tag{98}$$

It is easy to verify that :

$$\mathcal{S}\, D_a Z(J,\eta) = 0 \tag{99}$$

and that :

$$\Sigma_{phys}\, \partial_a Z(J,\eta)\Big|_{J=\eta=0} = \Sigma_{phys}\, D_a Z(J,\eta)\Big|_{J=\eta=0} \tag{100}$$

Since the Lowenstein action principles yields ;

$$D_a Z(J,\eta) = \frac{i}{\hbar}\left(\Delta_a^{S} \cdot Z\right)_4 (J,\eta) \tag{101}$$

for some insertion (D.V.O.) $\Delta_a^{(S)}$ of dimension four, we can write Eq.(99) and Eq.(100) in the form :

$$\mathcal{S}\left(\Delta_a^{(S)} \cdot Z\right)_4 (J,\eta) = 0 \tag{102}$$

and

$$\Sigma_{phys}\, \partial_a Z(J,\eta)\Big|_{J=\eta=0} = \frac{i}{\hbar}\Sigma_{phys}\left(\Delta_a^{(S)} \cdot Z\right)_4 (J,\eta)\Big|_{J=\eta=0} \tag{103}$$

In the following we shall call symmetrical an insertion satisfying Eq.(102). We shall also call an insertion Δ non physical if :

$$\sum_{\text{phys}} (\Delta . Z)_4 (J, \eta)\Big|_{J=\eta=0} = 0 \qquad (104)$$

Our aim is to prove that the symmetrical insertion $\Delta_\lambda^{(s)}$ (for $\lambda = \mathcal{K}, m_{\phi\pi}$) is non physical. (Eq.(100)). Now the set of the symmetrical insertions of our model is a linear space of dimension nine, since, for a prescribed Slavnov operator $\mathcal{S}$ (ρ fixed) our theory depends on nine parameters. Also, one can construct explicitly (as in reference [13] section 3-C) in terms of functional differential operators four independent non physical insertions :

$$\Delta_i^{(o,s)} \qquad i = 1, \dots 4 \qquad (105)$$

The remaining five independent symmetrical insertions :

$$\Delta_i^{(s)} \qquad i = 1, \dots . 5 \qquad (106)$$

are such that the matrix with columns :

$$\begin{aligned}
\Delta_{i,1}^{(s)} &= (\Delta_i^{(s)} . \Gamma)_{a_T a_T} (m) \\
\Delta_{i,2}^{(s)} &= \frac{\partial}{\partial p^2} (\Delta_i^{(s)} . \Gamma)_{a_T a_T} (p^2) \Big|_{p^2 = m^2} \\
\Delta_{i,3}^{(s)} &= (\Delta_i^{(s)} . \Gamma)_{\sigma\sigma} (M^2) \\
\Delta_{i,4}^{(s)} &= \frac{\partial}{\partial p^2} (\Delta_i^{(s)} . \Gamma)_{\sigma\sigma} (p^2) \Big|_{p^2 = M^2} \\
\Delta_{i,5}^{(s)} &= (\Delta_i^{(s)} . \Gamma)_{a_T a_T \sigma} (m^2, m^2, M^2)
\end{aligned} \qquad (107)$$

has non vanishing determinant :

$$\det \left\| \Delta_{i,j}^{(s)} \right\| \neq 0 \qquad (108)$$

This can be easily proved by constructing explicitly the five insertions in the tree approximation. Now the insertions (105) and (106) are a basis of the linear space of symmetrical insertions. Hence in particular :

$$\Delta_\lambda^{(s)} = \sum_1^4{}_i\, c_{\lambda,i}\, \Delta_i^{(o,s)} + \sum_1^5{}_j\, d_{\lambda,j}\, \Delta_j^{(s)} \tag{109}$$

and :

$$\partial_\lambda \Gamma(\psi,\eta) = -\frac{\partial_\lambda \rho}{\rho}\, \Lambda\Gamma(\psi,\eta) + \frac{i}{\hbar}\left(\Delta_\lambda^{(s)} \cdot \Gamma\right)(\psi,\eta) \tag{110}$$

The independence from λ of the physical parameters m, M, z_a, z_1, ε is expressed by the system :

$$\begin{aligned}
&(\partial_\lambda \Gamma)_{a_T a_T}(m^2) = 0\\
&(\partial_\lambda \partial_{p^2} \Gamma)_{a_T a_T}(m^2) = 0\\
&(\partial_\lambda \Gamma)_{\sigma\sigma}(M^2) = 0\\
&(\partial_\lambda \partial_{p^2} \Gamma)_{\sigma\sigma}(M^2) = 0\\
&(\partial_\lambda \Gamma)_{a_T a_T \sigma}(m^2\, m^2\, M^2) = 0
\end{aligned} \tag{111}$$

which, taking into account Eq.(110) is equivalent to :

$$\sum_1^5{}_j\, d_{\lambda,j}\, \Delta_{j,i}^{(s)} = 0 \qquad i = 1, \ldots 5 \tag{112}$$

which completes the proof.

Let us now give an outline of the connection between the Slavnov symmetry and the unitarity of the restriction of the S operator to the physical space [12,13].

We shall first introduce a new coordinate frame for the unphysical degrees of freedom of our model. Namely we shall replace the longitudinal photon and π fields by their independent linear combinations g^α (the gauge function) and :

$$\bar{g}_\alpha = \frac{\rho\pi_\alpha - \partial_\mu a^\mu_\alpha}{2\rho\,\Gamma_\gamma(m^2_{\phi\pi})} \tag{114}$$

where :

$$\Gamma_\gamma(p^2) = \frac{\delta}{\delta\tilde{\bar{c}}_\alpha(p)}\,\frac{\delta}{\delta\gamma^\alpha(0)}\,\Gamma(\psi,\eta)\Big|_{\psi=\eta=0} \tag{115}$$

Owing to the Slavnov symmetry the g-$\bar{g}$ propagator has a simple pole at $p^2 = m^2_{\phi\pi}$ with the same residue as the c-$\bar{c}$ propagator. Is is also easy to verify that the Fourier transformed g-g propagator is independent from the momentum.

We shall also put ourselves in the restricted t'Hooft gauge defined by the normalization condition :

$$i p_\mu \frac{\delta}{\delta\tilde{a}_{\mu\alpha}(p)}\,\frac{\delta}{\delta\pi_\beta(0)}\,\Gamma(\psi,\eta)\Big|_{\substack{\psi=\eta=0\\ p^2=m^2_{\phi\pi}}} = 0 \tag{116}$$

The main advantage of the restricted t'Hooft gauge is the absence of double poles in the propagators of the ghost fields. This greatly simplifies the analysis of the ghost contribution to the unitarity sum.

Introducing the complete system of asymptotic fields

$$\{\psi^{in}\} = \{\sigma^{in}, a^{T\,in}_{\mu\alpha}, g^{in}_\alpha, \bar{g}^{in}_\alpha, c^{in}_\alpha, \bar{c}^{in}_\alpha\}$$

and the corresponding wave operators K the scattering operator of our model is given by the L.S.Z. reduction formulae [6], namely :

$$\begin{aligned} S &= \Sigma\, Z(J,\eta)\big|_{J=\eta=0} \\ &= :\exp\int dx\,dy\,\Big(\psi^{in}(x)\,K_{xy}\,\frac{\delta}{\delta J(y)}\Big):\,Z(J,\eta)\big|_{J=\eta=0} \end{aligned} \tag{117}$$

Now the Slavnov identity can be translated into a symmetry property for S . Indeed introducing the functional differential operator :

$$\bar{\mathcal{S}} = \int dx \left[\mathcal{J}_{\bar{g}}^{\alpha} \frac{\delta}{\delta \xi_{\alpha}} - \bar{\xi}^{\alpha} \frac{\delta}{\delta \mathcal{J}_{g}^{\alpha}} \right](x) \tag{118}$$

which corresponds to the field transformations :

$$\begin{aligned} \delta c^{\alpha} &= \delta\lambda \bar{\mathcal{S}} c^{\alpha} = \delta\lambda g^{\alpha} \\ \delta \bar{g}^{\alpha} &= \delta\lambda \bar{\mathcal{S}} \bar{g}^{\alpha} = \delta\lambda \bar{c}^{\alpha} \\ \delta \bar{c}^{\alpha} &= \delta g^{\alpha} = \delta\sigma = \delta a_{\mu\alpha}^{T} = 0 \end{aligned} \tag{119}$$

the Slavnov identity for the amputated Green functions on the mass shell simplifies to

$$\sum \bar{\mathcal{S}}\, Z(\mathcal{J}, \eta) \Big|_{\mathcal{J}=\eta=0} = 0 \tag{120}$$

The essential meaning of this equation is that on the mass shell the g-$\bar{g}$ and c-$\bar{c}$ pairs are equally coupled. Since, owing to the anticommuting character of the Φ.Π. fields, their contribution to the unitarity sum is opposite in sign, it turns out that the total contribution to the physical unitarity of states involving g-$\bar{g}$ and (or) c-$\bar{c}$ pairs vanish. This is the whole contribution coming from the unphysical states since the g field is decoupled from the physical states (Eq.(25)) and the g-g propagator is momentum independent.

5. CONCLUSION

The gauge field models are characterized by the fulfilement of Slavnov identities. If the gauge group is semisimple and all the fields are massive an algebraic analysis of the possible quantum corrections to the action principle shows that, indeed, the Slavnov identities can be fulfilled in the absence of the Adler-Bardeen anomaly.

We have also discussed in the SU(2) Higgs-Kibble model the connection between the Slavnov identity and the unitarity and gauge invariance of the physical S operator.

The analysis of gauge independent local operators, although not touched here, should also be based on their Slavnov invariance [12].

Concerning the possible extensions of our results, Lowenstein will show at the end of his lectures how the infrared problems connected with the gauge field models can be overcome at least in the extreme case (pure Yang-Mills models) in which all the particules are massless [2,18] . The extension to G.F.M. with non semisimple gauge group (i.e. for example the Weinberg models) can be worked out supplementing the algebraic analysis of the anomalies by a power counting analysis similar to that used in reference [7] to deal with the invariant-abelian anomalies in the lagrangian models with broken symmetry.

REFERENCES

[1] J. SCHWINGER
Phys. Rev. 82, 918 (1951).

C.S. LAM
Nuovo Cimento 38, 1754 (1965).

[2] J. LOWENSTEIN
Lectures given at this School.

[3] E. SPEER
Lectures given at this School.

P. BREITENLOHNER
Lectures given at this School.

P. BREITENLOHNER, D. MAISON
Dimensional Renormalization and the Action Principle
MPI-PAE/PTh25/74 (May 1975).

[4] Y.M.P. LAM
Phys. Rev. D6, 2145 (1972).

[5] W. ZIMMERMANN
Ann. Phys. 77, 536-570 (1973).

J. LOWENSTEIN
Lectures given at this School.

[6] R. STORA
Lectures given at this School.

[7] C. BECCHI, A. ROUET, R. STORA
Renormalizable Theories with Symmetry Breaking
Marseille Preprint 75/P.734 (June 1975).

[8] A complete bibliography about the theory of gauge fields can be found for instance in :

E.S. ABERS, B.W. LEE
Phys. Reports, 9C n°1 (1973).

M. VELTMAN
Invited talk presented at the International Symposium on Electron and Photons at High Energies, Bonn, 27-31 August 1973.

J. ZINN-JUSTIN
Lectures given at the International Summer Institute

for Theoretical Physics, Bonn 1974.

[9] P. HIGGS
Phys. Letters 12, 132 (1964) ;
Phys. Rev. 145, 1156 (1966).

T.W.B. KIBBLE
Phys. Rev. 155, 1554 (1967).

F. ENGLERT, R. BROUT
Phys. Rev. Letters, 13, 321 (1964).

[10] L.D. FADDEEV, V.N. POPOV
Phys. Letters, 25B, 29 (1957).

[11] A.A. SLAVNOV
Teor. i Mat. Fiz. 10, 153 (1972).

J.C. TAYLOR
Nucl. Phys. B33, 436 (1971).

[12] C. BECCHI, A. ROUET, R. STORA
Phys. Letters 52B, 344 (1974).

C. BECCHI, A. ROUET, R. STORA
Commun. math. Phys. 42, 127 (1975).

[13] C. BECCHI, A. ROUET, R. STORA
Renormalization of Gauge Theories,
Marseille Preprint 75/P.723 (April 1975).

[14] J. GOLSTONE
Nuovo Cimento 19, 154 (1961).

[15] B. ZUMINO
in Lectures on Elementary Particles and Quantum Field Theory,
Brandeis University Summer Institute in Theoretical Physics ; vol.2 (the MIT Press, Cambridge Mass.)

[16] W.A. BARDEEN
Phys. Rev. 184, 1848 (1969).

[17] J. WESS, B. ZUMINO
Phys. Letters 37B, 95 (1971).

RENORMALIZABLE MODELS WITH BROKEN SYMMETRIES

C. BECCHI
University of Genova
Italy

A. ROUET
Max Planck Institut für Physik
und Astrophysik, München, Germany

R. STORA
Centre de Physique Théorique, CNRS
Marseille - France

Lectures given by R. STORA

1 - Introduction

These lectures are organized to make the transition between those on perturbative renormalization theory given by J.H. Lowenstein and W. Zimmermann, E.G. Speer and P. Breitenlohner, H. Epstein and R. Seneor, and those on gauge field models given by C. Becchi. The results are hopefully stated in a form which may survive in a non perturbative context when constructive field theory has reached a degree of sophistication that allows the construction of renormalizable models.

The material presented here is extracted from three articles:

[BRS 1] C. Becchi, A. Rouet, R. Stora

Renormalization of the Abelian Higgs Kibble Model, Commun.math.Phys. <u>42</u> , 127-162 (1975)

[BRS 2] C. Becchi, A. Rouet, R. Stora

Renormalization of Gauge Theories, Centre de Physique

G. Velo and A. S. Wightman (eds.), Renormalization Theory, 299–343.

Théorique, CNRS, Marseille, 75/P.723, April 1975, submitted for publication in Annals of Physics

[BRS 3] C. Becchi, A. Rouet, R. Stora

Renormalizable Theories with Symmetry Breaking (Preliminary version), Centre de Physique Théorique, CNRS, Marseille, 75/P.734, June 1975.

The main difficulties were connected with our incomplete knowledge of renormalization theory when vanishing mass parameters occur, in which case work is still in progress, as we have heard from previous speakers [L5][B7][S10] .

Whenever the symmetry under consideration does not force any of the mass parameters to vanish, the situation is easily handled e.g. by means of the BPHZ soft quantization procedure [LRSZ 4] Two interesting classes of symmetries violate however this condition : chiral symmetries in the presence of fermions, non abelian gauge symmetries. In many cases the breaking of these symmetries leads to a theory with no vanishing mass parameter which can in principle be tackled with the extensive body of knowledge that has been acquired.

The only reason why we have exclusively based our analysis on the BPHZ scheme [L 5] is that it is within this scheme that the main properties of the renormalized perturbation series were first rigorously derived : locality and power counting lead to the precise definition of composite operators whose properties are summarized in the renormalized quantum action principles thanks to which a finite number of finite dimensional algebraic problems gives the solution, in the renormalizable case which we shall exclusively consider. Of course, these algebraic problems are trivially solved by "intelligent" renormalization schemes such as the BPHZ soft quantization [LRSZ 4] , when it can be used, or the dimensional regularization scheme which has now been firmly constructed by E.G. Speer [S6] , P. Breitenlohner and D. Maison [B7] , in the absence of fermion anomalies, or when these come under control, whereas we have advocated brute force based on locality and power counting [EG 8][E 9][S10] summarized by the action principles, first derived by J.H. Lowenstein and Y.M.P. Lam [L5] .

However, even when perturbation theories involving vanishing mass parameters reach a state of perfection where the renormalized action principle is stated in all generality, it may very well be that the present situation will prevail, namely, no renormalization scheme will be universally adapted to all aspects of the theory - i.e. will trivialize all aspects simultaneously -. After all, a renormalization scheme, usually summarized with the help of a Lagrangian - be it an effective Lagrangian with finite coefficients in the BPHZ framework, or a Lagrangian with "infinite" counterterms in one of the BPH versions - is nothing else than a representative within an equivalence class of algorithms which yield a perturbation theory constrained by locality and power counting. One has a good experience in mathematics whereby choosing a representative within an equivalence class conflicts with universality. It is worth recalling at this point the difficulties experts meet in simultaneously exploiting locality and spectrum properties in "axiomatic" field theory, or unitarity and crossing in S matrix theory, and it is not totally crazy to fear similar conflicts in the present context.

At any rate whether there does not exist in principle a renormalization scheme that makes transparent simultaneously all aspects of perturbation theory is irrelevant since this seems to be the case at any given time.

We shall therefore advocate here maximum stupidity, brutality, and ignorance, as opposed to ingenuity, i.e. we shall rely on very few general properties (locality and power counting combined into the action principles) and precisely not worry about specifically adapted renormalization schemes.

For the reasons we have mentioned, the most severe restriction we have met concerns the non occurrence of vanishing mass parameters. The generality of our reasoning can be exploited as soon as the action principle is established in full generality for theories involving massless quanta. This is of course urgent since spontaneous global symmetry breaking is at the moment excluded, as well as the study of gauge theories involving massless quanta.

These lectures are organized into three main sections :

Section 2 summarizes the results of renormalized perturbation theory (in the absence of massless quanta).

Section 3 is devoted to global symmetry breaking, and a discussion of the associated currents in terms of the coupling with a classical Yang Mills field. This discussion generalizes the classical work of K. Symanzik [S11] to arbitrary symmetry groups and its highlight is the exhibition of a compact formula for the Adler Bardeen [A12] anomaly which is the only possible obstruction to the fulfillment of the local Ward identity which generalizes the integrated Ward identity that defines the broken symmetry. At this point, the reader should keep in mind that, as was stressed by K. Symanzik, once it has been proved that a Ward identity can be fulfilled, it can conveniently be used as a starting point throughout any renormalization scheme of his choice, to relate the parameters allowed by power counting - i.e. the coefficients of the Lagrangian - and insure that the results of practical calculations are consistent with the exact

- Section 1 - References -

[BRS 1] C. BECCHI, A. ROUET, R. STORA
"Renormalization of the Abelian Higgs Kibble Model", Commun.math.Phys. 42, 127-162 (1975)

[BRS 2] C. BECCHI, A. ROUET, R. STORA
"Renormalization of Gauge Theories", CNRS, Centre de Physique Théorique, Marseille, 75/P.723, April 1975 (Submitted for publication in Annals of Physics)

[BRS 3] C. BECCHI, A. ROUET, R. STORA
"Renormalizable Theories with Symmetry Breaking" (Preliminary Version), CNRS, Centre de Physique Théorique, Marseille, 75/P.734, June 1975

[LRSZ 4] J.H. LOWENSTEIN, A. ROUET, R. STORA, W. ZIMMERMANN
"Renormalizable Models with Broken Symmetries" in "Renormalization and Invariance in Quantum Field Theory", Capri, 1-14 July 1973, E. Caïanello Ed., Nato Advanced Study Institute Series, Physics Vol.5, Plenum Press 1974

[L 5] J.H. LOWENSTEIN , this volume

[S6] E.G. SPEER , this volume

[B7] P. BREITENLOHNER , this volume

[EG 8] H. EPSTEIN, V. GLASER
Ann.Inst.Henri Poincaré, Section A , Vol. XIX, n°3, 1973

[E 9] H. EPSTEIN, this volume

[S10] R. SENEOR , this volume

[S11] K. SYMANZIK
"Renormalization of Theories with Broken Symmetry" in "Cargèse Lectures in Physics 1970" , D. Bessis Ed., Gordon & Breach, NY 1972

[A12] S. ADLER , in "Lectures on Elementary Particles and Quantum Field Theory", 1970 Brandeis University Summer Institute in Theoretical Physics, Vol. 1 , S. Deser, M. Grisaru, H. Pendleton Eds., MIT Press, Cambridge, USA 1970

or broken symmetry under consideration. The relevance of Section 3 to C. Becchi's lectures should also become clear once one realizes that some of the algebraic problems appearing in fully quantized gauge theories have already appeared in the discussion of the perturbative analog of current algebra summarized by the local Ward identity.

Section 4 is devoted to a discussion of gauge theories at the classical level. Some steps look quite arbitrary, a sign that the geometry of gauge fields is not well understood, and deserves more attention. It is most likely that the natural set up should be the theory of fiber bundles and that making a choice of field coordinates makes the situation obscure. The Slavnov symmetry which characterizes gauge field theories is at the moment mysterious and its meaning ought to be clarified in the future. An attempt in this direction is made here.

Conspicuously absent is a discussion of supersymmetric models [PS 13][FP 14][PR 15] , where it seems that the BPHZ scheme has allowed progress where the dimensional scheme meets some difficulties [16] .

[PS 13] O. PIGUET, M. SCHWEDA
CERN TH 1980 , Feb. 1975

[FP 14] S. FERRARA, O. PIGUET
CERN TH 1995, March 1975

[PR 15] O. PIGUET, A. ROUET
Max Planck Institute PAE P Th/181/75

[16] P. BREITENLOHNER
Private communication

- Acknowledgement -

The lecturer bears the full responsibility for opinions which may not be shared by his coworkers, and also for the emphasis put on the selection of topics from the articles quoted in the beginning.

2 - Results from Renormalization Theory

We shall from the start limit ourselves to renormalizable theories. In this framework, all - equivalent - versions of renormalization theory have the following properties which have been completely proved to hold within the BPHZ scheme [1], at least when no massless field is involved. In the latter case, work is still in progress as well as within other renormalization schemes [2] (which involve different ultraviolet and adiabatic regularizations, dimensional or other whereas the BPHZ scheme involves only one which serves both purposes).

The object under consideration is the renormalized Green functional $Z(\underline{J},\underline{\eta})$ whose arguments are sources [3] linear coupled to the canonically quantized fields $\underline{\varphi}$, and external fields $\underline{\eta}$ [3] coupled to composite local operators (local polynomials of φ, and its derivatives). The connected Green functional $Z_c(\underline{J},\underline{\eta})$ is defined through

$$Z(\underline{J},\underline{\eta}) = \exp\frac{i}{\hbar} Z_c(\underline{J},\underline{\eta}) \tag{1}$$

where $\hbar$ is Planck's constant occurring in the canonical quantization of $\underline{\varphi}$. The vertex functional $\Gamma(\underline{\varphi},\underline{\eta})$ is defined through the Legendre transform

$$\Gamma(\underline{\varphi},\underline{\eta}) = Z_c(\underline{J},\underline{\eta}) - \int dx\, \underline{J}(x)\left[\underline{\varphi}(x)+\underline{F}\right]\Bigg|_{0=} \tag{2}$$

$$\left(\underline{F} = \frac{\delta Z_c}{\delta \underline{J}}\Bigg|_{\underline{J}=\underline{\eta}=0}\right) \qquad = \frac{\delta Z_c}{\delta \underline{J}} - (\underline{\varphi}+\underline{F})$$

(distributional sense)

Equivalently

$$Z_c(\underline{J},\underline{\eta}) = \Gamma(\underline{\varphi},\underline{\eta}) + \int dx\, J(x)\left[\underline{\varphi}(x)+\underline{F}\right]\Bigg|_{0=} \tag{3}$$

$$= \frac{\delta\Gamma}{\delta\underline{\varphi}} + \underline{J}$$

(distributional sense)

All functionals involved have kernels which are translation invariant temperate distributions and have to be understood as formal power series in their arguments, chosen from the proper Schwartz spaces. Besides, they are computable, in the formal power series sense in increasing powers of $\hbar$ and of the coefficients occurring in an interaction Lagrangian $\mathcal{L}_{int}(\underline{\varphi},\underline{\eta})$ through an algorithm whose output is a set of renormalized Feynman amplitudes (an effective interaction Lagrangian in the BPHZ scheme, a Lagrangian with regularization dependent counter terms, in other schemes). A total Lagrangian is then defined as

$$\mathcal{L} = \mathcal{L}_0 + \mathcal{L}_{int} \tag{4}$$

where $\mathcal{L}_0$ is the quadratic Lagrangian from which $\underline{\varphi}$ derives through canonical quantization. The mass and normalization parameters occurring in $\mathcal{L}_0$ also occur in normalization conditions fulfilled by the two point functions, so that the theory is interpretable as an operator theory within the Fock space defined by $\underline{\varphi}$, by means of the LSZ reduction formulae. Other normalization conditions involve coupling constants so that the coefficients of $\mathcal{L}_{int}$ are determined as formal power series in $\hbar$ by virtue of the implicit function theorem for formal power series [4] which will be of constant use in the sequel :

THO : Let $F_i(X_1,\cdots X_p;Y_1\cdots Y_q)$, $1\leq i\leq p$ be p formal power series in $X_1,\cdots X_p;Y_1,\cdots Y_q$ without constant coefficient, $LF_i(X_1,\cdots X_p;Y_1\cdots Y_q)$ their terms linear in $X_1,\cdots X_p$; if the formal power series

$$\det\left\|\frac{\partial LF_i}{\partial X_j}\right\|(Y_1,\cdots Y_q)$$ is invertible as a formal power series in $Y_1,\cdots Y_q$, i.e. has a non vanishing constant term, there exists one and only one formal power series solution

$$X_i = \varphi_i(Y_1,\cdots Y_q) \qquad 1\leq i\leq p$$

of the system $$F_i(X_1,\cdots X_p;Y_1,\cdots Y_q) = 0\,,\quad 1\leq i\leq p$$

In the present context, the parameters of the theory are those occurring in

$$\Gamma\Big|_{\hbar=0} = \int \mathcal{L}\,dx\,\Big|_{\hbar=0} \tag{5}$$

the classical Lagrangian underlying the quantum theory under investigation, and TH 0 is used in intermediate steps to determine some of the coefficients of $\mathcal{L}_{int}$ as formal power series of $\hbar$ and other variables of the problem.

The main features of renormalization theory - locality, power counting - are summarized in the renormalized action principle due to Lowenstein, and Lam [1] , a weak version of which - whose combination with TH 0 will be enough for our purpose - is the following :

Varying parameters.

$$\frac{\hbar}{i}\,\frac{\delta Z(\underline{J},\underline{\eta})}{\delta\underline{\eta}(x)} = \left(\Delta_{\underline{\eta}(x)}\cdot Z\right)(\underline{J},\underline{\eta}) \tag{6}$$

where $\Delta_{\eta(x)}\cdot$ is a linear operation which consists in inserting into the renormalized time ordered products from which $Z(\underline{J},\underline{\eta})$ is constructed the local operator

$$\text{"}\,\frac{\delta}{\delta\underline{\eta}(x)}\int dy\;\mathcal{L}(y)\,\text{"} \tag{7}$$

The meaning of the quotation marks is similar to that shortly given (cf. Eq. 9).

Varying a numerical parameter η may pose a problem of adiabatic limit in the zero mass case, as it amounts to inserting

$$\frac{d}{d\eta}\int dy\;\mathcal{L}(y)$$

Varying fields

Let

$$\delta\underline{\varphi}(x) = \underline{M}(\underline{\varphi},D\underline{\varphi})(x)\,\delta\lambda(x) \tag{8}$$

be a local field variation, where $\underline{M}(\underline{\varphi},D\underline{\varphi})$ is a monomial in φ and its derivatives, $\delta\lambda(x)$, an infinitesimal space time dependent parameter, then the Green functional is stationary in the following sense :

$$0 = \left(\left[\text{"} \frac{\delta}{\delta\lambda(x)} \int dy\, \mathcal{L}(y) \text{"} + \underline{J}(x) \underline{M}(\underline{\varphi}, D\underline{\varphi})(x) \right] \cdot Z \right)(\underline{J}, \underline{\eta}) \quad (9)$$

where the insertion of $\underline{M}$ is to be renormalized according to a prescription attached to the definition of $\underline{M}$ (a normal ordered product with dimension $d(\underline{M})$ larger than or equal to $d_{can}(M) = \sum_{\varphi \in M} d_\varphi + \#_M \partial$, where d_φ is the φ field canonical dimension computed according to the power counting index of its propagator : 1 for scalar fields or vector fields with indefinite metric, $\frac{3}{2}$ for Dirac fields, and $\#_M \partial$ is the number of first order derivative symbols in M). The insertion represented by the first term in Eq.9 has the following simple structure

$$\text{"} \frac{\delta}{\delta\lambda(x)} \int dy\, \mathcal{L}(y) \text{"} = \left[\frac{\delta}{\delta\lambda(x)} \int dy\, \mathcal{L}(y) \right]^{\text{naïve}} + \hbar\, Q(x) \quad (10)$$

where $Q(x)$ is a local polynomial with dimension smaller than or equal to $\sup_\varphi d(\mathcal{L}) - d(\varphi) + d(M) = \sup_\varphi 4 - d_\varphi + d(M)$.
$\hbar Q$ lumps together all radiative corrections to the naïve variation involved in the first term of Eq.10 . Of course, the coefficients appearing in Q depend on the renormalization prescriptions. Again, problems of adiabatic limit may arise if $\delta\lambda(x) = \delta\lambda$, space time independent. Once this general form of the renormalized action principle is understood, the coefficients can be recovered by application to particular cases (e.g. in the BPHZ scheme, zero momentum vertex functions). Note the generalization of Eq.5

$$\left[\frac{\delta}{\delta\lambda(x)} \int dy\, \mathcal{L}(y) \right]^{\text{naïve}} \cdot \Gamma = \frac{\delta}{\delta\lambda(x)} \int dy\, \mathcal{L}(y) + O(\hbar) \quad (11)$$

which will be frequently used.

Finally, if a particle interpretation is doable, performing mass renormalizations properly, one may use the reduction formuma :

$$S = :\exp\int dx\, dy\, \underline{\varphi}_{in}(x)\, K_{xy} \frac{\delta}{\delta \underline{J}(y)} : Z(\underline{J}, \underline{\eta})\Big|_{\underline{J}=0} \tag{12}$$

Here φ_{in} is the canonically quantized free field which can be derived from the asymptotic Lagrangian obtained as follows : keep from the vertex functional Γ terms quadratic in φ ; then retain from the p-space two point function the lowest non vanishing term of the Taylor expansion around the mass shell. Then K is the Euler Lagrange differential operator corresponding to the asymptotic Lagrangian.

- Section 2 - References and Footnotes -

[1] [LRSZ4] and [5] of Section 1

[2] [S10], [B7] , of Section 1
W. ZIMMERMANN, this volume

[3] We shall, for convenience use the functional language which is most convenient to cast the theory in a form that is most similar to that of classical quantum field theory, see e.g. [S11] of Section 1 and

[IIM3] J. ILIOPOULOS, C. ITZYKSON, A. MARTIN
Rev. Mod. Phys. 47, 165, 1975
Sources coupled to Fermi type operators are to be chosen from an infinite Grassmann algebra.

[4] N. BOURBAKI
Eléments de Mathématiques XI Algèbre CH4 §5 n° 9 .
We thank R. Seiler for this reference.

3 - Global Symmetry Breaking

a) The Classical Picture

As an introduction, let us consider a hydrogen atom in a homogeneous electric field, with Hamiltonian

$$H = H^s + \vec{E}.\vec{r}$$
$$H^s = \frac{\vec{p}^{\,2}}{2m} + V_{Coulomb}(|\vec{r}|) \qquad (1)$$

$\vec{p}$ and $\vec{r}$ are respectively the momentum and position of the electron, $\vec{E}$ is the electric field. H^S is invariant under O(3) , the rotation group in three dimensions. This symmetry is broken down to O(2) , which leaves $\vec{E}$ invariant, but the symmetry breaking term $\vec{E}.\vec{r}$ can be described in more details : under a transformation of O(3), $\vec{r}$, and therefore $\vec{E}.\vec{r}$ behaves according to the vector representation of O(3). The same would be true if $\vec{E}.\vec{r}$ were replaced by $\vec{E}.\vec{r}(\vec{r}^{\,2})$, which is more "singular" so that one may want to characterize the breaking by a dimension index. Of course, we have field theory in mind, and we wish to define notions which are meaningful in the framework of renormalized renormalizable perturbation theories. The recursive procedure which has to be used there forbids the characterization of the symmetry breaking in terms of the group whereas [1] it allows an infinitesimal description in terms of the corresponding Lie algebra. Thus, the consideration of the symmetry group will not survive the classical level. The degree of singularity of the breaking will be controlled by the theory of power counting, and its covariance under the symmetry group that is being broken is best characterized by introducing an external field coupled to the system via the breaking term. In the case of the hydrogen atom, let us couple the system to a time dependent electric field $\vec{\mathcal{E}}(t)$ so that the Hamiltonian becomes

$$H_{\varepsilon} = H^s + \left(\vec{\mathcal{E}}(t) + \vec{E}\right).\vec{r} \qquad (2)$$
$$H = H_{\varepsilon}\Big|_{\vec{\mathcal{E}}=0}$$

Now H_ε is invariant under the simultaneous transformation

$$\begin{aligned} \vec{r} &\to \overrightarrow{R.r} \\ \vec{p} &\to \overrightarrow{R.p} \qquad , \quad R \in O(3) \\ \vec{\mathcal{E}}(t)+\vec{E} &\to \overrightarrow{R.\,(\mathcal{E}(t)+E)} \end{aligned} \tag{3}$$

or equivalently

$$\begin{aligned} \vec{r} &\to \overrightarrow{R.r} \\ \vec{p} &\to \overrightarrow{R.p} \\ \vec{\mathcal{E}}(t) &\to \overrightarrow{R.\,(\mathcal{E}(t)+E)} - \vec{E} \end{aligned} \tag{4}$$

Here $(\vec{r}, \vec{p})$ denote the dynamical variables $\vec{\mathcal{E}}(t)$ an external field, $\vec{E}$ a fixed parameter.

By analogy we describe the field theory situation as follows :

Let G be a compact Lie group,

$\mathcal{G}$ its Lie algebra,

$\underline{\varphi}$ a field multiplet transforming under the unitary, completely reduced representation $\overset{\circ}{D}$ of G, acting in the vector space V

$$\varphi \to {}^{g}\varphi = \overset{\circ}{D}(g^{-1})\,\varphi \ , \quad \begin{matrix} g \in G \\ \varphi \in V \end{matrix} \tag{5}$$

with infinitesimal version

$$\delta_X \varphi = -\overset{\circ}{t}(X)\,\varphi \qquad X \in \mathcal{G} \tag{6}$$

where $X \to \mathring{t}(X)$ is obtained by differenciating $\mathring{D}$ in the neighbourhood of the identity of G.

Let $\underline{\mathring{b}}$ be a space time independent vector within another multiplet $\mathcal{V}$ characterized by a representation $\mathring{\mathcal{D}}$ of G, with differential $\mathring{\theta}$, $\underline{\beta}$ a classical field within the same multiplet.

We shall say that a Lagrangian density $\mathcal{L}(\underline{\varphi}, \mathring{b})$ describes the breaking of G along $\underline{\mathring{b}}$ if the corresponding action

$$\mathring{\Gamma}(\underline{\varphi}, \underline{\mathring{b}}) = \int dx \; \mathcal{L}(\underline{\varphi}, \underline{\mathring{b}})(x) \tag{7}$$

is invariant under

$$\begin{aligned} \varphi &\to {}^{g}\varphi = \mathring{D}(g^{-1})\varphi \\ \mathring{b} &\to {}^{g}\mathring{b} = \mathring{\mathcal{D}}(g^{-1})\,\mathring{b} \end{aligned} \qquad g \in G \tag{8}$$

Then $\mathring{\Gamma}(\underline{\varphi}, \underline{\beta} + \underline{\mathring{b}})$ is invariant under

$$\begin{aligned} \varphi &\to {}^{g}\varphi = \mathring{D}(g^{-1})\varphi \\ \beta &\to {}^{g}\beta = \mathring{\mathcal{D}}(g^{-1})(\underline{\beta} + \underline{\mathring{b}}) - \mathring{b} \end{aligned} \qquad g \in G \tag{9}$$

and consequently under the infinitesimal variation

$$\begin{aligned} \delta_X \varphi &= -\,\mathring{t}(X)\varphi \\ \delta_X \beta &= -\,\mathring{\theta}(X)(\underline{\beta} + \underline{\mathring{b}}) \end{aligned} \tag{10}$$

The analogy with our example is clear :

$$\begin{aligned} (\vec{r}, \vec{p}) &\leftrightarrow \varphi \\ \vec{E} &\leftrightarrow \underline{\mathring{b}} \\ \vec{\mathcal{E}}(t) &\leftrightarrow \underline{\beta}(x) \end{aligned} \tag{11}$$

$$O(3) \leftrightarrow G$$

$$H \leftrightarrow \mathring{\Gamma}$$

Finally, the degree of singularity of the breaking can be made precise in the framework of renormalizable theories : $\mathring{\Gamma}(\varphi, \beta + \mathring{b})$ only contains terms with dimension smaller than or equal to four, with the usual dimension assignment for φ and derivatives and a prescribed dimension assignment for β (not $\mathring{b}$)

$$d_\beta = 4 - d_B \tag{12}$$

where d_B is the dimension of the breaking. $\mathring{\Gamma}$ is a polynomial if and only if it is one for $\beta = 0$, i.e. one sticks to the renormalizable framework, and

$$0 < d_\beta \ (< 4) \tag{13}$$

which is Symanzik's criterium [2]. In the limiting case

$$d_\beta = 0 \tag{14}$$

a locality assumption insures that $\mathcal{L}(\varphi, \mathring{b})$ is analytic in $\mathring{b}$ and it is not true in general that the most general renormalizable $\mathcal{L}$ invariant under

$$\begin{aligned} \varphi &\to {}^{g}\varphi \\ \mathring{b} &\to {}^{g}\mathring{b} \end{aligned} \tag{15}$$

is the most general $\mathcal{L}$ invariant under the residual symmetry group $H(\mathring{b})$, the stability group of $\mathring{b}$ in G . This is due to the fact that some of the irreducible representations of G occurring in the tensor powers of $\mathcal{D}$ allowed by the renormalizability requirement may never occur in any of the tensor powers of $\mathring{\mathcal{D}}$ so that the trivial representations of $H(\mathring{b})$ contained in those will never appear.

Example

$$G = U(1) \qquad \varphi = (\varphi, \varphi^*) \qquad {}^g\varphi = \left(e^{i\omega(g)}\varphi, e^{-i\omega(g)}\varphi^*\right)$$

$$\mathring{b} = (\mathring{u}, \mathring{u}^*) \qquad {}^g b = \left(e^{4i\omega(g)}\mathring{u}, e^{-4i\omega(g)}\mathring{u}^*\right)$$

$$H(\mathring{b}) = \{e\}$$

Tensor powers of $\mathring{\mathcal{D}}$ only contain $e^{4ip\omega(g)}$ where p is a positive or negative integer or zero, and in particular never contains $e^{\pm 2i\omega(g)}$. The most general allowed Lagrangian is then :

$$\mathcal{L} = Z\left(\partial_\mu\varphi_1\,\partial^\mu\varphi_1 + \partial_\mu\varphi_2\,\partial^\mu\varphi_2\right) - M^2(\varphi_1^2 + \varphi_2^2)$$
$$- \lambda(\varphi_1^4 + \varphi_2^4) - \mu\,\varphi_1^2\varphi_2^2 - \nu(\varphi_1^3\varphi_2 - \varphi_1\varphi_2^3)$$

where we have put $\varphi = \varphi_1 + i\varphi_2$.

Although this example is at most amusing it may be worth keeping in mind this phenomenon which we shall now leave to consider only superrenormalizable breakings.

As a final remark before casting our observations into a form likely to go through renormalization one should point out that the breaking is not really characterized by $\mathring{b}$, but only by the stratum [3] of $\underline{\mathring{b}}$ under G in $\mathcal{V}$, i.e. the set of $\underline{\mathring{b}}$'s with conjugate stability groups.

If one wishes to interpret the theory in terms of particles, it is often necessary to perform a field translation

$$\underline{\varphi} \rightarrow \underline{\varphi} + \underline{\mathring{F}} \tag{16}$$

with

$$\left.\frac{\delta\mathring{\Gamma}}{\delta\varphi}\right|_{\varphi = \mathring{F},\ \beta = 0} = 0 \tag{17}$$

so that, in terms of the translated variables $\mathring{\mathcal{J}}(\underline{\varphi},\underline{\beta}) = \mathring{\Gamma}(\underline{\varphi}+\underline{\mathring{F}}, \underline{\beta}+\underline{\mathring{b}})$ is invariant under

$$\delta_x \underline{\varphi} = -\mathring{t}(x)(\underline{\varphi}+\underline{\mathring{F}})$$
$$\delta_x \underline{\beta} = -\mathring{\theta}(x)(\underline{\beta}+\underline{\mathring{b}}) \tag{18}$$

as expressed by the integrated Ward identity

$$0 = \mathring{W}(x)\,\mathring{\mathcal{J}}(\underline{\varphi},\underline{\beta})$$
$$= -\int dx \left[\frac{\delta \mathring{\mathcal{J}}}{\delta \underline{\varphi}}\, \mathring{t}(x)(\underline{\varphi}+\underline{\mathring{F}}) + \frac{\delta \mathring{\mathcal{J}}}{\delta \underline{\beta}}\, \mathring{\theta}(x)(\underline{\beta}+\underline{\mathring{b}}) \right](x) \tag{19}$$

The mass matrix is

$$\mathring{m} = \left.\frac{\delta^2 \mathring{\mathcal{J}}}{\delta \underline{\varphi}\,\delta \underline{\varphi}}\right|_{\varphi=\beta=0} = \left.\frac{\delta^2 \mathring{\Gamma}}{\delta \varphi\,\delta \varphi}\right|_{\underline{\varphi}=\mathring{F},\,\underline{\beta}=0} \tag{20}$$

Assuming that the kinematical part of $\mathcal{L}$ is non degenerate, the non degeneracy of the mass matrix insures the non occurrence of massless quanta, and it is then easy to check that $\underline{\mathring{F}}$ is a covariant function of $\underline{\mathring{b}}$, i.e.

$$\mathring{t}(x)\,\underline{\mathring{F}}(\underline{\mathring{b}}) = \frac{\partial \underline{\mathring{F}}}{\partial \underline{\mathring{b}}}\, \mathring{\theta}(x)\,\underline{\mathring{b}} \tag{21}$$

The same is true for $\mathring{m}$ and the other coefficients occurring in $\mathring{\mathcal{J}}$, due to their polynomial character in $\mathring{F}$ and $\mathring{b}$, but one has to be careful and observe in view of a particle interpretation of the renormalized theory that if $\mathring{m}$ is not the most general covariant constructed with $\overset{\circ}{b}$, i.e. the limitations due to the renormalizability criterium imply mass rules, the renormalized theory that follows will not lead to an operator theory in a fixed Fock space.

Also in view of the construction of the complete renormalized quantum theory based on the present classical one, one may object that the Ward identity (19) will possibly get renormalized in a way consistent with power counting so that one may wish to extend our definition of symmetry breaking in a way where only power counting and algebraic properties are involved. At the classical level the question is to find the most general action $\overset{\circ}{\Gamma}(\underline{\varphi}, \underline{\beta})$ without a term linear in $\underline{\varphi}$, invariant under the infinitesimal transformation

$$\begin{aligned} \delta_X \underline{\varphi} &= -\left[t(X)\underline{\varphi} + \underline{F}(X) \right] \\ \delta_X \underline{\beta} &= -\left[\Theta(X)\underline{\beta} + \underline{b}(X) \right] \end{aligned} \tag{22}$$

where $t(X)$, $\Theta(X)$, $F(X)$, $b(X)$ are linear in X , subject to

$$\delta_X \delta_Y - \delta_Y \delta_X = \delta_{[X,Y]}$$

In other words, the model is defined by the integrated Ward identity

$$\begin{aligned} 0 &= W(X)\, \overset{\circ}{\Gamma}(\varphi, \beta) \\ &\equiv - \int dx \left[\frac{\delta \overset{\circ}{\Gamma}}{\delta \underline{\varphi}} \left(t(X)\underline{\varphi} + \underline{F}(X) \right) + \frac{\delta \overset{\circ}{\Gamma}}{\delta \underline{\beta}} \left(\Theta(X)\underline{\beta} + \underline{b}(X) \right) \right](x) \end{aligned} \tag{23}$$

with

$$[W(X), W(Y)] = W([X,Y]) \tag{24}$$

and

$$\left. \frac{\delta \overset{\circ}{\Gamma}}{\delta \varphi} \right|_{\varphi = \beta = 0} = 0 \tag{25}$$

This algebraic constraint implies

$$\begin{aligned} \left[t(X), t(Y) \right] &= t([X,Y]) \\ \left[\theta(X), \theta(Y) \right] &= \theta([X,Y]) \end{aligned} \tag{26}$$

which means that $t(\cdot)$ and $\theta(\cdot)$ are representations of $\mathcal{G}$, and, the cocycle conditions

$$\begin{aligned} t(X)\underline{F}(Y) - t(Y)\underline{F}(X) &= \underline{F}([X,Y]) \\ \theta(X)\underline{b}(Y) - \theta(Y)\underline{b}(X) &= \underline{b}([X,Y]) \end{aligned} \tag{27}$$

If $\mathcal{G}$ is semi-simple, these cocycle conditions imply the existence of $\underline{F}$ and $\underline{b}$ such that

$$\begin{aligned} \underline{F}(X) &= t(X)\underline{F} \\ \underline{b}(X) &= \theta(X)\underline{b} \end{aligned} \tag{28}$$

However, if $\mathcal{G}$ has an abelian part, (23) can only be shown to have reasonable solutions (renormalizable, with non degenerate kinematical part) under the assumption that

$$\underline{b}(X) = \theta(X)\underline{b} \quad \text{for some } \underline{b} \tag{29}$$

which conforms with our philosophy of symmetry breaking. Under such conditions, $t(\cdot)$ and $\theta(\cdot)$ can be shown to be fully reducible and the cocycle $\underline{F}(\cdot)$ to be also trivial. This is our first contact with the cohomological difficulties [4] which go along with abelian groups and it will be no surprise that in the course of the renormalization procedure one meets in these cases anomalies - which can fortunately be eliminated, due to the renormalizability requirement - whose occurrence would mean that the classical representation of the Lie algebra becomes renormalized into one which cannot be lifted to the compact group it supposedly refers to.

b) Quantization

Given a classical model as previously described, i.e. where $\mathring{\Gamma}(\underline{\varphi},\underline{\beta})$ fulfills the Ward identity

$$
\begin{aligned}
0 &= \mathring{W}(X)\,\mathring{\Gamma}(\underline{\varphi},\underline{\beta}) \\
&\equiv -\int dx \left\{ \frac{\delta\mathring{\Gamma}}{\delta\underline{\varphi}}\, \mathring{t}(X)\left(\underline{\varphi}+\underline{\mathring{F}}\right) + \frac{\delta\mathring{\Gamma}}{\delta\underline{\beta}}\, \mathring{\theta}(X)\left(\underline{\beta}+\underline{\mathring{b}}\right) \right\}(x)
\end{aligned}
\tag{30}
$$

where $\mathring{t}(\cdot)$, $\mathring{\theta}(\cdot)$ are fully reducible representations of $\mathcal{G}$, find a quantum extension $\Gamma(\underline{\varphi},\underline{\beta})$ fulfilling

$$
\begin{aligned}
0 &= W(X)\,\Gamma(\underline{\varphi},\underline{\beta}) \\
&\equiv -\int dx \left\{ \frac{\delta\Gamma}{\delta\varphi}\left(t(X)\varphi+F(X)\right) + \frac{\delta\Gamma}{\delta\beta}\left(\theta(X)\beta+b(X)\right) \right\}(x)
\end{aligned}
\tag{31}
$$

i.e. an effective action

$$
\Gamma_{eff} = \mathring{\Gamma} + \hbar\,\Delta\Gamma \tag{32}
$$

leading to Γ, with

$$
\begin{aligned}
t(X) &= \mathring{t}(X) + O(\hbar) \\
\theta(X) &= \mathring{\theta}(X) + O(\hbar) \\
\underline{F}(X) &= \mathring{t}(X)\,\underline{\mathring{F}} + O(\hbar) \\
\underline{b}(X) &= \mathring{\theta}(X)\,\underline{\mathring{b}} + O(\hbar)
\end{aligned}
\tag{33}
$$

and the constraints

$$
[W(X), W(Y)] = W([X,Y]) \tag{34}
$$

$$\left.\frac{\delta\Gamma}{\delta\varphi}\right|_{\varphi=\beta=0} = 0 \tag{35}$$

We shall in fact see that the problem always has a solution with

$$\begin{aligned} &t = \overset{\circ}{t} \\ &\Theta = \overset{\circ}{\Theta} \\ &F(x) = \overset{\circ}{t}(x)\,F \qquad \underline{F} = \overset{\circ}{\underline{F}} + O(\hbar) \ \text{fixed by (35)} \\ &\underline{b}(x) = \overset{\circ}{\Theta}(x)\,\underline{b} \qquad \underline{b} = \overset{\circ}{\underline{b}} + O(\hbar) \ \text{arbitrary} \end{aligned} \tag{36}$$

The quantum extension of the classical theory will be chosen such that $\underline{b}$ lies in the stratum of $\overset{\circ}{\underline{b}}$. Once this restriction has been imposed, the most general solution is obtained by field amplitude (finite) renormalizations.

We shall only sketch the proof whose details are given in [BRS 3] .

First assume that one tries to fulfill (31, 36). Starting from an arbitrary Γ_{eff} of the form (32) and applying the action principle for the variation

$$\begin{aligned} \delta_x \underline{\varphi} &= -\overset{\circ}{t}(x)\,(\underline{\varphi} + \underline{F}) \\ \delta_x \underline{\beta} &= -\overset{\circ}{\Theta}(x)\,(\underline{\beta} + \underline{b}) \end{aligned} \qquad x \in \mathcal{G} \tag{37}$$

yields :

$$W(x)\,\Gamma = \Delta(x).\,\Gamma \tag{38}$$

where $\Delta(x)$ is a dimension four insertion of the form

$$\Delta(X) = \delta_X \Gamma_{eff} + \hbar Q(X) \tag{39}$$

where $Q(X)$ lumps together the radiative corrections to the naïve form of the action principle.

Then use the algebraic constraint (34) :

$$W(X)\Delta(Y).\Gamma - W(Y)\Delta(X).\Gamma = \Delta([X,Y]).\Gamma \tag{40}$$

which is of the form

$$W(X)\Delta(Y) - W(Y)\Delta(X) - \Delta([X,Y]) = O(\hbar\Delta) \tag{41}$$

Expanding $\Delta(X)$, Γ_{eff} , $Q(X)$ in terms of a basis which reduces the representation induced by (37) on monomials of dimension smaller than or equal to four :

$$\Delta(X) = \underline{\delta}_\natural(X).\underline{\Delta}_\natural + \underline{\delta}_\flat(X).\underline{\Delta}_\flat = \underline{\delta}(X).\underline{\Delta}$$

$$\Gamma_{eff} = \underline{c}_\natural.\underline{\Delta}_\natural + \underline{c}_\flat.\underline{\Delta}_\flat = \underline{c}.\underline{\Delta} \tag{42}$$

$$Q(X) = \underline{q}_\natural(X).\underline{\Delta}_\natural + \underline{q}_\flat(X).\underline{\Delta}_\flat = \underline{q}(X).\underline{\Delta}$$

where $\underline{\Delta}_\natural$ spans the invariant part and $\underline{\Delta}_\flat$ spans the non invariant part of this fully reducible representation $T(X)$, we get

$$T(X)\underline{\delta}(Y) - T(Y)\underline{\delta}(X) - \underline{\delta}([X,Y]) = O(\hbar\delta) \tag{43}$$

There is a delicacy on the way which is due to the possible occurrence of a field independent part in $\Delta(X)$ to which $W(X)$ is not a priori applicable, but can be shown to drop out from (43) . (43) is a quantum perturbation of a cocycle condition, with $T(\cdot)$ fully reducible.
Restricting (43) to its $\underline{b}$ component yields

$$\underline{\delta}_b(X) = T(X) \frac{1}{\{T,T\}_b} \{T, \underline{\delta}_b\} + O(\hbar\delta) \quad (44)$$

where $\{\ \ \}$ is an invariant quadratic form on the dual of $\mathcal{G}$, so that $\{T, T\}$ commutes with $T(X)$, and is invertible on the b component. Now, we may choose T_{eff} in such a way that

$$\{T, \underline{\delta}_b\} = 0 \quad (45)$$

which is equivalent to

$$\underline{C}_b + \hbar \frac{\{T, q_b\}}{\{T, T\}_b} = 0 \quad (46)$$

This system is soluble for $C_b - \mathring{C}_b$ in terms of $C_{\mathcal{G}} - \mathring{C}_{\mathcal{G}}$, $\underline{F} - \underline{\mathring{F}}$, $\underline{b} - \underline{\mathring{b}}$ since $\underline{q}_b$ is a formal power series in these variables and $\hbar$.

Similarly, for $X \in \mathcal{S}$, the semi-simple part of $\mathcal{G}$, the non degeneracy of the Killing form of $\mathcal{S}$ implies that

$$\underline{\delta}_{\mathcal{G}}(X) = O(\hbar\delta) \qquad X \in \mathcal{S} \quad (47)$$

However, for $X \in \mathcal{A}$, the abelian part of $\mathcal{G}$, (43) yields no information on $\underline{\delta}_{\mathcal{G}}(X)$. Showing that also

$$\underline{\delta}_{\mathcal{G}}(X) = O(\hbar\delta) \qquad X \in \mathcal{A} \quad (48)$$

can be done by a more subtle argument which involves a detailed study of the behaviour of the theory under scaling transformations and can be found in [BRS 3] . Putting then together (44), (46), (47), (48) shows that

$$\underline{\delta}(x) = O(\hbar\,\delta) \tag{49}$$

hence

$$\underline{\delta}(x) = 0 \tag{50}$$

i.e. the Ward identity

$$\begin{aligned} 0 &= W(x)\,\Gamma \\ &\equiv -\int dx \left\{ \frac{\delta\Gamma}{\delta\underline{\varphi}} \mathring{t}(x)(\underline{\varphi}+\underline{F}) + \frac{\delta\Gamma}{\delta\underline{\beta}} \mathring{\theta}(x)(\underline{\beta}+\underline{b}) \right\}(x) \end{aligned} \tag{51}$$

is fulfilled. There, $\underline{F}$ is determined by the condition that Γ_{eff} has no term linear in $\underline{\varphi}$, i.e. the linear part of $\underline{C}_4 \cdot \underline{\Delta}_4$ cancels that of $\underline{C}_b \cdot \underline{\Delta}_b$, which is easily seen to be soluble for $\underline{F} - \mathring{\underline{F}}$ provided that the mass matrix $\mathring{m}$ is invertible, which was assumed from the start. The theory then depends on the parameters $\underline{C}_4$ and $\underline{b}$, which can be adjusted by imposing on the Green's functions the same normalization conditions as those which specify $\mathring{\underline{C}}_4$, $\mathring{\underline{b}}$ at the classical level, in an invertible way.

There remains now to check that (31, 36) provides up to field amplitude renormalizations the most general quantum perturbation of (30). In order to do so, one has to check that the most general solution of (31), (34) is given by

$$\begin{aligned} t(x) &= Z\,\mathring{t}(x)\,Z^{-1} \\ \theta(x) &= z\,\mathring{\theta}(x)\,z^{-1} \\ \underline{F}(x) &= t(x)\,\underline{F} \\ \underline{b}(x) &= \theta(x)\,\underline{b} \end{aligned} \tag{52}$$

This is again a cohomological problem which raises no difficulty for $X \in \mathcal{S}$. Scaling fields through $\mathcal{Z}$, $\mathfrak{z}$, $W(X)$ then reduces to (31, 36) except for $X \in \mathfrak{a}$, but precisely the deviations of $t(X)$, $\theta(X)$, $F(X)$, from $\mathring{t}(X)$, $\mathring{\theta}(X)$, $t(X)F$, for $X \in \mathfrak{a}$ contribute to $\Delta_4(X)$, $X \in \mathfrak{a}$, up to $O(\hbar\Delta)$ and can be shown to vanish by virtue of the argument leading to (48). This concludes our sketchy discussion.

c) Currents and the Local Ward Identity

Given a model defined by the integrated Ward identity (19) one may now construct the corresponding -dimension 3- Noether current [5] and its correlation functions by looking for an extension $\mathring{\Gamma}(\underline{\varphi}, \underline{\beta}, \underline{a}_\mu)$ of $\mathring{\Gamma}(\underline{\varphi}, \underline{\beta})$ ($\mathring{\Gamma}(\varphi, \beta) = \mathring{\Gamma}(\underline{\varphi}, \underline{\beta}, \underline{a}_\mu)\big|_{\underline{a}=0}$) where $\underline{a}_\mu$ is a classical dimension one external Yang Mills field (i.e. with values in $\mathcal{G}$) such that invariance under local gauge transformations holds :

$$\delta_\omega \varphi(x) = - \mathring{t}(\omega_x)(\underline{\varphi}_x + \underline{\mathring{F}})$$
$$\delta_\omega \beta(x) = - \mathring{\theta}(\omega_x)(\underline{\beta}_x + \underline{\mathring{b}}) \tag{53}$$
$$\delta_\omega a_\mu(x) = \partial_\mu \omega_x - [\omega_x, a_{\mu x}]$$

Here ω belongs to $\mathcal{G} \times$ Schwartz space, and parametrizes the gauge Lie algebra, with the bracket

$$[\omega, \omega'](x) = [\omega(x), \omega'(x)] \tag{54}$$

The corresponding Ward identity is

$$0 = \mathring{\mathcal{W}}(\omega)\, \mathring{\Gamma}(\varphi, \beta, a_\mu) \equiv \tag{55}$$

$$\equiv \int dx \left\{ \frac{\delta \mathring{\Gamma}}{\delta a_\mu} \left(\partial_\mu \omega - [\omega, a_\mu] \right) - \right.$$

$$\left. - \frac{\delta \mathring{\Gamma}}{\delta \varphi} \mathring{t}(\omega)(\underline{\varphi} + \underline{\mathring{F}}) - \frac{\delta \mathring{\Gamma}}{\delta \beta} \mathring{\theta}(\omega)(\underline{\beta} + \underline{\mathring{b}}) \right\}(x) \tag{55}$$

It is readily fulfilled by replacing in $\mathring{\Gamma}$ derivatives by covariant derivatives :

$$\begin{aligned} \partial_\mu \underline{\varphi} &\rightarrow D_\mu \varphi = \partial_\mu \varphi + \mathring{t}(a_\mu)(\underline{\varphi} + \underline{\mathring{F}}) \\ \partial_\mu \underline{\beta} &\rightarrow \Delta_\mu \beta = \partial_\mu \beta + \mathring{\theta}(a_\mu)(\underline{\beta} + \underline{\mathring{b}}) \end{aligned} \tag{56}$$

and include gauge invariant terms constructed from the curvature tensor

$$G_{\mu\nu} = \partial_\mu a_\nu - \partial_\nu a_\mu - [a_\mu, a_\nu] \tag{57}$$

At the quantum level however, the same procedure as was previously applied yields a priori to the local Ward identity

$$\mathcal{W}(\omega)\, \Gamma(\underline{\varphi}, \underline{\beta}, \underline{a}_\mu) = \Delta(\omega) \cdot \Gamma(\underline{\varphi}, \underline{\beta}, \underline{a}_\mu) \tag{58}$$

where $\mathcal{W}$ is obtained from $\mathring{\mathcal{W}}$ by replacing $\underline{\mathring{F}}$ by $\underline{F}$, $\underline{\mathring{b}}$ by $\underline{b}$. The commutation relation

$$[\mathcal{W}(\omega), \mathcal{W}(\omega')] = \mathcal{W}([\omega, \omega']) \tag{59}$$

implies the Wess Zumino [6] consistency condition :

$$\mathcal{W}(\omega)\Delta(\omega') - \mathcal{W}(\omega')\Delta(\omega) - \Delta([\omega, \omega']) = O(\hbar \Delta) \tag{60}$$

and the question is whether $\Gamma_{eff}(\underline{\varphi}, \underline{\beta}, \underline{a}_\mu)$ can be found in such a way that

$$\Delta(\omega) = 0 \tag{61}$$

The answer is in general no : there may survive an anomaly, the so called Adler Bardeen anomaly [7] , which only depends on the external gauge field, and has the following form :

$$\Delta(\omega) = (\partial_\mu K^\mu)(\omega) \tag{62}$$

where

$$K_\mu(\omega) = \epsilon_{\mu\nu\rho\sigma}\left[D(\omega, \partial^\nu a^\rho, a^\sigma) + F(\omega, a^\nu, a^\rho, a^\sigma)\right] \tag{63}$$

where ϵ is the totally antisymmetric tensor, D is a completely symmetric trilinear invariant form on $\mathcal{G}$ and F a quadrilinear invariant form on $\mathcal{G}$ defined by

$$F(X,Y,Z,T) = \frac{1}{12}\Big[D(X,Y,[Z,T]) + D(X,Z,[T,Y]) + D(X,T,[Y,Z])\Big] \tag{64}$$

We refer the reader to the rather lengthy details given in [BRS 3] where the first local cohomology of the gauge Lie algebra is computed up to dimension four and this obstruction is computed. This anomaly is well known to occur only when Fermion fields are involved, with pseudoscalar couplings.

- Section 3 - References and Footnotes -

[1] We are indebted to H. Epstein for this remark.

[2] [S 11] of Section 1.

[3] For a condensed exposition of basic facts on actions of compact groups we refer to :

L. MICHEL : in "Proceedings 3rd GIFT Seminar in Theoretical Physics", p. 49-131, Madrid 1972

L. MICHEL : in "Proceedings of 1974 Warsaw Symposium in Mathematical Physics

L. MICHEL : in "Proceedings 4th International Colloquium on Group Theoretical Methods in Physics, Nijmegen 1975".

[4] For a review of the cohomology of Lie Algebras, one may consult :

R. HERMANN : Vector Bundles in Mathematical Physics, Vol. II, Mathematical Physics Monograph Series, W.A. Benjamin N.Y. 1970 ,
Séminaire Sophus Lie 1 - 1954-1955, Théorie des Algèbres de Lie, Topologie des Groupes de Lie, Secrétariat Mathématique, 11 rue Pierre Curie, Paris 5e.

[5] This was first done in detail for the σ model with nucleons in :

C. BECCHI : Commun.math.Phys. 39 , 329 (1975)

[6] J. WESS, B. ZUMINO
Phys. Lett. 37B, 95 (1971)

[7] W.A. BARDEEN
P.R. 184 , 1848 (1949)

4 - Classical Gauge Theories

a) Introduction

Renormalizable gauge theories [1] are based on the so called Faddeev Popov [2] Lagrangian which involves a matter field φ , Yang Mills fields - gauge fields - a_μ , Faddeev Popov ghost fields c , $\bar{c}$, the latter three taking their values in the Lie algebra $\mathcal{G}$ underlying the gauge symmetry. This Lagrangian has a complicated looking structure which depends on a gauge function $g(\underline{\varphi}, \underline{a})$ with value in $\mathcal{G}$, and its construction relies on a beautiful heuristic argument due to Faddeev and Popov [2] , based on the consideration of the Feynman functional integral, and subsequently developed by Fradkin and Tyutin [3] and 't Hooft and Veltman [1] . Although it is currently accepted that the Faddeev Popov ghosts are only needed at the quantum level, we shall indicate here [4] that they also naturally appear at the classical level whenever the gauge degrees of freedom are not eliminated from the start so that only physical degrees of freedom appear in the Lagrangian. Such an elimination leads however to the "unitary gauge" which does not lead to a renormalizable theory and thus falls out of our framework. Here, we shall limit ourselves to Lorentz invariant, renormalizable gauges.

b) The Faddeev Popov Lagrangian and the Slavnov Symmetry

We shall first review the geometrical situation we have been dealing with in the past two sections, with slight notational changes.

G is a compact Lie group, $\mathcal{G}$ its Lie algebra. The associated gauge group is $\underline{G} = G \times \mathcal{S}$ where $\mathcal{S}$ is the Schwartz space over Minkowski space M_4 , with the composition law

$$(\underline{g} \cdot \underline{g}')(x) = \underline{g}(x)\, \underline{g}'(x) \qquad \underline{g}, \underline{g}' \in G \times \mathcal{S}(R^4) \tag{1}$$

and its Lie algebra is $\underline{\mathcal{G}} = \mathcal{G} \times \mathcal{S}(R^4)$ with the composition law

$$[\underline{\omega}, \underline{\omega}'](x) = [\underline{\omega}(x), \underline{\omega}'(x)] \quad , \quad \underline{\omega}, \underline{\omega}' \in \mathcal{G} \times \mathcal{S}(R^4) \tag{2}$$

We consider a matter field $\underline{\varphi} \in \underline{V} \equiv V \times \mathcal{S}(R^4)$ where V is a representation space of G . Let D be the corresponding representation and t the corresponding representation of $\mathcal{G}$. $\underline{G}$ acts on $\underline{\varphi}$ according to

$$\underline{\varphi} \to {}^{\underline{g}}\underline{\varphi} : {}^{\underline{g}}\underline{\varphi}(x) = D(\underline{g}^{-1}(x))\, \underline{\varphi}(x) \qquad \underline{g} \in \underline{G} , \quad \underline{\varphi} \in \underline{V} \tag{3}$$

with the infinitesimal version

$$\delta_{\underline{\omega}} \underline{\varphi} : \quad \delta_{\underline{\omega}} \underline{\varphi}(x) = -t(\underline{\omega}(x))\, \underline{\varphi}(x) \qquad \underline{\omega} \in \underline{\mathcal{G}} = \mathcal{G} \times \mathcal{S}(R^4) \tag{4}$$

Since the dynamics of $\underline{\varphi}$ is to be derived from a Lagrangian involving the space time derivatives of $\underline{\varphi}$, one needs to compare the actions of G on $\underline{\varphi}$ at different space time points, which requires the introduction of a connexion form [5]

$$\underline{a} : \qquad \underline{a}(x, dx) = \underline{a}_\mu(x)\, dx^\mu \tag{5}$$

with values in $\mathcal{G}$. $\underline{a}_\mu$ is usually referred to as the gauge field or Yang Mills field. $\underline{G}$ acts on $\underline{a}$ as follows :

$$\underline{a} \to {}^{\underline{g}}\underline{a} = d\underline{g} \cdot \underline{g}^{-1} + \underline{g} \cdot \underline{a} \cdot \underline{g}^{-1} \tag{6}$$

where the indicated products are pointwise in M_4 taken in G . The last term in (6) involves the adjoint representation of G in $\mathcal{G}$. The infinitesimal version of (6) is :

$$\delta_{\underline{\omega}} \underline{a} = d\underline{\omega} + [\underline{\omega}, \underline{a}] \tag{7}$$

Thus, the covariant differential

$$\mathcal{D}\underline{\varphi} = d\underline{\varphi} - t(\underline{a})\underline{\varphi} \tag{8}$$

transforms as it should :

$$\mathcal{D}\underline{\varphi} \to {}^{g}\mathcal{D}\varphi = D(\underline{g}^{-1})\mathcal{D}\underline{\varphi} \tag{9}$$

Furthermore, the curvature form [4]

$$\underline{F} = d\underline{a} - [\underline{a}, \underline{a}] \tag{10}$$

transforms according to the adjoint representation

$$\underline{F} \to {}^{g}\underline{F} = \underline{g}\,\underline{F}\,\underline{g}^{-1} \tag{11}$$

$$\delta_{\underline{\omega}} \underline{F} = [\underline{\omega}, \underline{F}] \tag{12}$$

A gauge invariant Lagrangian density can then be constructed

$$\mathcal{L}_{inv} = -\frac{1}{4}\{\underline{F}_{\mu\nu}, \underline{F}^{\mu\nu}\} + \mathcal{L}(\underline{\varphi}, \mathcal{D}\underline{\varphi}) \tag{13}$$

where $\mathcal{L}(\underline{\varphi}, \mathcal{D}\underline{\varphi})$ is pointwise invariant under the action of G, and $\{\ ,\ \}$ is a nondegenerate symmetric invariant quadratic form on $\mathcal{G}$ whose choice fixes the coupling constants.

The geometrical situation at hand can be described in terms of a fiber bundle of base M_4, with structure group G, the fibers consisting of $\mathcal{F} = V \oplus \mathcal{G} \times M_4$, G acting according to (3), (6). Actually one has to enrich the fibers so as to include space time derivatives of $\underline{\varphi}$ and $\underline{a}$, but we refer the reader to [5] for the details. Due to gauge invariance, however the extremum problem derived from (13) has orbits of solutions obtained by the action of $\underline{G}$. The first way out is to eliminate the gauge degrees of freedom and define the physical degrees of freedom in the "unitary gauge" which is not renormalizable [1]. Alternatively, one may extremize

$$\Gamma_{inv}(\underline{\varphi},\underline{a}) = \int dx\, \mathcal{L}_{inv}(\underline{\varphi},\underline{a})(x) \tag{14}$$

subject to a gauge constraint

$$\underline{g}(\underline{\varphi},\underline{a}) = 0 \tag{15}$$

where $\underline{g}$ is a suitable local, relativistic invariant functions of $\underline{\varphi}$, $\underline{a}$, with values in $\mathcal{G}$. This procedure is definitely not gauge invariant unless the hypersurface (15) in "$\mathcal{F} \times M_4$" is gauge invariant in the following sense: "$\mathcal{F} \times M_4$" is actually meant to be a quotient space

$$\text{"}\mathcal{F} \times M_4\text{"} = (\mathcal{F} \times G \,/\, \text{action of } G \text{ on } \mathcal{F}) \times M_4 \tag{16}$$

and so far we have expressed this quotient by choosing coordinates ($\underline{\varphi}$, $\underline{a}$, e = identity of G) in $\mathcal{F} \times G$. In order to express invariantly variations of functions of $\underline{\varphi}$, along the fibers, we need a connexion on G, which is the well known Maurer Cartan left invariant canonical form [6] on G, ω, with value in $\mathcal{G}$, fulfilling the Maurer Cartan structure equation

$$\delta\omega = -[\omega,\omega] \tag{17}$$

where δ denotes the exterior differenciation on G.
For historical reasons, we shall denote

$$\omega\big|_e = \bar{c} \qquad e = \text{identity of } G \qquad (18).$$

$\bar{c}$ will turn out to be one of the Faddeev Popov ghosts.
It appears here naturally as embedded in the Grassmann algebra of differential forms on G whose left invariant ideal is isomorphic with the exterior algebra of $\mathcal{G}$.

Then, the gauge condition (15) is invariant along the fibers if

$$\delta \underline{g} = \left(\bar{c}, \frac{\delta_\omega \underline{g}}{\delta \underline{\omega}}\right) \equiv m\bar{c} = 0 \qquad (19)$$

This means in general that the invariant connexion $\bar{c}$ on G has to be considered as a field coupled to $\Phi = \{\underline{\varphi}, \underline{a}\}$ in the same way as in the first step, the Yang Mills field $\underline{a}$ had to be considered as coupled with $\underline{\varphi}$.

The minimization of Γ_{inv} (14) subject to the constraints (15), (19) can now be performed by introducing Lagrange multipliers θ, c (the second Faddeev Popov ghost) which leads to the action

$$\Gamma(\underline{\varphi}, \underline{a}, c, \bar{c}, \theta) = \Gamma_{inv}(\underline{\varphi}, \underline{a}) - (\underline{\theta}, \underline{g}) - (m\bar{c}, c) \qquad (20)$$

The Faddeev Popov charge may be identified with the grading associated with forms and vector fields on G : assigning charge -1 to forms $(\bar{c})$, $+1$ to vector fields (c) zero to functions ($\underline{\varphi}$, $\underline{a}$ and thus $\underline{\theta}$, $\underline{g}$) Γ is $\Phi.\Pi.$ neutral.

Now, since

$$\delta m\bar{c} = \delta^2 g = 0 \qquad (21)$$

$$\delta \Gamma = -(\theta, m\bar{c}) \tag{22}$$

one has the expression of the "Slavnov symmetry" [7] :

$$s \Gamma = 0 \tag{23}$$

with the definition

$$\begin{aligned} s \Phi &= \delta \Phi \\ s c &= \theta \\ s \theta &= 0 \end{aligned} \tag{24}$$

where s is an antiderivation [8] of degree 1 on the Grassmann algebra generated by c and $\bar{c}$ in which the scalar product of a form with a vector field is defined as being antisymmetric in its argument [9] . Thus, by construction s^2 is a derivation and actually

$$s^2 = 0 \tag{25}$$

Going back to (23) , one may write

$$s \Gamma \equiv \frac{\delta \Gamma}{\delta \phi} s \Phi + \frac{\delta \Gamma}{\delta \bar{c}} s \bar{c} + \frac{\delta \Gamma}{\delta c} \theta = 0 \tag{26}$$

where some care has to be exercised in computing $\delta\Gamma/\delta\bar{c}$, $\delta\Gamma/\delta c$.

In this framework, the gauge function has to be chosen in such a way that the degeneracy of the quadratic part of Γ is non degenerate, e.g.

$$\underline{g} = \partial \underline{a} + \cdots \tag{27}$$

and the equations of motion derived from Γ show that θ is a massless field, a characteristic feature of the Landau type gauges. Now, of course, the Slavnov symmetry is preserved if one adds to Γ an arbitrary function of θ alone :

$$\Gamma \to \Gamma + \mathcal{Q}(\theta) \tag{28}$$

and θ can be eliminated from the equation

$$\underline{g} = \frac{\delta \mathcal{Q}}{\delta \underline{\theta}} \tag{29}$$

provided $\mathcal{Q}$ contains a quadratic part, by performing a Legendre transformation with respect to θ [1d] :

$$\Gamma(\Phi, c, \bar{c}, R) = \Gamma(\Phi, c, \bar{c}) - (g, \theta) + \mathcal{Q}(\theta) + (R, \theta)\Big|_{-g + \frac{\delta \mathcal{Q}}{\delta \theta} + R = 0} \tag{30}$$

and putting $R = 0$. This operation corresponds to the choice of a Feynman like gauge - a Feynman gauge if $\mathcal{Q}$ is purely quadratic -. The Slavnov symmetry assumes now a new form

$$\mathcal{S}\Gamma = 0 \tag{31}$$

with

$$\mathcal{S}\Phi = s\Phi \tag{32}$$

$$\mathcal{S}\bar{c} = \theta(g, R) \qquad \text{deduced from } -g + \frac{\delta \mathcal{Q}}{\delta \theta} + R = 0$$

$$\mathcal{S}R = 0$$

which allows to consider R as an external field. Now,

$$\mathcal{S}^2 \neq 0 \tag{33}$$

because the new transformation corresponds to

$$s\theta \neq 0$$

The simplest case corresponds to a purely quadratic Q which falls within the renormalizable framework for dimension two gauge functions, θ and R being assigned dimension two. Then,

$$\Gamma(\Phi, c, \bar{c}, R) = \Gamma(\Phi, c, \bar{c}) - \frac{1}{2}\left[(g-R),(g-R)\right]$$

$$\mathcal{S}C = \frac{\delta\Gamma}{\delta R} \tag{34}$$

where $[\quad]$ denotes the inverse of Q.

In order to cast the Slavnov identity into a form likely to pass through renormalization, it is convenient to introduce external fields coupled to the Slavnov variations of the basic fields, carrying the correct $\Phi\Pi$ charge and dimension so that Γ is $\Phi\Pi$ neutral, and has dimension four :

$$\Gamma(\Phi, c, \bar{c}, R, \gamma, \zeta) = \Gamma(\Phi, c, \bar{c}, R) + (\gamma, \mathcal{S}\Phi) + (\zeta, \mathcal{S}\bar{c}) \tag{35}$$

The Slavnov invariance now reads

$$\int dx \left(\frac{\delta\Gamma}{\delta\Phi}\frac{\delta\Gamma}{\delta\gamma} + \frac{\delta\Gamma}{\delta\bar{c}}\frac{\delta\Gamma}{\delta\zeta} + \frac{\delta\Gamma}{\delta C}\frac{\delta\Gamma}{\delta R} \right)(x) = 0 \tag{36}$$

In terms of the connected Green's functional the Legendre transform of Γ with respect to sources linearly coupled to the quantized fields Φ , c , $\bar{c}$:

$$Z_c(J, \bar{\xi}, \xi, R, \gamma, \zeta) = \Gamma(\Phi, c, \bar{c}, R, \gamma, \zeta) + + J\Phi + \bar{\xi}c + \bar{c}\xi \Bigg|_{\substack{\frac{\delta\Gamma}{\delta\Phi} + J = 0 \\ \frac{\delta\Gamma}{\delta c} - \bar{\xi} = 0 \\ \frac{\delta\Gamma}{\delta\bar{c}} - \xi = 0}} \tag{37}$$

one has [1d] :

$$\mathcal{S} Z_c(J, \xi, \bar{\xi}, R, \gamma, \zeta) \equiv \int dx \left(J \frac{\delta Z_c}{\delta\gamma} - \xi \frac{\delta Z_c}{\delta\zeta} - \bar{\xi} \frac{\delta Z_c}{\delta R} \right)(x) = 0 \tag{38}$$

In the case of linear gauges

$$\mathcal{G} = (g, \Phi) \tag{39}$$

there is no need to introduce the source R coupled to the Gauge function, since one may substitute

$$\frac{\delta\Gamma}{\delta R} = (g, \Phi) - R \tag{40}$$

and put $R = 0$.

In this case, the Slavnov identity (38) reads :

$$\mathcal{S}Z_c(J,\bar{\xi},\xi,\gamma,\zeta) = \int dx \left(J\frac{\delta Z_c}{\delta\gamma} - \xi\frac{\delta Z_c}{\delta\gamma} - \bar{\xi}\left(g,\frac{\delta}{\delta J}\right)\right) Z^c(x) = 0 \tag{41}$$

Before leaving the subject, it is worthwhile pointing out that whereas (26) and (36) are equivalent, (41), in the linear case is not equivalent to them and allows in some abelian cases (cf. [BRS 1, 2]) the addition of some mass terms which allow for instance to consider as gauge theories massive quantum electrodynamics in the Stueckelberg gauge and the non spontaneously broken Higgs Kibble model which is void of physical interpretation. If one chooses (36) to define gauge theories, it is tempting to look for its general solution for which a conjecture has been made by J. Zinn Justin [10] .

c) Physical Interpretation

We shall insist here on the case where $\mathcal{L}_{inv}(\underline{\varphi},\underline{a})$ is invariant under G but where this symmetry is spontaneously broken, namely, the particle interpretation requires a field translation

$$\varphi \rightarrow \varphi + F \tag{42}$$

where

$$\left.\frac{\delta \Gamma_{inv}}{\delta \varphi}\right|_{\underline{\varphi}=\underline{F},\,\underline{a}=0} = 0 \tag{43}$$

In terms of the translated variable

$$\left.\frac{\delta \mathcal{T}_{inv}}{\delta \underline{\varphi}}\right|_{\underline{\varphi}=\underline{a}=0} \tag{44}$$

Let us decompose $\mathcal{G}$ into the stabilizer of $\underline{F}$ and its complement

$$\mathcal{G} = \mathcal{G}_F + \mathcal{G}_F^{\perp} \tag{45}$$

$$t(x)\,\underline{F} = 0 \qquad x \in \mathcal{G}_F \tag{46}$$

and let us decompose V into the orbit of $\underline{F}$ and its complement :

$$V = O_F + O_F^{\perp} \tag{47}$$

O_F is generated by $t(x)\,\underline{F}$, $x \in \mathcal{G}_F^{\perp}$.

The quickest way to proceed is to write down the quadratic part of the most general solution of the local Ward identity (55) of section 3 - c.

$$\Gamma^{(2)}_{inv}(\underline{\varphi},\underline{a}) = \text{quadratic part of}$$

$$-\frac{1}{4}\left\{\underline{F}_{\mu\nu},\underline{F}^{\mu\nu}\right\} + \frac{1}{2}\left(\mathcal{D}_\mu(\underline{\varphi}+\underline{F}),\mathcal{D}^\mu(\underline{\varphi}+\underline{F})\right) + V(\underline{\varphi}+\underline{F}) \tag{48}$$

where $\{\ \}$, $(\ ,\ )$ and V are invariant under $\mathcal{G}$. Before the translation is performed, this quadratic form is degenerate since $\underline{\varphi}$ and $\underline{a}$ are decoupled and $\{\partial_\mu a_\nu - \partial_\nu a_\mu\, ,\, \partial^\mu a^\nu - \partial^\nu a^\mu\}$ is degenerate. Although the $\underline{\varphi}$ field translation couples $\underline{\varphi}$ and $\partial\underline{a}$, this situation will prevail. Mass terms develop, with the following structure :

$$m^{inv}_{\varphi\varphi} = \begin{array}{c|cc} & O_F & O_F^{\perp} \\ \hline O_F & 0 & 0 \\ O_F^{\perp} & 0 & M \end{array} \tag{49}$$

(as can easily be derived from the integrated Ward identity which yields $[m^{inv}_{\varphi\varphi}, t(X)] = 0$ for $X \in \mathcal{G}_F$

$$m^2_{aa} = \begin{array}{c|cc|} & \mathcal{G}_F & \mathcal{G}_F^\perp \\ \mathcal{G}_F & 0 & 0 \\ \mathcal{G}_F^\perp & 0 & m_a^2 \end{array} \tag{50}$$

(as can be derived by differenciating the local Ward identity with respect to $\underline{a}$ at $\underline{a} = \underline{\varphi} = 0$, and with respect to momentum at 0), and the general form of $\Gamma^{(2)}_{INV}$ is :

$$\begin{aligned} \Gamma^{(2)}_{INV}(\underline{\varphi}, \underline{a}) = -\frac{1}{4}\left\{\partial_\mu \underline{a}_\nu - \partial_\nu \underline{a}_\mu, \partial^\mu \underline{a}^\nu - \partial^\nu \underline{a}^\mu\right\} \\ + \frac{1}{2}\left(\partial_\mu \varphi, \partial^\mu \varphi\right) + \left(\varphi, t(\partial a)F\right) \\ + \frac{1}{2}\left(t(a)F, t(a)F\right) - \frac{1}{2}\left[\varphi_{O_F^\perp} M^2 \varphi_{O_F^\perp}\right] \end{aligned} \tag{51}$$

where $\{\ ,\ \}$, $(\ ,\ )$, are bilinear forms invariant under $\mathcal{G}$ whereas $[\ \ ]$ is invariant under $\mathcal{G}_F$. From this general form, one observes that :

$a^T|_{\mathcal{G}_F^\perp}$ acquires a mass (the index T indicates the transverse part).

$\varphi|_{O_F^\perp}$ is in general massive

$a|_{\mathcal{G}_F}$ remains massless

$\partial a|_{\mathcal{G}_F^\perp}$ and $\varphi|_{O_F}$ couple within a massless channel,

the "Goldstone channel".

The degeneracy of the quadratic form is limited to the last two components and its removal through the introduction of the gauge term yields gauge dependent particle interpretations.

The situation is summarized in the following table :

<u>Physical degrees of freedom</u> (unaffected by the choice of the gauge function) :	
$a^T\|_{G_F^\perp}$	massive spin 1 particles
$\varphi\|_{O_F^\perp}$	massive
$a^T\|_{G_F}$	massless helicity 1 particles
<u>Unphysical degrees of freedom</u> (affected by the choice of the gauge function)	
$a^L\|_{G_F}$	massless longitudinal and scalar quanta
$\partial a\|_{G_F^\perp}$, φ_{O_F}	longitudinal "photons" and Goldstone bosons

We shall now see how the choice of the gauge function specifies the particle interpretation of the non physical degrees of freedom. Obviously, only the linear part of the gauge function is involved :

$$g^L = (g, \Phi) = g(\partial a) + f(\varphi) \tag{52}$$

where we have limited ourselves to relativistically invariant gauges. Although the t'Hooft gauges which respect the same decoupling as in $\Gamma^{(2)}_{inv}$ are simpler, we shall treat the general case and show that the physical content (masses) is gauge independent and that the unphysical degrees of freedom are pairwise degenerate in mass, provided physical and unphysical field combinations have been suitably defined.

We thus look at

$$\Gamma^{(2)}_{inv}(\underline{\varphi}, \underline{a}) - \frac{1}{2}\left\{ g(\partial a) + f(\varphi), g(\partial a) + f(\varphi) \right\} \tag{53}$$

Using the non degeneracy and invariance of $\{\ ,\ \}$, $(\ ,\ )$ we may choose the corresponding matrices to be unity, and define $m = tF$ as a matrix with one index in $\mathcal{G}_F^\perp$, and one in $\mathcal{O}_F$, so that m^2 leaves $\mathcal{G}_F^\perp$ stable.

The equations of motion derived from (52) are :

$$\mathcal{O}_F^\perp:\ (\Box + M^2)\varphi_{\mathcal{O}_F^\perp} = -\{g\,\partial a + f\varphi, f_{\mathcal{O}_F^\perp}\} \qquad \text{(a)}$$

$$\mathcal{O}_F:\ \Box\varphi_{\mathcal{O}_F} + m\,\partial a_{\mathcal{G}_F^\perp} = -\{g\partial a + f\varphi, f_{\mathcal{O}_F}\} \qquad \text{(b)}$$

(54)

$$\mathcal{G}_F^\perp:\ (\Box g_{\mu\nu} - \partial_\mu\partial_\nu)\,a^\nu_{\mathcal{G}_F^\perp} + (\partial_\mu\varphi_{\mathcal{O}_F} m)_{\mathcal{G}_F^\perp} + m^2 a^\nu_{\mathcal{G}_F^\perp} \qquad \text{(c)}$$
$$= -\partial_\mu\{g\partial a + f\varphi, g_{\mathcal{G}_F^\perp}\}$$

$$\mathcal{G}_F:\ (\Box g_{\mu\nu} - \partial_\mu\partial_\nu)\,a^\nu_{\mathcal{G}_F} = -\partial_\mu\{g\partial a + f\varphi, g_{\mathcal{G}_F}\} \qquad \text{(d)}$$

Taking the transverse part of the last two equations yields

$$\mathcal{G}_F^\perp\ \ (\Box g_{\mu\nu} - \partial_\mu\partial_\nu \mathbb{1} + m^2 g_{\mu\nu})\,a^{\nu T}_{\mathcal{G}_F^\perp} = 0 \qquad \text{(a)}$$

(55)

$$\mathcal{G}_F\ \ (\Box g_{\mu\nu} - \partial_\mu\partial_\nu)\,a^{T\nu}_{\mathcal{G}_F} = 0 \qquad \text{(b)}$$

So, the transverse photons have mass m^2 along $\mathcal{G}_F^\perp$, 0 along $\mathcal{G}_F$.

Taking the divergence of the last two equations in (54) yield :

$$\mathcal{G}_F^\perp:\ \Box\varphi_{\mathcal{O}_F} m + m^2(\partial a)_{\mathcal{G}_F^\perp} = -\Box\{g\partial a + f\varphi, g_{\mathcal{G}_F^\perp}\} \qquad \text{(a)}$$

(56)

$$\mathcal{G}_F:\ \Box\{g\partial a + f\varphi, g_{\mathcal{G}_F}\} = 0 \qquad \text{(b)}$$

Combining (55b), (56b) yields :

$$\Box \left\{ g\partial a + f\varphi ,\, g_{\mathcal{G}_F} \right\} = 0 \tag{57}$$

i.e. $\{ g\partial a + f\varphi ,\, g_{\mathcal{G}_F} \}$ is a massless field.
Combining (52b) and (56a) yields :

$$\Box \left\{ g\partial a + f\varphi ,\, g_{\mathcal{G}_F^{\perp}} \right\} + m \left\{ g\partial a + f\varphi ,\, f_{O_F} \right\} = 0 \tag{58}$$

which we shall later compare to one of the $\Phi\Pi$ equations of motion along $\mathcal{G}_F^{\perp}$. Using (57) (58), one can determine

$$\begin{aligned} \varphi'_{O_F^{\perp}} = \varphi_{O_F^{\perp}} + t_{O_F^{\perp}}{}^{\mathcal{G}_F} \left\{ g\partial a + f\varphi ,\, g_{\mathcal{G}_F} \right\} \\ + u_{O_F^{\perp}}{}^{\mathcal{G}_F^{\perp}} \left\{ g\partial a + f\varphi ,\, g_{\mathcal{G}_F^{\perp}} \right\} \end{aligned} \tag{59}$$

i.e. show the existence of the matrices $t_{O_F^{\perp}}{}^{\mathcal{G}_F}$, $u_{O_F^{\perp}}{}^{\mathcal{G}_F^{\perp}}$, so that

$$(\Box + M^2)\, \varphi'_{O_F^{\perp}} = 0 \tag{60}$$

Now, looking at the quadratic part of the $\Phi\Pi$ Lagrangian

$$\left\{ c ,\, \Box g \bar{c} + f m \bar{c} \right\} \tag{61}$$

we see that (57) (58) has the same structure as the equation of motion for the c ghost with c replaced by the gauge function. Since system (54) derives from a Lagrangian, there is another field combination $\bar{g}$ for which the field equation is

$$\Box \left\{ \bar{g} ,\, g \right\} + \left\{ \bar{g} ,\, f_{O_F} \right\} m = 0 \tag{62}$$

So that the coupled g , $\bar{g}$ system as well as the c , $\bar{c}$ system derive from the same Lagrangian which yields the following differential operator :

$$\begin{array}{c|c|} & \mathcal{G} \\ \hline \mathcal{G} & \Box g \end{array} \; + \; \begin{array}{c|c:c|} & \mathcal{G}_F & \mathcal{G}_F^{\perp} \\ \hline \mathcal{G}_F & 0 & \\ & & m f_{\varphi} \\ \mathcal{G}_F^{\perp} & 0 & \end{array} \qquad (63)$$

and its transposed. Clearly, the mass matrix in (63) has $\dim \mathcal{G}_F$ nul eigenvalues, which corresponds to the massless fields in (57) , and all masses occur by degenerate pairs as they do for the $c\ \bar{c}$ ghosts. The parameters have of course to be chosen in such a way that the masses come out real.

Other consequences of gauge invariance can be directly deduced from the Slavnov identity, e.g.

$$\frac{\delta^2 Z_c}{\delta R(x)\, \delta R(y)} = 0 \qquad (64)$$

which means that at the tree level, the two point function of the gauge operator is a δ function. For the gauge independence of the S operator, one derives from the Slavnov identity that the gauge function has vanishing matrix elements between physical states. We refer to C. Becchi's lectures for illustration of these facts at the quantum level.

- Section 4 - References and Footnotes -

[1] a) E.S. ABERS, B.W. LEE
Physics Reports 9c n° 1 (1973)

b) M. VELTMAN
in "International Symposium on Electrons and Photons at High Energies", Bonn 27-31, August 1973

c) G.'t HOOFT, M. VELTMAN
"Diagrammar" CERN-TH/73/9 yellow report, September 1973

d) J. ZINN JUSTIN
in "Trends in Elementary Particle Theory", International Summer Institute on Theoretical Physics, Bonn 1974, Lecture Notes in Physics, vol. 37, Springer Verlag New York 1975

[2] L.D. FADDEEV, V.N. POPOV
Phys. Lett. 25B, 29 (1967)

[3] E.S. FRADKIN, IV. TYUTIN
PRD 2, 284, (1970)

[4] The possibility of making the construction indicated here has been suggested in the course of a discussion between C. Becchi and B. Zumino in Cargèse, July 1975.

[5] Ref. [4] of Section 3

[6] a) J. DIEUDONNE
Eléments d'Analyse, vol. 4 , Gauthiers-Villars, Paris 1971

b) S. STERNBERG
Lectures on Differential Geometry, Prentice Hall 1964

[7] In A.A. SLAVNOV , Teor. Mat. Fiz. 10 , 153 (1972), the Slavnov identities first appear in functional form. The identities written there are restricted to $\Gamma(\varphi, a, c=\bar{c}=0)$ and derived on the basis of a heuristic argument which relies on the formal properties of the Feynman functional integral (cf. [1d]). Considering the full $\Gamma(\varphi, a, c, \bar{c})$ yields the form of the Slavnov invariance.

[8] E. NELSON

Tensor Analysis, Mathematical Notes, Princeton University Press, Princeton N.J., 1967,
and 6b).

[9] This formulation avoids the introduction of the anticommuting parameter $\delta\lambda$ introduced in [BRS 1,2] .
The construction given in the text may not look very compelling at a stage where c and $\bar{c}$ appear linearly in Γ . However, as soon as external fields coupled to $s\Phi$, $s\bar{c}$ are introduced (cf. Eq. 35) $\bar{c}$ naturally appears as embedded in the exterior algebra of left invariant differential forms on G , and an invariance property of Γ can only be expressed in terms of the action principle as done in [BRS 1,2] if c is also embedded into the exterior algebra of vector fields on G and a skew product $(\bar{c}, c) = -(c, \bar{c})$ is defined.

[10] J. ZINN JUSTIN

Private communication.

RENORMALIZED PERTURBATION THEORY: ACHIEVEMENTS, LIMITATIONS AND OPEN PROBLEMS

B.SCHROER

Institut für Theoretische Physik
Freie Universität Berlin, Germany

Lectures presented at the International School of Mathematical Physics "Ettore Majorana", Centre for Scientific Culture.
Erice, Italy Sept. 18-30, 1975

G. Velo and A. S. Wightman (eds.), Renormalization Theory, 345–369.

I. Quantization of Classical Lagrangians and the Problem of Parametrization.

Perturbative renormalization theory (P.R.T.) of renormalizable Lagrangians has in recent years reached an unprecedented perfection and sofistication. There are now so many renormalization techniques available that one cannot conceive of any renormalizable Lagrangian deserving its name which cannot be treated by one or the other method[1]. This state of affairs contrasts for example with the situation concerning nonabelian gauge theories some years ago.

The spectacular success of QED up to the shortest distances measured so far and the impressive application of nonabelian gauge theories to a unified picture of electromagnetism, weak interactions and perhaps also strong interaction constitute the most convincing achievements of P.R.T. Obviously a series of three lectures can not do justice to discuss these achievements; there are however many beautifully written reviews and conference proceedings[2] which should convince even sceptical minds.

We will concentrate in the following on achievements and problems of P.R.T. which can be demonstrated and discussed in simple field theoretical models and which up to now have not attracted a lot of attention.
Any of the many methods of P.R.T usually start from the Gell-Mann Low formula either in its conventional operator - or functional-form. For the A^4-theory (to which we will confine ourselves for explaining matters of principle) the unrenormalized Gell-Mann Low formula reads[3]

$$< T X_u > = C^{-1} < T X_{uo} \exp i\int \mathcal{L}_{int,u} d^D x > \quad (1)$$

with $C = < T \exp i\int \mathcal{L}_{int,u} d^D x >$, $X_u = \prod_1^N A_u(x_i)$

index u = unrenormalized

or in the (euclidean) functional form:

$$< X_u >_e = C^{-1}\int d[A] X[A] \exp.\int \mathcal{L}[A] d^D x \quad (2)$$

$$C = \int d[A] \exp.\int \mathcal{L}[A] d^D x$$

Here the total Lagrangian $\mathcal{L}$ is in some way split up into a bilinear part and the rest: i.e. for a A^4-theory:

$$\mathcal{L} = \frac{1}{2}\partial_\mu A_u \partial^\mu A_u - \frac{1}{2} m_o^2 A_u^2 - \frac{g_o}{4!} A_u^4 \qquad (3)$$

$$\mathcal{L} = \mathcal{L}_{u,o} + \mathcal{L}_{int,u}$$

The fields $A_{uo}(x)$ are free fields belonging to the unrenormalized mass m_o. The $\mathcal{L}_{int,u}$ in the exponential is understood as a function of the free fields A_{uo}.

In order to have a finite expression one has to introduce a cutoff Λ or a regularization.

It is easy to see (for example by formal functional manipulation[4]) that the Gell-Mann Low formula does not change if we make another split:

$$\mathcal{L} = \hat{\mathcal{L}}_{bil.} + \hat{\mathcal{L}}_{int} \qquad (4)$$

as long as the free fields in X_o are those referring to the linear part (i.e. with the same mass). Multiplicative renormalization consists in introducing

$$A = Z^{-1/2} A_u \qquad (5a)$$

$$< T X > = Z^{-N/2} < T X_u > \qquad (5b)$$

In evaluating renormalized Green functions it is convenient to split:

$$\begin{aligned}\mathcal{L} &= \frac{1}{2}\partial_\mu A \partial^\mu A - \frac{m^2}{2} A^2 - \frac{g_o Z^2}{4} A^4 \\ &- \frac{1}{2}(Z-1)(\partial_\mu A \partial^\mu A - m^2 A^2 + \delta m^2 A^2) + \frac{\delta m^2}{2} A^2\end{aligned} \qquad (6)$$

We also use the additive notation $(g_o Z^2 \equiv g Z_1)$:

$$\mathcal{L} = \frac{1}{2}\partial_\mu A \partial^\mu A - \frac{m^2}{2} A^2 - \frac{g-c}{4!} A^4 + \frac{1}{2} b\, \partial_\mu A \partial^\mu A - \frac{1}{2} a A^2 \qquad (7)$$

where the additive counter terms a b c are related

to the Z, Z_1 and $\delta\, m^2$. The main statement of multiplication renormalization theory is that Z, Z_1 and $\delta\, m^2$ respectively a b and c may by choosen as such (divergent) functions of the cutoff Λ that the Green function<TX> approach a finite limit $\Lambda\to\infty$. In order to not only have finite Green functions but also to uniquely specify them as functions of finite parameters m^2 and g , one usually introduces normalization conditions. Rather stating them in terms of Green functions it is more convenient to work with one particle irreducible functions[1] often called <u>vertex functions</u> $\Gamma^{(N)}(p_1..p_N)$. The simplest choice is the choice of "<u>intermediate</u>" <u>normalizations</u>:

$$\Gamma^{(2)}\Big|_{p=o} = -\,i\,m^2 \;,\quad \frac{d}{dp^2}\,\Gamma^{(2)}\Big|_{p=o} = i \tag{8}$$

$$\Gamma^{(4)}\Big|_{p=o} = -\,i\,g$$

on the right hand side appears always the lowest order expression for the corresponding vertex functions.

Let $\Gamma_n^{(N)}$ be the sum of nth order perturbation expression which has been obtained by iterative determination of counter-terms and subsequent elimination of the cut-off $\Lambda\to\infty$ (via normalization conditions). Then we have the following <u>statement</u>[5]:

$$\lim_{\Lambda\to\infty} \Gamma_n^{(N)}(p_1...p_N) = F.P.\sum_{G_n}\int I_{G_n}(\underline{p},\underline{k})\,d\underline{k} \tag{9}$$

The notation is the following:

G_n = n^{th} order Feynman graph without counter-terms

F.P. = Finite part operation

= Momentum space subtraction with Zimmermann's forest combinations[1] i.e.

$$P.F.\; I_{G_n} \equiv R_{G_n} \tag{10}$$

where R_{G_n} is the renormalized integrand[1].

Since on the level of the renormalized integrand R_{G_n}

there was no use of counter-terms (we have only done zero momentum Taylor-subtractions on unrenormalized integrands), we say that in the intermediate normalization (8) the effective interaction Lagrangian has the form:

$$\mathcal{L}_{int} = -\frac{g}{4!} A^4 \tag{11}$$

The usual normalizations in high energy physics are the mass shell conditions:

$$\Gamma^{(2)}\Big|_{p^2=m^2} = 0\ , \quad \frac{\partial}{\partial p^2}\Gamma^{(2)}\Big|_{p^2=m^2} = i$$

$$\Gamma^{(4)}\Big|_{s.p.(m^2)} = -ig \tag{12}$$

In that case the previous statement (9) looks still the same, however now the unrenormalized integrand also contains those vertices coming from finite a-b-c counter-terms. Their values are determined by enforcing the normalization conditions iteratively. We therefore say that the effective interaction has now the form:

$$\mathcal{L}_{int} = -\frac{g-c}{4!} A^4 + \frac{b}{2}\partial_\mu A\, \partial^\mu A - \frac{a}{2} A^2 \tag{13}$$

For many applications one wants a m^2, g-parametrization of Γ's which admit a smooth limit for zero mass: $m \to 0$. This requirement necessitates the introduction of at least another normalization parameter μ. The two most prominent schemes are:

(a) The Gell-Mann Low parametrization:

$$\Gamma^{(2)}\Big|_{p=0} = i\, m^2 \quad (\text{or } \Gamma^{(2)}\Big|_{p^2=m^2} = 0)$$

$$\frac{\partial}{\partial p^2}\Gamma^{(2)}\Big|_{p^2=-\mu^2} = i\ , \quad \Gamma^{(4)}\Big|_{s.p.(-\mu^2)} = -ig \tag{14}$$

In order to have absolutly convergent Feynman integrands in (9) and a finite form (i.e. finite counterterms) of the effective Lagrangian in the limit $m \to o$, one must interpret the F.P. as a Taylor-operator not around $p = o$ but rather around finite euclidean momenta with length μ^2. This framework has been recently worked out by P.K. Mitter[6].

b) The Kadanoff parametrization:

This parametrization works with a variable mass. The vertex-functions depend as in the Gell-Mann Low approach on μ, m, g, but in the normalization conditions one leaves the physical values for Γ and treats m as a variable parameter in the normalization conditions[7] i.e. one works with unphysical normalizations

$$\Gamma^{(2)}\Big|_{p=o,\ m=o} = 0$$

$$\frac{\partial}{\partial p^2}\Gamma^{(2)}\Big|_{p=o,\ m=\mu} = i \ , \quad \frac{\partial}{\partial m^2}\Gamma^{(2)}\Big|_{p=o,\ m=\mu} = -i \qquad (15)$$

$$\Gamma^{(4)}\Big|_{p=o,\ m=\mu} = -ig$$

The subtraction operators in the forest formula for R_{G_n} are constructed in such a way that the first condition is automatically fulfilled without use of counter-terms.

In high energy physics the advantage of the Kadanoff-parametrization is that the leading asymptotic behaviour for nonexceptional momenta (in the sense of Symanzik) is equal to its zero mass limit:

$$\lim_{m\to o}\Gamma^{(N)}(p_1\ldots p_N) = \Gamma^{(N)}_{as.}(p_1\ldots p_N) \qquad (16)$$

The Gell-Mann Low parametrization does not lead to such an equality even though each side exists separately. The main advantage of this parametrization lies in the form of the parametric differential equations it leads to:

$$\left\{ 2\mu^2\frac{\partial}{\partial\mu^2} + 2\delta m^2\frac{\partial}{\partial m^2} + \beta\frac{\partial}{\partial g} - N\gamma \right\}\Gamma^{(N)} = 0 \qquad (17a)$$

$$\frac{\partial}{\partial m^2}\Gamma^{(N)} = \Delta_o\,\Gamma^{(N)}$$

Δ_o = mass insertion operation.

δ β and γ are only functions of g. From the normalization condition it easily follows that:

$$2\,\delta = \gamma_{A^2} \tag{18}$$

γ_{A^2} = "would-be" anomaleous dimension of A^2.
The homogeneous differential equation (together with Euler's homogeneity equation for $p,\mu,m \to \lambda p, \lambda\mu, \lambda m$) tells us that a scale change in the momenta may be absorbed into a change of the coupling constant, the "scaling" mass m and a multiplicative finite wave function renormalization.

Such a statement in the presence of an infrared-stable zero of would immediatly lead to the differential version of Kadanoff's scaling law, which was first introduced in the statistical mechanics treatment of critical phenomena[8].

The subtraction method leading to the Kadanoff parametrization is very similar to a method which was recently worked out by Lowenstein and Zimmermann[9] for the direct renormalization of zero mass theories. It was successfully applied to nonabelian gauge theories by Lowenstein[1].

Let us briefly explain the Kadanoff parametrization in the "softness" parameter language of Lowenstein and Zimmermann. Choose

$$\mathcal{L}_{eff} = \frac{1}{2}\,\partial_\mu A\,\partial^\mu A - \frac{1}{2}\,m^2(s,\sigma)\,A^2 - \frac{g}{4}\,A^4 \tag{19}$$

$$\text{with} \quad m^2(s,\sigma) = \sigma^2\mu^2 + s^2 m^2$$

and define subtraction operators τ

$$\tau^{(o)}\,F(p;\,s,\sigma) = t^{(o)}_{p,s,\sigma-1}\,F(p;\,s,\sigma) \tag{20a}$$

$$(1-\tau^{(2)})\,F(p;\,s,\sigma) = (1-t^{(o)}_{p,s,\sigma})(1-t^{(2)}_{p,s,\sigma-1})F(p;s,\sigma) \tag{20b}$$

Here $t^{(2\delta)}_{p,s,\sigma}$ is the Taylor-subtraction operator up to 2 δ around p = o = s = σ. The σ and s-variables which enter together with the quadratic mass terms into the Feynman denominators carry canonical dimension two[10], and therefore derivatives with respect to s and σ corresponding to odd p-derivatives will be absent; the highest s, σ derivative is therefore of order δ. The generalization to higher degrees then two for the purpose of defining arbitrary composite fields in our Kadanoff-parametrized A^4-model is identical to that given by Lowenstein[11].

From the relation:

$$\frac{\partial}{\partial s}(1-\tau^{(2)})\,F = (1-\tau^{(o)})\,\frac{\partial}{\partial s}\,F \tag{21}$$

it then follows that the differentiation

$$\frac{\partial}{\partial s}\,\Gamma^{(N)}_{s,\sigma} = m^2\,\frac{\partial}{\partial m^2}\,\Gamma^{(N)}_{s,\sigma} \tag{22}$$

can also be written as a mass insertion:

$$m^2\,\Delta_o\,\Gamma^{(N)}_{s,\sigma} \tag{23}$$

For the physical values: $\Gamma^{(N)} = \Gamma^{(N)}_{1,o}$ this leads to (17b) and therefore establishes the validity of the renormalized action principle with respect to m-differentiation:

$$\begin{aligned}\frac{\partial}{\partial m^2} < T\,X > &= i < T\,X\,\frac{\partial}{\partial m^2}\,\mathcal{L}_{eff}(x)\,d^4x > \\ &= -\frac{i}{2} < T\,X\,\frac{\partial}{\partial m^2}\int N_4\,[m^2A^2]\,d^4x > \\ &= -\frac{i}{2} < T\,X\int N_2\,[A^2]\,d^4x >\end{aligned} \tag{24}$$

We will not attempt to formulate the equivalence statement corresponding to formula (9). This could be achieved by splitting the self-mass term in the conventional cut-off formulation into a μ and m-dependent part.

The homogeneous differential equation (17a) can be derived in the standard fashion[7]. Although the most useful applications of homogeneous differential equation (17a) involving an effective mass m are in the theory of critical phenomena, there have been recently some attempts to utilize such equations in high energy phenomenology[12].

Note that the Kadanoff parametrization is the only one which does not work perturbativly (for A^4) in $D < 4$ even though we are talking about a massive theory. The difficulty is that the first requirement presupposes a smooth zero mass limit in perturbation theory. On the other hand we know that the approach to zero mass is equivalent to the critical limit and therefore requires nonperturbative arguments. In this case one may however make a compromise between the perturbative starting point and the desired scaling structure of the non-perturbative critical limits by demanding instead of the first normalization condition

$$\Gamma^{(2)}\Big|_{p=o,\ m=\mu} = -\, i\, \mu^2 \tag{25}$$

In that case we obtain a differential equation in the form:

$$\{2\,\mu^2 \frac{\partial}{\partial\mu^2} + 2(\delta_1 + \frac{\mu^2}{m^2}\,\delta_2) m^2 \frac{\partial}{\partial m^2} + \beta\,\frac{\partial}{\partial g} - N\gamma_A\}\;\Gamma^{(N)} = 0 \tag{26}$$

$$\frac{\partial}{\partial m^2}\,\Gamma^{(N)} = \Delta_o\,\Gamma^{(N)} \tag{27}$$

From the normalization properties one obtains

$$\delta_1 + \delta_2 = \gamma_A \quad , \quad 2\,\delta_1 = \gamma_{A^2} \tag{28}$$

The scaling property forthe effective mass is:

$$\hat{m}^2(\kappa^2) = \frac{1}{\kappa^2}\,\overline{m}^2(\kappa^2)$$

where $\overline{m}^2$ obeys the inhomogeneous differential equation

$$\frac{d\,\overline{m}^2\,(\kappa^2)}{d\,\ell n\,\kappa^2} = \delta_1\,\overline{m}^2(\kappa^2) + \mu^2\,\kappa^2\,\delta_2 \tag{29}$$

The solution fulfilling the boundary condition $m^2(\kappa^2)\Big|_{\kappa^2=1} = m^2$ is (assuming that we are at a zero of β i.e. constant δ's)

$$\overline{m}^2\,(\kappa^2) = \kappa^4\,\frac{\delta_2}{2-\delta_1}\,\mu^2 + (m^2 - \frac{\delta_2}{2-\delta_1}\,\mu^2\,)(\kappa^2)^{\delta_1} \tag{30}$$

Therefore $\hat{m}^2(\kappa^2)$ has for small κ^2 the desired scaling property ($2\,\delta_1 = \gamma_{A^2} < 2$)

$$\hat{m}^2\,(\kappa^2) \sim (m^2 - \frac{\delta_2}{2-\delta_1}\,\mu^2)(\kappa^2)^{\delta_1 - 1} \tag{31}$$

However the coefficient is unfortunatly not m^2 i.e. zero mass does not mean that $m^2 = o$ but rather that a function of m^2 and μ^2 plays the role of mass near $\kappa^2 \to o$ (long distances)[13]. This example shows the limitations of perturbation theory in the problem of infrared behaviour in superrenormalizable theories. Even for strictly renormalizable theories in $D = 4$ the question of a "good" parametrization of a model is in many cases an open problem. Consider two typical examples:

1. Problem of introducing gauge invariant coupling strength into nonabelian Yang-Mills theory. In this case the β-function must be independent of the gauge parameter α. It has been suspected for some time that the t'Hooft "evaluator procedure" (i.e. renormalization without imposing normalization conditions) based on dimensional regularization leads to such a desired situation. A proof to 2nd order of perturbation theory has been given by Caswell and Wilszek[14]. It is not clear whether one can obtain such a simple parametrization from normalization conditions. In such a parametrization the discussion of gauge invariant composite fields is expected to be simpler.

2. Consider the SU(n) version of the massive Thirring model. This model is known (by generalized Fierz identities) to be reducible to a 3 parametric Lagrangian[16])

$$\mathcal{L} = i\,\bar{\psi}\gamma^{\mu}\partial_{\mu}\psi + m\,\bar{\psi}\psi - \frac{1}{2}g\,\bar{\psi}\gamma_{\mu}\psi\bar{\psi}\gamma^{\mu}\psi - \frac{1}{2}g_{\nu}\,\bar{\psi}\vec{\lambda}\gamma_{\mu}\psi\bar{\psi}\vec{\lambda}\gamma^{\mu}\psi - \frac{1}{2}g_{s}\,\bar{\psi}\psi\bar{\psi}\psi \tag{32}$$

Out of this 3 parametric manifold we may select a one parametric "Thirring" manifold C by demanding tha absence of any hard breaking terms in the iso-scalar and iso-vector axial currents:

$$\partial^{\mu} j_{\mu 5} \xrightarrow{m\to o} 0 \quad ; \quad \partial^{\mu} j^{a}_{\mu 5} \xrightarrow{m\to o} 0$$

This is an asymptotically scale invariant one parametric subset. From this Thirring curve (Fig. 1) there extends a two dimensional surface S with soft broken isoscalar axial current (but hard breaking of $j^{a}_{\mu 5}$):

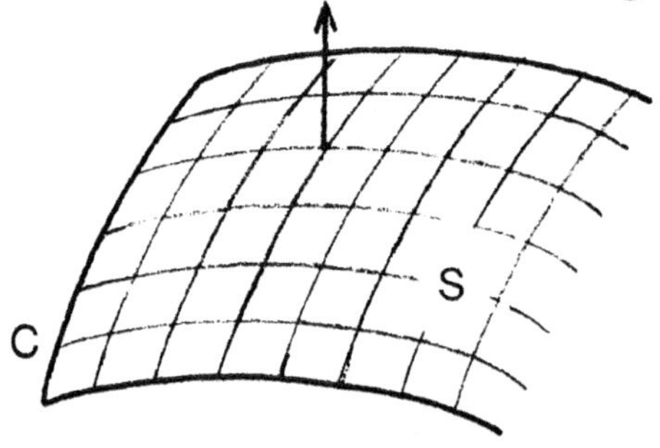

Fig. 1

C. curve on which both currents are asymptotically (m = o)conserved.
S: surface on which only $j_{\mu 5}$ is asymptotically (m=o) conserved.

Going into the direction of the arrow leads to a hard breaking of all axial symmetries.

Unfortunatly there does not seem to be a renormalization scheme in which the Thirring curve will simply read as: $g_v=o=g_s$, g arbitrary; and S as: $g_s=o, g, g_v$ arbitrary. Perhaps the dimensional renormalization could achieve this "classical" parametrization directly[15].

This last example also suggests a very interesting general type of problem of renormalization theory: Develop a P.T. around a nontrivial scale invariant field theory!

The conventional renormalized perturbation theory may be viewed as the special case of perturbation around free massless fields. Strictly renormalizable couplings lead to theories which are either "short distance" or "long distance free" i.e.

$$\dim \mathcal{L}_{Pert} = D \quad \begin{cases} \longrightarrow \text{s.d. - free} \\ \longrightarrow \text{l.d. - free} \\ \dashrightarrow \text{remains scale invariant (changed dimensions)} \end{cases} \qquad (33)$$

The third case namely the persistence of scale invariance with changed dimension is an exceptional case and it only seems to happen in the Thirring[17] model and the closely related Baxter model[18]. The derivation of the infrared respectively ultraviolet "asymptotic freedom" (33) for strictly renormalizable Lagrangians in 4-dimensional P.R.T. (i.e. for perturbation on the free field fixed point) should be considered as one of the major achievements of P.R.T. in recent years[2]. Asymptotic freedom for short distances severely limits the application of P.R.T. for long distances and vice versa.

For superrenormalizable perturbations we encounter infrared-divergences (uncurable in P.R.T.) which force us to "resumm" the generated mass. This is simply done by doing a new perturbation theory on a massive free Lagrangian.

For perturbations on non-canonical scale invariance field theory strictly renormalizable coupling (dim $\mathcal{L}_{\text{Pert}}$ = D) are the exception and the infrared difficulties with superrenormalizable interactions appear to be insurmountable. An explicit "resummation" (which probably is tantamount to an explicite solution) even for the simplest case of a mass perturbation on the Thirring model does not seem to be feasable with present field theoretical techniques. However in this particular example there may be very special circumstances (infinitly many conservation laws) which facilitate the construction of an explicit solution. We will come back to this in the next section.

II. Limitation of Renormalized Perturbation Theory

Some limitations of P.R.T were already discussed in the last section. In the following we look at another type of limitation.

For small value of the coupling constant in a conventional Lagrangian field theory we expect all bound states to disappear and the remaining particles to be in one-to one correspondence with the fields. We say that perturbation theory is consistent with this "naive picture" if the perturbation unitarity equations (i.e. the perturbative version of the Glaser-Lehmann-Zimmermann equations[19]) can be established in every order. Gauge theories which introduce zero mass terms are the most prominent exceptions (besides theories with unstable particles which become stable in only the limit of zero coupling). For abelian gauge theories (as QED) our picture is that charged particles can not be idealized by a discrete eigenstate of the mass operator[20]. The LSZ limits in perturbation theory do not exist (outside of perturbation theory they vanish), but asymptotic observables corresponding to cross sections with finite resolution do exist and can be evaluated in renormalized perturbation theory. In the nonabelian case even a small coupling is expected to lead to a more violent change: the true particle spectrum is presumably non-perturbative around $g = 0$; the appearant formal particles suggested by the structure of the Lagrangian may be "confined"[21]. In both cases Green's functions (properly normalized) exist in perturbation theory, but can not be used to check the unitarity GLZ relations.

Apart from these exceptions we have gotten accustomed to the "naive picture" and it came as a surprise to some of us that there are counter-examples provided by conventional theories.
Let us look at one.
Consider the A^4_2 theory.
The Wick-ordered (renormalized) Lagrangian:

$$\mathcal{L} = \frac{1}{2} : \partial_\mu A \, \partial^\mu A : - \frac{m^2}{2} : A^2 : - \frac{g}{4} : A^4 : \quad (34)$$

yields via the Gell-Mann Low formula a finite perturbation theory. Hence we would expect an elementary particle for small g and for $\frac{g}{m^2} > \gamma_c$ (from analogy to statistical mechanics) a first order phase transition. This picture has been indeed confirmed outside of perturbation theory by methods of constructive field theory[22].

It is formally easy to design a new parturbation theory in the phase transition regime. In this version we start with a mass term of the opposite sign (Higgs potential) and look for minima of the classical effective potential[23])

$$U(A) = -\frac{\mu^2}{2}A^2 + \frac{\lambda}{4}A^4 \tag{35}$$

With

$$A_{min} = \pm\frac{\mu}{\sqrt{\lambda}} \tag{36}$$

and the shifted field: $A' = A - \frac{\mu}{\sqrt{\lambda}}$ we obtain:

$$\begin{aligned}\mathcal{L} &= \frac{1}{2}\partial_\mu A\,\partial^\mu A + \frac{\mu^2+a}{2}A^2 - \frac{\lambda}{4}A^4 \\ &= \frac{1}{2}\partial_\mu A'\,\partial^\mu A' - \frac{2\mu^2-a}{2}A'^2 - \mu\sqrt{\lambda}A'^3 - \frac{\lambda}{4}A'^4 \\ &\quad + \delta A' + c\end{aligned} \tag{37}$$

We now Wick-order our interaction terms with respect to the mass $m = 2\mu^2$; this leads to a change of the mass-counter-term a. The linear term is choosen in such a way that the vacuum-expectation value of A' vanishes and without loss of much generality we may choose $a = 0$ i.e.

$$\mathcal{L}_{int} = -\mu\lambda :A'^3: - \frac{\lambda}{4} :A'^4: + \delta A' \tag{38}$$

$$\text{with } \delta = \sqrt{\lambda}^3\,\delta_1 + \text{higher terms.}$$

It is known that the corresponding classical Lagrangian (i.e. no counterterms)

$$\mathcal{L}_{c\ell} = \frac{1}{2}\partial_\mu A\,\partial^\mu A - U(A) \tag{39}$$

has in addition to the "vacuum" solutions(36) other stationary classical solutions[23]) ("kinks")

$$A_{c\ell} = \pm\frac{\mu}{\sqrt{\lambda}}\tanh\frac{\mu(x-a)}{\sqrt{2}} \tag{40}$$

What do these solutions correspond to in the quantum theory? Goldstone and Jackiw[23]) and other authors[24]) have given very convincing arguments that these solutions are the quasi-classical manifestations of heavy particle in the weak coupling approximation $\lambda \to 0$. In other words in addition to the "mesons" which we expect on the basis of our "naive" picture there are stable "baryons": The state $|p_1 \ldots p_n\ k \ldots k_m >^{in}$ is according to Goldstone and Jackiw the most general state of the broken A^4-theory for small λ (apart from possible bound states of baryons and mesons). So in this situation the perturbative (small λ) unitarity check of the naive picture would be missleading. Baryons of the above type unfortunatly do not notify their presence via a manifest disease of naive renormalized perturbation theory, they do not couple (as pairs) in the perturbative intermediate state structure.

The situation may, however, not be quite as bad as it appears from this consideration. The new particles which indicate their appearance as "kink" solutions of the nonlinear classical equations have the general property that they lead to superselection sectors[25]. In the A_2^4 case we have in addition to the vacuum sector just one additional sector. This suggests that a possible local "baryon" field which links the two sectors should be like a Majorana field (whose even functions have the same quantum numbers as the vacuum). In the quasiclassical weak coupling limit this is hard to see. An independant argument comes from the rigorous study of $D = 2$ critical behaviour at the end of the phase transition region for $\mu\lambda^{-\frac{1}{2}} \to o$. It has been recently established [26] that the scale invariant limit of the $D = 2$ Lenz-Ising model can be described by a relativistic field theory in terms of a two-component Majorana field. The situation is illustrated in the following diagram:

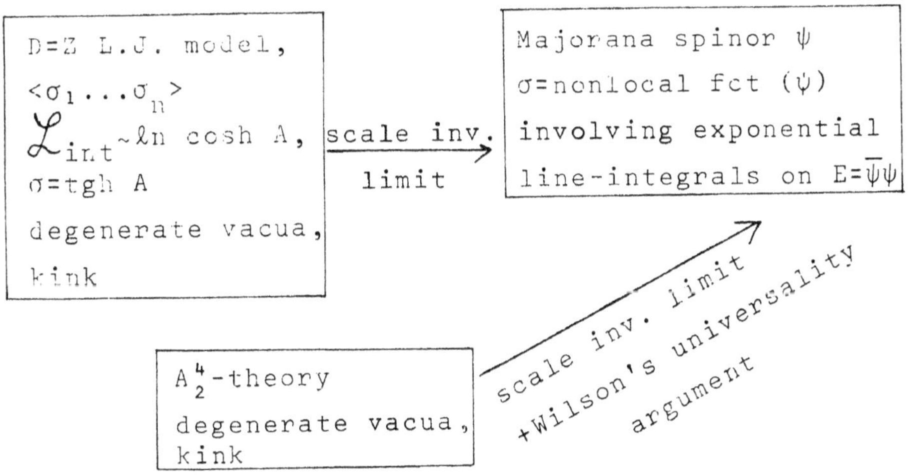

It is very suggestive to interprete the Majorana-field as the relict of the kink baryon if one approaches the critical point from the first order phase transition region. The A_2^4-theory in the first order phase transition region has presumably an equivalent local Lagrangian description which contains an explicit Majorana-field: $\mathcal{L}(\psi, A)$. If such a (presently very speculative) picture could be established in the future, the appearant difficulties of nonlinear quasiclassical states with the naive perturbative unitarity picture could be resolved. Let us consider another model in which such a picture has been established recently[26].

Consider the D = 2 Sine-Gordon equation:

$$\mathcal{L}_{c\ell} = \frac{1}{2} \partial_\mu \phi \partial^\mu \phi - \frac{m^4}{\lambda} [\cos \frac{\lambda}{m} \phi - 1] \tag{41}$$

In that case the kink solution leads to an additive charge superselection rule (soliton situation). A convincing but nevertheless somewhat formal argument has been given to the extent that the soliton operators correspond to the Dirac-field of the massive Thirring-model[26]:

$$\mathcal{L}(\phi) \longleftrightarrow \mathcal{L}(\psi) = i \bar{\psi} \gamma^\mu \partial_\mu \psi + m \bar{\psi} \psi - \frac{g}{2} j_\mu j^\mu \tag{42}$$

The equivalence proofs rely essentially on certain short-distance identities in the Thirring-model[27], the same identities which bring the Thirring energy momentum tensor of the massless Thirring-model into the Sugawara form

$$\Theta_{\mu\nu}^{Th} = \pi (: j_{\mu 5} j_{\nu 5} : - \frac{1}{2} g_{\mu\nu} : j_{\kappa 5} j^{\kappa 5} :) \tag{43}$$

From the derivation of the equivalence it is therefore clear that there is no relation to be expected between the classical versions of the two models. In the Thirring version in which the sector generating operators (solitons) have been incorporated into the Lagrangian the naive perturbation unitarity picture is in agreement with known reality. Note that the perturbative renormalization aspects of the massive Thirring-model are also much simpler[28] since we are talking about a polynomial strictly renormalizable interaction instead of a nonpolynomial interactionwhose power series expansion gives superrenormalizable terms.

There is another extremly interesting aspect regarding renormalized perturbation theory: the classical Sine-Gordon equation is an integrable dynamical system and leads to infinitely many conservation laws (Noether currents involving nonpolynomial functions of the field). Perturbative investigations of Flume[29] indicate that in the Sine-Gordon picture these conservation laws, although fulfilled on the tree-level, develop anomalies on the one loop level. These currents, converted to the Thirring language, will be ordinary polynomial currents. In showing their conservation, one would have to use algebraic short distance singularities of a similar type as the ones used for the equivalence[27]. A recent perturbation check on the level of trees of the Thirring S-matrix established the absence of pair creation[30]. The status of pair creation on the loop level is presently under intensive investigation. If certain polynomial currents (of arbitrarily high degree) in the Thirring-model turn out to be free of quantum anomalies one would have the strange example (not possible in D = 4) of an interacting local field theory (p-dependent phase shift) without pair creation. In this situation, the old Dirac picture based on negative energy states instead of antiparticles, would not lead to chaotic instabilities in the presence of the Thirring interaction. Indeed the old Thirring-Glaser formulation of the massless model was based on the negative energy treatment in which the ψ annihilates the pseudo-vacuum $|\Omega>$:

$$\psi \mid \Omega > = 0 \qquad (44)$$

For the massive version, Berezin and Sushko[31] have investigated the relativistic quantum theory (δ-function interaction) based on (44). The fact that a "diseased" theory leads to a reasonable relativistic unitary S-matrix fulfilling a restricted version of crossing and which possesses an interesting spectrum of bound states is already quite remarkable[32].

Even if the massive Thirring-model leads to pair creation via quantum loops, I have the feeling that the S-matrix of the model based on the pseudo-vacuum (44) may have some useful relation to the massive Thirring-model. There are several reasons to expect this model to be the first explicitly soluble quantum field theory with an LSZ interpretation and a nontrivial p-dependent S-matrix. So at least for this special case there is some hope to understand a superrenormalizable perturbation of a non-canonical Lagrangian theory:

$$\mathcal{L} = \mathcal{L}_{Th} + m\, N\, [\bar{\psi}\, \psi] \qquad (45)$$

which in a naive perturbation (in m) approach creates incurable infrared-singularities. This situation in which an interaction becomes very strong in the infrared region is interesting from the point of view of recent "confinement" discussions[21].

III. Global Operator Expansions and P.R.T.

Let A and B be two local fields from the denumerable set $C_n(x)$ of such local composite fields which can not be written as total derivations of others. In the case of free fields the set of local fields consist of the basic field and all local Wick-polynomials including derivative inside but excluding a derivation in front.

A global operator expansion on the vacuum is a formula of the type

$$A(x)\ B(y)|o\rangle = \sum_n \int K^{A,B}_{c_n}(x,y,z)\ C_n(Z)d^4z\ |o\rangle \qquad (46)$$

In contradistinction to a short distance[33] (Wilson) expansion, which in general (i.e. apart from matrix elements between finite energy momentum states), converges only asymptotically for short (say space-like) distances, a global expansion is supposed to converge for all points (this is of course meant in the sense of distribution theory). It is in a certain sense the "correctly summed up" short distance expansion. The latter would follow by expanding the kernel $\tilde{K}_n(x,y,p)$ in a Taylor series around $p = o$.

Global operator expansion would convert the algebra of composite fields of a Wightman theory into an algebra closed under multiplication. On the vacuum, multilocal states could be always resolved in terms of integrals over local states. Clearly such a property is closely linked to the property of completness of $C_n(x)\ |\ o\rangle$:

$$\mathcal{H} = \overline{\left\{ \int d^D x C_n(x)\ |\ o\rangle \right\}} \qquad (47)$$

In fact if one wants to demonstrate this completness for states in a given field theoretical model, one most probably would take the road via establishing (46). The validity of global expansions in the explained sense has first been seen by studying conformal invariant models[34,35,36]. In that case one would expect that one can restrict the C_n's to conformal tensors of the first type, which are those which cannot be written as a total derivative

$$C_n = \partial\ C_m \qquad (48)$$

(These are also the only ones which have a simple infinitesimal conformal transformation property.) In $D = 4$ massless free field theory the canonical dimension and the Lorentz-indices uniquely characterize such C_n and, barring accidental degeneracies of the dimensional spectrum, one would expect this state of affairs in the general case. The two point functions of the C_n fulfill in this case orthogonality relations:

$$< C_n(x)\ C_m(y) > = \delta_{n,m}\ F_n(x-y) \qquad (49)$$

Therefore by multiplying relation (46) from the left with C_n and taking the vacuum expectation value, one can project out the n th term:

$$< C_n(z)\ A(x)\ B(y)> = \int K^{AB}_{C_n}(x,y,z')\ F_n(z'-z)\ dz' \qquad (50)$$

Hence in the conformal invariant case the vacuum expansion is term-wise conformal invariant and the general form of K^{AB}_{C} which is closely related to three-point functions ("shadow" 3-point function) can be written down. For example if A, B and $C_n = C$ are scalar fields we obtain for the scalar 3-point function:

$$K^{AB}_{C}(x_1,x_2,x_3) = N_{ABC} \prod_{i<j} (-x^2_{ij})^{-[\lambda_{ij},\ \xi_{ij}]} \qquad (51)$$

$$\xi_{23} = o,\ \xi_{12} = \lambda_{12},\ \xi_{13} = \xi_a - \lambda_{12},$$

Notation:

$$(-x^2)^{-[\lambda,\xi]} = (-x^2_+)^{-\lambda+\xi}(-x^2_-)^{-\xi},$$

$$x^2_\pm = (x_o \pm i\varepsilon)^2 - x^2 \qquad (52)$$

$\lambda_{12} = \frac{1}{2}(d_a + d_b - d_c)$, cyclic; d_c is replaced by the "shadow dimension"): $d^*_c = 4 - d_c$.

It is very tempting to suggest that the validity of such operator expansions does not hinge on conformal invariance but also holds in Lagrangian P.R.T.

A prerequisite for such a property, namely the validity of (46) for massive free fields and their composites can be checked by realizing that the method of reference 36 goes through with slightly changed (massive) kernels K. The algebraic structure (i.e. the selction rules) is the same as in the zero mass case. This ceases to be so in P.R.T. There, depending on the perturbation order, one obtains i.e. many scalar contributions in the decomposition of the basic scalar field (say in A -coupling):

$$A(x)\,A(y)\,|\,o> = \int K_{A^2}(x,y,z)\;N[A^2](z)\,|\,o> \qquad (53)$$

$$+ \int K_{A^4}(x,y,z)\;N\,[A^4](z)\,|\,o> + \ldots.$$

$N[\mathcal{O}]$ = normal products representing the composite $\mathcal{O}$
However in each finite perturbative order, there will be only a finite number of contributions and the multiplication of wich these composites from the left will lead to a system of integral equations relating the 3-point and two-point functions

$$< N[A^2](z)\;A(x)A(y) > ,\; < N[A^4](z)A(x)A(y) > ,\ldots(54)$$

$$<N[A^2](x)N[A^2](y)> ,\; <N[A^2](x)N[A^4](y) ,$$

$$N[A^4](x)N[A^4](y) ,\ldots.$$

to the kernels K_{A^2}, K_{A^4},...
Since the three- and two-point functions involving arbitrary composite fields can be computed in P.R.T., one would expect the "shadow" functions K to be determinable. Such a program has not yet been carried out.
There are two reasons why we think that a solution of this open problem in P.R.T. may be important.
The first has to do with the structure of Lagrangian field theories. Given a set of time-ordered Green-functions, the only known relation to a positive definit Hilbert-space structure (i.e. Wightman's positivity[37]) goes via the LSZ asymptoties (GLZ-unitarity equations[19]). It would be desirable to dissociate the issue of stable particles from the positive metric structure of the state-space. The global expansion achieves this by resolving any multilocal state to a state generated by C_n (47). Instead of studying positivity for Wightman functions with arbitrary many arguments one just has to investigate the positivity of the infinite matrix of two point functions:

$$\{ < C_n(x)\;C_m(y) > \} \qquad (55)$$

The special interest in conformal invariant theories[35] stems from the fact that this matrix simplifies to a diagonal matrix because of the afore-mentioned selection rules (49).

The second reason has to do with the current interest in the asymptotic structure of correlation functions. The approach based on Callan-Symanzik equations and Wilson short distance expansion is limited to certain highly inclusive processes (deep inelastic e-p scattering). The Regge-limit or the rather interesting on mass shell wide angle limit which may be relevant to the discussion of confinement[37] is outside the handling power of the present formalism. Here we expect global expansions together with differential equations and perhaps some reasonable assumptions to be useful. It has already been stressed by Poliakov[38] and Mack[39] that a crossing symmetric version of "conformal partial wave" expansion of the 4-point functions leads to a situation which resembles the structure of dual models. The main difference to dual models is that instead of an exchange of particles on Regge-trajectories we obtain an exchange of composite fields on (s,d)-trajectories where s is the spin and d the (anomaleous) dimension. In contrast to the dual model situation where byond Born approximation the duality includes cuts, the crossing symmetry on the level of intermediate composite fields is an exact property. In order to have more than just a formal analogy one needs a relation between composite fields and composite particles.

Acknowledgement:

I profited from discussion with J. Lowenstein (1st section). I would like to point out that G. Mack arrived recently at similar conclusions concerning the viability of global Minkowski expansions outside of conformal invariance.

References and Footnotes

1) See the lecture notes and references of the other lectures of this summer-school in particular those of J.H. Lowenstein and E. Speer.
2) See for example the Proceedings of 17th Conference in High Energy Physics, London 1974.
3) The derivation of the Gell-Mann Low formula on a formal level can be found in most textbooks on quantum field theory i.e. the formal aspects have been reviewed by S. Gasiorowicz, "Elementary particle Theory". John Wiley and Sons, Inc. New York 1966.
4) Functional techniques go back to Feynman's work and have been extensively used for example in papers of Fadejev and collaborators. See also J. Zinn Justin in Lecture Notes in Physics, page 2, Vol. 37, 1975 Springer Verlag.
5) This statement is the prototype of equivalence theorems connecting old fashioned R.P.T. with the modern method of Bogoliubov-Parasiuk-Hepp-Zimmermann and derivates thereof[1]. It has been called "Folk Theorem" in Wightman's orientation lecture. The proof of such theorem can probably be found in some forthcoming University of Maryland lecture notes of P.K. Mitter.
6) P.K. Mitter, see forthcoming University of Paris preprints.
7) M. Gomes and B. Schroer, Phys. Rev. D10, (1974) and references quoted therein. The form of the differential equations in the Kadanoff parametrization is closely linked to the differential equations studied in G. t'Hooft, Nucl. Phys. B61, 455 (1973) and S. Weinberg, Phys. Rev. D8, 3497 (1973).
8) The connection of this form of the differential equations with the Kadanoff scaling laws and the "scaling field" concept of Riedel and Wegner has been discussed for example in B. Schroer, Revista Brasileira Fisica, Vol. 4, 323 (1974).
9) J. Lowenstein and W. Zimmermann, Nucl. Phys. B86, 77 (1975).
10) Our σ and s corresponds to the square of the softness parameters in 8).
11) J. Lowenstein, NYU preprint TR 3/75. See also the last section of the lecture notes of this summer-school by the same author.
12) W. Ernst and I. Schmitt, University of Bielefeld preprints 75/01/03/07.

13) Formula (31) replaces the incorrect expression for $\overline{m}(\kappa)$ in reference 7). I am indepted to H. Wagner for pointing this out.
14) W. Caswell and F. Wilszek, Phys. Lett. B49 (1974) 291.
15) I owe this remark to R. Stora. For the most recent state of art concerning dimensional renormalization look at the lecture notes of this school by E. Speer and P. Breitenlohner.
16) P.K. Mitter and P.H. Weisz, Phys. Rev. D8 (1973) 4410.
17) B. Klaiber, 1967 Boulder Lectures in Physics Vol. 10 A: Quantum Theory and Statistical Theory, edited by W.E. Brittin et al. (Gordon and Breach, New York 1968) p. 141.
See also earlier references quoted therein.
18) R.J. Baxter, Phys. Rev. Lett. 26, 832 (1971).
The relation of the scale invariant limit of the Baxter model to the Thirring model has been worked out in:
B. Schroer, FU preprint HEP 6/75.
See also:
A. Luther and I. Peschel, Harvard University preprint 1975.
19) V. Glaser, H. Lehmann and W. Zimmermann, N.C. 6, 1122 (1957).
20) This infra-particle phenomenon can be rigorously studied in two dimensional field theories.
21) S. Weinberg, Phys. Rev. Lett. 31 (1973) 494.
22) J. Glimm, A. Jaffe and T. Spencer: "Phase Transitions for ϕ^4_2 Quantum Fields", Harvard University preprint 1975.
23) J. Goldstone and R. Jackiw, Phys. Rev. D11, 1486 (1973).
24) R. Dashen, B. Hasslacher and A. Neveu, Phys. Rev. D10, 4114 (1974).
25) The superselection rules are the quantum version of the classical "kink stability". See the review by R. Rajaraman, Institute for Advanced Study preprint 1975.
26) B. Berg and B. Schroer, FU-Berlin preprint HEP 75/8
See also:
B. Schroer, DESY 75/22.
27) S. Coleman, Phys. Rev. D11, 2088 (1975).
28) This shows up very clearly in the equivalence proof given by the author FU-Berlin preprint HEP 75/5. This treatment as however the same infrared desease as ref. 27) and will be superseeded by a treatment working directly with the massive Thirring model.

29) See for example:
M. Gomes and J.H. Lowenstein, Nucl. Phys. B45, 252 (1972).
30) R. Flume, DESY preprint. This paper relies on the ordinary perturbation theory for the Sine-Gordon equation, a perturbation theory which is quite different from that of the Thirring model (where quantum solitions are present). It is not clear whether the Sine-Gordon perturbation theory reproduces known facts as the softness of the trace of the energy momentum tensor (i.e. β = o in the Callan-Symanzik equations).
31) B. Berg, M. Karowski and H.-J. Thun, FU-Berlin preprint HEP 75/14.
32) F.A. Berezin and V.N. Sushko, JETP 21, 865 (1965). The bound states and the crossing structure of the N-body S-matrix will be discussed in a forthcoming publication with T. Truong.
33) K.G. Wilson, Cornell Report (1964) unpublished. W. Zimmermann, Brandeis lectures (1970), Ann. Phys. 77, (1973) 536.
34) S. Ferrara, R. Gatto and A. Grillo, N.C. Lett 2, 1369 (1971). Here a formally summed up expression was written down for K.
35) The euclidean aspect of conformal partial wave expansion of the 4-point function were studied by G. Mack and M. Lüscher.For recent developments and references see:
V.K. Dobrev, V.B. Petkova, S.G. Petrova and I.T. Todorov, Institute for Advanced Study preprint, July 1975.
36) The first systematic investigation on the basis of global Minkowski conformal invariance was carried out in:
B. Schroer, J.A. Swieca and A.H. Völkel, Phys. Rev. D11, 1509 (1973).
37) A.S. Wightman, Phys. Rev. 101, 860 (1956).
38) See J.M. Cornwall and G. Tiktopoulos, UCA/75/TEP/1 where an interesting conjecture concerning the on-shell-infrared structure was made.
39) A.M. Polyakov, JETP 39, 10 (1974).
40) G. Mack in "Proceedings of the 11e Conference Internationale sur les Particules Elementaires", Aix-en-Provence 1973. Sup.aux Journal de Physique, Vol. 34, CI-99 (1973).

QUANTUM SINE-GORDON EQUATION AND QUANTUM SOLITONS IN TWO SPACE-TIME DIMENSIONS

Jürg Fröhlich

Department of Mathematics, Princeton University, Princeton, New Jersey 08540, U.S.A.

CONTENTS:

I. INTRODUCTION: THE PROGRAM OF THESE LECTURES

First a warning: The main emphasis of the "International School of Mathematical Physics, 1975" at Erice was clearly on renormalized perturbation theory in the framework of renormalizable quantum field theories which, one hopes, may be relevant for the physics (not only the mathematics) of quantized fields and elementary particles. Participants of this school wanted to learn something non-trivial about: ultraviolet renormalization, infrared power counting, the strong adiabatic limit (i.e. the construction of the scattering matrix), composite fields, Yang-Mills theories, etc. In my lectures, however, I study mainly the construction and the detailed properties of a new model of a self-interacting, relativistic scalar Bose field in two space-time dimensions: the quantum sine-Gordon equation (≡ s-G theory). It is formally defined by the following field equation:

$$(\Box + m_o^2)\ \phi\ (\vec{x},t) = -\ \varepsilon\lambda\ :\sin\ (\varepsilon\phi\ (\vec{x},t) + \theta):_1, \qquad \text{(I.1)}$$

where m_o is the bare mass, λ an arbitrary real coupling constant,

G. Velo and A. S. Wightman (eds.), Renormalization Theory, 371–414.

ε a real parameter with $\varepsilon^2 \le 4\pi$, in general, and θ is an angle in $[0,2\pi)$. The colons $: - :_1$ denote Wick ordering with respect to bare mass 1.

The s-G theory is almost trivial from the point of view of renormalization theory: If $\varepsilon^2 < 4\pi$ Wick ordering removes all ultraviolet divergencies. For $4\pi < \varepsilon^2 < 8\pi$ this theory does have ultraviolet divergencies which are so serious that I do not understand them well enough to say something non-trivial about them.

My warning is thus, not to expect too much enlightment from my contribution to this school concerning its main themes.

Nevertheless I hope my lectures are a rather useful illustration of the success and the limitations of renormalized perturbation theory:

For $\varepsilon^2 < 4\pi$, $m_o > 0$ and $|\lambda|$ small the s-G theory can be constructed by means of a convergent perturbation expansion in λ. This theory is the first and in some sense presumably the only example of a relativistic field theory with a convergent perturbation series.

Moreover perturbation theory in λ is asymptotic to the scattering amplitudes of the s-G theory, and scattering is non-trivial, provided $\varepsilon^2 < 4\pi$ is small enough.

For $4\pi < \varepsilon^2 < 8\pi$ the s-G theory has ultraviolet divergencies which can perhaps be renormalized, but generically not on the level of perturbation theory in λ. For $\varepsilon^2 = 4\pi$ the ultraviolet divergencies are super renormalizable, but there is no scattering. For $\varepsilon^2 \ge 8\pi$ the theory is meaningless.

For $\varepsilon^2 < 4\pi$ and $m_o = 0$ the s-G theory has infrared divergencies; perturbation theory breaks down. The nature of the (infrared) singularity at $\lambda = 0$ can be determined explicitly.

For $m_o = 0$, $\varepsilon^2 < 16/\pi$ the theory has super-selection (soliton -) sectors which cannot be approximately constructed by standard perturbation theory.

Other phenomena encountered in the analysis of the s-G theory may be vaguely related with the main topics of this school: Fields with anomalous dimensions, scaling equations, non-uniqueness of the vacuum ... .

Finally our results on the s-G theory yield a construction of massive QED in two dimensions which is the first model where the problems associated with indefinite metric can be understood rigorously.

All this sounds very interesting. Yet the s-G theory is so simple and easy to analyze compared to realistic field theories that if I asserted it tells us something relevant about the main topics of this school, e.g. renormalized perturbation theory for a renormalizable field theory in four space-time dimensions, I would find myself in the rôle of the philosopher in one of Kafka's stories who gets very excited when he discovers children playing with a top, watches it for a while and then tries to catch the top; for he believes that one should find the truth about this world in its simplest phenomena.

Keeping this warning always in mind I now want to summarize some of the field theoretic concepts and programs which I hope can be somewhat clarified by studying simple models such as the sine-Gordon theory, the Yukawa model in two dimensions (see Seiler's lectures) or the more difficult ϕ^4 model in three dimensions (see the contribution of Feldman).
Programs of the type summarized below, in particular, Programs 1-3 and 5 have been advocated for a number of years, notably by A. Wightman, J. Glimm and A. Jaffe; see [GJS1, Part I] .

Program 1:

Given a formal Lagrangian construct a relativistic quantum field theory satisfying the Wightman axioms the dynamics of which is determined by this formal Lagrangian in a sense to be made precise later.

Program 1 requires:

(1a) Ultraviolet renormalizations; investigation of infrared singularities.

(1b) Infinite volume limit, e.g. stability, (which is finiteness of the renormalized vacuum energy density), and convergence of space-time cutoff Wightman functions, as the cutoff is removed.

(1c) Verification of the axioms.

Program 2:

If the existence of a field theory can be established, investigate whether it is non-trivial, e.g. whether the scattering matrix S is different from $\mathbb{1}$.

Program 3:

Given the existence of a theory determine whether perturbation theory makes accurate predictions on the theory for weak coupling; e.g.

(3a) does the perturbation expansion of, say, the Euclidean Green's ($\equiv$ Schwinger) functions converge, or is it e.g. Borel summable, or is it asymptotic?

(3b) is perturbation theory asymptotic to the S-matrix? If yes, this yields in general an answer to Program 2.

Program 4:

Find rigorous connections between long range forces, non-uniqueness of the vacuum, confinement of quantum numbers, etc.

Program 5:

Construct field theories showing a phase transition, as some coupling constant is varied, having a broken symmetry, having a Goldstone boson.

Program 6:

Find field theories having non-trivial super-selection rules. Subprograms:

(6a) The quantum soliton.

(6b) Connections between Programs 5 and 6.

Program 7:

Construct gauge theories, such as massive QED in (two or) three space-time dimensions and formulate Osterwalder-Schrader (physical) positivity [OS1] for the Euclidean Green's functions in a (super -) renormalizable gauge with indefinite metric.

This list should be completed with

Program 8:

Suppose we succeed with Programs 1-7 in the framework of

simple models in two or three space-time dimensions. Can we extract some "general wisdom" about four dimensional theories?

In the following we want to describe what we may learn about Programs 1-4, 6 and 7 from an analysis of the sine-Gordon theory. Other (more sophisticated) illustrations of Programs 1 and 3 can be found in the contributions of Seiler and Feldman. For an illustration of Program 5 we refer to the deep results of Glimm, Jaffe and Spencer [GJS 2] who prove existence of a phase transition in ϕ_2^4. An analysis of the connection between Programs 5 and 6 and an illustration of Program 6 for various quantum field models can be found in [Fr 5]. Program 7 becomes hard in more than two dimensions, so that this and Program 8 are a task for the future.

II. GENERALITIES ABOUT THE EUCLIDEAN STRATEGY IN THE ANALYSIS OF PROGRAMS 1-3

For the construction of interacting relativistic quantum fields in two and three dimensions the Euclidean approach to quantum field theory has turned out to be particularly advantageous. It will be the basis of this as well as of the articles of Seiler and Feldman. This is our motivation to give (as a service to the reader) a very brief summary of the Euclidean formulation of quantum field theory and an outline of the strategy presently used for the construction of models of quantum fields. We limit our considerations to the case of scalar Bose fields. (In two dimensions this is no serious lack of generality, since in this case statistics is a matter of convention, [Ch, Col, Fr S]). However Erhard Seiler will describe a very successful approach (Euclidean Matthews-Salam formalism) to the Euclidean formulation of theories of interacting Fermi and Bose fields.

(See Wightman's article for an overall view of the Euclidean strategy).

II.1 Notations:

The number of (space-time dimensions is denoted by d. A point in $\mathbb{R}^d$ is denoted $x = (\vec{x},t)$. The Schwartz space over $\mathbb{R}^d$ is denoted by $\mathcal{S}_{\mathbb{C}} = \mathcal{S}(\mathbb{R}^d)$ and $\mathcal{S}$ is the real part of $\mathcal{S}_{\mathbb{C}}$; $\mathcal{S}_{\mathbb{C}}'$ ($\mathcal{S}'$) is the space of (real-valued) tempered distributions, and

$$\underline{\mathcal{S}} = \bigoplus_{m=0}^{\infty} \mathcal{S}_{\mathbb{C}}^{\otimes m} \quad (\text{with } \mathcal{S}_{\mathbb{C}}^{\otimes 0} = \mathbb{C})$$

is the Borchers tensor algebra over $\mathscr{S}_{\mathbb{C}}$, [J]. A scalar Bose field is denoted by ϕ (x).

II.2 What is a Relativistic Quantum Field Model?

A model for (a theory of) a relativistic quantum field ϕ (x) is defined as a state (positive semi-definite linear functional) on $\underline{\mathscr{S}}$:

$$\omega = \left\{W_n (x_1,\ldots,x_n)\right\}_{n=o}^{\infty},$$

where W_n is the n-point Wightman distribution (vacuum expectation value) and the W_n's are supposed to satisfy the Wightman axioms with the possible exception of the cluster property (uniqueness of the vacuum). In this case we call ω a Wightman state on $\underline{\mathscr{S}}$. It determines uniquely a sequence

$$\underline{S} = \left\{S(x_1,\ldots,x_n)\right\}_{n=o}^{\infty}$$

of Euclidean Green's (or Schwinger) functions (EGF's) which have a certain number of properties, notably:

Euclidean invariance, physical positivity, symmetry under permutation of arguments, etc., } (II.1)

see [OS1], and which are obtained from the Wightman distributions by analytic continuation in the time variables to the Euclidean region.

Osterwalder and Schrader [OS1] have proven that the following axioms on $\underline{S}$ guarantee that $S(x_1,\ldots,x_n)$ is the n-point EGF obtained from an n-point Wightman distribution, for all n:

(E0) A distribution property (which restricts the order and the size of the distribution $S(x_1,\ldots,x_n)$ independence of n).

(E1) Euclidean invariance.

(E2) Osterwalder-Schrader ($\equiv$ physical) positivity; (the Euclidean form of Wightman positivity, i.e. positivity of the metric in the physical Hilbert space).

(E3) Symmetry (under permutation of arguments).

The connection between Wightman's and the Euclidean formulation of quantum field theory can be summarized in:

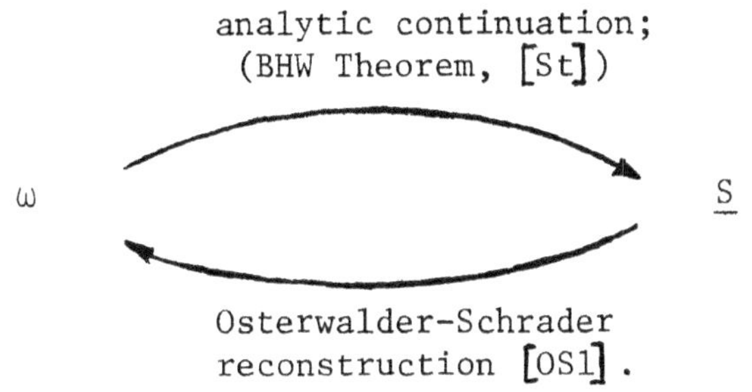

For more details see [OS1], also [G, J].

II.3 Euclidean Field Theory:

For the purposes of constructive quantum field theory the following additional structure discovered by Symanzik and Nelson has turned out to be particularly useful:

A theory (a Wightman state, the EGF's, resp.) is (are) called Nelson-Symanzik (NS) positive [N1,Sy] if and only if

[(a) the EGF's $S(x_1,\ldots,x_n)$ are locally integrable on $\mathbb{R}^{dn}$, for all n, and]

(b) $\underline{S}$ is a state on $\mathcal{S}$.

From (a), (b) and (II.1) it follows that the EGF's are the expectation values of products of Euclidean covariant, scalar random fields also denoted by $\phi(x)$. So far there is no satisfactory general condition guaranteeing that a Wightman state ω is NS positive, but there are many examples of NS positive theories: The free, scalar field [N2], all the models (in some sense even the Yukawa model discussed in Seiler's article) constructed so far. References [GJS1, S1, N1] are convincing witnesses for the success of NS positivity in constructive field theory.

Given [(a) and](b) one can find necessary and sufficient conditions for the EGF's to be the moments of a unique, Euclidean invariant probability measure $d\mu(\phi)$ on $\mathcal{S}'$, [Bo]. Since the measure $d\mu(\phi)$ is derived from a Wightman state, it has very special properties, notably a positivity property derived from Osterwalder-Schrader positivity. These properties are discussed in [Fr2]. We call such a measure a physical measure.

NS positivity and the existence of physical measures are the basic reasons for the existence of connections between quantum field theory and classical statistical mechanics, [GRS2], and provide a rigorous version of Feynman's history integrals.

II.4 The Simplest NS Positive Quantum Field Theory: The Free Scalar Field

The Wightman distributions and the EGF's of a free, scalar field $\phi(x)$ (with mass m_o) are determined completely by the one - and the two point function. (The one-point function can be made = 0 by shifting ϕ). NS positivity of the theory of a free, scalar field can therefore be read off the (Källen-Lehmann representation of its) two-point function [N2, S1]. This theory is thus characterized by a physical measure $d\mu_o\ (\phi)$ the moments of which are the EGF's. Heuristically $d\mu_o$ is the Gaussian measure defined by

$$d\mu_o\ (\phi) = N \prod_x \left\{ e^{-\frac{1}{2}\left[(\mathrm{grad}\ \phi)^2 + m_o^2\ \phi^2\right](x)}\ \mathcal{D}\phi\ (x) \right\}, \quad \text{(II.2)}$$

where N is an infinite normalization factor. The expression (II.2) is well known, but it is purely formal. Mathematically a Gaussian measure is determined completely by $S(x,y) \equiv \int d\mu_o\ (\phi) \times \phi\ (x)\ \phi\ (y)$.

The formal expression (II.2) tells us that $S(x,y)$ is the kernel of the operator $(-\Delta + m_o^2)^{-1}$, i.e. the Euclidean propagator of a free, scalar field. By Minlos' theorem [Mi] $d\mu_o\ (\phi)$ can be defined as the Gaussian measure on $\mathcal{S}'$ with mean 0 and covariance ($\equiv$ Euclidean propagator) $(-\Delta + m_o^2)^{-1}$. It is also determined by its characteristic functional $\int d\mu_o\ (\phi)\ e^{i\phi(f)} = e^{-\frac{1}{2}(f,(-\Delta + m_o^2)^{-1}f)}$, $f\varepsilon\mathcal{S}$.

Presently interacting fields are usually constructed by perturbing the free field measure $d\mu_o\ (\phi)$ in a way specified by the Euclidean Gell Mann - Low formula; (see Wightman's lecture). Often it is however more convenient, not to start from $d\mu_o\ (\phi)$ but from other closely related Gaussian measures: Let C be a positive quadratic form on $\mathcal{S}$. The Gaussian measure on $\mathcal{S}'$ with mean 0 and covariance C is denoted by $d\mu_C$, the expectation with respect to $d\mu_C$ by $\langle\cdot\rangle_C$. The characteristic functional of $d\mu_C$ is

$$\langle e^{i\phi(f)}\rangle_C = e^{-\frac{1}{2}(f,Cf)}, \quad f\varepsilon\mathcal{S}, \quad \text{(II.3)}$$

and Wick ordering with respect to covariance C is defined by

$$:e^{i\phi(f)}:_C = e^{i\phi(f)} e^{\frac{1}{2}(f,Cf)} \tag{II.4}$$

Equation (II.4) also determines Wick ordering of polynomials in ϕ.

We now summarize some of the most frequent choices for C

(i) $C = C_{m_o} \equiv (-\Delta + m_o^2)^{-1}$. In this case we speak of free (F) boundary conditions.

(ii) Let Γ be a collection of (d-1) dimensional (continuous, piece-wise smooth) surfaces in $\mathbb{R}^d$; define

$$C = C_\Gamma (\beta,t) = (-\Delta_{\Gamma,\beta} + m_o^2)^{-1} e^{-t(-\Delta_{\Gamma,\beta} + m_o^2)}, \tag{II.5}$$

where $\Delta_{\Gamma,\beta}$ is the d-dimensional Laplacian with some b.c. β on Γ which have the property that $C_\Gamma(\beta,t)$ is a continuous, positive quadratic form on $\mathscr{S}$, for all $t \geq 0$. (The factor $e^{-t(-\Delta_{\Gamma,\beta} + m_o^2)}$ provides an ultraviolet cutoff used in Feldman's lectures; in the following $t = 0$, and $C_\Gamma(\beta,0) \equiv C_\Gamma(\beta)$).

Example: Γ = a lattice of planes (lines) parallel to coordinate planes; β = 0-Dirichlet data (D), or Neumann b.c. (N), or periodic b.c. (P), or a combination of such b.c..

Covariances of the form (II.5), or convex combinations of them with D, N or P b.c. provide cutoffs which are useful for the construction of interacting fields and which will be removed in various steps when the renormalizations have been done.

References: [N2, S1, GRS2, GJS1, FeO1].

II.5 The Construction of Interacting Fields:

With a formal interaction Lagrangian $\mathcal{L}_I(\phi)$ and some b.c. (Γ,β) of the type introduced in II.4 we associate a Euclidean action

$$U_\Lambda (C_\Gamma(\beta)) = \int_\Lambda dx :\mathcal{L}_I(\phi):_{C_\Gamma(\beta)} (x), \tag{II.6}$$

where Λ is a space-time cutoff (e.g. a rectangle). A relativistic quantum field model with interaction $\mathcal{L}_I$ is presently generally constructed in three (or four) steps:

(1) For every bounded rectangle, construct a cutoff interacting measure (according to the Euclidean GM-L formula)

$$d\mu^{\Lambda}_{C_\Gamma(\beta)}(\phi) \equiv \left[Z^{\Lambda}_{C_\Gamma(\beta)}\right]^{-1} e^{-U_\Lambda(C_\Gamma(\beta))}\, d\mu_{C_\Gamma(\beta)}(\phi), \tag{II.7}$$

where $Z^{\Lambda}_{C_\Gamma(\beta)} = \langle e^{-U_\Lambda(C_\Gamma(\beta))}\rangle_{C_\Gamma(\beta)}$.

In more than two dimensions the Euclidean action $U_\Lambda(C_\Gamma(\beta))$ is, in all non-trivial cases, not a $d\mu_{C_\Gamma(\beta)}$- measurable function. This is another way of saying that Wick ordering does not remove all ultraviolet divergencies. Hence, in order to construct $d\mu^{\Lambda}_{C_\Gamma(\beta)}(\phi)$ one has to do the necessary ultraviolet renormalizations! [Gℓ1,GJ3,Fe01,MS].

(2) Show that the measures $\{d\mu^{\Lambda}_{C_\Gamma(\beta)}\}$ converge, as $C_\Gamma(\beta) \to C_{m_0}$, $\Lambda \nearrow \mathbb{R}^d$, e.g. in the sense that their characteristic functionals $\{\langle \exp i\phi(f)\rangle^{\Lambda}_{C_\Gamma(\beta)}\}$ converge on $\mathcal{S}$. Usually this involves a proof of bounds on the vacuum energy density ($\equiv$ "pressure" [GRS2])

$$p^{\Lambda}_{C_\Gamma(\beta)} \equiv \frac{1}{|\Lambda|} \log Z^{\Lambda}_{C_\Gamma(\beta)}, \tag{II.8}$$

that are uniform in Λ and (Γ,β).

(3) Show that the moments of the limiting measure $d\mu$ satisfy e.g. the Osterwalder-Schrader axioms (E0) - (E3), i.e. that they determine a Wightman state.

((4) Try to construct translation covariant super-selection sectors orthogonal to the vacuum sector reconstructed from $d\mu$).

Interpretation of this procedure: The connection of this three step procedure and of the Euclidean Gell Mann-Low formula with the Hamiltonian formalism and the formal field equation derived from the Lagrangian $\mathcal{L}_I$ is given by Nelson's Feynman-Kac formula [N1]; see also [SeS3]. Typically, in two dimensions, one is able to show that the quantum field obtained from $d\mu$ by Osterwalder-Schrader reconstruction satisfies the expected field equations. In general (e.g. for $d>2$) it is however more convenient and conceptually more important to establish the connection between the formal Lagrangian and the theory reconstructed from the limiting measure $d\mu$ by proving that standard perturbation theory (in $\mathcal{L}_I$) is e.g. asymptotic to the EGF's or the S-matrix obtained from $d\mu$. (See [Di2, EMS, FrS, EEF]).

III. SUMMARY OF RESULTS FOR THE SINE-GORDON THEORY

For the rest of my lectures the number of space-time dimensions is equal to 2.

The sine-Gordon theory is defined by the formal field equation

$$(\Box + m_o^2)\ \phi(\vec{x},t) = -\varepsilon\lambda\ :\sin(\varepsilon\ \phi(\vec{x},t) + \theta):_1 \qquad \text{(III.1)}$$

which is derived from the interaction Lagrangian

$$\begin{aligned}\mathcal{L}_I(\phi) &= -\lambda\ :\cos(\varepsilon\ \phi\ (\vec{x},0) + \theta):_1 \\ &= -\lambda\ m_o^{\alpha}\ :\cos(\varepsilon\ \phi\ (\vec{x},0) + \theta):_{m_o}\ ,\end{aligned} \qquad \text{(III.2)}$$

where λ is an arbitrary, real coupling constant,

$$\alpha \equiv \varepsilon^2/4\pi < 1,\ \theta \in [0,2\pi) \text{ and } :-:_m = :-:_{C_m}.$$

The Euclidean action we are going to analyze is

$$U_\Lambda(C_1) = -\lambda \int_\Lambda dx\ :\cos(\varepsilon\ \phi\ (x) + \theta):_1 \qquad \text{(III.3)}$$

We will describe the sine-Gordon theory in terms of the chiral fields

$$\chi_\vartheta(x) = :\cos(\varepsilon\ \phi\ (x) + \vartheta):_1,\ \ \vartheta \in [0,2\pi); \qquad \text{(III.4)}$$

the field ϕ is used merely for mathematical convenience at various places.

Theorem 1: (See Program 1)

(a) [Fr4] For all $m_o \geq 0$, $\varepsilon^2 < 4\pi$, $\lambda \in \mathbb{R}$, $\theta \in [0,2\pi)$

$$p(\lambda, \alpha, m_o) = \lim_{\Lambda \nearrow \mathbb{R}^2} \frac{1}{|\Lambda|} \log \langle e^{-U_\Lambda(C_1)} \rangle_{C_{m_o}} < \infty$$

exists and is positive. For $m_o = 0$

$$p(\lambda,\alpha) \equiv p(\lambda,\alpha,0) = F(\alpha)\,\lambda^{2/(2-\alpha)} \tag{III.5}$$

for some positive convex function $F(\alpha)$ with $F(0) = 0$. As a function of

$$\rho = \lambda \frac{\partial p(\lambda,\alpha)}{\partial \lambda}$$

$$p(\lambda(\rho), \alpha) = \rho\left(1 - \frac{\alpha}{2}\right) \tag{III.5^1}$$

For $\varepsilon^2 \geq 4\pi$, $p(\lambda,\alpha,m)$ is infinite, and the theory does not exist without ultraviolet renormalizations.

(b) [Fr3] For all $m_o \geq 0$, $\varepsilon^2 < 16/\pi$, $\lambda \in \mathbb{R}$, $\theta \in [0,2\pi)$, infinite volume EGF's for the fields χ_ϑ exist and satisfy the Osterwalder-Schrader axioms (E0) - (E3).*) (The vacuum in the physical Hilbert space need not be unique).

Remark: Equation (III.5) proves that for $\alpha < 1$ and $m_o = 0$ perturbation theory in λ at $\lambda = 0$ breaks down; the divergencies in the perturbation series are easily identified as infrared divergencies. It follows from equations (III.5) and (III.5^1) (see also [Co1,Fr3]) that the theory is meaningless for $\alpha > 2$. Taking equation (III.5) at face value for $1 < \alpha < 2$, the singularity of $p(\lambda,\alpha)$ at $\lambda = 0$ in this range of values of α is caused by ultraviolet divergencies. If we postulate that any acceptable renormalization of the s-G theory for $1 < \alpha < 2$ should preserve equations (III.5) and (III.5^1) we conclude that this theory has no renormalized perturbation series, except possibly for $\alpha = \alpha_n = 2 - 2/n$, $n = 1,2,3,\ldots$. We conjecture that the s-G theory may be renormalized only if $\alpha = \alpha_n$, (for some n).

*) This is based on $P(\phi)_2$- results of [GRS2]; (also [Fr1]).

Theorem 2: (See Program 2)

(a) [Fr2, Fr3] For all $m_0 > 0$, $\varepsilon^2 < 16/\pi$, $\lambda \neq 0$, $\theta \in [0,2\pi)$ the sine-Gordon theory constructed in Theorem 1, (b) is not in the Borchers class of the (generalized) free field.

(b) [FrS, EEF] For all $m_0 > 0$, $\varepsilon^2 < \varepsilon_c^2 \leq 4\pi$, $0 < |\lambda/m_0^2| < \lambda_1(\varepsilon)$, $\theta \in [0,2\pi)$, there is non-trivial scattering.

Theorem 3: (See Program 3)

(a) [FrS] For all $m_0 > 0$, $\varepsilon^2 < 4\pi$, $|\lambda/m_0^2| < \lambda_0(\varepsilon)$, the infinite volume EGF's have a convergent perturbation expansion in λ at $\lambda = 0$. The energy-momentum spectrum of the theory reconstructed from these EGF's has a positive mass gap.*)

(b) [FrS, EEF] Under the assumptions of Theorem 2, (b) there exist one particle states of discrete mass (hence a Haag-Ruelle scattering theory)*), and perturbation theory in λ at $\lambda = 0$ is asymptotic to the scattering amplitudes.

In order to describe our results concerning Programs 4 and 7 we must explain Coleman's isomorplism [Col] : Let $\psi(x)$ be a Dirac two spinor field in two dimensions. Let $\mathcal{L}_0(\psi)$ be the free Lagrangian with bare mass 0. Define

$$j^{\mu}(x) = :\bar{\psi}\,\gamma^{\mu}\,\psi:(x) \tag{III.6}$$

Consider now the following interaction Lagrangian

$$\left.\begin{aligned} &\mathcal{L}_I(\psi) = \mathcal{L}_1(g, e, \psi) + M:\bar{\psi}\psi:, \\ &\text{where } \mathcal{L}_1(g, e, \psi) = g/2 :j^{\mu} j_{\mu}: + e^2: j^{o}\, V_c * j^{o}:, \end{aligned}\right\} \tag{III.7}$$

and $V_c(x) = (\pi/2)|x|$ is the one dimensional Coulomb potential.

The formal Lagrangian $\mathcal{L}_0(\psi) + \mathcal{L}_I(\psi)$ with $\mathcal{L}_I(\psi)$ given by (III.7) defines the so called massive Thirring-Schwinger model (QED_2).

---

*) These results are based on powerful expansions of [GJS1, Sp2]

Theorem 4: [Col, LE, FrS]

For $g > o$ the massive Thirring-Schwinger model on the charge 0 sector and the sine-Gordon theory on the vacuum sector obtained in Theorems 1, (b) and 3, (a) are equivalent. The equivalence identifies

$$\left.\begin{array}{l} j^{\mu} \text{ with } -\frac{\varepsilon}{2\pi}\ \varepsilon^{\mu\nu}\partial_{\nu}\phi \\ \\ :\bar{\psi}\,\psi : \text{ with } : \cos\,(\,\varepsilon\,\phi + \theta)\ :_{1},\ \theta \in [0,\ 2\pi) \\ \\ :j^{o}\ V_{c} * j^{o}: \text{ with } :\phi^{2}: \text{ (Schwinger mechanism [Sw])} \end{array}\right\} \quad \text{(III.8)}$$

and sets $\varepsilon^2/4\pi = (1 + g/\pi)^{-1}$, $m_o^2 = e^2\ (1 + g/\pi)^{-1}$, $\lambda = M$.(III.9)

Notice that in (III.8) θ remains undetermined The value of θ, in the language of QED_2, is determined by the value of a constant, universal c - number electric field which is not fixed by the Lagrangian (III.7). See [Co 2] .

The charge operator is defined by

$$Q = \int d\vec{x}\ j^{o}\ (\vec{x},t) = -\frac{\varepsilon}{2\pi}\int d\vec{x}\ \partial_{\vec{x}}\ \phi(\vec{x},t) \qquad \text{(III.10)}$$

Theorem 5: (See Program 4)

Under the conditions of Theorem 2, (b) all physical states of the sine-Gordon theory $\equiv QED_2$ have charge 0; [LS, Fr S] . The scattering matrix $S = S\ (\theta)$ depends on θ in a non-trivial way; [Fr3, EEF] . For $\theta = 2\pi n$, $n \in \mathbb{Z}$, $S(\theta) = S(0)$: This can be interpreted, in the language of QED_2, as charge shielding; see [Co 2] .

This theorem asserts the remarkable fact that there are formal Lagrangians (e.g. (III.7)) which do not determine one unique, but rather (uncountably) many, physically different, relativistic quantum field theories which are labelled by a parameter (in our example : θ) not explicitly occurring in the Lagrangian that specifies properties of the physical vacuum; see also [Co 2] .

Theorem 6: [Fr 3, Fr 5] (See Program 6)

For $m_0 = 0$, $\varepsilon^2 < 16/\pi$, all real λ and $\theta \in [0, 2\pi)$ the sine-Gordon theory has infinitely many disjoint super-selection (soliton -) sectors labelled by the eigenvalues $m \in \mathbb{Z}$ of the charge operator Q. The sector associated with m = 0 is the vacuum sector obtained in Theorem 1, (b).

Remarks:

This theorem tells us that if the electric charge e is 0, i.e. if there is no Coulomb force, then there exist super-selection sectors of non-vanishing Q-charge, i.e. the confinement of the Q-charge asserted in Theorem 5, for $e \neq 0$, disappears when $e = 0$.

Heuristically the super-selection sector with Q-charge m is obtained by applying the intertwining operator $\exp i\pi(g)$ to the vacuum (charge 0) sector. Here π is the momentum operator canonically conjugate to ϕ and g is a smooth function with

$$\lim_{\vec{x} \to -\infty} g(\vec{x}) = 0, \lim_{\vec{x} \to +\infty} g(\vec{x}) = \frac{2m\pi}{\varepsilon} \tag{III.11}$$

See [Fr3, Fr5] for a rigorous construction of the charged sectors. We conjecture that the sectors with charge ± 1 contain one particle states (the charged particles of the massive Thirring model, [Col]).

Massive QED_2: (See Program 7)

Let A_μ be a neutral vector field with bare mass $m > 0$. Massive QED_2 is defined by the following formal Lagrangian

$$\mathcal{L}(\psi, A_\mu) = \mathcal{L}_o(\psi) + \mathcal{L}_o(A_\mu)$$
$$- \left[g/2 : j^\mu j_\mu : + e\, j^\mu A_\mu + M{:}\bar{\psi}\psi{:} \right] \tag{III.12}$$

On the charge 0 sector and for $g > 0$ the equivalence ($j^\mu \to -\varepsilon/2\pi\ \varepsilon^{\mu\nu}\partial_\nu \phi, \ldots$) asserted in Theorem 4 can be extended to this theory, so that we may rewrite it as a theory of a neutral, scalar field ϕ and the vector field A_μ, (as long as we only consider the currents j^μ and $:\bar{\psi}\psi:$, but not the fields ψ, $\bar{\psi}$).

We are then able to prove

Theorem 7:

In the Stückelberg gauge and for $g > \pi\cdot(\pi^2/4 - 1)$, (i.e. $\varepsilon^2 < 16/\pi$), the infinite volume EGF's of the fields j^μ, $:\bar{\psi}\psi:$ and A_μ exist and satisfy axioms (E0), (E1) and (E3). Osterwalder-Schrader positivity (E2) is restored if the Euclidean field A^μ is smeared out only with transverse test functions $(f_0, f_1) = (\partial_{\vec{x}} g, i\partial_t g)$.

The theory has infinitely many charged super-selection sectors labelled by the integer eigenvalues of the charge operator $Q = \int d\vec{x}\, j^0(\vec{x},t)$. These sectors are constructed as in the Remark after Theorem 6; see also [Fr3, Fr5].

Presumably this theorem presents the first rigorous solution of problems associated with indefinite metric, (even though the model considered here is admittedly somewhat too simple).

IV. AN OUTLINE OF THE PROOF OF THEOREM 1

References for Section IV: [Fr3, Fr4].

IV.1 Step (1), II.5, of the proof:

(a) Preliminaries:

We consider the Euclidean action $U_\Lambda(C_1)$ of the s-G theory defined in (III.3), where Λ is some bounded rectangle. In step (1), II.5, we must show that $\exp -\mathcal{U}_\Lambda(C_1)$ is a measurable, $d\mu_{C_\Gamma(\beta)}$-integrable function on $\mathscr{S}'$, and we shall need to consider two cases: $C_\Gamma(\beta) = C_{m_0}$ or $= C_{\partial\Lambda}(D)$. It turns out that in Step (1) it suffices to consider the case where $C_\Gamma(\beta) = C_{m_0}$.*) For $\varepsilon^2 < 4\pi$ measurability of $\exp -\mathcal{U}_\Lambda(C_1)$ is shown by standard arguments. We assert that, for $\alpha \equiv \varepsilon^2/4\pi < 1$, $U_\Lambda(C_1)$ is in $L^2(\mathscr{S}', d\mu_{C_{m_0}})$: It is shown by a simple computation, using (II.3), (II.4), that

$$<U_\Lambda(C_1)^2>_{C_{m_0}} = \frac{\lambda^2}{2} m_0^\alpha \int_{\Lambda\times\Lambda} dx dx' \left[\cos\theta\, e^{-\varepsilon^2 C_{m_0}(x-x')} + e^{\varepsilon^2 C_{m_0}(x-x')}\right] \quad (IV.1)$$

Since

*) See also Lemma IV.10.

$$C_{m_o}(x) = -\frac{1}{4\pi}\ \ell n\ (C\ m_o^2\ x^2) + 0\ (x^2),\ \text{for}\ x \approx 0, \qquad (IV.2)$$

$\langle U_\Lambda(C_1)^2\rangle_{C_{m_o}}$ is finite if and only if $\alpha < 1$, by (IV.1); (C is some numerical constant). Thus for $\varepsilon^2 < 4\pi$

$$e^{\pm U_\Lambda(C_1)} = \sum_{n=o}^{\infty} \frac{1}{n!}\ (\pm\ U_\Lambda(C_1))^n, \qquad (IV.3)$$

holds $d\mu_{C_{m_o}}$ - almost everywhere.

Clearly

$$\begin{aligned}\langle e^{-U_\Lambda(C_1)}\rangle_{C_{m_o}} &\leq 2\ \langle \text{Cosh}\ U_\Lambda(C_1)\rangle_{C_{m_o}} \\ &= 2\sum_{n=o}^{\infty} \frac{1}{(2n)!}\ \langle U_\Lambda(C_1)^{2n}\rangle_{C_{m_o}}\end{aligned} \qquad (IV.4)$$

by (IV.3). As in (IV.1) one shows by a computation that $\langle U_\Lambda(C_1)^{2n}\rangle_{C_{m_o}}$ takes its maximum value if $\theta = 0$. Therefore we set $\theta = 0$ throughout the proof of step (1).

We define

$$\left.\begin{aligned}\gamma\ (\varepsilon,C;h) &= :\cos\varepsilon\phi:_C(h) \\ \gamma\ (\varepsilon;h) &= \gamma(\varepsilon,C_1;h)\end{aligned}\right\} \qquad (IV.5)$$

Thus for the proof of Step (1), II.5, it suffices to show that

$$\langle\gamma\ (\varepsilon;\chi_\Lambda)^{2n}\rangle_{C_{m_o}} \leq K(\Lambda)^n \left[(2n)!\right]^{1-\delta}, \qquad (IV.6)$$

for some $\delta > 0$ and a constant $K(\Lambda) < \infty$, by (IV.4) and (IV.5).

(b) Connection with classical statistical mechanics (CSM):

Let $C(x,x')$ be the kernel of a covariance C, and suppose for the moment that $\exp\left[pC(x,x')\right]$ is locally Lebesgue-integrable for all finite p.

Lemma IV.1:

$$< \prod_{j=1}^{n} : e^{i\varepsilon_j \phi} :_C (x_j) >_C = e^{-\frac{1}{2} \sum_{i \neq j} \varepsilon_i \varepsilon_j C(x_i, x_j)} \tag{IV.7}$$

Proof:

This follows directly from the definitions of $<\cdot>_C$ and $: - :_C$, see (II.3), (II.4).

Q.E.D.

The r.h.s. of (IV.7) is obviously the Gibbs density of a classical gas of n particles with charges $\varepsilon_1, \ldots, \varepsilon_n$ and a pair potential $C(x,x')$.

Lemma IV.1 permits us to evaluate $<\gamma(\varepsilon;\chi_\Lambda)^{2n}>_{C_{m_0}}$ explicitly, an important step in the proof of (IV.6). We define

$$Z_n^q (C,h) = \int \prod_{i=1}^{n-q} h(x_i) dx_i \prod_{j=1}^{n+q} h(x_j') dx_j' \tag{IV.8}$$

$$\times \exp \left[-\varepsilon^2 \left(\sum_{1 \leq i < j \leq n-q} C(x_i, x_j) + \sum_{1 \leq i < j \leq n+q} C(x_i', x_j') - \sum_{\substack{i=1 \ldots n-q \\ j=1 \ldots n+q}} C(x_i, x_j') \right) \right]$$

This is the <u>classical partition function</u> of a gas of n-q particles with charge + ε, n+q particles with charge - ε, pair potential C, in an external potential - ℓn h, at inverse temperature 1.

From (IV.5), (IV.7) and (IV.8) we now conclude

$$< \gamma(\varepsilon, C; h)^{2n} >_C = 2^{-2n} \sum_{q=-n}^{n} \binom{2n}{n-q} Z_n^q (C,h) \tag{IV.9}$$

Equations (IV.7)-(IV.9) and other relations described below are the basis for the following isomorphism, [Fr3, Fr4] : <u>The sine-Gordon Euclidean field theory is isomorphic to CSM of the two-component Yukawa-, for $m_0 > 0$, and the two-component, neutral Coulomb gas, for $m_0 = 0$, in two space dimensions in the grand canonical ensemble, at inverse temperature 1; λ is identified with the fugacity of the gas</u>. (See also [A1]).

This observation recalls useful estimates first proven in CSM. As an important example we mention the following inequality which is presum-ably well known in CSM and the proof of which can be found in [Fr4] , (by some straight forward modifications).

Lemma IV.2: ("Inverse conditioning", [Fr4])

Let C^1 and C^2 be two covariances with $C^1 \geq C^2$ (as quadratic forms on $\mathscr{S}$) and set

$$h^{12}(x) = \exp \frac{\varepsilon^2}{2}\left[C^1 - C^2\right](x,x) \qquad \text{(IV.10)}$$

Let h be a real-valued function in $L^1(\mathbb{R}^2) \cap L^p(\mathbb{R}^2)$, some $p > 1$. Then $< \gamma(\varepsilon,C^1;h)^{2n}>_{C^1} \leq < \gamma(\varepsilon,C^2;h^{12}h)^{2n}>_{C^2}$

(c) The "conditioning" inequality:

Next we state an inequality going in the opposite direction of Lemma IV.2.

Lemma IV.3: ("Conditioning")

Let C^1, C^2 and h be as in Lemma IV.2.

Then $<\gamma(\varepsilon,C^1;h)^{2n}>_{C^1} \geq <\gamma(\varepsilon,C^2;h)^{2n}>_{C^2}$.

Proof:

This is a direct consequence of the fact that $C^1-C^2 \geq 0$, i.e. C^1-C^2 is a covariance, and of the Hölder inequality: see [Fr4] for details. This Lemma is based on earlier results of [GRS2] .

Q.E.D.

Conditioning and inverse conditioning are the basic ingredients for the proof of (IV.6), Step (1). By the definition of Wick ordering (II.4) and (IV.2)

$$\gamma(\varepsilon;h) = m_o^{\alpha} \cdot \gamma(\varepsilon,C_{m_o};h) \qquad \text{(IV.11)}$$

By (IV.2) $\exp \frac{\varepsilon^2}{2}\left[C_{m_o} - C_1\right](x,x) = m_o^{-\alpha}$, (IV.12)

and of course $C_{m_o} \geq C_1$, for $m_o < 1$.

Thus inverse conditioning gives

$$<\gamma(\varepsilon;h)^{2n}>_{C_{m_o}} = m_o^{2n\alpha} < \gamma(\varepsilon,C_{m_o};h)^{2n}>_{C_{m_o}}$$

$$\leq <\gamma(\varepsilon;h)^{2n}>_{C_1}, \qquad \text{(IV.13)}$$

for all $m_o \leq 1$.

Lemma IV.4: [W,Fr4]

$$\lim_{m_o \searrow 0} m_o^{2n\alpha} <\gamma(\varepsilon,C_{m_o};h)^{2n}>_{C_{m_o}}$$

$$= \frac{(2n)!}{(n!)^2\, 2^{2n}} \int \prod_{j=1}^{n} h(x_j)\, h(x_j')\, dx_j\, dx_j'$$

$$\times\, e^{-U_{G_o}(x_1,\ldots,x_n;\, x_1',\ldots,x_n')} > 0,$$

where $G_o(x) = \frac{1}{2\pi} \ln \frac{C}{|x|}$, and

$$U_{G_o}(x_1,\ldots,x_n,x_1' \ldots) = \varepsilon^2 \Big[\sum_{1\leq i<j\leq n} G_o(x_i - x_j) + G_o(x_i' - x_j')$$

$$- \sum_{i,j=1}^{n} G_o(x_i - x_j') \Big] \qquad \text{(IV.14)}$$

Proof:

$$\lim_{m_o \searrow 0} \Big[C_{m_o} + \frac{1}{4\pi} \ln m_o^2 \Big](x) = G_o(x) \qquad \text{(IV.15)}$$

Therefore, by Lemma IV.1 and ReWick-ordering (II.4), (IV.11)

$$\lim_{m_o \searrow 0} < \prod_{j=1}^{n} :e^{i\varepsilon_j\phi}:_1 (x_j) >_{C_{m_o}}$$

$$= \lim_{m_o \searrow 0} m_o^{\frac{1}{4\pi}\sum_1^n \varepsilon_j^2} \langle \prod_{j=1}^n : e^{i\varepsilon_j \phi} :_{m_o} (x_j) \rangle_{C_{m_o}}$$

$$= \left[\lim_{m_o \searrow 0} m_o^{\frac{1}{4\pi}(\sum_1^n \varepsilon_j)^2} \right] e^{-\sum_{1\le i<j\le n} \varepsilon_i \varepsilon_j G_o (x_i - x_j)} ,$$

which obviously vanishes, unless $\Sigma_1^n \varepsilon_j = 0$; (neutrality!). This and (IV.9) complete the proof.

Q.E.D.

By (IV.13) and Lemma IV.4 it suffices now to estimate $\langle \gamma(\varepsilon;h)^{2n} \rangle_{C_1}$. This requires a change of boundary conditions.

(d) Change of boundary conditions:

By scaling we may choose for Λ a unit square and by Euclidean invariance of $d\mu_{C_{m_o}}$ we may suppose that Λ is centered at $(\vec{x},t) = (1,0)$ with sides parallel to the coordinate axes. We let S be the circle of radius two centered at (0,0).

Fig. 1

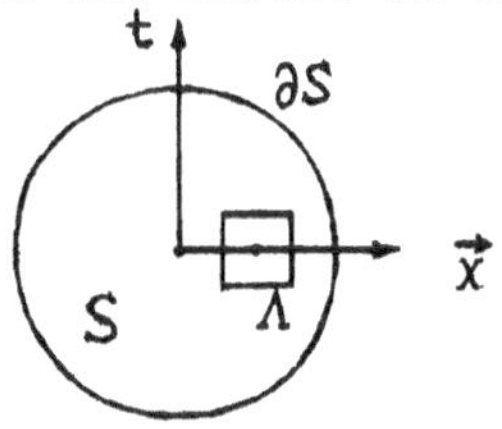

Clearly dist $(\Lambda, \partial S) = 2 - \frac{\sqrt{10}}{2} > 0$

Let $C_{\partial S}(D) = (-\Lambda_{\partial S,D} + 1)^{-1}$, and

$h^{12}(x) = \exp \frac{\varepsilon^2}{2} \left[C_1 - C_{\partial S}(D) \right] (x,x)$

Since dist $(\Lambda, \partial S) > 0$,

$$|h^{12}(x)\, \chi_\Lambda(x)| \le w < \infty \tag{IV.16}$$

Lemma IV.5: [Fr4]

Let h be a real function in $L^p(\Lambda, dx)$, (some $p = p(\varepsilon) > 1$), and let $C^o_{\partial S} = (-\Delta_{\partial S,D})^{-1}$. Then

$$\langle \gamma(\varepsilon;h)^{2n} \rangle_{C_1} \le \langle \gamma(\varepsilon, C^o_{\partial S}; w\, h)^{2n} \rangle_{C^o_{\partial S}}$$

Proof: Using Sobolev spaces it is easy to see that $C_1 \ge C_{\partial S}(D)$, as quadratic forms on $L^2(S,dx)$, [GRS2]. Since $\Lambda \subset S$, supp $h \subseteq \Lambda$, and by (IV.16) inverse conditioning gives

$$<\gamma(\varepsilon;h)^{2n}>_{C_1} \leq <\gamma(\varepsilon, C_{\partial S}(D);\ w\,h)^{2n}>_{C_{\partial S}(D)}.$$

Because of 0 - Dirichlet data at ∂S $C^o_{\partial S}$ is a bounded, positive operator on $L^2(S,dx)$, hence a covariance. Clearly $C^o_{\partial S} \geq C_{\partial S}(D)$ as quadratic forms (and as operators on $L^2(S,dx)$). Therefore conditioning gives

$$< \gamma(\varepsilon, C_{\partial S}(D);\ w\cdot h)^{2n}>_{C_{\partial S}(D)} \leq < \gamma(\varepsilon, C^o_{\partial S};\ w\cdot h)^{2n}>_{C^o_{\partial S}} \tag{IV.17}$$

Q.E.D.

Fortunately the r.h.s. of (IV.17) can be calculated and estimated explicitly:

With $x = (\vec{x},t)$ we associate complex numbers $z = \vec{x} + it$ and $\hat{z} = 4/\bar{z}$, (the reflection of z at ∂S). By the method of image charges one shows that

$$C^o_{\partial S}(x,x') = -\frac{1}{4\pi}\{\ell n|z-z'| + \ell n|\hat{z}-\hat{z}'| - \ell n|\hat{z}-z'| - \ell n|z-\hat{z}'|\} \tag{IV.18}$$

Given $x_1,\ldots,x_n$, and $x_1',\ldots,x_n'$ we define

$$z_j = \begin{cases} z_j, & j=1,\ldots,n \\ \hat{z}'_{j-n}, & j=n+1,\ldots,2n, \end{cases} \quad \text{and similarly } z_j',\ j=1,\ldots,2n.$$

Now

$$< \gamma(\varepsilon, C^o_{\partial S};\ w\cdot h)^{2n}>_{C^o_{\partial S}} \leq <|\gamma(\varepsilon, C^o_{\partial S};\ w\cdot h)|^{2n}>_{C^o_{\partial S}}$$

$$= w^{2n}\int \prod_{j=1}^{n} h(x_j)h(x_j')\,dx_j\,dx_j' \frac{\prod_{1\leq i<j\leq 2n} |z_i-z_j|^{\alpha}|z_i'-z_j'|^{\alpha}}{\prod_{i,j=1}^{2n} |z_i - z_j'|^{\alpha}}$$

$$\times \prod_{j=1}^{n} |z_j - z'_{j+n}|^{\alpha}\,|z_j' - z_{j+n}|^{\alpha},$$

and $\alpha = \varepsilon^2/4\pi$, by (IV.7) and (IV.18). See [Fr4] for details.

By a lemma of Cauchy's (see e.g. [DL])

$$\frac{\prod_{1\le i<j\le 2n} |z_i - z_j|^\alpha \, |z_i' - z_j'|^\alpha}{\prod_{i,j=1}^{2n} |z_i - z_j'|^\alpha} = \left|\mathrm{Det}\left(\frac{1}{z_i - z_j'}\right)\right|^\alpha . \qquad \text{(IV.19)}$$

By the geometry of Λ and ∂S, Fig. 1

$$\max_{\substack{z_j,\, z_j' \text{ in } \Lambda \\ j=1,\cdots,n}} \left\{ \prod_{j=1}^{n} |z_j - z_{j+n}'|^\alpha \, |z_j' - z_{j+n}|^\alpha \right\} \le K^n , \qquad \text{(IV.20)}$$

for some finite constant K independent of $\alpha \in [0,1]$.

From (IV.19) and (IV.20) and the Hölder inequality with respect to $\prod_{j=1}^{n} \chi_\Lambda(x_j)\, \chi_\Lambda(x_j')\, dx_j\, dx_j'$ we conclude

Theorem IV.6: [Fr4]

Let $\infty > p > (1-\alpha)^{-1}$ and $h \in L^p(\Lambda, dx)$. Then

$$\langle \gamma(\varepsilon,h)^{2n} \rangle_{C_1} \le \langle \gamma(\varepsilon, C^o_{\partial S}; w\cdot h)^{2n} \rangle_{C^o_{\partial S}} \le K_p^n \left[(2n)!\right]^{\frac{p-1}{p}} ||h||_p^{2n}$$

(The details of the proof can be found in [Fr4]).

Corollary IV.7:

For all $m_o \ge 0$ and with $U_\Lambda(C_1)$ given by (III.3)

$$d\mu^\Lambda_{C_{m_o}}(\phi) = \left[Z^\Lambda_{C_{m_o}}\right]^{-1} e^{-U(C_1)}\, d\mu_{C_{m_o}}(\phi)$$

is a well defined probability measure on $\mathscr{S}'$, i.e. step (1), II.5, is complete.

Proof: For $m_o > 0$ the corollary follows from Theorem IV.6, (IV.13) and subsection IV.1(a), in particular (IV.4) and (IV.6).

Let the field χ_ϑ be given by (III.4). Then by (IV.13) and Lemma IV.4

$$\lim_{m_o \searrow 0} \int d\mu_{C_{m_o}}^{\Lambda}(\phi) \exp i\,\chi_\vartheta(f) \equiv \langle \exp i\,\chi_\vartheta(f)\rangle_{C_o}^{\Lambda} \tag{IV.21}$$

exists and is a functional of positive type which is normalized and continuous on $\mathscr{S}$. By Minlos' theorem [Mi] it is the Fourier transform of some probability measure on $\mathscr{S}'$, denoted by $d\,\mu_{C_o}^{\Lambda}(\phi)$.

Q.E.D.

Note that the fields $\phi(f)$, $f \in \mathscr{S}$, are not well defined random variables for $d\,\mu_{C_o}^{\Lambda}$, unless $\tilde{f}(k=0) = 0$.

IV.2 Step (2), II.5: Infinite volume limit for the pressure

(a) Stability:

A basic estimate for Step (2) in the construction of the s-G theory is stability:

$$e^{-O(|\Lambda|)} \le Z_{C_{m_o}}^{\Lambda} \le e^{O(|\Lambda|)} \tag{IV.22}$$

By Jensen's inequality

$$Z_{C_{m_o}}^{\Lambda} = \langle e^{m_o^{\alpha} \lambda \gamma(\varepsilon, C_{m_o}; \chi_\Lambda)} \rangle_{C_{m_o}} > e^{m_o^{\alpha} \lambda \langle \gamma(\varepsilon, C_{m_o}; \chi_\Lambda)\rangle} = 1 \tag{IV.23}$$

Concerning the upper bound we notice that it suffices to consider the case, where $m_o = 1$ (by (IV.13) and Cor. IV.7). Let $\mathfrak{C}_\Lambda$ be a covering of Λ by unit squares. We decompose each $\Delta \in \mathfrak{C}_\Lambda$ into four squares $\Delta(1), \ldots, \Delta(4)$ of equal size. By the Hölder inequality and the translation invariance of $\langle\cdot\rangle_{C_1}$

$$\langle e^{-U_\Lambda(C_1)}\rangle_{C_1} \le \prod_{i=1}^{4} \langle \exp\Big[4\lambda \sum_{\mathfrak{C}_\Lambda} \gamma(\varepsilon;\chi_{\Delta(i)})\Big]\rangle_{C_1}^{1/4}$$

$$= \langle \exp\Big[4\lambda \sum_{\mathfrak{C}_\Lambda} \gamma(\varepsilon;\chi_{\Delta(1)})\Big]\rangle_{C_1} \tag{IV.24}$$

This inequality separates adjacent squares.

Next we cover $\mathbb{R}^2$ with a cubic grid Γ of mesh 1 in such a way that, for each $\Delta \in \mathcal{C}_\Lambda$, $\Delta(1)$ is in the center of one of the primitive cells $\bar{\Delta}$ of Γ, and dist $(\Delta(1), \partial\bar{\Delta}) = \frac{1}{2}$. The following idea (Neumann b.c.) for the proof of (IV.22) is inspired by Guerra, Rosen and Simon [GRS3] :

We consider the covariance $C_\Gamma(N)$ with Neumann b.c. at Γ. It is well known that $C_\Gamma(N)$ decouples different cells of Γ, i.e.

$$C_\Gamma(N) = \bigoplus C_{\partial\bar{\Delta}}(N), \tag{IV.25}$$

where the direct sum is over all cells $\bar{\Delta}$ of Γ.

We assert that for all λ and all $\alpha < 1$

$$\langle \exp\left[4\lambda\gamma(\varepsilon, C_{\partial\bar{\Delta}}(N); \chi_{\Delta(1)})\right]\rangle_{C_{\partial\bar{\Delta}}(N)} \text{ is finite.} \tag{IV.26}$$

Proof: Since dist $(\Delta(1), \partial\bar{\Delta}) = \frac{1}{2}$,

$$h^{12} \equiv \exp\left[\varepsilon^2 \max_{x\,\varepsilon\Delta(1)} \left[C_{\partial\bar{\Delta}}(N) - \chi_{\bar{\Delta}} C_1 \chi_{\bar{\Delta}}\right](x,x)\right] < \infty$$

It is well known [GRS3] that $C_{\partial\bar{\Delta}}(N) - \chi_{\bar{\Delta}} C_1 \chi_{\bar{\Delta}} \geq 0$, (IV.27)

as quadratic forms on $L^2(\bar{\Delta}, dx)$.

Thus inverse conditioning (Lemma IV.2) gives

$$\langle \exp\left[4\lambda\, \gamma(\varepsilon, C_{\partial\bar{\Delta}}(N); \chi_{\Delta(1)})\right]\rangle_{C_{\partial\bar{\Delta}}(N)}$$

$$\leq 2 \sum_{n=0}^{\infty} \frac{(4\lambda)^{2n}}{(2n)!} \langle \gamma(\varepsilon, C_{\partial\bar{\Delta}}(N); \chi_{\Delta(1)})^{2n}\rangle_{C_{\partial\bar{\Delta}}(N)}$$

$$\leq 2 \sum_{n=0}^{\infty} \frac{(4\lambda h^{12})^{2n}}{(2n)!} \langle \gamma(\varepsilon; \chi_{\Delta(1)})^{2n}\rangle_{C_1}$$

$$\leq 2 \sum_{n=0}^{\infty} \frac{(4\lambda h^{12})^{2n}}{(2n)!} K_p^n \left[(2n)!\right]^{\frac{p-1}{p}} \left(\frac{1}{4}\right)^{\frac{2n}{p}} \leq K < \infty,$$

provided we choose $(1-\alpha)^{-1} < p < \infty$, by Theorem IV.6, which proves (IV.26). It is well known that $C_\Gamma(N) \geq C_1$, as quadratic forms [GRS3]. Thus conditioning (Lemma IV.3) gives

$$<e^{-U_\Lambda(C_1)}>_{C_1} \leq <\exp\left[4\lambda \sum_{\mathfrak{C}_\Lambda} \gamma\,(\varepsilon,\chi_{\Delta(1)})\right]>_{C_1}$$

$$\leq <\exp\left[4\lambda \sum_{\mathfrak{C}_\Lambda} \gamma\,(\varepsilon,C_\Gamma(N);\chi_{\Delta(1)})\right]>_{C_\Gamma(N)}$$

$$= \prod_{\Delta\in\mathfrak{C}_\Lambda} <\exp\left[4\lambda\, \gamma\,(\varepsilon,C_{\partial\bar\Delta}(N);\chi_{\Delta(1)})\right]>_{C_{\partial\bar\Delta}(N)},$$

by (IV.25)

$< K^{|\mathfrak{C}_\Lambda|}$; where $|\mathfrak{C}_\Lambda|$ is the number of unit squares in $\mathfrak{C}_\Lambda$. Let $|\partial\Lambda|$ be the length of $\partial\Lambda$; if $|\partial\Lambda|^2 \leq K_1 |\Lambda|$, for some finite K_1, then $|\mathfrak{C}_\Lambda| = 0\,(|\Lambda|)$ and $K^{|\mathfrak{C}_\Lambda|} = e^{O(|\Lambda|)}$. This completes the proof of (IV.22).

(b) Guerra's theorem, [Gu] :

Theorem IV.8:

Let $\ell \times T = [-\ell/2, \ell/2] \times [-T/2, T/2]$. Then for all λ, all $\alpha < 1$ and all $m_0 \geq 0$

$$p(\lambda,\alpha,m_0) \equiv \lim_{\ell,T\to\infty} \frac{1}{\ell T} \log <e^{-U_{\ell\times T}(C_1)}>_{C_{m_0}} \tag{IV.28}$$

exists and is positive; (see Theorem 1,(a), Section III). For $m_0 = 0$ the identities (III.5) and (III.5^1) hold, in particular $p(\lambda,\alpha) \equiv p\,(\lambda,\alpha,0) = F(\alpha)\,\lambda^{2/2-\alpha}$.

Remarks: (IV.28) is Guerra's theorem; (for its sine-Gordon version, see [Fr4]). It is a consequence of the fact that, by (IV.22) and (IV.13)

$$\frac{1}{\ell T} \log <e^{-U_{\ell\times T}(C_1)}>_{C_{m_0}}$$

is bounded uniformly in ℓ,T and $m_o \in [0,1]$.

The second part of the theorem follows from the well known scaling properties of the Coulomb potential G_0; see Lemma IV.4. The proof can be found in [Fr4]. (Earlier heuristic results were obtained in [M]).

(c) Uniform bounds on the generating functional of the EGF's: Theorem IV.8 can of course be extended to the case where

$$U_{\ell xT} = -\lambda\, \gamma(\varepsilon;\chi_{\ell xT}) - f \cdot \chi_{\vartheta}(\chi_{\ell xT}),$$

where $\chi_{\vartheta}(x) = :\cos(\varepsilon\phi(x) + \vartheta):_1$.

The corresponding infinite volume pressure is denoted by $p(\{f,\vartheta\},\lambda,\alpha,m_o)$, and

$$\langle\cdot\rangle^{\Lambda}_{C_{m_o}} = \int_{\mathcal{S}'} \cdot \; d\mu^{\Lambda}_{C_{m_o}}(\phi);$$ see (II.7) and Cor. IV.7.

Theorem IV.9:

$$\lim_{\ell\to\infty} \sup \lim_{T\to\infty} \langle e^{\chi_{\vartheta}(f)\,\ell xT} \rangle_{C_{m_o}}$$

$$\leq e^{\int dx\, [p(\{\mathrm{Re} f(x),\vartheta\},\lambda,\alpha,m_o) - p(\lambda,\alpha,m_o)]},$$

and the r.h.s. in bounded uniformly in $m_o \in [0,1]$, for $f \in L^1(\mathbb{R}^2) \cap L^q(\mathbb{R}^2)$ with $q > (1-\alpha)^{-1}$.

Remarks: Thanks to [GRS1] this theorem is nowadays a rather straight-forward corollary of Theorem IV.8. The proof is identical to the one of analogous bounds in the $P(\phi)_2$ - models which is by now standard. See [GRS1]. The estimate on q follows from Theorem IV.6; see [Fr4]. For the s-G theory results of the type of Theorem IV.9 are derived in [Fr3].

IV.3 Steps (2) and (3), (II.5): The lattice approximation [GRS2], the infinite volume limit and verification of axioms

In this subsection we first prove the necessary correlation inequalities which permit us then to construct the infinite volume EGF's of the sine-Gordon theory in close analogy to Nelson's construction of the infinite volume EGF's in $P(\phi)_2$, [N 3]. In this subsection m_0 is positive. By scaling we may as well set $m_0 = 1$. The EGF's for $m_0 = 0$ are later obtained as limits of the ones for $m_0 > 0$, as $m_0 \searrow 0$. Technically it is easier to derive this limit from Theorem 3(a), Section III, by using weak coupling boundary conditions [GJ4]. References for this and the next subsection are [Fr3, FrS, N3, GRS2, GJ4].

(a) A digression on periodic boundary conditions:

We let $C_{\ell T}(\beta_1,\beta_2)$ be the covariance $(-\Delta+1)^{-1}$ with b.c. β_1 at $\vec{x} = \pm\ \ell/2$ and b.c. β_2 at $t = \pm\ T/2$.

Lemma IV.10:

For all λ and all $\alpha < 1$

$$1 \leq \ \langle e^{-U_{\ell x T}(C_{\ell T}(D,D))}\rangle_{C_{\ell T}(D,D)} \leq e^{O(\ell\cdot T)}$$

$$1 \leq \ \langle e^{-U_{\ell x T}(C_{\ell+1,T}(D,F))}\rangle_{C_{\ell+1,T}(D,F)} \leq e^{O(\ell\cdot T)}$$

Proof: Again the lower bounds follow from Jensen's inequality; see (IV.22). From [GRS2] we infer

$C_{\ell T}\ (D,D) \leq \chi_{\ell x T}\ C_1\ \chi_{\ell x T}$, and

$C_{\ell+1,T}\ (D,F) \leq \chi_{\ell+1,T}\ C_1\ \chi_{\ell+1,T}$, as quadratic forms.

Therefore Lemma IV.10 follows from (IV.22) (see also Theorem IV.8) by conditioning (Lemma IV.3).

Q.E.D.

Definition: Let F_ℓ be the symmetric Fock space over $L^2([-\ell/2,\ \ell/2]\ ;\ dx)$, Ω_0 the bare vacuum in F_ℓ, and $\langle\cdot\ ,\cdot\rangle$ the scaler product of F_ℓ.

By Nelson's Feyman-Kac formula [N1, GRS2] there exists a self-adjoint operator $H_{\ell D}(\lambda,\alpha)$ on $F_{\ell+1}$ such that

$$<e^{-U_{\ell xT}(C_{\ell+1,T}(D,F))}>_{C_{\ell+1,T}(D,F)} = <\Omega_o, e^{-TH_{\ell D}(\lambda,\alpha)}\Omega_o> \quad (IV.29)$$

From Lemma IV.10 and (IV.29) we obtain by standard arguments (see e.g. [Fr1])

$$H_{\ell D}(\lambda,\alpha) \geq -0(\ell), \quad (IV.30)$$

for all λ and all $\alpha < 1$.

Let $H^o_{\ell D} = H_{\ell D}(0,0)$ be the free Hamiltonian on $F_{\ell+1}$ (with 0-Dirichlet data at $\vec{x} = \pm \frac{\ell+1}{2}$). It is well known (see e.g. [Ho]) that the covariance $C_{\ell+1\ T}(D,P)$ is the two point ("temperature ordered") EGF of a gas of free, relativistic scalar Bosons of mass 1 in a box with rigid walls ($\approx$ 0-Dirichlet data) at $\vec{x} = \pm \frac{\ell+1}{2}$ with Hamiltonian $H^o_{\ell D}$, at inverse temperature T (in the canonical ensemble). From this fact and the Feynman-Kac formula [Ho] we conclude

$$<e^{-U_{\ell xT}(C_{\ell+1,T}(D,F))}>_{C_{\ell+1,T}(D,P)} = \mathrm{Tr}\,(e^{-TH^o_{\ell D}})^{-1}$$
$$\times\ \mathrm{Tr}\,(e^{-TH_{\ell D}(\lambda,\alpha)}). \quad (IV.31)$$

The r.h.s. of (IV.31) can be estimated by means of the <u>Golden-Thompson inequality</u>:

If $e^{-T(A+B)}$ satisfies the hypotheses for the Trotter product formula the Golden-Thompson inequality applies to $e^{-T(A+B)}$ and gives

$$\mathrm{Tr}\,(e^{-T(A+B)}) \leq \mathrm{Tr}\,(e^{-TA}e^{-TB}) \leq ||e^{-TB}||\,\mathrm{Tr}\,(e^{-TA}),$$

where $||e^{-TB}||$ is the operator norm of e^{-TB}. Therefore, setting $A = \frac{1}{2} H^o_{\ell D}$ and $B = \frac{1}{2} H_{\ell D}(2\lambda,\alpha)$, we obtain

$$\mathrm{Tr}\,(e^{-TH_{\ell D}(\lambda,\alpha)}) \leq \mathrm{Tr}\,(e^{-\frac{T}{2}H^o_{\ell D}})\,||e^{-\frac{T}{2}H_{\ell D}(2\lambda,\alpha)}||$$
$$\leq e^{0(\ell\cdot T)}, \quad (IV.32)$$

by (IV.30) and the fact that $1 < \mathrm{Tr}\,(e^{-T/2\,H^{o}_{\ell D}}) \leq e^{O(\ell\cdot T)}$.
By (IV.31) and the finiteness of re Wick-ordering this yields

$$< e^{-U_{\ell x T}(C_{\ell+1,T}(D,P))} >_{C_{\ell+1,T}(D,P)} \leq e^{O(\ell\cdot T)}$$

We let $C_T(P)$ be the covariance $(-\Delta+1)^{-1}$ with periodic b.c. at $t = \pm T/2$. A by now standard application of <u>inverse conditioning</u> (Lemma IV.2) gives

$$< e^{-U_{\ell x T}(C_T(P))} >_{C_T(P)} \leq 2 \sum_{n=0}^{\infty} \frac{1}{(2n)!} < U_{\ell x T}(C_T(P))^{2n} >_{C_T(P)}$$

$$\leq 2 \sum_{n=0}^{\infty} \frac{(h^{12})^{2n}}{(2n)!} < U_{\ell x T}(C_{\ell+1,T}(D,P))^{2n} >_{C_{\ell+1,T}(D,P)}$$

$$= 2 < \mathrm{Cosh}\left[h^{12}\, U_{\ell x T}(C_{\ell+1,T}(D,P))\right] >_{C_{\ell+1,T}(D,P)}$$

$$\leq e^{O(\ell\cdot T)}, \text{ where} \tag{IV.33}$$

$$h^{12} = \exp\left[\varepsilon^2 \max_{x\varepsilon\, \ell x T} \left[C_T(P) - C_{\ell+1,T}(D,P)\right](x,x)\right] < \infty$$

Again by the Feynman-Kac formula there exists a selfadjoint operator $H_{TP}(\lambda,\alpha)$ on F_T such that

$$< e^{-U_{\ell x T}(C_T(P))} > = < \Omega_o, e^{-\ell H_{TP}(\lambda,\alpha)}\, \Omega_o > ,$$

and by (IV.33) $H_{TP}(\lambda,\alpha) \geq -O(T)$. (IV.34)

We note that $1 < \mathrm{Tr}\,(e^{-\ell H^{o}_{TP}}) \leq e^{O(\ell\cdot T)}$ (IV.35)

Applying again the Golden-Thompson inequality, using (IV.34) and (IV.35) and doing finite re Wick-ordering we finally get

Theorem IV.11:

$$<e^{-U_{\ell xT}(C_{\ell T}(P,P))}>_{C_{\ell T}(PP)} \leq e^{O(\ell\cdot T)}$$

Remark: By Lemma IV.10 the analogous stability estimate also holds for Dirichlet b.c. and can be proven for Neumann b.c., using inverse conditioning, (at least if $\varepsilon^2 < 2\pi$; however we will not use that).

(b) Isomorphism between s-G-theory and Ising model:

Next we describe an isomorphism between Euclidean field theory and classical statistical mechanics discovered by Guerra, Rosen and Simon [GRS2] (who considered the $P(\phi)_2$ -models). For the s-G theory it is:

The Euclidean sine-Gordon theory is isomorphic to a two dimensional, generalized, continuous spin, ferromagnetic Ising model.

This isomorphism provides the necessary correlation inequalities for taking the infinite volume limit of the EGF's of the s-G theory à la Griffiths-Nelson [N3] . Let n_1 and n_2 be integers and consider the rectangle $\Lambda = n_1 \times n_2$. Let $h_1, \ldots, h_n$ be arbitrary functions in $L^2(\Lambda, dx)$. Then

Lemma IV.12:

The periodic box cutoff EGF's of the s-G theory

$$S_P^{\Lambda}(h_1,\ldots,h_n) = \int_{\mathscr{S}'} \prod_{j=1}^{n} \phi(h_j)\, d\mu^{\Lambda}_{C_{\partial\Lambda}(P)}(\phi)$$

exist; (here $C_{\partial\Lambda}(P) = C_{n_1 n_2}(P,P)$).

Proof:

This follows from the Schwarz inequality with respect to $d\mu_{C_{\partial\Lambda}(P)}$, Theorem IV.11 and the fact that $\int \prod_{j=1}^{n} \phi(h_j)\, d\mu_{C_{\partial\Lambda}(P)}(\phi)$ is well defined.

Q.E.D

In order to make the isomorphism between the s-G theory and a generalized, ferromagnetic Ising model precise we must introduce lattice approximations to the EGF's S_p^{Λ} and identify the EGF's in the lattice approximation as Ising correlation functions. For this purpose we cover Λ with a cubic grid (lattice) Γ_ℓ^{Λ} of mesh $\delta_\ell = 2^{-\ell}$. Points on this lattice are denoted $J = (i,j)$, with $i = 1,2,\ldots, 2^\ell n_1 + 1$, and $j = 1,2,\ldots, 2^\ell n_2 + 1$.

Since we are interested in periodic b.c. at $\partial\Lambda$ we must identify $(1,j)$ with $(2^\ell n_1 + 1,j)$ and $(i,1)$ with $(i, 2^\ell n_2 + 1)$. At each point $J \in \Gamma_\ell^\Lambda$ we are given a random variable $\hat{\phi}_J$ with distribution

$$\exp\left[- \delta_\ell^2 (\hat{\phi}_J^2 + \lambda_\ell \cos(\varepsilon \hat{\phi}_J + \theta))\right] d\hat{\phi}_J$$

(up to a normalization factor); λ_ℓ is chosen such as to achieve the correct Wick ordering when we let $\ell\to\infty$, later. Since we consider periodic b.c. we may shift $\hat{\phi}_J$ for convenience: $\phi_J = \hat{\phi}_J + \varepsilon^{-1}\theta$. The distribution of ϕ_J is thus

$$\exp\left[-\delta_\ell^2 (\phi_J^2 + 2\varepsilon^{-1}\theta\phi_J + \lambda_\ell \cos(\varepsilon\phi_J))\right] d\phi_J, \qquad \text{(IV.36)}$$

(up to a normalization factor).

Next we introduce nearest neighbor couplings: The joint distribution of all random variables $\{\phi_J\}_{J\in\Gamma_\ell^\Lambda}$ is

$$d\mu_\ell^\Lambda(\{\phi_J\}) \propto e^{2\sum_{|J-J'|=1}\phi_J\phi_{J'}}$$

$$\times \prod_{J\in\Gamma_\ell^\Lambda} e^{-(\delta_\ell^2+8)\phi_J^2 + 2\delta_\ell^2\varepsilon^{-1}\theta\phi_J + \delta_\ell^2\lambda_\ell\cos(\varepsilon\phi_J)} \, d\phi_J$$

(The factors δ_ℓ^2 in the exponent will permit us to pass to the limit $\ell = \infty$). The normalization is chosen such that $d\mu_\ell^\Lambda$ is a probability measure on $\mathbb{R}^{2^\ell n_1 \cdot 2^\ell n_2}$. The measure $d\mu_\ell^\Lambda$ is ferromagnetic in the sense of Guerra, Rosen and Simon [GRS2]. Up to the factors $e^{2\delta_\ell^2\varepsilon^{-1}\theta\phi_J}$ it is even. Without loss of generality we may assume hence-forth that $\varepsilon^{-1}\theta$ is positive. Therefore $d\mu_\ell^\Lambda$ defines a generalized, continuous spin, ferromagnetic Ising model for which the following is known, [GRS2]:

Theorem IV.13:

Let $S_\ell^\Lambda(J_1,\ldots,J_n) = \langle \prod_{j=1}^n \phi(J_j)\rangle_\ell^\Lambda$

$$\equiv \int \prod_{j=1}^n \phi(J_j)\, d\mu_\ell^\Lambda(\{\phi_J\}).$$

Then $S_\ell^\Lambda(J_1,\ldots,J_n) \geq 0$, (first Griffiths inequality), and $S_\ell^\Lambda(J_1,\ldots,J_n) \geq S_\ell^\Lambda(J_1,\ldots,J_k)\, S_\ell^\Lambda(J_{k+1},\ldots,J_n)$, (second Griffiths inequality).

If F and G are monotone increasing functions of $\{\phi_J\}_{J\varepsilon\Gamma_\ell^\Lambda}$ then

$\langle F\cdot G\rangle_\ell^\Lambda > \langle F\rangle_\ell^\Lambda \langle G\rangle_\ell^\Lambda$, (Fortuin-Kasteleyn-Ginibre inequality).

The correlation inequalities of Theorem IV.13 supply almost all information needed to complete the proof of Theorem 1, (b), Section III, for $m_0 > 0$. The only missing steps in the proof are

(i) $S_\ell^\Lambda \to S_P^\Lambda$, as $\ell \to \infty$,

(i.e. convergence of the lattice approximation [GRS2]),

and

(ii) replace periodic b.c. at $\partial\Lambda$ by 0-Dirichlet data; see [N3, GRS2] for motivation.

(c) Convergence of the lattice approximation:

We first note that

$$d\mu_{0,\ell}^\Lambda(\{\phi_J\}) \propto e^{2\sum_{|J-J'|=1}\phi_J\phi_{J'}} \prod_{J\varepsilon\Gamma_\ell^\Lambda} e^{-(\delta_\ell^2+8)\phi_J^2} \qquad \text{(IV.37)}$$

is the Gaussian measure on $\mathbb{R}^{2^\ell n_1\cdot 2^\ell n_2}$ with mean 0 and covariance

$$C_{\partial\Lambda}^\ell = (-\Delta_{\Lambda,\ell}+1)^{-1},$$

where $\Delta_{\Lambda,\ell}$ is the finite difference approximation to the Laplacian with periodic b.c. at $\partial\Lambda$.

Let P^{Λ} be the momentum space lattice whose first Brillouin zone is Λ. One can show by some straightforward estimates [GRS2] :

$$0 < \tilde{C}^{\ell}_{\partial\Lambda}(k) < f_{\Lambda}^{2} \frac{\pi^2}{4} (\vec{k}^2 + (k^o)^2 + 1)^{-1}, \tag{IV.38}$$

for all $k \in P^{\Lambda}$, where $f_{\Lambda} \to 1$, as $\Lambda \nearrow \mathbb{R}^2$. It is assumed that $\tilde{C}^{\ell}_{\partial\Lambda}(k) = 0$ if k is not in the first Brillouin zone associated with Γ^{Λ}_{ℓ}. Obviously (IV.38) yields an inequality between quadratic forms on $\ell^2(P^{\Lambda})$.

As long as we only want to estimate $d_{\ell}\mu^{\Lambda}_{0,\ell}$ -integrals of translation invariant functions on $\mathbb{R}^{2^{\ell}n_1 \cdot 2^{\ell}n_2}$, we may interpret $\tilde{C}^{\ell}_{\partial\Lambda}(k)$ as the covariance of a Gaussian measure on $\mathscr{S}'$ with mean 0 and a covariance whose x-space kernel is

$$C^{\ell}_{\partial\Lambda}(x-y) = \frac{1}{2\pi} \sum_{k\varepsilon P^{\Lambda}} \frac{4\pi^2}{n_1 \cdot n_2} e^{ik(x-y)} \tilde{C}^{\ell}_{\partial\Lambda}(k), \tag{IV.39}$$

(since periodic b.c. guarantee total momentum conservation). By (IV.38)

$$0 \leq C^{\ell}_{\partial\Lambda} \leq f^2_{\Lambda} \frac{\pi^2}{4} C_{\partial\Lambda}(P), \tag{IV.40}$$

as quadratic forms on $\mathscr{S}$.

Next we choose the coupling constant λ_{ℓ} - see (IV.36) - by defining the lattice action

$$U_{\Lambda}(C^{\ell}_{\partial\Lambda}) = -\lambda \sum_{J\varepsilon\Gamma^{\Lambda}_{\ell}} \delta^2_{\ell} \left[:\cos(\varepsilon\phi_J):_{C^{\ell}_{\partial\Lambda}} + 2\theta\, \varepsilon^{-1} \phi_J \right]$$

$$\text{(by (IV.39),} \quad = -\lambda \int_{\Lambda} dx \left[:\cos \varepsilon\phi:_{C^{\ell}_{\partial\Lambda}}(x) + 2\theta\, \varepsilon^{-1} \phi(x) \right] \text{)}.$$

The basic step in the proof of convergence of the lattice approximation is to show that

$$\int d\mu^{\Lambda}_{o,\ell}(\{\phi_J\})\, e^{-U_{\Lambda}(C^{\ell}_{\partial\Lambda})} = \langle e^{-U_{\Lambda}(C^{\ell}_{\partial\Lambda})} \rangle_{C^{\ell}_{\partial\Lambda}}$$

is bounded uniformly in ℓ. (The equation follows from (IV.39)). Using (IV.40) and conditioning we obtain

$$<e^{-U_\Lambda (C^\ell_{\partial\Lambda})}>_{C^\ell_{\partial\Lambda}} \leq <e^{-U_\Lambda (\frac{1}{4} f_\Lambda{}^2 \pi^2 C_{\partial\Lambda}(P))}>_{\frac{1}{4}f_\Lambda{}^2\pi^2 C_{\partial\Lambda}(P)} \tag{IV.41}$$

Thus for $\varepsilon^2 < \frac{16}{\pi}$ there exists a bounded rectangle Λ so big that $\frac{1}{4} f_\Lambda{}^2 \pi^2 \varepsilon^2 < 4\pi$. But then the r.h.s. of (IV.41) is finite, by Theorem IV.11. Using inequality (IV.41) we may now prove as in [A2]

Theorem IV.14:

Let $\varepsilon^2 < \frac{16}{\pi}$ and let Λ be so large that $f^2_\Lambda < {}^{16}/_{\pi\varepsilon^2}$. Let $h_1, \ldots, h_n$ be continuous, periodic functions on Λ. Then

$$\lim_{\ell\to\infty} \sum_{\substack{J_i \varepsilon \Gamma^\Lambda_\ell \\ i=1,\ldots,n}} \prod_{i=1}^n \delta^2_\ell\, h_i\, (J_i)\, S^\Lambda_\ell\, (J_1,\ldots,J_n) = S^\Lambda_P\, (h_1,\ldots,h_n).$$

Theorem IV.14 is satisfactory, provided we can replace periodic b.c. by 0-Dirichlet data and do Wick ordering with respect to C_1. (These are the requirements that must be met if we want to apply the Griffiths-Nelson method [N3] to construct the EGF's in the limit $\Lambda = \mathbb{R}^2$). Fortunately this is possible. Let $\bar\Lambda \subset \Lambda$ be such that dist $(\bar\Lambda, \partial\Lambda) > 0$; $\varepsilon^2 < {}^{16}/_{\pi}$.

By Theorem IV.11

$$<e^{-U_{\ell xT} (C_1)}>_{C_{\ell T}(P,P)} = <e^{-h^{12} U_{\ell xT} (C_{\ell T} (P,P))}>_{C_{\ell T}(P,P)}$$

$$\leq e^{O(\ell\cdot T)}, \tag{IV.42}$$

where $h^{12} = \exp \left[\varepsilon^2 (C_{\ell T} (P,P) - C_1) (0)\right] < \infty$,

Conditioning gives

$$< e^{-U_{\bar\Lambda} (C_1)}>_{C_{\partial\bar\Lambda}(D)} \leq < e^{\gamma(\varepsilon;h^{12})}>_{C_1}, \tag{IV.43}$$

where now $h^{12} = \exp\left[\varepsilon^2 (C_1 - C_{\partial\bar{\Lambda}}(D))(x,x)\right] \chi_{\bar{\Lambda}}(x)$. For $\varepsilon^2 < 16/\pi$ ($< 2\pi$) the r.h.s. is finite as a consequence of Theorem IV.6. Thanks to inequalities (IV.42) and (IV.43) we may now pass from periodic b.c. and Wick ordering with respect to $C_\Lambda(P)$ to 0-Dirichlet data and Wick ordering with respect to C_1:

Doing Wick ordering with respect to C_1 we may use the second Griffiths inequality to increase the local bare mass on $\Lambda\backslash\bar{\Lambda}$ and eventually let it become ∞.

This procedure yields a decreasing sequence of EGF's which converges to $S_D^{\bar{\Lambda}}$ (= EGF's with 0-Dirichlet data at $\partial\bar{\Lambda}$) and it obviously preserves all correlation inequalities of Theorem IV.13. The method described here has been developed in detail in [Fr1] for the $P(\phi)_2$-models. It yields

Corollary IV.15:

For b.c. $\beta = P$ or D and $\varepsilon^2 < 16/\pi$ the space-time cutoff Euclidean sine-Gordon theory satisfies the two Griffiths and the FKG inequalities; (see Theorem IV.13). Using ϕ-bounds and their Euclidean form [Fr1, Fr2] , (they are similar to the bounds established in Theorem IV.9), we derive upper bounds on the EGF's $S_D^{\bar{\Lambda}} > 0$ that are uniform in Λ. By the second Griffiths inequality the EGF's $S_D^{\bar{\Lambda}}$ are monotone increasing in $\bar{\Lambda}$. This is Nelson's argument [N3], see also [GRS2, Fr3] . We conclude :
(d) Convergence of EGF's as $\Lambda \nearrow \mathbb{R}^2$:

Theorem IV.16:

Let $\varepsilon^2 < 16/\pi$, $m_0 > 0$, λ and θ arbitrary. Let $h_1, \ldots, h_n$ be in $L^1(\mathbb{R}^2) \cap L^2(\mathbb{R}^2)$. Then, for all n,

$$S_{m_0}(h_1,\ldots,h_n) = \lim_{\bar{\Lambda}\nearrow\mathbb{R}^2} S_D^{\bar{\Lambda}}(h_1,\ldots,h_n)$$

$$\equiv \lim_{\bar{\Lambda}\nearrow\mathbb{R}^2} \int_{\mathscr{S}'} \prod_{i=1}^{n} \phi(h_i)\, d\mu^{\bar{\Lambda}}_{C_{\partial\bar{\Lambda}}(D)}(\phi)$$

exists and is Euclidean invariant.

The distributions $\{S_{m_0}(x_1,\ldots,x_n)\}$ satisfy the Osterwalder-Schrader axioms (E0) - (E3)°.

Remarks: Axioms (E1) - (E3) are obviously satisfied; ((E2) and (E3) are already true for bounded, time-reflection invariant regions $\bar{\Lambda}$, and (E1) (Euclidean invariance) follows from monotomicity, [N3]). Finally (E0) (Distribution property) follows from the ϕ-bounds; see [Fr1] . These ϕ-bounds also imply that

$$S_{m_o}(h_1,\ldots,h_n) = \int_{\mathscr{S}'} \prod_{j=1}^{n} \phi(h_j)\, d\mu_{m_o}(\phi), \qquad \text{(IV.44)}$$

where $d\mu_{m_o}$ is a physical measure on $\mathscr{S}'$, [Fr1, Fr3] . Using a slight generalization of the cos $\varepsilon\phi$ - bounds of Theorem IV.9 [Fr3] and ϕ - bounds it is easy to prove that (with $\chi_\vartheta(x) = :\cos(\varepsilon\phi + \vartheta):_1(x)$)

$$\left\langle \prod_{j=1}^{n} \chi_{\vartheta_j}(h_j)\right\rangle_{m_o} \equiv \int \prod_{j=1}^{n} \chi_{\vartheta_j}(h_j)\, d\mu_{m_o}(\phi) \qquad \text{(IV.45)}$$

are well defined EGF's satisfying axioms (E0) - (E3). This completes the proof of Theorem 1 (b), Section III, for all $m_o > 0$.

IV.4 The Limit, as $m_o \searrow 0$

References: [Fr3,H]

(a) Monotomicity of EGF's in m_o:

It is a standard consequence of the second Griffiths inequality that $S_{m_o}(x_1,\ldots,x_n) \geq 0$ is monotone increasing in $1/m_o$. (IV.46)

This looks promising, because now the problem of proving the existence of a limit of the EGF's, as $m_o \searrow 0$, is reduced to deriving upper bounds on the EGF's $S_{m_o}(x_1,\ldots,x_n)$ that are uniform in $m_o \in (0,1]$.

We suppose that $\theta \neq 0$ and we let $m(\lambda,\alpha,m_o)$ denote the physical mass of the s-G theory reconstructed from the EGF's $\{S_{m_o}\}$. The basic (but so far unproven) assumption of subsection IV.4 (a) is that

$$m(\lambda,\alpha,m_o) \geq m_* > 0, \text{ for all } m_o \in (0,1]. \qquad \text{(IV.47)}$$

For $\theta \neq 0$ this seems to be a reasonable assumption. For $\theta = 0$ there may be a phase transition (and long range order) if m_o is small enough. This is predicted by the naive Goldstone picture. In this case we may however decompose the theory into its pure phases, see [Fr2] , and (IV.47) should still be true in all pure phases below the critical point, i.e., for m_o small enough. Indeed, using the isomorphism between the s-G theory at $\theta=m_o=0$ and the classical, neutral Coulomb gas in two dimensions we may interpret (IV.47) as Debye screening (m is the inverse correlation length!) which is generally believed to take place; see also [Da].

Theorem IV.17:

Assume (IV.47). Then there exists a Schwarz norm $|\cdot|_{\mathscr{S}}$ such that

$$|S_{m_o}(h_1,\ldots,h_n)| \leq n! \prod_{i=1}^{n} |h_i|_{\mathscr{S}}, \qquad \text{(IV.48)}$$

uniformly in $m_o \in (0,1]$, and

$$S(x_1,\ldots,x_n) = \lim_{m_o \searrow 0} S_{m_o}(x_1,\ldots,x_n) \qquad \text{(IV.49)}$$

exists (in the distributional sense). The EGF's S are the moments of a physical measure, i.e.

$$S(h_1,\ldots,h_n) = \langle \prod_{i=1}^{n} \phi(h_i)\rangle \qquad \text{(IV.50)}$$

Proof: For $m_o > 0$ it is relatively straightforward to show that the s-G theory is a canonical quantum field theory if formulated in terms of the field ϕ. The bounds (IV.48) are a consequence of the canonical commutation relations and the positivity of the mass gap, $m(\lambda,\alpha,m_o) \geq m_* > 0$, by a general, axiomatic result of Herbst [H]. Convergence (IV.49) follows from (IV.46) and (IV.48), and (IV.50) is a simple consequence of (IV.48) and (IV.49).

Q.E.D.

Remarks: Using integration by parts on function space, see e.g. [GJS1, Fr1] we may replace the field ϕ by the fields $:\cos(\varepsilon\phi+\theta):_1$ and $:\sin(\varepsilon\phi+\theta):_1$. As a consequence of Theorem IV.17 plus technicalities and the scaling properties of the two dimensional classical, neutral Coulomb gas we obtain

Corollary IV.18: [Fr3]

Let $u_\lambda(x-y) \equiv \langle :\sin(\varepsilon\phi+\theta):_1(x)\ :\sin(\varepsilon\phi+\theta):_1(y)\rangle^T$ and let $d\rho_\lambda(a)$ denote the measure occurring in the Källen-Lehmann spectral representation of u_λ. Then

(a) $\vartheta^{-2\alpha} u_{\vartheta^{2-\alpha}\lambda}(x) = u_\lambda(\vartheta x)$, $(\alpha \equiv \varepsilon^2/4\pi)$

(b) $m(\lambda,\alpha) \equiv m(\lambda,\alpha,0) = G(\alpha)\,\lambda^{\frac{1}{2-\alpha}}$

for some positive function G. All eigenvalues of the mass operator depend on λ like $\lambda^{1/2-\alpha}$.

(c) Spectral function sum rules:

$$\lambda \int \frac{d\,\rho_\lambda(a)}{a} = \langle :\cos(\varepsilon\phi + \theta):_1 (0)\rangle, \text{ and}$$

$$\varepsilon^2\,\lambda^2 \int \frac{d\,\rho_\lambda(a)}{a^2} = 1.$$

Remarks:

(a) follows from scaling, [Fr3]. (b) is a simple consequence of (a), integration by parts on function space and of the FKG inequalities (Theorem IV.13). See [S1] (and [Fr4] for the case of the s-G theory). Finally (c) follows from integration by parts on function space, the Källen-Lehmann representations of $\langle\phi(x)\phi(y)\rangle$ and $u_\lambda(x-y)$ and from the positivity of $m(\lambda,\alpha)$. (See also [Fr3]). The deep results of [Fad, Da] and Corollary IV.18 suggest that certain quantities of the sine-Gordon theory such as the mass spectrum, or e.g. the two-particle scattering amplitude can be calculated explicitly. (Presumably one would have to use a quantized version of [Fad]).

(b) A (complete) proof for the existence of an $m_0 = 0$ limit: By a lengthy argument we can prove that the generating functionals for the EGF's of the fields χ_ϑ satisfy the bounds

$$|\langle e^{\chi_\vartheta(f)}\rangle_{m_0}| \leq \exp\left[\int dx\,\{p(\{\mathrm{Re} f(x)\},\vartheta,\lambda,\alpha,m_0) - p(\lambda,\alpha,m_0)\}\right] \qquad \text{(IV.51)}$$

(see Theorem IV.9, Section IV.2 (c)) which are uniform in $m_0 \varepsilon (0,1]$.

Unfortunately these bounds do not trivially follow from Theorem IV.9, since we have used 0-Dirichlet data (rather than free b.c.) for the construction of the expectation $\langle\cdot\rangle_{m_0}$. Indeed, our proof of the bounds (IV.51) is technically rather complicated: We must make a detour over Theorem 3(a), Section III (see [FrS]), and then use "weak coupling boundary conditions" [GJ4] to construct $\langle \cdot \rangle_{m_0}$. We therefore refer the reader to [Fr3] .

Obviously the bounds (IV.51) suffice to select a subsequence $\{m_k\}_{k=1}^{\infty}$ of bare masses converging to 0 such that

$$\lim_{k \to \infty} \langle \prod_{i=1}^{n} \chi_{\vartheta_i} (h_i) \rangle_{m_k} \equiv \langle \prod_{i=1}^{n} \chi_{\vartheta_i} (h_i) \rangle$$

exists, for all n and $h_1, \ldots, h_n$ in $L^1(\mathbb{R}^2) \cap L^q(\mathbb{R}^2)$ with $q > (1-\alpha)^{-1}$ (as in Theorem IV.9). For a proof, use the bounds (IV.51) and Cantor's diagonal procedure. Of course the limiting expectation $\langle \cdot \rangle$ may depend on the choice of $\{m_k\}_{k=1}^{\infty}$, but in any event the EGF's $\langle \prod_{i=1}^{n} \chi_{\vartheta_i}(h_i) \rangle$ satisfy the Osterwalder- Schrader axioms (E0) - (E3), and our results suffice to construct the soliton-sectors of the s-G theory for $m_0 = 0$, (Theorem 6, Section III). See [Fr3,Fr5] .

V. Massive QED_2 (Program 7, Section I)

In this section we merely explain the formal steps required for a proof of Theorem 7, Section III. The experienced reader will then be able to complete the proof of Theorem 7 as a slightly non-trivial exercise in applying the results of [Fr3,Fr4,FrS] and Section IV.

Recall that the formal Lagrangian of massive QED_2 is given by

$$\mathcal{L}(\psi, A_\mu) = \mathcal{L}_0(\psi) + \mathcal{L}_0(A_\mu) - \Big[g/2 :j^\mu j_\mu: + e\, j^\mu A_\mu + M :\bar{\psi}\psi: + C \Big],$$

where C is an appropriately chosen, scalar counterterm. Using Theorem 4, Section III (which we here take for granted, but see [Col,FrS]) we may rewrite this Lagrangian in terms of A_μ and of a neutral, scalar Bose field ϕ of bare mass 0. One is left with the following interaction Lagrangian:

$$- q\, \varepsilon^{\mu\nu} \partial_\nu \phi A_\mu + M: \cos(\varepsilon\phi + \theta):_1 + C, \qquad \text{(V.1)}$$

where $q = e \frac{\varepsilon}{2\pi}$, $j^\mu = - \frac{\varepsilon}{2\pi} \cdot \varepsilon^{\mu\nu} \partial_\nu \phi, \ldots,$ ε and θ as in Theorem 4, Section III.

For the Euclidean propagator of the vector field A_μ we choose

$$\tilde{C}_{\mu\nu}(k) = \frac{g_{\mu\nu}}{k^2 + m^2}, \quad - g_{00} = g_{11} = 1, \; g_{01} = g_{10} = 0 \qquad \text{(V.2)}$$

(The terms $k_\mu k_\nu (m^2(k^2 - m^2))^{-1}$ in the relativistic propagator can be ignored, since A_μ is coupled to a current of the form $\varepsilon^{\mu\nu}\partial_\nu \phi$. This is obviously special about two space-time dimensions). Note that, for transverse $(f^o, f^1) = (\text{grad}g,\ i\partial_t g)$ with g a test function with support in $\{x = (\vec{x},t)/t>0\}$ and with $f^\mu_\vartheta(\vec{x},t) = f^\mu(\vec{x}, -t)$, etc.

$$\langle f^\mu_\vartheta, C_{\mu\nu} f^\nu \rangle_{L^2} = m^2 \langle g_\vartheta, (-\Delta + m^2)^{-1} g\rangle_{L^2} \geq 0,$$

(by the Osterwalder-Schrader positivity of the free, scalar field). For $M = 0$ the theory obtained from the formal Lagrangian (V.1) and the propagator (V.2) and massive QED_2 can be solved explicitly and are equivalent, and the case where $M > 0$ is covered by Theorem 4. The formal Euclidean action obtained from the Lagrangian (V.1) is

$$U(\phi, A_\mu) = - q \int dx \, \{ [(\text{grad}\phi) A_o + (i\partial_t \phi) A_1](x) + M : \cos(\varepsilon\phi + \theta) :_1 (x) + C\} \qquad \text{(V.3)}$$

Let $(f^o, f^1) = (\text{grad}g,\ i\partial_t g)$, and $g \in \mathscr{S}$; ((f^o, f^1) is <u>transverse</u> in the Euclidean sense). We propose to calculate the generating functional

$$Z^{-1} \langle e^{-U(\phi, A_\mu)} e^{iA_\mu(f^\mu)} \rangle_A ,$$

where $\langle\cdot\rangle_A$ denotes functional integration over A_μ and $Z = \langle \exp - U(\phi, A_\mu)\rangle$ is the total partition function. After properly choosing the counterterm C we find

$$Z^{-1} \langle e^{-U(\phi, A_\mu)} e^{iA_\mu(f^\mu)} \rangle_A$$

$$= Z^{-1} e^{-\int \{M : \cos(\varepsilon\phi + \theta) :_1 (x) + \frac{q^2}{2} :\phi^2: (x)\} dx}$$

$$\times\, e^{\frac{m^2 q^2}{2} \int dx\, dy\, \phi(x)\, C_m(x-y)\, \phi(y)}$$

$$\times\, e^{-iq\,\{\phi(g) - m^2\, \phi(C_{m^*}\, g)\}} \tag{V.4}$$

$$\times\, e^{\frac{1}{2}\int dx g(x)^2 - \frac{m^2}{2}\int dx\, dy\, g(x)\, C_m(x-y)\, g(y)}$$

The first two exponentials on the r.h.s. of (V.4) can be interpreted as an effective Euclidean action for the field ϕ. This observation permits us to reduce the proof of stability of massive QED_2 as defined by (V.1) - (V.3) by conditioning to the stability of the s-G theory (inequality (IV.22), Theorem IV.8).

We can absorb a term proportional to $\exp\left[-q/2 \int dx{:}\phi^2{:}(x)\right]$ in the free measure $d\mu_{C_0}(\phi)$ and assume hence forth that ϕ is a scalar field of bare mass q with free measure $d\mu_{C_q}(\phi)$. Next, let χ be another, independent, free, scalar Bose field of bare mass m, and consider the effective action

$$U_{eff.}(\phi,\chi) = \int dx\, \{mq\, \phi(x)\chi(x) + M{:}\cos(\varepsilon\phi + \theta){:}_1(x)\} \tag{V.5}$$

Let $\langle\cdot\rangle_\chi$ denote functional integration over χ, and $Z' = \langle \exp - U_{eff.}(\phi,\chi)\rangle$ the partition function. Then one may easily verify that

$$Z^{-1} \langle e^{-U(\phi,A_\mu)}\, e^{iA_\mu(f^\mu)}\rangle_A$$

$$= e^{\frac{1}{2}\int dx g(x)^2}\, e^{iq\phi(g)}\, (Z')^{-1} \langle e^{-U_{eff.}(\phi,\chi) - im\,\chi(g)}\rangle_\chi \tag{V.6}$$

Therefore the generating functional of the EGF's of massive QED_2 on the charge 0-sector is formally given by

$$Z\,(w, v_\mu; f^\mu) = e^{\frac{1}{2}\int dx g(x)^2}\, (Z')^{-1} \int_{\mathscr{S}' \times \mathscr{S}'} d\mu_{C_q}(\phi)\, d\mu_{C_m}(\chi)$$

$$\times\, e^{i:\cos(\varepsilon\phi+\theta):_1(w)}\, e^{i\phi(\mathrm{grad}\, v_o - \partial_t v_1)}\, e^{iq\phi(g)}$$

$$\times\, e^{-im\chi(g)}\, e^{-U_{\mathrm{eff.}}(\phi,\chi)} \qquad \text{(V.7)}$$

The factor $e^{\frac{1}{2}\int dx\ g(x)^2}$ could be omitted. It is a contact term and therefore irrelevant for the reconstruction of Wightman distributions and quantum fields from the functional $\mathcal{Z}$.

We note that the measure

$$(Z')^{-1}\, e^{-U_{\mathrm{eff.}}(\phi,\chi)}\, d\mu_{C_q}(\phi)\, d\mu_{C_m}(\chi)$$

is <u>ferromagnetic</u>:

Let $U_1(\phi) = M \int dx\ :\cos(\varepsilon\phi + \theta):_1(x)$, and

$$Z_1 = \langle e^{-U_1(\phi)} \rangle_{C_q}$$

Then $Z_1^{-1}\, e^{-U_1(\phi)}\, d\mu_{C_q}(\phi)\ d\mu_{C_m}(\chi)$ is, formally, a ferromagnetic measure, by the results of Section IV.3, (a) - (c). By the $\chi \to -\chi$ symmetry of $d\mu_{C_m}(\chi)$ we can assume that $q < 0$. Then the term

$$(Z')^{-1}\, Z_1\, e^{-mq\int dx\phi(x)\chi(x)}$$

introduces a <u>ferromagnetic nearest neighbor coupling</u> between (two independent, generalized, ferromagnetic Ising models with) measures $Z_1^{-1}\, e^{-U_1(\phi)}\, d\mu_{C_q}(\phi)$ and $d\mu_{C_m}(\chi)$, which proves our assertion. Therefore we have correlation inequalities, in particular the <u>Griffiths inequalities</u> which are one essential ingredient for taking the infinite volume limit à la Griffiths and Nelson.

These observations plus cos $\varepsilon\phi$- and grad ϕ-bounds (because of (V.4) no χ-bounds are needed) combined with the results of Section IV and of [Fr3] suffice to give a rigorous proof of Theorem 7.

Remarks: The verification of axioms (E0) - (E3) follows from the facts that for $(f^o, f^1) = (\text{grad}g, i\partial_t g)$ the functional $\mathcal{Z}(w, v_\mu, f^\mu)$ obviously satisfies Osterwalder-Schrader positivity in the form of [Fr1, Fr2] and is Euclidean invariant; (E0) follows from cos $\varepsilon\phi$- and grad ϕ-bounds. The construction of the charged sectors proceeds as in the sine-Gordon theory ($m_0 = 0$). The essential ingredient for the construction of the charged sectors (of massive QED) is the invariance of the r.h.s. of (V.4) under shifts $\phi \rightarrow \phi + \frac{2\pi n}{\varepsilon}$, for $g = 0$; see [Fr3, Fr5].

We are aware of the fact that our presentation of massive QED_2 is rather formal (or "non-standard") and sketchy. Yet, we hope that the reader may grasp the main ideas by reading this section carefully. Detailed proofs of the assertions of this section and the construction of the charged sectors will presumably be presented elsewhere.

VI. Some concluding remarks

For reasons of space we were forced to omit all proofs of Theorems 2-6, Section III, and detailed illustrations of Programs 2-6, Section I. We have decided to emphasize Theorems 1 (existence) and 7 (massive QED_2), since the other results have been discussed more extensively in the available literature, or follow easily from known results. It would have been in the spirit of this school to emphasize Theorem 3 (convergence of perturbation series of EGF's, perturbation theory asymptotic to scattering amplitudes). Since a more or less self-contained proof of Theorem 3 would easily cover one hundred pages we could not present it here.

Detailed investigations of Programm 6 (super-selection sectors, quantum solitions) can be found in [Ch, DHR, Fr5] and refs. given there. For a proof of Theorem 6, see [Fr3, Fr5]. A brief summary of these matters is contained in the authors contribution to the 1975 Colloquium on Mathematical Methods of Quantum Field Theory in Marseille.

Concerning Programm 5, we refer the reader to the exciting paper of Glimm, Jaffe and Spencer [GJS2] proving a phase transition for $\phi^4{}_2$

Acknowledgements: I have greatly profited from numerous interesting, partly crucial discussions with S. Coleman, E. Lieb, E. Seiler and A. Wightman. I wish to thank them all.

NON-PERTURBATIVE RENORMALIZATION IN THE YUKAWA MODEL IN TWO DIMENSIONS

E. Seiler

Institute for Advanced Study, Princeton
and
Max-Planck-Institut für Physik und Astrophysik, München†

I. INTRODUCTION

For any genuine renormalizer the two-dimensional Yukawa (Y_2) model must look very trivial, since its only primitively divergent diagram is  , leading to an infinite boson mass renormalization. But if we want to go beyond perturbation theoretic methods, we have to be very modest, at least for the time being. After two-dimensional polynomially self-interacting fields (which have no ultraviolet divergences) have been treated successfully by the constructive field theorists (to mention just a few significant names: Glimm, Jaffe, Spencer, Nelson, Guerra, Rosen, and Simon - consult the 1973 Erice notes for more information [VW]) the next model in line is Y_2 , the "first" one to require some real renormalization. The next step after that, ϕ_3^4 , which is even more singular, will be discussed by J. Feldman. The major complication in Y_2 , and at the same time the reason why it can be handled, as we shall see later, is the appearance of fermions.

Glimm (1967)[Gl 1] was the first one to study rigorously the Y_2 Hamiltonian (with a spatial cutoff); he showed that the renormalized Hamiltonian is bounded from below. This Hamiltonian approach proved to be very fruitful: Based on it, Glimm and Jaffe constructed the infinite volume theory in the sense of local observables [GJ 2], proving all the Haag-Kastler axioms except Lorentz invariance; this was later supplied by McBryan and Park [McP]. Dimock [Di 1] proved that renormalized field equations hold; Schrader [Sch] proved a linear lower bound on the Hamiltonian and showed that the infinite volume states that can be constructed by a compactness argument are "locally Fock".

G. Velo and A. S. Wightman (eds.), Renormalization Theory, 415–433.

Most of these bounds were also obtained by Federbush and Brydges [Br,BrF] by a different ("semi-Euclidean") method. What was still missing was the right physical representation of the infinite volume field algebra, generated by a Lorentz invariant state (the physical vacuum).

In these lectures I will describe a different approach that hopefully will lead some day to the solution of that problem. Actually this approach is very old dating back to the heroic times of quantum field theory. The idea is to construct the (Euclidean) Green's functions (also known as Schwinger functions) for the theory starting with a heuristic formula proposed by Matthews and Salam [MaS]*). From the work of Osterwalder and Schrader [Os 1] we have learned that constructing these functions and verifying a number of properties (the Osterwalder-Schrader axioms) is actually sufficient to guarantee the existence of a corresponding Wightman theory by analytic continuation. So far this approach has not yet been carried through completely, but the chances are pretty good, at least for the case of weak coupling, that this can and will be done soon.

In these lectures I will try to emphasize the point of view of renormalization theory; that is, both the similarities with and the differences from perturbative renormalization.

II. RENORMALIZATION IN A FINITE VOLUME

1. The formulae of Matthews and Salam [MaS]

Matthews and Salam looked at the expressions for the Green's functions in terms of a "sum over histories" as given by Feynman [Fey]:

$$G(x_1,\dots,x_n;y_1,\dots,y_n;z_1,\dots,z_m) = \tag{1}$$
$$= 1/Z \int \mathcal{D}\psi\mathcal{D}\psi^{\dagger}\mathcal{D}\phi\; e^{-i\int\mathcal{L}(\phi,\psi,\bar{\psi})dx} \prod_i \psi_i(x_i)\psi_i^{\dagger}(y_i) \prod_k \phi_k(z_k)$$

where Z is a normalization factor: $Z = \int\mathcal{D}\psi\mathcal{D}\bar{\psi}\mathcal{D}\phi\; e^{-i\int\mathcal{L}dx}$).

If one imagines doing the ψ-"integration" first, ϕ plays the role of an external field; the remaining integration just "averages" over this external field. Now (for $\mathcal{L}_I = \lambda\bar{\psi}\Gamma\psi\phi$) $\mathcal{L}$ is bilinear in $\psi,\bar{\psi}$; the ψ, $\psi^{\dagger}$-integrations can be "done" and

*) That this could be a useful starting point has been emphasized, for instance, by Guerra, Osterwalder, and Wightman on different occasions.

yield

$$G = 1/Z \int \mathcal{D}\phi e^{-i \int \mathcal{L}_0(\phi)dx} \det(1-i\lambda K(\phi))$$
$$\det((S_F'(x_i,y_k;\phi))^n_{i,k=1} \prod_\ell \phi(z_\ell) \tag{2}$$

where K is an integral operator with kernel

$$S_F(x-y)\,\Gamma\,\phi(y) \qquad (\Gamma = 1 \text{ or } i\gamma_5)\ ; \tag{3}$$

the first det denotes the corresponding Fredholm determinant; S_F is the free fermion two-point function and S_F' the fermion two-point function in the external field ϕ. The formulae for the external field case were also given by Schwinger [Sw]. So far, everything is purely formal. But motivated by the success of Euclidean methods in $P(\phi)_2$ it is tempting to "Euclideanize" formula (2) and give it some precise mathematical meaning. One arrives at the following type of expression for the Schwinger functions

$$S_g(x_1,\ldots,x_n;y_1,\ldots,y_n;z_1,\ldots,z_m) =$$
$$= 1/Z \int d\mu_0(\phi)\det_{ren}(1-\lambda K(\phi))$$
$$\det((S_F'(x_i,y_k;\phi))^n_{i,k=1} \prod_\ell \phi(z_\ell) \tag{4}$$

This requires some explanation:

a) $d\mu_0(\phi)$ denotes the Gaussian measure for the free Euclidean Bose field which can be characterized in all its essential aspects by its Fourier transform

$$\int d\mu_0(\phi)\ e^{i\phi(f)} = e^{-1/2(f,(-\Delta+\mu^2)^{-1}f)} \tag{5}$$

According to Minlos' theorem [GV,Mi] it can be realized on the Schwartz space $\mathcal{S}'$.

b) $\det_{ren}$ denotes some "renormalized" version of the Fredholm determinant which will be described below.

c) K has the kernel

$$S_F(x-y)\,\Gamma\phi(y)g(y) \tag{6}$$

where S_F now denotes the free Euclidean fermion two-point function, g is a space-time cutoff.

It should be remarked that an ultraviolet cutoff version of (4) can be "derived", either by using Euclidean Fermi fields and their connection with the Hamiltonian (see Osterwalder and Schrader [OS 2]), or more directly by computing explicitly expressions like $(\Omega_0, e^{-Ht}\Omega_0)$ (Ω_0 is the Fock vacuum), which turns out to be equal to the Euclidean Z with a space-time cutoff g that is of the form $\chi_{[0,T]}(t)\ \chi_\ell(x)$, or like

$$(\Omega_0, e^{-t_0 H}\psi^{\#}\ e^{-Ht_1} \dots \psi^{\#} e^{-Ht_N}\ \Omega_0) \quad (\psi^{\#} \text{ either } \psi, \bar{\psi}, \text{ or } \phi)$$

which produce the numerator in (4). For $T \to \infty$ we obtain what we are interested in:

$$(\Omega_\ell,\ e^{-t_0 H_\ell}\ \psi^{\#} \quad \dots \quad \psi^{\#}\ e^{-H_\ell t_N}\ \Omega_\ell) \tag{7}$$

where Ω_ℓ is the ground state for H_ℓ (this requires $(\Omega_0,\Omega_\ell) \neq 0$). Compare [GJ 1],[SeS 3],[Mc 3]. This derivation is described in more detail in [SeS 3]. But I want to outline the proof for the simplest case, since it is closely related to a standard field theoretic method (the "Dyson expansion"): If we write $H = H_0+V$ and assume enough cutoffs to have $\mathrm{Ran}(e^{-tH_0}) \subset D(V)$ for all $t > 0$ and $\int_0^1 ||Ve^{-tH_0}||dt < \infty$, then the so-called Phillips (iterated Duhamel) expansion holds:

$$e^{-Ht} = \sum_{n=0}^{\infty} (-\lambda)^n \int_{0 \le t_1 \le \dots \le t_n \le t} dt_1 \dots dt_n\ e^{-t_1 H_0}\ Ve^{-(t_2-t_1)H_0} \dots e^{-(t_n-t_{n-1})H_0} Ve^{-(t-t_n)H_0} \tag{8}$$

If we introduce $V(it) = e^{-H_0 t}\ Ve^{H_0 t}$ this can be written as

$$e^{-Ht} = T \exp[-\lambda \int_0^t V(it)dt]\ e^{-H_0 t}$$

(where T is a time ordering symbol). If we compute $(\Omega_0, e^{-Ht}\Omega_0)$ using (8), we obtain just the classical Fredholm expansion of $\det(1-\lambda K)$ integrated over $d\mu_0$.

2. The renormalized determinant

That the determinant of $1-\lambda K$ is divergent can be seen by using the well-known formula

$$\det(1-\lambda K) = \exp \mathrm{Tr} \ln(1-\lambda K) \tag{9}$$

and expanding in λ. $\mathrm{Tr}K^n$ corresponds to a closed fermion loop with n corners and the bose field "sitting at the corners".

TrK and TrK^2 are divergent. Now TrK disappears if we simply Wick-order the fermions in the basic interaction. The expression we obtain by deleting all terms TrK is called $\det_2$; it can be defined directly by

$$\det{}_2(1+A) = \det((1+A)e^{-A}) \qquad (10)$$

(this $\det_2$ has a long history involving, among many others, Poincaré, Hilbert, and Schwinger*)). Now we could just delete Tr K^2 as well, but this arbitrary procedure would not correspond to a renormalization! From renormalization theory we know that we should have a counterterm of the form $\delta\mu^2\phi^2$; choosing $\delta\mu^2$ cutoff-dependent in the right way we can indeed make Tr $K^2 - \delta\mu^2\phi^2$ finite, as long as we assume that ϕ is nice enough. That is how far standard renormalization theory guides us and that is sufficient for the external field case (as has been noted e.g. by Schwinger [Sw], cf. also Matthews and Salam [MaS].

But obviously there is another problem in the interacting theory: ϕ is in general a distribution and it is not even clear a priori whether our kernel defines any reasonable operator. Now people have studied the carrier of the measure $d\mu_0$ in detail (see, e.g. Cannon [Ca], or Colella and Lanford [CoL]) and they found that the generic ϕ is not as bad as it could be (it is "almost" a function almost certainly), so there is some hope.

Actually, the first statements of that kind of which I know can be found in the very interesting discussion at the end of the second paper by Matthews and Salam [MaS]; they were, however, too optimistic. The problem is that a Gaussian measure in general is not supported where its defining quadratic form is finite. For instance, if we consider the measure $d\nu = \prod_{i=1}^{\infty} 1/\sqrt{\pi}\, e^{-1/2\, x_i^2}\, dx_i$ on $\mathbb{R}^\infty$, the subset $\ell^2 = \{x \in \mathbb{R}^\infty \mid \sum_{i=1}^{\infty} x_i^2 < \infty\}$ has measure zero.

First we have to set up a Hilbert space on which K is supposed to act, we choose $\mathcal{H} = \mathbb{C}^2 \otimes \mathcal{H}_{\frac{1}{2}}$ where $\mathcal{H}_{\frac{1}{2}}$ consists of (real) functions on $\mathbb{R}^2$ such that $\int |\tilde{f}(p)|^2 \sqrt[2]{p^2+m}\, d^2p < \infty$ ($\tilde{f}$ Fourier transform).

*) See, e.g., B. Simon's recent "Notes on infinite determinants" of Hilbert space operators [Si 2].

Next we define Banach spaces of operator-valued random variables

$$C_{p;q} = \{A: \mathcal{S}' \to \mathcal{L}(\mathcal{H}) \mid \int d\mu_0 (\mathrm{Tr}(A^*A)^{\frac{p}{2}})^{\frac{q}{p}} < \infty\} \qquad (11)$$

with the norm

$$||A||_{p;q} = (\int d\mu_0 (\mathrm{Tr}(A^*A)^{\frac{p}{2}})^{\frac{q}{p}})^{\frac{1}{q}} \qquad (12)$$

(completeness of these spaces follows by standard arguments, cf. Halmos [Ha]).

Now, if we introduce an ultraviolet cutoff κ in the Bose field ϕ, it is easy to see that then $K_\kappa \in C_{p;q}$ for all $4 \le p \le \infty$; $1 \le q < \infty$, because ϕ_κ is a C^∞ function. Furthermore, if we look at $||K_\kappa - K_{\kappa'}||_{4;4}$, we realize that this leads to differences of Feynman diagrams like

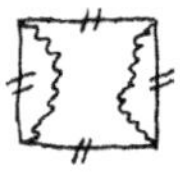 or 

with different cutoffs on the boson lines (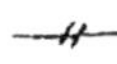 denotes the Fourier transform of $1/\sqrt{p^2+m^2}$) and goes therefore to 0 as $\kappa,\kappa' \to \infty$. So K can be defined as $C_{4;4}$ limit (a refinement of this argument shows that 4 can be replaced by $2+\varepsilon$ for any $\varepsilon > 0$, see [SeS 1]).

It is not difficult to see now that

$$\det_3(1+\lambda K) = \det((1+\lambda K)e^{-K+\frac{1}{2}K^2}) \qquad (13)$$

can be defined a.e. $d\mu_0$ and is finite a.e. $d\mu_0$ (see[Se]). Since we want to use only local counterterms, we define

$$\det_{ren}(1+ K) = \det(1+\lambda K)e^{\lambda^2 \mathrm{Tr}:K^2: - \delta\mu^2 :\phi^2:} \qquad (14)$$

Here : : denotes Wick-ordering of the bosons (that eliminates the divergent vacuum graph ) and the last factor is to be interpreted as a $L^2(d\mu_0)$ limit (which is easily seen to exist) of similar expressions with cutoffs σ on the fermions and κ on the bosons.

If we expand (13) in a power series in λ, insert it in our Euclidean Matthews-Salam formula and integrate term by term, we just get the renormalized perturbation expansion. The counterterm $\delta\mu^2:\phi^2:$ in (13) contains in closed form all the subtractions that are necessary whenever  appears as a subgraph.

3. Existence of the finite volume Schwinger functions.

So far we have succeeded in making sense of the integrand in (4) (S_F' has to be interpreted essentially as the resolvent kernel to K - up to a factor S_F). The next question is obvious: Is the renormalized determinant integrable?

The idea for that proof is taken over from Nelson's work on $P(\phi)_2$ [N 1, 3]: We introduce a cutoff κ in and show that the corresponding integrand*)

$$|I_\kappa| \le v_\kappa \le e^{c\,\ell n \kappa} \tag{15}$$

($I_\kappa = \det_3(1 - \lambda K_\kappa)\exp(-\frac{\lambda^2}{4}\operatorname{Tr}:(K_\kappa + K_\kappa^*)^2:)$ whereas the measure of the sets

$$\{\phi \in \mathscr{S}' : \ell n|I| - \ell n\, v_\kappa \ge 1\} \text{ is } O(\kappa^{-\epsilon})$$

for some $\epsilon > 0$. (15) follows from the following simple inequality:

$$|\det_2(1+A)| \le e^{\frac{1}{2}\operatorname{Tr}A^*A} \tag{16}$$

(Schwinger [Sw] gives a one line proof:

$$|\det_2(1+A)|^2 = \det(1 + A + A^* + AA^*)$$

$$e^{-\operatorname{Tr}(A+A^*)} \le e^{\operatorname{Tr}AA^*}$$

because for selfadjoint $B \in \mathcal{C}_1$ $\det(1+B) \le e^{\operatorname{Tr}B}$.) Of course $\operatorname{Tr}K_\kappa^* K_\kappa$ is infinite, but $\operatorname{Tr}(K_\kappa^* K_\kappa + K_\kappa^2)$ is finite and $O(\ell n \kappa)$. The estimates on the measure require some technical inequalities on the quotients of determinants (for details see [Se]) but finally everything boils down to convergent Feynman graphs in which at least one line has momentum higher than κ.

Alternatively we can write

$$I = \sum_{n=1}^{\infty} (I_{\kappa_n} - I_{\kappa_{n-1}})$$

and estimate each term in this sum by some determinant inequality which leads, after integration over $d\mu_0$, to a bound of the form

*) After we have separated a Gaussian part using Hölder's inequality.

$$\kappa_n^{-\epsilon} e^{\lambda^2 c \ell n(1+\frac{\kappa_n}{\mu})} .$$

For small λ and a suitable sequence κ_n the sum over n converges (O. A. McBryan, private communication; he also has a method to remove the restriction on small λ). Some people might worry about possible singularities of $S_F'(x, y;\phi)$ which is essentially the kernel of $(1-\lambda K)^{-1}$. The point is that possible poles get just cancelled by corresponding zeros of $\det_{ren}$; actually the numerator ZS can be controlled by essentially the same estimates that control Z (see [Se 1]). Furthermore it turns out that $\det_{ren}$ is always ≥ 0, a remarkable and somewhat mysterious result - it expresses the so-called Symanzik-Nelson positivity for the effective Boson interaction and leads to the obviously important result $Z > 0$ [Se 1].

Remark. J. Bellissard has informed me that he has obtained results of the last two sections independently (unpublished).

4. Remarks on finite mass renormalizations.

In the preceding discussion we assumed a certain fixed regularization of the divergent graph  (e.g. subtraction at zero momentum). This is not necessary, or stated differently, we can add an arbitrary term $\pm M^2\phi^2$ to the interaction density - as has been shown by McBryan [Mc 1] and Seiler and Simon [SeS 1]. The method is to treat the part of the interaction coming from low Fermion momenta separately; the high momentum part then has a negative quadratic term in ϕ that gets more and more negative as we restrict ourselves to higher and higher momenta: therefore it can balance any term $M^2{:}\phi^2{:}$.

III. THE PROBLEM OF THE THERMODYNAMIC LIMIT

1. What is the problem?

In perturbative renormalization theory the thermodynamic limit for the Schwinger functions or Green's functions is rather trivial because vacuum graphs, which show, of course, volume divergences, do not appear in the perturbation expansion of Green's functions. If we formally expand the denominator Z in our basic formula (4) we just obtain the sum of all vacuum graphs; but these graphs also appear in the expansion of the numerator and disappear if we divide the two formal power series, as everybody knows.

The lesson is, of course, that we expect numerator and denominator of (4) to diverge, as the cutoff function g tends to 1, and the problem is to exhibit the cancellation of these divergences that works so beautifully in perturbation theory. This problem is of course exactly analogous to the problem of the thermodynamic limit of the correlation functions in statistical mechanics.

In $P(\phi)_2$ and weakly coupled $(\phi^4)_3$, where the analogy to statistical mechanics is even closer, this problem has been solved by methods which either come directly from statistical mechanics (such as correlation inequalities which can be taken over from the Ising model - see Nelson [N 3], Guerra, Rosen and Simon [GRS 2]) or are inspired by it such as the "cluster-expansion" of Glimm, Jaffe and Spencer [GJS 1], which has been successfully extended to the case of ϕ_3^4 by Feldman and Osterwalder [FeO 1, 2] and by Magnen and Sénéor [MS].

The situation in Y_2 is not yet as satisfactory, since the presence of fermions creates additional difficulties; but it is a safe bet that for the weakly coupled theory convergence of the cluster expansion will be established soon.

But there are many things that are known about the volume dependence of the Y_2 theory and I will discuss them briefly.

2. The basic stability bounds.

The first thing that always has to be established in statistical mechanics is that the numerator and denominator of expressions analogous to (4) do not grow worse than exponentially in the volume. This has been established for Y_2 independently by McBryan [Mc 2] and Seiler and Simon [SeS 2] (it could also be derived from Schrader's linear lower bound on H_ℓ [Sch]). I will sketch briefly the ideas behind the second proof. It is inspired by Nelson's proof of the linear lower bound in $P(\phi)_2$.

The crucial input is the following theorem of Guerra, Rosen and Simon which stems from Nelson's "best hypercontractive bounds" for the measure $d\mu_0$ (see [N 2], [N 3]).

Theorem [GRS 2]: Cover $\mathbb{R}^2$ with unit squares labeled by α; let u_α be (measurable) functions of the free Bose field in the cube α ($\alpha \in \mathbb{Z}^2$). Then

$$\int d\mu_0 \left|\prod_\alpha u_\alpha\right| \le \prod_\alpha \left(\int |u_\alpha|^\beta d\mu_0\right)^{\frac{1}{\beta}} \tag{17}$$

with

$$\beta = \frac{4}{1 - e^{-\mu}}.$$

If our $\det_{ren}(1 - \lambda K)$ were of the form $\prod_{\alpha\in\Lambda} u_\alpha$ with each u_α depending only on fields in the square α, a bound $Z \le e^{a|\Lambda|}$ would follow (the integrability proof gives $\det_{ren} \in \bigcap_{1\le p<\infty} L^p$ at once). But things are not quite as simple, so we try the next best thing: we try to prove an inequality of the form $\det_{ren}(1 - \lambda K) \le \prod_{\alpha\in\Lambda} u_\alpha$. This is unfortunately rather technical, so I cannot describe the details here. But the main idea is to split K into a "diagonal part" $D = \sum_{\alpha\in\Lambda} D_\alpha$ with $D_\alpha D_\beta = 0$ for $\alpha \ne \beta$ and a correction term B that links different squares. The point is that the diagonal part does have the factorizing property

$$\det_{ren}(1 - \lambda D) = \prod_{\alpha\in\Lambda} \det_{ren}(1 - \lambda D_\alpha)$$

whereas B is a much nicer operator that does not produce any divergences.*) The two contributions are separated by a determinant inequality which yields

$$\det_{ren}(1 + D + B) \le \prod_{\alpha\in\Lambda} J(D_\alpha) e^{\frac{1}{2}\mathrm{Tr} B^* B} \tag{18}$$

The second factor can be bounded by a product v_α because of the exponential falloff of the propagators which link different squares. We have simplified the discussion, because in general we also have to separate out a third piece C, a low momentum contribution, to allow for an arbitrary finite mass renormalization.

The numerator ZS can be handled by the same methods if we discuss a generating functional and then use Cauchy estimates

*) It turns out to be Hilbert-Schmidt a.e. ($d\mu_0$).

to bound its derivatives.*) The bound so obtained looks as follows:

$$|ZS(f_1,\cdots,f_k;g_1,\cdots,g_k;h_1,\cdots,h_n)| \le \le C_1^{|\Lambda|} C_2^{n+k} \prod_\alpha \sqrt{n_\alpha !} \prod_{r=1}^{n} \|h_r\|_{-1} \prod_{i=1}^{k} \|f_i\|_{-\frac{1}{2}} \|g_i\|_{-\frac{1}{2}} \tag{19}$$

where n_α is the number of h_r with support in the square α (assume for simplicity that all h_r are supported in one of the unit squares).

The bound (19) is sufficient to recover Schrader's linear lower bound for the Hamiltonian if we use the connection between the Hamiltonian picture and the Matthews-Salam picture established e.g. in [SeS 3]; there it is shown for instance that

$$\int d\mu_0 \det_{ren}(1 - \lambda K_{\ell T}) = (\Omega_0, e^{-H_\ell T}\Omega_0) \tag{20}$$

(with a slight finite change in the counterterms [see below]), and more generally for a total set of vectors in Fock**) space (so-called Jost states)

$$(\psi, e^{-H_\ell T}\psi) = ZS_{\ell T}(\underline{h};\underline{f},\underline{g})$$

with suitable test functions $\underline{h}$, $\underline{f}$, $\underline{g}$ determined by ψ.

The direct meaning of the bounds $e^{a|\Lambda|}$ is therefore a bound on the vacuum energy density; by analogy with statistical mechanics the bound $\frac{1}{|\Lambda|}\ell n\, Z_\Lambda \le a$ is called a bound on the "pressure."

*) This leads just to a slight modification of K and an additional factor $\exp\phi(f)$ - see below.

**) That H_ℓ can be constructed in Fock space is well known from the work of Glimm and Jaffe [GJ 1]; it also has been shown by Euclidean techniques in [SeS 3].

3. Local scalar counterterms

That the counterterms can be chosen as local operators is one basic result of renormalization theory. In this connection normally nobody worries about c-number counterterms because a) a c-number is always "local" and b) all c-numbers drop out of the expressions for the Green's functions.

So what is our point here? Of course it is still true that c-numbers cancel in numerator and denominator of (4). But we renormalized so that both are finite and for this we employed a scalar counterterm corresponding to the Feynman-graph  (this was hidden in our prescription to Wick order $\mathrm{Tr}K^2$ in eq. (14). Note that at each end of  there is a space-time cut-off function g so that the whole term is of the form

$$\int F(x-y)g(x)g(y)dxdy.$$

This is obviously nonlocal in g in the sense that it is not of the form

$$c\int g(x)^2 dx.$$

But of course we learn from renormalization theory that we can make  finite by subtracting a local quantity, if we consider g(x) as an additional (classical) field (that is, we interpret  as ). In the simple case at hand this can be seen directly without any difficulty.

The replacement of  by something linear in T is actually necessary in order to make sense of eq. (20) (with a time-independent Hamiltonian). Using the good local counterterm in addition preserves Euclidean covariance of Z which leads to the so-called "Nelson symmetry"

$$(\Omega_0, e^{-H_\ell T}\Omega_0) = (\Omega_0, e^{-\ell H_T}\Omega_0) \tag{21}$$

and its far-reaching consequences (see below).

The method of local scalar counterterms is also very useful in $(\phi^4)_3$ (as has been demonstrated by Seiler and Simon [SeS 3]), where, however, the situation is more complicated because of the appearance of divergent surface terms. I will not discuss this subject any further.

Quite generally, it can be said that the local scalar counter-terms are "the right ones" in constructive field theory.

4. Thermodynamic limit for the "pressure" and vacuum energy density (Guerra's theorem).

Although the existence of the thermodynamic limit for the Schwinger functions has not yet been established (as I said before), it can be proven that

$$\lim_{\ell, t \to \infty} \frac{1}{\ell t} \ln Z_{\ell t} \equiv a_\infty$$

exists and is equal to the vacuum energy density. For the $P(\phi)_2$ case this has been first proven by Guerra [Gu] in an ingenious way; we follow the somewhat different method of the later paper by Guerra, Rosen and Simon [GRS 2] with some necessary modifications.

To see the existence of the limit is easy (after [GRS 2]):

$$a_{\ell t} \equiv \frac{1}{\ell t} Z_{\ell t} = \frac{1}{\ell t} \ln(\Omega_0, e^{-H_\ell t}\Omega_0) = \tag{22}$$

$$= \frac{1}{\ell t} \ln(\Omega_0, e^{-H_t \ell}\Omega_0)$$

by (20) and (21). By the spectral theorem and Hölder's inequality or otherwise just by use of the Schwarz inequality and the bounds of the previous section, it follows that $\ln Z_{\ell t}$ is convex in ℓ and t separately; since $Z_{0t} = Z_{\ell 0} = 1$, it follows that $a_{\ell t}$ is increasing in ℓ and t; since it is bounded it has a (unique) limit.

Obviously

$$\lim_{t \to \infty} \frac{1}{t} \ln(\Omega_0, e^{-H_\ell t}\Omega_0) \equiv \tilde{E}_\ell$$

is the lowest energy contained in Ω_0. But is it really the ground state energy? The answer is yes; Ω_0 is not orthogonal to Ω_ℓ, the ground state of H_ℓ (that H_ℓ has a ground state is a result of

Glimm and Jaffe [GJ 1]). The proof uses Euclidean covariance (Nelson's symmetry) and physical (Osterwalder-Schrader) positivity as well as a density argument essentially identical to the Reeh-Schlieder theorem.

Consider the states created from the bare vacuum by applying field operators with imaginary time argument and space argument $|x| \geq \ell$ (so-called Jost states; cf. Jost's book [J]; the times have to be ordered). These states form a total set by the classical Reeh-Schlieder argument. So let η be such a vector with $(\eta, \Omega_\ell) \neq 0$. Then

$$(\eta, e^{-H_\ell t}\eta) = (\tilde{\eta}, e^{-H_t \ell}\Omega_0)$$

where $\tilde{\eta}$ is another Jost state, by Euclidean covariance (a 90° rotation). By the Schwarz inequality the right hand side is

$$\leq (\tilde{\eta}, e^{-H_t \ell}\tilde{\eta})^{\frac{1}{2}}(\Omega_0, e^{-H_t \ell}\Omega_0)^{\frac{1}{2}}$$

which in turn can be interpreted as

$$(\tilde{\tilde{\eta}}, e^{-H_\ell t}\tilde{\tilde{\eta}})^{\frac{1}{2}}(\Omega_0, e^{-H_\ell t}\Omega_0)^{\frac{1}{2}}$$

and is bounded by

$$\|\tilde{\tilde{\eta}}\| e^{-\frac{1}{2}E_\ell t} e^{-\frac{1}{2}\tilde{E}_\ell t}.$$

So we get

$$\lim_{t\to\infty} \frac{1}{t} \ln(\eta, e^{-H_\ell t}\eta) = -E_\ell \leq -\tfrac{1}{2}E_\ell - \tfrac{1}{2}\tilde{E}_\ell$$

or

$$E_\ell = \tilde{E}_\ell$$

This proof might be easier to understand pictorially:

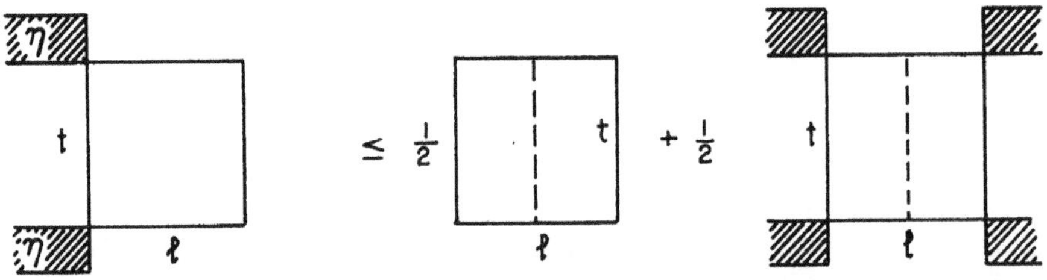

(The empty boxes represent the interaction, the shaded boxes Jost states; the diagrams are meant to include taking logarithms).

The results of this section have been obtained independently by McBryan [Mc 3] and Seiler and Simon [SeS 3].

5. Bounds on the normalized Schwinger functions.

We do not know yet whether the Schwinger functions (4) have a limit as $g \to 1$, but at least we know they are bounded uniformly, if we first let t and then ℓ go to infinity. So by picking a subsequence we can construct a limit; but unfortunately we do not know whether it is Euclidean invariant. In that respect the situation is somewhat similar to the one in Schrader's paper [Sch] based on the Hamiltonian approach.

The proof of these bounds (which are known as Fröhlich bounds) again uses critically Nelson's symmetry (21), so the use of the "right" counterterm is relevant.

We again deduce the bounds for the Schwinger functions from bounds on the generating functional defined as follows: Let $f \in \mathcal{D}(\Delta)$, $q = \sum_{n=1}^{N} g_n \otimes h_n$ ($g_n, h_n \in \mathcal{D}(\Delta)$) ($\Delta$ a unit square), then

$$F_g(f, q) = \sum_{n, m=0}^{\infty} \frac{1}{n!} \frac{1}{m!} \int S_g(x_1, \cdots, x_n; y_1, \cdots, y_n; z_1, \cdots, z_m)$$

$$f(z_1) \cdots f(z_n) q(x_1, y_1) \cdots q(x_m, y_m) d\underline{x} d\underline{y} d\underline{z} \tag{23}$$

The usefulness of generating functionals in terms of a bilinear source q in the case of a charged field has been emphasized by my teacher H. Mitter [Mr]. In terms of Euclidean fields ϕ, $\psi^{(1)}$, $\psi^{(2)}$ (see [OS 2])

$$F_{\ell+a,\, t+b}(f, q) = \frac{1}{Z_{\ell+a,\, t+b}} \int d\mu_0 \langle e^{-\lambda V_g + \int \psi^{(1)}(y) \psi^{(2)}(z) q(y, z) dy dz} \rangle e^{\phi(f)} \tag{24}$$

($g(x) = \chi_{\ell+a,\, t+b}(x)$) or in the Matthews-Salam framework

$$F_{\ell+a,\, t+b}(f, q) = \frac{1}{Z_{\ell+a,\, t+b}} \int d\mu_0 \det{}_{ren}(1 - \lambda K - C_q) e^{\phi(f)} \qquad (25)$$

where C_q is the (finite rank) operator with kernel $\int S_F(x - x')q(x', y)dx'$. $Z_{\ell+a,\, t+b}F_{\ell+a,\, t+b}$ can be interpreted as the (Fock space) scalar product of two vectors (generalized Jost states) in different ways, depending on which line we interpret as the (imaginary) time axis. By using the Schwarz inequality N times in t direction and M times in x direction and taking logarithms we obtain

$$\begin{aligned} \ln |Z_{\ell+a,\, t+b}F_{\ell+a,\, t+b}(f, q)| &\le (1 - 2^{-N}) \ln Z_{\ell+a,\, t} + \\ &+ 2^{-N}(1 - 2^{-M}) \ln Z_{\ell+a,\, t+2^N b} + \\ &+ 2^{-N-M} \ln |Z_{\ell+2^M a,\, t+2^N b} F_{\ell+2^M a,\, t+2^N b}(f_{MN}, q_{MN})| \end{aligned} \qquad (26)$$

where f_{MN}, q_{MN} are the objects arising from f, q by reflecting N times in x direction and M times in t direction. Pictorially the elementary step looks as follows:

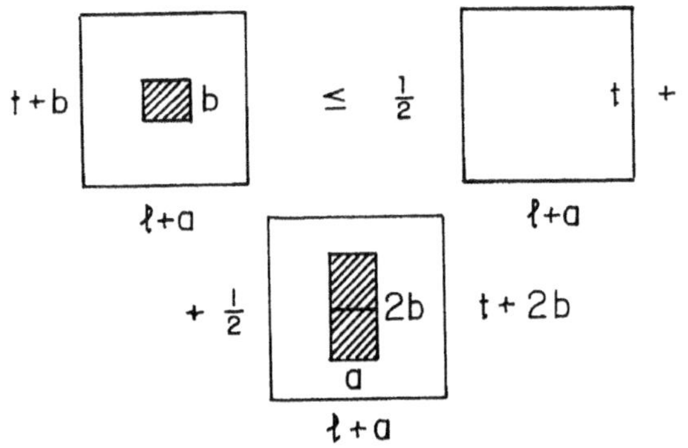

(The empty box denotes the interaction, the shaded box the interaction modified by f and q). The crucial point is that by the methods of section 2 (see [SeS 2]) we can easily see that the last term in (26) stays bounded for $N, M \to \infty$ (actually it can be shown to converge) by a constant c(f, q), whereas the second term converges by convexity of $\ln Z_{\ell+a,\, t+b}$ to $-E_{\ell+a}$. So we have

$$\ln |Z_{\ell+a,\, t+b}F_{\ell+a,\, t+b}(f, q)| \le \ln Z_{\ell+a,\, t} - E_{\ell+a} + c(f, q) \qquad (27)$$

Now again by convexity

$$\lim_{t\to\infty} (\ln Z_{\ell+a,\, t} - \ln Z_{\ell+a,\, t}) = E_{\ell+a} \tag{28}$$

(this is immediately clear from looking at the graph of $\ln Z_{\ell+a,\, t}$) and furthermore

$$E_{\ell+a} - E_\ell \le E_a - E_o = E_a \tag{29}$$

so

$$\overline{\lim_{t\to\infty}} \ln |F_{\ell+a,\, t+b}(f, q)| \le E_a + c(f, q) \tag{30}$$

which is the desired bound. It is easy to see that

$$c(f, q) \le c\|f\|_{-1} \sum_n \|g_n\|_{-\frac{1}{2}} \|h_n\|_{\frac{1}{2}} \tag{31}$$

where $\| \ \|_\alpha$ is the norm of the Sobolev space of order α. ($\|f\|_\alpha^2 = \int |\hat{f}(p)|^2 (p^2 + m^2)^\alpha dp$) Bounds on the Schwinger functions follow now by using Cauchy estimates. It should be remarked that the same method also gives Fröhlich bounds in $(\phi^4)_3$ (see [SeS 2]), but it is by no means the only one (J. Fröhlich had proven these and other bounds by a different method [private communication]).

Working a little harder one can get bounds of the form

$$\overline{\lim_{\ell\to\infty}}\ \overline{\lim_{t\to\infty}} |S_{\ell t}(\underline{f}, \underline{g}, \underline{h})| \le C \prod_{i,k} \|f_i\|_S \|g_i\|_S \|h_k\|_{S'}$$

where $\| \ \|_S$, $\| \ \|_{S'}$ are some Schwartz space norms (see [SeS 3]). O. A. McBryan [Mc 3] has obtained similar bounds by somewhat different methods.

6. Outlook

The bounds of the previous section allow the construction of infinite volume Schwinger functions by choosing convergent subsequences. But this unpleasant procedure has one big problem, as we remarked already: We do not know whether these Schwinger functions are Euclidean invariant. So a relativistic theory cannot be reconstructed. Therefore the existence and uniqueness of the thermodynamic limit remains an important open question.

In even $P(\phi)_2$ (allowing a linear term) and in $(\phi^4)_3$ the relation to the Ising model supplies correlation inequalities which show that the Schwinger functions with suitable boundary conditions increase with the volume (see [N 3] or [GRS 2]). Such a property seems to be missing in Y_2 (as well as in the general uneven $P(\phi)_2$ theories). But is there any other method of showing that the Schw inger functions converge to unique limit (which then would be automatically Euclidean invariant)?

There is, as I said, good reason to expect that an adaptation of the cluster expansion of Glimm, Jaffe and Spencer [GJS 1] (about which J. Feldman will say a little more) can prove this in the weak coupling case.

But there is another big question which makes the strongly coupled theory particularly interesting: In the pseudoscalar Y_2 theory there should be a phase transition for sufficiently large negative finite renormalization of the boson mass (which is the same as a large negative ϕ^2 term in the interaction) - together with a breakdown of parity symmetry (J. Preskill [Pr] has obtained some support for this belief by studying the one-loop approximation of the effective potential). The recent proof of the phase transition in $P(\phi)_2$ by Glimm, Jaffe and Spencer [GJS 2] gives some hope that this question can be answered for Y_2 too. Their announced "low temperature" cluster expansion [GJS 2] might be particularly helpful because it does not make use of correlation inequalities. So the chances for "understanding" Y_2 are not bad.

Let me add one short remark about more realistic theories, that is, more than two dimensions. The methods sketched here certainly do not work even in Y_3; what is needed there is at least as hard as, probably harder, than the phase-cell expansion for $(\phi^4)_3$ (see Glimm and Jaffe [GJ 3], and J. Feldman's lecture). But maybe some ingenious estimater will be able to push that through.

But of course we cannot claim that we really understand nonperturbative renormalization unless we construct a renormalizable theory, may it be Y_4, $(\phi^4)_4$ or some super-gauge theory. Unfortunately I have nothing more to say about that.

My part of the work described in these lectures was done while I was staying at the Institute for Advanced Study and supported in part by the National Science Foundation (Grants No. GP-16147A No.1 and GP 40768X); a smaller part was done at the Aspen Center for Physics. A large part of the work was done in cooperation with Barry Simon; his and Arthur Wightman's critical comments concerning these notes were of great help to me. And it should be said that my work in this field would not have been started without Arthur Wightman's stimulating influence. Furthermore I want to thank Mrs. Grace Anderson and Mrs. Barbara Drauschke for typing this manuscript under the pressure of a deadline.

---

†Present address: Department of Physics, Harvard University, Cambridge, Massachusetts 02138.

Supported in part by the National Science Foundation under Grant MPS 75-21212.

THE NON-PERTURBATIVE RENORMALIZATION OF $(\lambda\varphi^4)_3$

Joel S. Feldman[1, 2]

Jefferson Laboratory of Physics
Harvard University
Cambridge, Mass.

I THE MODEL

The $\lambda\varphi^4$ model in three space-time dimensions provides us with one of the few examples of the success of renormalization theory that goes beyond perturbation theory. The counterterms (over and above Wick ordering) required by the Euclidean Green's functions are summarized in the table below.

counterterm	order	degree of divergence	divergence cancelled	role in Hamiltonian theory
$\frac{1}{2}\langle(\int d^3x\lambda:\Phi^4:)^2\rangle$	λ^2	linear	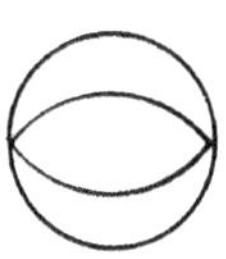	infinite vacuum energy and wave function renormalization
$-\frac{1}{6}\langle(\int d^3x\lambda:\Phi^4:)^3\rangle$	λ^3	logarithmic		infinite vacuum energy renormalization
$\frac{\lambda^2}{2}\delta m^2\int d^3x:\Phi^2(x):$	λ^2	logarithmic	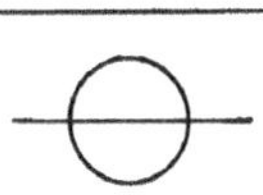	infinite mass renormalization

[1]Present address: Dept. of Mathematics, Mass. Inst. of Technology, Cambridge, Mass. 02139

[2]Supported in part by the National Science Foundation under grant MPS 73-05037

G. Velo and A. S. Wightman (eds.), Renormalization Theory, 435–460.

Hence the cutoff Euclidean Green's functions are given by

$$\langle\cdot\rangle_{m_0} = \text{norm. const.} \int \cdot\, e^{-\frac{1}{2}\int d^3x((\nabla\Phi)^2+m_0^2\Phi^2)} \prod_{x\in\mathbb{R}^3} d\Phi(x)$$

$$S^n_{\kappa,\Lambda}(f_1,\ldots,f_n) = Z^{-1}_{\kappa,\Lambda}\langle\Phi(f_1)\ldots\Phi(f_n)e^{-V_{\kappa,\Lambda}}\rangle_{m_0}$$

$$Z_{\kappa,\Lambda} = \langle e^{-V_{\kappa,\Lambda}}\rangle_{m_0}$$

$$V_{\kappa,\Lambda} = V_I + V_C$$

$$V_I = \lambda\int_\Lambda d^3x{:}\Phi_\kappa^4(x){:} - \mu\int_\Lambda d^3x\Phi_\kappa(x)$$

$$V_C = \frac{1}{2}\langle(\lambda\int_\Lambda {:}\Phi_\kappa^4{:})^2\rangle - \frac{1}{6}\langle(\lambda\int_\Lambda {:}\Phi_\kappa^4{:})^3\rangle_{m_0} + \frac{\lambda^2}{2}\delta m_\kappa^2\int_\Lambda {:}\Phi_\kappa^2{:}$$

The definition for $\langle\cdot\rangle_{m_0}$ given above is, of course, formal. We really mean the expectation with respect to the Gaussian measure on $\mathscr{S}'_{\mathbb{R}}(\mathbb{R}^3)$ with mean zero and covariance $(-\Delta+m_0^2)^{-1}$. We also introduce boundary conditions à la statistical mechanics into the $\{S^n_{\kappa,\Lambda}\}$ by imposing boundary conditions on the Laplacian. Placing zero Dirichlet data or periodic boundary conditions on $\partial\Lambda$, the surface of the region of interaction, has proven particularly useful.

For proper renormalization of theories with boundary conditions it is important that the Wick ordering and second order scalar counterterm remain matched to the covariance. (For example ${:}\Phi^2(x){:}_C = \Phi^2(x) - C(x,x)$.) This is optional for the other counterterms. As a matter of convenience we do match the third order scalar counterterm. When we want to match δm^2 to the covariance we use the counterterm

$$= \frac{\lambda^2}{2} \int_\Lambda d^3x \delta m_\kappa^2(x) : \Phi_\kappa^2(x):$$

$$\delta m_\kappa^2(x) = \int d^3y C_\kappa^3(x, y).$$

II THE RESULTS

There are four circumstances under which we have existence theorems.

	Coupling Restrictions	Boundary Conditions
W 1	$\lambda \geq 0$ small, $m_0 > 0$ large	free, periodic, Dirichlet
W 2	$\lambda \geq 0$, $m_0 > 0$, $\lvert\mu\rvert$ large	periodic
S 1	$\lambda \geq 0$, $m_0 > 0$, $\mu = 0$	Dirichlet
S 2	$\lambda \geq 0$, $m_0 > 0$, $\mu \neq 0$	Dirichlet

Theorem I ([MS], [FeO 1, 2])

The no-cutoff limit

$$S^n(f_1, \ldots, f_n) = \lim_{\Lambda \to \mathbb{R}^3} \lim_{\kappa} S^n_{\kappa,\Lambda}(f_1, \ldots, f_n)$$

exists in all the above cases. The $\{S^n\}$ satisfy the axioms for Euclidean Green's functions of [OS 1] (with the possible exception, in case S 1, of clustering). Hence they are the Euclidean Green's functions of a unique Wightman theory. (Again, in case S 1, the vacuum need not be unique.) They are also the moments of a unique probability measure on $\mathscr{S}'_{\mathbb{R}}(\mathbb{R}^3)$.

Theorem II ([MS], [FeO 1, 2])

This Wightman theory has a non-zero mass gap in cases W 1, W 2 and S 2.

Theorem III ([MS], [FeO 1, 2])

The $\{S^n\}$ are C^∞ in $\lambda \in [0, \lambda_0)$ and analytic in μ. Perturbation theory provides an asymptotic expansion for the $\{S^n\}$.

III THE INFINITE VOLUME LIMIT

The two principal tools used to control the infinite volume limit -- the cluster expansion and correlation inequalities -- both derive from analogous tools in statistical mechanics. The cluster expansion [GJS 1], [Sp 1] is a high temperature expansion and is used in the weak coupling cases W 1 and W 2. It is based on the observation that if there were no coupling between different space-time regions the infinite volume limit would be trivial. In particular suppose we partition space-time into disjoint unit cubes and place zero Dirichlet data on all surfaces of these unit cubes. Then our covariance no longer couples different cubes (i.e. $(-\Delta_D + m_0^2)^{-1}(x, y) = 0$ if x and y are in different cubes) and the Gaussian measure factors into a product over unit cubes. Hence if x_i is in the unit cube Δ_i

$$S^n(x_1 \ldots x_n) = \lim_{\Lambda \to \mathbb{R}^3} S^n_\Lambda(x_1 \ldots x_n)$$

$$= \lim_{\Lambda \to \mathbb{R}^3} Z_\Lambda^{-1} \langle \Phi(x_1) \ldots \Phi(x_n) e^{-V_\Lambda} \rangle_{m_0}$$

formally (since $V_{\Lambda, \kappa=\infty}$ does not exist)

$$= \lim_{\substack{\Lambda \to \mathbb{R}^3 \\ \cup \Delta_i \subset \Lambda}} Z^{-1}_{\cup \Delta_i} Z^{-1}_{\Lambda \sim \cup \Delta_i} \langle \Phi(x_1) \ldots \Phi(x_n) e^{-V_{\cup \Delta_i}} \rangle \langle e^{-V_{\Lambda \sim \cup \Delta_i}} \rangle$$

$$= S^n_{\cup \Delta_i}(x_1 \ldots x_n).$$

The cluster expansion perturbs about this completely decoupled theory. In the weak coupling cases W 1 and W 2 different cubes are strongly decoupled and the expansion converges, yielding the infinite volume limit.

Correlation inequalities [N 3], [GRS 2] are applied in cases S 1 and S 2. There are two steps to the argument. Firstly Nelson's monotonicity tells us that the Schwinger functions with zero Dirichlet data on $\partial\Lambda$ are increasing in Λ.

$$S^n_\Lambda(x_1 \ldots x_n)_{\text{Dirichlet}} \leq S^n_{\Lambda'}(x_1 \ldots x_n)_D \quad \text{if } \Lambda' \supset \Lambda, \quad \mu \geq 0.$$

(It is fortunate that for a φ^4 theory this inequality is still true for full Dirichlet boundary conditions - i.e. even when the Wick ordering depends on the covariance. Introducing a Λ dependence into the Wick ordering corresponds to introducing a Λ dependence into the single spin distribution of a continuous spin lattice system. Luckily, in our case, this just strengthens the inequality.) We now need only get an upper bound. We observe that

1) $S^n_\Lambda(x_1 \ldots x_n)_{\text{Dirichlet}} \leq S^n_\Lambda(x_1 \ldots x_n)_{\text{periodic}}$ if $\mu \geq 0$

2) $S^n_\Lambda(x_1 \ldots x_n)_{\text{periodic},\mu} \leq S^n_\Lambda(x_1 \ldots x_n)_{\text{periodic},\mu'}$

if $0 \leq \mu \leq \mu'$

and 3) $|S^n_\Lambda(f_1 \ldots f_n)| \leq (n!)^{1/2} |f_1| \ldots |f_n|$ if $|\mu'| >> 1$

by the cluster expansion of case W 2.

IV THE CUTOFF COVARIANCE

The frequent appearance of complicated boundary conditions in the cluster expansion suggests the use of an ultraviolet cutoff based on the Wiener measure representation of the bare (Euclidean) propagator.

$$(-\Delta+m_0^2)^{-1}(x,y) = \int_0^\infty dt \int \frac{d^3p}{(2\pi)^3} e^{ip\cdot(x-y)} e^{-t(m_0^2+p^2)}$$

$$= \int_0^\infty dt\, e^{-m_0^2 t} (4\pi t)^{-3/2} e^{-\frac{(x-y)^2}{4t}}$$

$$= \int_0^\infty dt\, e^{-m_0^2 t}\, p(t,x,y) \quad \text{(see [Gi] for example)}$$

$$= \int_0^\infty dt\, e^{-m_0^2 t} \int P_{xy}^t(d\omega)$$

Here $P_{xy}^t(d\omega)$ is the conditional Wiener measure on the set of all paths starting at x at time zero and ending at y at time t. The mass of this measure, $p(t,x,y)$, is just the total probability that a path travels from x to y in time t. To introduce a high (low) momentum cutoff we merely replace $\int_0^\infty dt$ by $\int_{t_m}^\infty dt$ $(\int_0^{t_M} dt)$. To introduce periodic boundary conditions we use the Wiener process on a torus rather than on $\mathbb{R}^3$. To introduce zero Dirichlet data on a surface Γ we integrate only over those paths that avoid Γ. We now discuss some properties [FeO 1] of the typical propagator

$$C_\Gamma(x,y) = \int_{t_m}^{t_M} dt\, e^{-m_0^2 t} \int P_{xy}^t(d\omega)\chi_\Gamma^t(\omega)$$
$$\chi_\Gamma^t(\omega) = \begin{cases} 0 & \text{if } \omega(s)\in\Gamma \text{ for some } s\in[0,t] \\ 1 & \text{otherwise.}\end{cases} \tag{4.1}$$

(P 1) $0 \le C_\Gamma \le C_\emptyset \equiv \int_{t_m}^{t_M} dt\, e^{-m_0^2 t}\, p(t,x,y) \le (-\Delta+m_0^2)^{-1}$ point-wise and as operators. The pointwise statement follows from the positivity of the integrand and the operator statement is a consequence of

$$\int P^t_{xy}(d\omega)\chi^t_\Gamma(\omega) = e^{-t\Delta_\Gamma} \leq e^{-t\Delta} = p(t, x, y).$$

(Δ_Γ is the Laplacian with zero Dirichlet data on Γ.)

(P 2) <u>t_m is an upper momentum cutoff,</u> as may be seen from

$$C_\Gamma(x, y) \leq \int_{t_m}^{\infty} dt\, e^{-m_0^2 t}\, p(t, x, y)$$

$$= \int \frac{d^3p}{(2\pi)^3}\, e^{ip(x-y)}\, \frac{e^{-t_m(m_0^2+p^2)}}{p^2 + m_0^2}\,.$$

Momenta larger than $t_m^{-1/2}$ are exponentially damped.

(P 3) <u>Decay properties</u>. Our propagator is bounded above by

$\int_{t_m}^{t_M} dt(4\pi t)^{-3/2} e^{-m_0^2 t} e^{-\frac{(x-y)^2}{4t}}$. The $(4\pi t)^{-3/2}$ determines its

ultraviolet (i.e. short distance) behavior while

$e^{-m_0^2 t} e^{-\frac{(x-y)^2}{4t}}$ determines its long distance behavior. Since

$$e^{-m_0^2 t} e^{-\frac{(x-y)^2}{4t}} \leq \begin{cases} e^{-\frac{m_0}{2}|x-y|} & \left(\text{consider } t \geq \frac{|x-y|}{2m_0} \text{ and } t \leq \frac{|x-y|}{2m_0} \text{ separately}\right) \\[2ex] e^{-\frac{|x-y|^2}{4t_M}} & \text{since } t < t_M \end{cases}$$

$C_\Gamma(x, y)$ really decays according to a scaled distance

$d_s(x,y) = \max(m_0|x-y|, t_M^{-1/2}|x-y|)$. For high momentum lines (i.e. t_M near zero) this decay is much stronger than one would expect from m_0. Rather it is commensurate with the energy carried by the propagator.

(P4) Smoothness Properties. Let us start by observing that in position space renormalization cancellations are implemented through smoothness properties of the propagator. In the following example we exhibit the cancellation between a mass bubble  and its counterterm ,

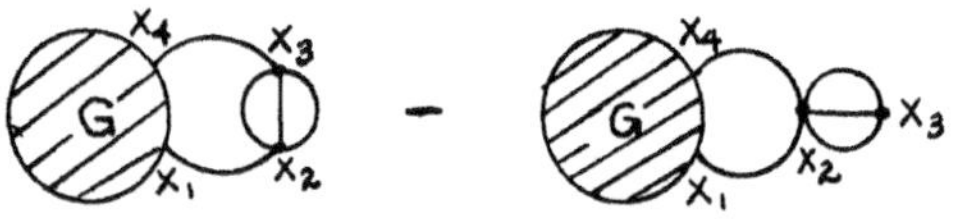

$$= \int dx_1 dx_2 dx_3 dx_4 G(x_1, x_4) C(x_1, x_2) C^3(x_2, x_3)[C(x_3, x_4) - C(x_2, x_4)].$$

By itself $C^3(x_2, x_3) \sim |x_2 - x_3|^{-3}$ and has a nonintegrable singularity at $x_2 = x_3$. However if C is Hölder continuous of index $\alpha > 0$, $|C(x_3, x_4) - C(x_2, x_4)| \sim |x_2 - x_3|^{\alpha}$. This cancellation moderates the singularity leaving it integrable.

Now let us prove some smoothness properties, starting with $C_\emptyset(x,y)$. Now $p(t,x,y)$ is obviously C^∞ in x and y. For definiteness we can use

$$|e^{-b^2} - e^{-a^2}| \leq O(1)|a-b|^{\alpha}(e^{-\frac{3}{4}b^2} + e^{-\frac{3}{4}a^2}) \quad 0 \leq \alpha \leq 1$$

to yield

$$|p(t,x,y) - p(t,x,y')| \leq O(1)\frac{|y-y'|^{\alpha}}{t^{\alpha/2}}[p(\tfrac{4}{3}t, x, y) + p(\tfrac{4}{3}t, x, y')]$$

$$\leq O(1)|y-y'|^{\alpha}[|x-y|^{-\alpha} + |x-y'|^{-\alpha}][p(2t,x,y) + p(2t,x,y')] \tag{4.2}$$

(since $x^{\alpha/2}e^{-1/4\,x^2} \leq O(1)$). So

$$|C_\emptyset(x,y)-C_\emptyset(x,y')| \le O(1)|y-y'|^\alpha[|x-y|^{-\alpha} + |x-y'|^{-\alpha}]\int_{2t_m}^{2t_M} dt' e^{-\frac{m_0^2}{2}t'}[p(t',x,y)+p(t',x,y')]$$

after the change of variable $t' = 2t$. Notice that $\int_{2t_m}^{2t_M} dt \ldots$ looks a great deal like the propagator we started with. Only m_0, t_m and t_M have changed by irrelevant constant factors.

Now we get to the propagator (4.1). According to the theory of elliptic operators $(-\Delta_\Gamma+m_0^2)^{-1}(x,y)$ is C^∞ except on the edges of Γ where we expect Hölder continuity. However we use only crude methods and so we only derive crude estimates - but they suffice.

$$C_\Gamma(x,y) = C_\emptyset(x,y) - \int_{t_m}^{t_M} dt\, e^{-m_0^2 t}\int P_{xy}^t(d\omega)(1-\chi_\Gamma^t)(\omega) \quad (4.3)$$

We have already dealt with the first term. Since $(1-\chi_\Gamma^t)(\omega)$ forces ω to cross Γ sometime in the interval $[0,t]$ we may express the second term using first hitting times.

$$\int P_{xy}^t(d\omega)(1-\chi_\Gamma^t)(\omega) = \int_{0\le t_1\le t} E_x\{\chi(\tau_\Gamma(\omega)\in dt_1)p(t-t_1,\omega(t_1),y)\} \quad (4.4)$$

where E_x is the Wiener expectation for paths starting at x at time zero.

$\chi(\tau_\Gamma(\omega)\in dt_1)$ is the characteristic function of the set of all paths whose first hitting time for Γ, τ_Γ, lies in dt_1.

Equation (4.4) writes the probability that a path starts at time zero passes through Γ and ends at y at time t as the integral over t_1 of the probability that the path starts at x at time zero, hits Γ for the first time at $\tau_\Gamma \in dt$, and in the remaining time, $t - t_1$, travels to y. In other words the right hand side of (4.4) just classifies paths according to when they first hit Γ.

We are looking for Hölder continuity in y and y appears only in $p(t-t_1, \omega(t_1), y)$. By (4.2)

$$
\begin{aligned}
&|p(t-t_1, \omega(t_1), y) - p(t-t_1, \omega(t_1), y')| \\
&\le O(1)|y-y'|^{\alpha}[|\omega(t_1)-y|^{-\alpha} + |\omega(t_1)-y'|^{-\alpha}][p(2(t-t_1), \omega(t_1), y) \\
&\qquad\qquad + p(2(t-t_1), \omega(t_1), y')] \\
&\le O(1)|y-y'|^{\alpha}[d(\Gamma, y)^{-\alpha} + d(\Gamma, y')^{-\alpha}][p(2(t-t_1), \omega(t_1), y) \\
&\qquad\qquad + p(2(t-t_1), \omega(t_1), y')]
\end{aligned}
$$

where $d(\Gamma, y)$ is the Euclidean distance from y to Γ. We may now substitute this into (4.4) and make a change of variables to get

$$
\begin{aligned}
&\left|\int_{t_m}^{t_M} dt\, e^{-m_0^2 t}\left[\int P_{xy}^{t}(d\omega)(1-\chi_{\Gamma}^{t})(\omega) - \int P_{xy'}^{t}(d\omega)(1-\chi_{\Gamma}^{t})(\omega)\right]\right| \\
&\le O(1)|y-y'|^{\alpha}[d(\Gamma, y)^{-\alpha}+d(\Gamma, y')^{-\alpha}]\left\{\int_{t_m}^{2t_M} dt\, e^{-\frac{m_0^2 t}{2}}\left[\int P_{xy}^{t}(1-\chi_{\Gamma})\right.\right. \\
&\qquad\qquad \left.\left. + \int P_{xy'}^{t}(1-\chi_{\Gamma})\right]\right\} \\
&\le O(1)|y-y'|^{\alpha}[d(\Gamma, y)^{-\alpha}+d(\Gamma, y')^{-\alpha}][C_{\emptyset}(\tfrac{m_0}{\sqrt{2}}; x, y)+C_{\emptyset}(\tfrac{m_0}{\sqrt{2}}; x, y')].
\end{aligned}
$$

To round out this section we will outline how one can get simultaneous smoothness estimates on both variables of $C_{\Gamma}(x, y)$.

$$
\begin{aligned}
C_{\Gamma}(x, y) &= \int dt\, e^{-m_0^2 t}\, e^{-t\Delta_{\Gamma}}(x, y) \\
&= \int dt\, e^{-m_0^2 t}\, \frac{1}{t}\int_0^t dt_1 dz\, e^{-t_1\Delta_{\Gamma}}(x, z)e^{-(t-t_1)\Delta_{\Gamma}}(z, y)
\end{aligned}
$$

We have just seen how to get a smoothness estimate on the x dependence of $e^{-t_1\Delta_\Gamma}(x, z)$ and the y dependence of $e^{-(t-t_1)\Delta_\Gamma}(z, y)$. Combining the two gives a simultaneous smoothness estimate.

(P 5) Factorization Properties. Recall that the cluster expansion perturbs about a theory which has no coupling between different unit cubes in space-time. This theory may be realized by placing zero Dirichlet data on the surfaces of all unit cubes. Then $C(x, y) = 0$, if x and y are in different cubes, even in the presence of momentum cutoffs (cf. (4.1)). As a result the Gaussian measure with covariance C factorizes:

$$d\mu_C = \prod_\Delta d\mu_C \upharpoonright \Delta.$$

Since the nonscalar part of the action is local the interacting measure also factorizes:

$$Z_\kappa^{-1}(\Lambda)e^{-V_\kappa(\Lambda)} d\mu_C = \prod_\Delta Z_\kappa^{-1}(\Delta)e^{-V_\kappa(\Delta)} d\mu_C \upharpoonright \Delta.$$

V THE METHODS - THE PHASE SPACE CELL EXPANSION (PSCE)

We use the PSCE [GJ 3] to control the ultraviolet limit. It is really a combination of many expansions, each associated with a different region of (Euclidean) space-time. Hence it is essential that we be able to introduce different momentum cutoffs in different space-time regions. To this end we define an auxiliary Gaussian field $\psi(t, x)$ $(t \in (0, \infty),\ x \in \mathbb{R}^3)$ whose two point function is given by

$$\langle\psi(t, x)\psi(s, y)\rangle = \delta(t-s)e^{-m_0^2 t} \int P^t_{xy}(d\omega)\chi^t_\Gamma(\omega).$$

The familiar free Euclidean field of mass m_0 is just

$\Phi(x) = \int_0^\infty dt\, \psi(t, x)$ while $\int_{t_m}^{t_M} dt\, \psi(t, x)$ is a field with upper and lower momentum cutoffs. We say that $\psi(t, x)$ has momentum $t^{-1/2}$ and position x and we refer to $(0, \infty) \times \mathbb{R}^3$ as phase space. (This terminology should, of course, not be taken too seriously.)

The PSCE for the $(\lambda\varphi^4)_3$ model is extremely complicated. So we will look instead at a simplified model which has a more transparent expansion. This simplified model ignores one of the more nasty problems of $(\lambda\varphi^4)_3$ - that of the momentum crossterms that couple different momentum regions.

Firstly we partition the cutoff phase space $[t_m, \infty) \times \Lambda$. ($t_m$ is an overall ultraviolet cutoff and Λ an overall space-time cutoff.)

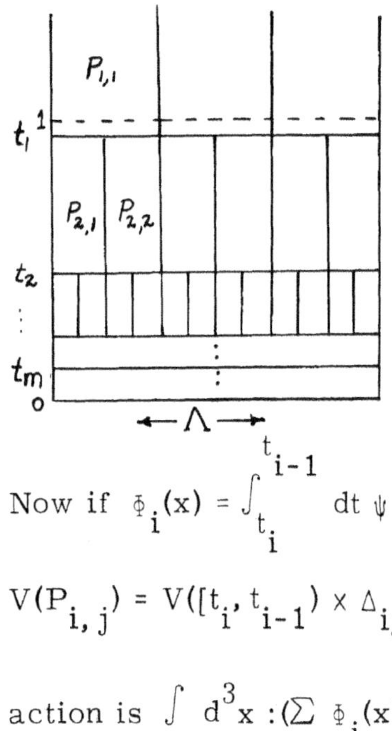

$$[t_m, \infty) = \bigcup_{i=1}^{N} [t_i, t_{i-1})$$

$$\text{where } t_0 = \infty,\ t_1 < 1,\ t_i = t_1^{(1+\nu)^{i-1}} \tag{4.5}$$

$$[t_i, t_{i-1}) \times \Lambda = \bigcup_j [t_i, t_{i-1}) \times \Delta_{i,j} = \bigcup_j P_{ij}$$

$$\text{where } |\Delta_{ij}| t_i^{-1} = 1$$

Now if $\Phi_i(x) = \int_{t_i}^{t_{i-1}} dt\, \psi(t, x)$ we define

$V(P_{i,j}) = V([t_i, t_{i-1}) \times \Delta_{ij}) \equiv \int_{\Delta_{ij}} d^3x : \Phi_i(x)^4 : +$ ctms. The real action is $\int_\Lambda d^3x : (\sum_i \Phi_i(x))^4 : +$ ctms but we will study $\prod_{i,j} \langle e^{-V(P_{ij})} \rangle$. The dominant feature here is that we have dropped the momentum crossterms $:\Phi_{i_1}(x)\Phi_{i_2}(x)\Phi_{i_3}(x)\Phi_{i_4}(x):$

(not all the i's equal) that couple different momentum regions $[t_i, t_{i-1})$. (Less significantly we have also decoupled different space-time regions by moving the $\prod_{i,j}$ outside the expectation.)

Expand each $\langle e^{-V(P_{ij})}\rangle$ in a Taylor series to fourth order and use $\frac{d^n}{d\lambda^n}\langle e^{-V(P_{ij})}\rangle\Big|_{\lambda=0} = 0$ for $n = 1$ (by Wick ordering), $n = 2$ (by the choice of the second order scalar counterterm) and $n = 3$ (by the choice of the third order scalar counterterm) to give

$$\langle e^{-V(P_{ij})}\rangle = 1 + \frac{1}{3!}\int_0^\lambda (\lambda-\lambda')^3 \frac{d^4}{d\lambda'^4}\langle e^{-V(P_{ij})}\rangle_{\lambda'}\, d\lambda'.$$

Now define $M = \sup_{\substack{\lambda'\in[0,\lambda]\\ i,j}} |\Delta_{ij}|^{-1} t_i^{-\epsilon} \left|\frac{d^4}{d\lambda'^4}\langle e^{-V(P_{ij})}\rangle_{\lambda'}\right|$ where

$\epsilon > 0$ is to be determined later. Then

$$\begin{aligned}\prod_{ij}\langle e^{-V(P_{ij})}\rangle &\le \prod_{ij}\left(1 + \frac{\lambda^4}{4!} M|\Delta_{ij}|t_i^\epsilon\right)\\ &\le \exp\left(\sum_{i,j}\frac{\lambda^4}{4!} M|\Delta_{ij}|t_i^\epsilon\right)\\ &\le \exp\left(\frac{\lambda^4}{4!} M|\Lambda| \sum_{i=1}^\infty t_i^\epsilon\right).\end{aligned}$$

Now $\sum_{i=1}^\infty t_i^\epsilon = \sum_{i=1}^\infty t_i^{\epsilon(1+\nu)^{i-1}} < \infty$ (by (4.5)) so that an ultraviolate uniform estimate on $\prod \langle e^{-V(P_{ij})}\rangle$ will follow from $M < \infty$ or, in other words from

$$\left|\frac{d^4}{d\lambda'^4}\langle e^{-V(P_{ij})}\rangle_{\lambda'}\right| \leq O(1)\,|\Delta_{ij}|\,t_i^{\varepsilon}.$$

This estimate is proven in four steps.

Step 1: The P-C Expansion. In this step we evaluate the derivatives $\frac{d}{d\lambda'}$. If G is a polynomial in the fields that has arisen from earlier differentiations

$$\frac{d}{d\lambda'}\langle Ge^{-V(P_{ij})}\rangle = \langle(\frac{d}{d\lambda'}G)e^{-V(P_{ij})}\rangle - \langle\int_{\Delta_{ij}} dx{:}\Phi_i^4(x){:}Ge^{-V}\rangle$$

$$- \text{ counterterms}$$

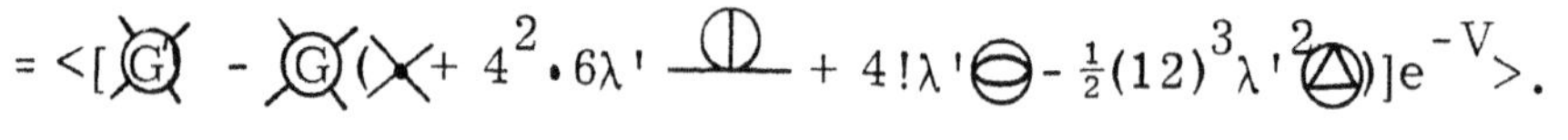

(4.6)

To obtain decent estimates we obviously must perform some renormalization cancellations (cf. ). The divergence of perturbation theory prevents us from cancelling to all orders. Fortunately, as we shall see, it suffices to cancel the lowest order divergences each time we take a derivative. We use the integration by parts (contraction) formula [GJ 3], [GJS 1]

$$\langle\psi(t,x)F(\psi)\rangle = \int dy\, e^{-m_0^2 t}\int P_{xy}^t(d\omega)\chi_\Gamma^t(\omega)\langle\frac{\partial F}{\partial\psi(t,y)}\rangle$$

to give

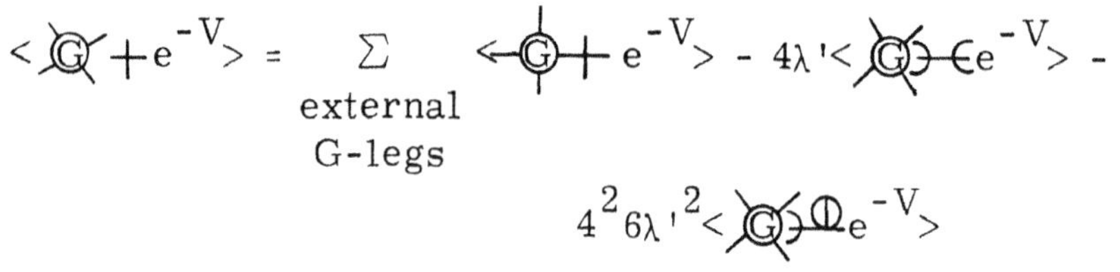

(4.7)

Now rewrite 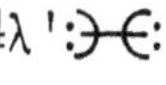 as a sum of Wick ordered monomials

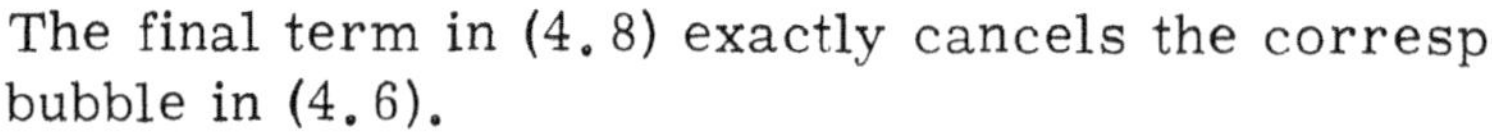

(4.8)

The final term in (4.8) exactly cancels the corresponding bubble in (4.6).

We still have lots of divergences left. However if we continue in this manner we can express the right hand side of (4.6) as a sum of diagrams each of which has a convergent piece and (possibly) some logarithmic divergences. (Note that  is the only non-logarithmic divergence in the theory and it is always exactly cancelled). For example  has a convergent cancelled mass subdiagram () and a divergent uncancelled mass counterterm (). This residual divergence will not cause any problems. When we evaluate some vacuum graph containing , the divergent bubble will indeed contribute the divergent factor $\ln t_i^{-1}$. However  is convergent and its lines carry only high momenta. (i.e. those lines have lower momentum cutoff $t_{i-1}^{-1/2}$; all lines have upper momentum cutoff $t_i^{-1/2}$.) As a result,  contributes a factor of t_{i-1}^{ε} and this is sufficient to control the $\ln t_i^{-1}$. (cf.

$$\Delta \text{ } \Delta \leq \int_{x\in\Delta} dxdy\, dt_1 dt_2 (4\pi t_1)^{-3/2} (4\pi t_2)^{-3/2} e^{-m_0^2(t_1+t_2)} e^{-\frac{(x-y)^2}{4}\left(\frac{1}{t_1}+\frac{1}{t_2}\right)}$$

$$\leq O(1)|\Delta| \int_{t_i}^{t_{i-1}} dt_1 dt_2 (t_1+t_2)^{-3/2} \leq O(1)|\Delta| t_{i-1}^{1/2}.)$$

There are two additional complications in the real PSCE. Firstly, because of the crossterms, not all legs in  carry the same momentum and in fact some may carry momentum which is very low compared to that in the divergent pieces. Hence we must be careful to ensure that there are always sufficiently many high momentum legs in the convergent pieces to supply the required convergence factors.

Secondly the coupling between different space-time regions means that we get $\sum_{j'}$ 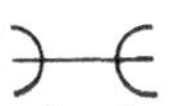 $\Delta_{ij}\Delta_{ij'}$ rather than 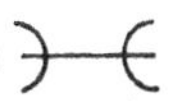 $\Delta_{ij}\Delta_{ij}$ whenever we perform an integration by parts. However this is easily controlled by the exponential decay of the propagator joining Δ_{ij} and $\Delta_{ij'}$. For example

$$\sum_{j'} \left| \begin{matrix} \Delta_{ij} \\ \text{(diagram)} \\ \Delta_{ij'} \end{matrix} \right| \le \text{const} \sum_{j'} e^{-\frac{1}{2} d_s(\Delta_{ij}, \Delta_{ij'})}$$

$$\le \text{const} \sum_{j'} e^{-\frac{1}{2} |\Delta_{ij'}|^{-\frac{1}{2(1+\nu)}} d_e(\Delta_{ij}, \Delta_{ij'})}$$

$$\text{since } t_i = |\Delta_{ij'}| \text{ and } t_{i-1} = t_i^{\frac{1}{1+\nu}}$$

$$\approx \text{const} \int \frac{d^3\vec{x}}{|\Delta|} \exp[-\tfrac{1}{2} |\Delta|^{-\frac{1}{2(1+\nu)}} |\vec{x}|]$$

$$\le O(1) \text{ independent of } |\Delta|.$$

Step 2: Squaring. At this stage we have

$$\frac{d^4}{d\lambda'^4} \langle e^{-V(P_{ij})} \rangle = \sum_k \langle G_k e^{-V(P_{ij})} \rangle$$

We wish to bound $e^{-V(P_{ij})}$ using the formal positivity of $V(P_{ij})$. Before employing any inequality we must replace G_k by something positive. That's easy.

$$|G_k| \le \tfrac{1}{2}(\delta_k^2 + \frac{1}{\delta_k^2} G_k^2) \tag{4.9}$$

(The choice $\delta_k = 1$ is obviously inadequate since we are trying to prove that $\langle G_k e^{-V} \rangle$ is small.)

Step 3: The Wick Expansion. Since the Wick constant, $\langle \phi_i(x)^2 \rangle$, for $V(P_{ij})$ obeys

$$<\Phi_i(x)^2> \leq \int_{t_i}^{t_{i-1}} dt\,(4\pi t)^{-3/2} \leq O(1)t_i^{-1/2}$$

we have $\quad :\Phi_i(x)^4: \geq -O(1)t_i^{-1}$

and $\quad V(P_{ij}) \geq -O(1)|\Delta_{ij}|t_i^{-1} \geq -O(1) \qquad (4.10)$

independent of i and j. This is obviously why we coupled the cube size to the momentum cutoff (cf. (4.5)). Notice that we have used the Wick bound independently in each phase space cell. This becomes drastically more complicated when the crossterms intervene. Then the formally positive $V(\sum \Phi_i)$ does not decompose into the sum $\sum V(\Phi_i)$ of formally positive polynomials.

Step 4: Graph Estimates. We now have

$$\left|\frac{d^4}{d\lambda'^4}<e^{-V(P_{ij})}>\right| \leq \sum c_e |<G_e>|$$

where the constants c_e have arisen from the δ_k's in (4.9) and from the Wick bound (4.10). Since G_e is a monomial in the fields $<G_e>$ is the usual sum of vacuum diagrams and

$$\left|\frac{d^4}{d\lambda'^4}<e^{-V(P_{ij})}>\right| \leq \sum c_r |V_r|. \qquad (4.11)$$

The vacuum graph V_r is a big complicated multiple integral. To estimate it we need to split it up into manageable pieces.

We associate with each (not necessarily vacuum) graph the kernel of an integral operator. (A vacuum graph will be an operator on $L_2(\mathbb{R}^0)$ i.e. a number.) A graph G consists of

$V(G) = \{$all vertices in G$\}$
$I(G) = \{$all internal lines in G$\}$
$E(G;v) = \{$all external legs hooked to $v \in V(G)\}$
$E(G) = \bigcup_{v\in V(G)} E(G;v)$.

We associate with each vertex a cutoff function $h_v(x_v)$, with each internal line a propagator and with each external leg the operator square root of a propagator (i.e. half a line). We split $E(G)$ into two disjoint subsets $E_i(G)$ and $E_f(G)$, called initial and final legs respectively, and then

$$K(G)(z,y) = \int \prod_{v\in V(G)} \{d^3x h_v(x_v) \prod_{\ell\in E_f(G;v)} \ell^{1/2}(z_\ell, x_v) \prod_{\ell\in E_i(G;v)} \ell^{1/2}(x_v, y_\ell)\} \prod_{\ell_{ij}\in I(G)} \ell_{ij}(x_i, x_j) \tag{4.12}$$

is the kernel of an operator mapping L_2 of the initial variables (i.e. the y's) into L_2 of the final variables (i.e. the z's). (In a graphical representation of G initial (final) legs point to the left (right).) Of course when G is a vacuum graph $K(G) = G$ and $|G| = \|G\| \equiv$ the operator norm of (4.12).

We may now decompose G by writing $V(G) = \bigcup_j V_j$ with the different V_j disjoint and then defining G_j

$$V(G_j) = V_j$$

$$I(G_j) = \{\text{all lines in } I(G) \text{ connecting two vertices in } V_j\}$$

$$E_i(G_j) = \bigcup_{v\in V_j} E_i(G;v) \cup \{\text{all lines in } I(G) \text{ connecting a vertex in } V_j \text{ with one in } V_{j'} \text{ with } j' < j\}$$

$$E_f(G_j) = \bigcup_{v\in V_j} E_f(G;v) \cup \{\text{all lines in } I(G) \text{ connecting a vertex in } V_j \text{ with one in } V_{j'} \text{ with } j' > j\}.$$

It is clear that the operator G is a product of operators of the form $G_j \otimes I$ (the I is needed to match the various $L_2(\mathbb{R}^s)$

spaces) and

$$\|G\| \leq \prod_j \|G_j\|. \tag{4.13}$$

Returning to (4.11), we just cut V_r up into pieces and use

$\|$ 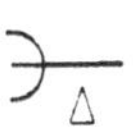 $\| \leq O(1)|\Delta|^{1/4} t_M^{1/16} \ell n^{1/2} t_m$

$\|$  $\| \leq O(1)|\Delta|^{1/4} t_M^{1/128}$

$\|$ 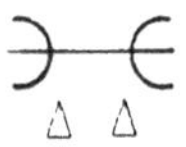 $\|_{H.S.} \leq O(1)|\Delta|^{1/2} t_M^{1/8} \ell n\, t_m$ (4.14)

$\|$  $\|_{H.S.} \leq O(1)|\Delta|^{1/2} t_M^{1/8} \ell n\, t_m$

$\|$ 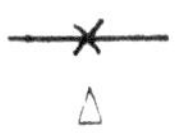 $\|_{H.S.} \leq O(1)|\Delta|^{1/3} t_M^{1/16}$

where $\| \ \|_{H.S.}$ refers to the Hilbert-Schmidt norm.

(Proof: $\|$  $\|^4 = \|$ 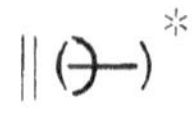 $\|^2 = \|$  $\|^2 \leq \|$  $\|^2_{H.S.}$ = 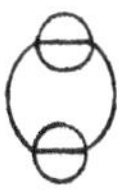

$\leq O(1)|\Delta| t_M^{1/4} \ell n^2 t_m$ etc.) This yields

$$|V_r| \leq O(1)(t_M^{1/128} \ell n\, t_m |\Delta_{i,j}|^{1/4})^{\text{number of vertices in } V_r}$$

Recall that we took four derivatives $\frac{d}{d\lambda'}$ so, before squaring (4.9), each term had at least four vertices. Hence if we choose $\delta_k^2 = t_M^{1/32}|\Delta_{i,j}|$ we have

$$C_r|V_r| = \tfrac{1}{2}\delta_k^2 \leq \tfrac{1}{2} t_M^{1/32} |\Delta_{i,j}| \leq \tfrac{1}{2} t_i^{1/32(1+\nu)} |\Delta_{ij}| \tag{4.15}$$

if V_r arose from the first term of (4.9) and

$$
\begin{aligned}
C_r |V_r| &\le O(1)\delta_k^{-2} t_M^{1/16} \ell n^8 t_m |\Delta_{i,j}|^2 \\
&\le O(1) t_M^{1/32} \ell n^8 t_m |\Delta_{ij}| \\
&\le O(1) t_{i-1}^{1/32} \ell n^8 t_i |\Delta_{ij}| \\
&\le O(1) t_i^{1/40(1+\nu)} |\Delta_{ij}| \qquad (4.16)
\end{aligned}
$$

if V_r arose from the second term of (4.9) (and hence had at least eight vertices.) The substitution of (4.15) and (4.16) into (4.11) completes the proof that $M < \infty$.

Those wishing to delve deeper into the PSCE might try reading [GJ 3], [Fe], [FeO 1] and [MS]. For earlier work on $(\lambda\varphi^4)_3$ see [Gl 2] and the references in [FeO 1].

Bibliography for the lectures of J. Feldman, J. Fröhlich and E. Seiler.

[A 1] S. Albeverio and R. Høegh-Krohn, Comm. Math. Phys. 30, 171, (1972).

[A 2] ______, J. Funct. Anal. 16, 39, (1974)

[Bo] H. Borchers, contribution in the Proceedings of the International Colloquium on Mathematical Methods of Quantum Field Theory, Marseille (1975).

[Br] D. Brydges, J. Math. Phys., 16, 1649 (1975) and "Boundedness Below for Fermion Model Theories, part II" University of Michigan preprint (1974).

[Br F] __________ and P. Federbush, J. Math. Phys. 15, 730 (1974).

[Ca] J. T. Cannon, Comm. Math. Phys. 35, 215 (1974).

[Ch] N. H. Christ and T. D. Lee, "Quantum Expansion of Soliton Solutions", Columbia University, Preprint CO-2271-55, (1975); see also refs. therein.

[Co L] P. Colella and O. Lanford, "Sample Field Behavior for the Free Markov Field", in Constructive Quantum Field Theory, eds. G. Velo and A. S. Wightman, Springer-Verlag (1973).

[Co 1] S. Coleman, Phys. Rev. D11, 2088, (1975).

[Co 2] ______, R. Jackiw and L. Susskind, "Charge Shielding and Quark Confinement in the Massive Schwinger Model", Ann. Phys., to appear.

[Da] R. Dashen, B. Hasslacher and A. Neveu, "The Particle Spectrum in Model Field Theories from Semiclassical Functional Integral Techniques", Phys. Rev., to appear.

[DL] C. Deutsch and M. Lavaud, Phys. Rev. A9, 2598, (1974).

[DHR] S. Doplicher, R. Haag and J. E. Roberts, Comm. Math. Phys. 23, 199, (1971) and 35, 49, (1974).

[Di 1] J. Dimock, Ann. Phys. 72, 177 (1972).

[Di 2] ______, Comm. Math. Phys. 35, 347, (1974).

[EEF] J.-P. Eckmann, H. Epstein and J. Fröhlich, "Asymptotic Perturbation Expansion for the S-Matrix and the Definition of Time-Ordered Functions in Relativistic Quantum Field Models", University of Geneva, Preprint, (1975).

[EMS] J.-P. Eckmann, J. Magnen and R. Seneor, Comm. Math. Phys. 39, 251 (1975).

[Fad] L. Faddeev and L. A. Takhtajan, Theor. Math. Phys. 21, 160, (1974) and refs. given there; see also: L. Faddeev, "Quantization of Solitons", IAS, Preprint, (1975).

[Fe] J. Feldman, Comm. Math. Phys. 37, 93 (1974) and "Thesis", Harvard University (1974).

[FeO 1] __________ and K. Osterwalder, "The Wightman Axioms and the Mass Gap for Weakly Coupled $(\Phi^4)_3$ Quantum Field Theories", Annals of Physics, to appear.

[FeO 2] ______, "The Construction of $\lambda\varphi_3^4$ Quantum Field Models", Proceedings of the International Colloquium on Mathematical Methods of Quantum Field Theory, Marseille, (1975) and work in preparation.

[Fey] R. P. Feynman, Rev. Mod. Phys. 20, 367 (1948).

[Fr 1] J. Fröhlich, Ann. Inst. Henri Poincaré, A21, 271, (1974), and "Schwinger Functions and Their Generating Functionals, II, ..." to appear in Adv. Math., (1975).

[Fr 2] ______, "The Pure Phases, the Irreducible Quantum Fields and Dynamical Symmetry Breaking in Symanzik-Nelson positive Quantum Field Theories", Princeton University, Preprint, (1975).

[Fr 3] ______, Phys. Rev. Letters 34, 833, (1975), and paper in preparation.

[Fr 4] ______, "Classical and Quantum Statistical Mechanics in One and Two Dimensions: Two-Component Yukawa- and Coulomb Systems", to appear in Comm. Math. Phys.

[Fr 5] ______, "New Super-Selection Sectors ('Soliton-States') in Two dimensional Bose Quantum Field Models", Princeton University, Preprint, (1975).

[Fr S] ______ and E. Seiler, "The Massive Thirring-Schwinger Model: Convergence of Perturbation Theory in the Mass", Preprint to appear.

[GV] I. Gel'fand and N. Vilenkin, Generalized Functions, Vol. 4, Applications of Harmonic Analysis, Academic Press, New York, 1964.

[Gi] J. Ginibre, "Some applications of Functional Integration in Statistical Mechanics" in Statistical Mechanics and Quantum Field Theory, eds. C. DeWitt and R. Stora, Gordon and Breach (1971).

[G] V. Glaser, Comm. Math. Phys. 37, 257, (1974).

[Gl 1] J. Glimm, Comm. Math. Phys. 5, 343 (1967) and 6, 61 (1967).

[Gl 2] ______, Comm. Math. Phys. 10, 1 (1968).

[GJ 1] ______ and A. Jaffe, Ann. Phys. 60, 321 (1970).

[GJ 2] ______, J. Funct. Anal. 7, 323 (1971).

[GJ 3] ______, Fortsch Physik 21, 327 (1973).

[GJ 4] ______, On the Approach to the Critical Point", Harvard University, Preprint, (1974).

[GJS 1] ______ and T. Spencer, "The Particle Structure of the Weakly Coupled $P(\varphi)_2$ Model and Other Applications of High Temperature Expansions," in: Constructive Quantum Field Theory, ed. G. Velo and A. Wightman, Springer Lecture Notes in Physics, 1973.

[GJS 2] ______, "Existence of Phase Transitions for φ_2^4 Quantum Fields", Proceedings of the International Colloquium on Mathematical Methods of Quantum Field Theory, Marseille (1975) and "Phase Transitions for φ_2^4 Quantum Fields" Comm. Math. Phys. to appear and "A Cluster Expansion in the Two Phase Region" in preparation.

[Gu] F. Guerra, Phys. Rev. Lett. 28, 1213, (1975).

[GRS 1] F. Guerra, L. Rosen and B. Simon, Comm. Math. Phys. 27, 10, (1972) and 29, 233, (1973).

[GRS 2] ______, Ann. Math. 101, 111, (1975).

[GRS 3] ______, "Boundary Conditions in the $P(\varphi)_2$ Euclidean Quantum Field Theory", Princeton University,

Preprint, to appear.

[Ha] P. R. Halmos, Measure Theory, Van Nostrand, New York (1950).

[H] I. Herbst, "Remarks on Canonical Quantum Field Theory", paper in preparation.

[Ho] R. Høegh-Krohn, Comm. Math. Phys. 38, 195, (1974).

[J] R. Jost, "The General Theory of Quantized Fields", American Math. Soc., Publ. 1965, Providence, R.I.

[LS] J. Lowenstein and A. Swieca, Ann. Phys. (N. Y.) 68, 172, (1971).

[LE] A. Luther and V. J. Emery, Phys. Rev. Lett., 33, 589, (1974).

[MS] J. Magnen and R. Seneor, "The Infinite Volume Limit of the φ_3^4 Model", to appear in Ann. Inst. Henri Poincaré.

[Ma S] P. T. Matthews and A. Salam, Nuovo Cim. 12, 563 (1954) and 2, 120 (1955).

[M] R. May, Physics Letters 25A, 282, (1967).

[Mc 1] O. A. McBryan, "Finite Mass Renormalizations in the Euclidean Yukawa_2 Field Theory", Comm. Math. Phys., to appear.

[Mc 2] ______, "Volume Dependence of Schwinger Functions in the Yukawa_2 Quantum Field Theory", Comm. Math. Phys., to appear.

[Mc 3] ______, "Convergence of the Vacuum Energy Density, φ-bounds and Existence of Wightman Functions for the Yukawa Model", Rockefeller University preprint (1975).

[MP] ______ and Y. M. Park, J. Math. Phys. 16, 104 (1975).

[Mi] R. Minlos, Tr. Mosk. Mat. Obs. 8, 471, (1959), see also [GV] vol. 4.

[Mr] H. Mitter, Z. f. Naturforschung 20a, 1505 (1965).

[N 1] E. Nelson, "Quantum Fields and Markov Fields", in Proceedings of the Summer Inst. on Part. Diff. Equ., Berkeley 1971, American Math. Soc., Publ. 1973, Providence, R. I.

[N 2] ______, J. Funct. Anal. 12, 211, (1973).

[N 3] ______, "Probability Theory and Euclidean Field Theory", in Constructive Quantum Field Theory, G. Velo and A. S. Wightman (eds.), Lecture Notes in Physics 25, Springer-Verlag, Berlin-Heidelberg-New York, (1973).

[OS 1] K. Osterwalder and R. Schrader, Comm. Math. Phys. 31, 83, (1973) and 42, 281, (1975).

[OS 2] ______, Helv. Phys. Acta 46, 272, (1973).

[Pa] Y. Park, "Lattice Approximation of the $(\varphi^4-\mu\varphi)_3$ Field Theory" to appear in J. Math. Phys., and "The $\lambda\varphi_3^4$ Euclidean Quantum Field Theory in a Periodic Box", Yonsei University preprint.

[Pr] J. Preskill, Princeton University Senior Thesis (1975).

[Sch] R. Schrader, Ann. Phys. 70, 412, (1972).

[Sw] J. Schwinger, Phys. Rev. 93, 615 (1953) and 128, 2425, (1962).

[Se] E. Seiler, Comm. Math. Phys. 42, 163 (1975).

[SeS 1] E. Seiler and B. Simon, "On Finite Mass Renormalizations in the Two-dimensional Yukawa Model", J. Math. Phys., to appear.

[SeS 2] ______, "Bounds in the Yukawa_2 Quantum Field

Theory: Upper Bound on the Pressure, Hamiltonian Bound and Linear Lower Bound", Comm. Math. Phys., to appear.

[SeS 3] ______, "Nelson's Symmetry and All That in the $Yukawa_2$ and $(\varphi^4)_3$ Field Theories", Princeton University preprint (in preparation).

[S 1] B. Simon, "The $P(\varphi)_2$ Euclidean (Quantum) Field Theory", Princeton Series in Physics, Princeton University Press, (1974).

[Si 2] ______, "Notes on Infinite Determinants of Hilbert Space Operators", Princeton University preprint (1975).

[Sp 1] T. Spencer, Comm. Math. Phys. 39, 63 (1974).

[St] R. Streater and A. Wightman, PCT, Spin, Statistics and All That, Benjamin, New York, 1964.

[Sy] K. Symanzik, J. Math. Phys. 7, 510, (1966), and "Euclidean Quantum Field Theory", in Local Quantum Theory, Int. School of Physics 'Enrico Fermi', Course 45, R. Jost (ed.), Academic Press, New York-London (1969).

[W] A. S. Wightman, in Cargèse Lectures in Theor. Physics, Gordon & Breach, New York-London-Paris, (1967).

NON-RENORMALIZABLE QUANTUM FIELD THEORIES

K. Pohlmeyer

II. Institut für Theoretische Physik
der Universität Hamburg, Germany

I. INTRODUCTION

Since the early days of renormalization theory there have been repeated attempts to quantize certain non-renormalizable classical local relativistic Lagrangian field theories formally i.e. giving the respective Green's functions as formal power series in the coupling constants, their logarithms etc.[1)–6)]. The real difficulty encountered lies in the uniqueness and not in the existence of a solution[7),8)]. So far none of these endeavours provided an undisputed answer to the problem of how to associate formally a quantum theory to a given non-renormalizable classical theory in a "natural" way.

Field theories are classified as super-renormalizable, renormalizable and non-renormalizable within the context of perturbation theory in integer powers of the coupling constants. The classification is based on the power counting theorem[9),10)]. An interaction in ν space-time dimensions is called non-renormalizable if it contains a coupling constant with negative (mass)dimension. For instance consider the interaction term involving b Bose fields and f Dirac spinor fields which may be written symbolically as $g\,"D^r\phi^b\psi^f"$ where the r derivatives D^r are applied in a specific way to the various fields and where r, b and f are non-negative integers. Such an interaction is non-renormalizable if

$$\omega = \delta - \nu > 0$$

where

$$\delta = b\frac{\nu-2}{2} + f\frac{\nu-1}{2} + r$$

since the dimension of the coupling constant g is $-\omega$.

The significance of this classification within the context of perturbation theory becomes clear if we consider the contribution

G. Velo and A. S. Wightman (eds.), Renormalization Theory, 461–482.

of an internal vertex v of type $g\,"D^r\phi^b\psi^f"$ (Wick-ordered) to the degree of superficial divergence ω_G of some graph G: upon contraction, the b propagators Δ_F and the f propagators S_F contribute to ω_G a share $-2b - f$, $(b + f)$ integrations over internal momenta a share $(b + f)\cdot\nu$, the δ-function which ensures energy-momentum conservation at the vertex v, contributes a share $-\nu$ and the r derivatives a share $+ r$ (out of the first three contributions only half of them is due to the vertex v). Consequently, the net contribution of v to ω_G is

$$\tfrac{1}{2}(-2b - f + b\nu + f\nu) - \nu + r = \delta - \nu = \omega$$

The degree of superficial divergence of graphs G which are built of V vertices, exclusively of type $g\,"D^r\phi^b\psi^f"$, is

$$\omega_G = \omega(V - C)$$

where C is a number determined by the external structure of G.

Keeping the external structure fixed while letting the perturbation theoretical order V grow, we find that the degree of superficial divergence of the respective graphs grows beyond all limits. Hence the definition of the Feynman amplitudes corresponding to these graphs requires more and more undetermined subtraction constants. This can also be seen from the fact that the corresponding regularized Feynman amplitudes behave, to leading order in the cut-off Λ, as

$$[\Lambda^{\omega}]^{V-C}$$

with the same V - independent number C as before.

The subtractions can be incorporated into the Lagrangian as counterterms [11),12)]. The unlimited number of subtractions implies an infinite variety of counterterms corresponding to new vertices i.e. interactions. This raises doubts concerning the extent to which one is still dealing with the original interaction.

The fact that the number of super-renormalizable and renormalizable theories (they are essentially [13)] the only ones that formally can be quantized) is so limited and excludes, in addition, such prominent theories like Einstein's theory of pure gravity [14)], Fermi's current x current universal weak interaction model [15)] and Weinberg's non-linear σ-model [16)], has caused misgivings all along. The above examples have the common feature that their effective coupling constants gE^{ω} are much smaller than unity for a sizable range of the energy scale E. Thus, at least in this range of E, they ought to be predestinated for a perturbation theoretical approach. The failure to quantize them at least formally seems, therefore, especially deplorable.

However, from a phenomenological point of view non-renormalizable theories are not necessarily void of predictive or descriptive power if one is willing to make the following assumption: the undetermined subtraction constants do not jeopardize order of magnitude estimates i.e. the neglected contributions from higher orders in the effective coupling constant gE^{ω} are small as long as gE^{ω} remains small. Accurate low energy theorems follow for the examples above from the tree approximations corresponding to the respective classical theories [17]. (In case of the Fermi interaction we restrict ourselves to pure lepton reactions the masses of the leptons taken to be zero.) The one-loop contributions can be calculated for quantum gravity without any remaining arbitrariness [18),19] and for the Fermi interaction and the non-linear σ-model up to two and one undetermined subtraction constants respectively [20),21]. As long as the energies involved are much smaller than $g^{-1/\omega}$, the one-loop contributions give small corrections to the respective tree approximations. However, when unitarized with the help of rational fractions à la Padé or with the help of effective range expansions, a low energy description of a large number of reactions ranging up to energies of order $g^{-1/\omega}$ is obtained in terms of zero, two and one free parameters respectively. Nevertheless, it is true that the contributions from more and more loops introduce more and more - though at each stage finitely many - free parameters into the phenomenological description.

From a less pragmatic, more fundamental point of view this seeming indeterminacy of a quantum theory associated with a given "reasonable" non-renormalizable classical theory is unacceptable. It may just reflect the inadequacy of the perturbation theoretical approach to this problem and the link between classical and quantum field theory may become transparent only outside of conventional perturbation theory.

Invariances of the theory are welcome since the number of undetermined subtraction constants is restricted by invariance requirements. However, for a finite parameter symmetry group the number of undetermined parameters in perturbation theory will still be infinite. Under favourable circumstances local gauge theories may evade the problem of non-renormalizability by providing a "renormalizable" gauge albeit with indefinite metric [22)-25].

Alternatively, instead of flooding the original (classical) interaction with an infinite variety of new interactions in the form of counterterms, one may introduce a cut-off Λ into the theory and renormalize the cut-off theory in order to insure the correct connection between interpolating and asymptotic fields and in order to insure that the couplings have specified strengths. One may then sum (partially) over the regularized renormalized perturbation theory and prove that the corresponding Green's functions tend to well-defined limits which are the Green's functions of an acceptable, in particular ghost-free field theory. In addition, one should prove that the limits are independent of the way in which the cut-off was introduced. The difficulties mentioned above are likely to arise from taking limits in the wrong order, first Λ going to infinity, then summing the perturbation expansions. This idea is by no means new. Already,

T.D. Lee [3] and Pais and Feinberg [2] pursued a program of this kind.

Some time ago, Symanzik encountered a similar situation while studying the limit "physical mass m tending to zero" of the super-renormalizable ϕ^4- theory in $(4-\varepsilon)$ dimensions, $\varepsilon > 0$ [26]. In finite order perturbation theory this limit does not exist, whereas Wilson's ε-expansion [27-31], in the context of the theory of critical phenomena, and attempts to construct a conformal invariant ϕ^4-theory in $(4-\varepsilon)$ dimensions [32] suggest that the limit $m \downarrow 0$ does exist. The observation of formal analogies between the limit $m \downarrow 0$ for the super-renormalizable ϕ^4 -theory in $(4-\varepsilon)$ dimensions and the limit $\Lambda \uparrow \infty$ for the non-renormalizable massless ϕ^4 -theory in $(4+\varepsilon)$ dimensions, $\varepsilon > 0$, recently led Symanzik to study the non-renormalizability problem of the massless ϕ^4 -theory in $(4+\varepsilon)$ dimensions [5], [6]. Symanzik's approach relies on a careful analysis of the cut-off dependence of the regularized theory.

The non-renormalizability problem may be reduced as follows: one need not sum the perturbation expansion of the individual renormalized regularized Green's functions as a whole. It simplifies matters considerably to separate the dependence of the Green's functions on the cut-off Λ from their dependence on the momenta $\underline{p}$ by factorization and to set up a new perturbation expansion for the momentum dependent factors. It suffices to sum only the perturbation series for the Λ-dependent factors and to take the limit $\Lambda \uparrow \infty$. Actually, one can do even better, as will be explained in the first part of these lectures [33] : instead of separating the Λ-and $\underline{p}$ - dependence for the individual Green's functions, one introduces an effective Lagrangian first for the unrenormalized regularized and later for the renormalized regularized theory. These effective Lagrangians furnish for the respective Feynman amplitudes corresponding to any given finite order graph the asymptotic expansions for large values of the cut-off Λ. They are themselves asymptotic series whose individual terms consist of a Λ-dependent coefficient times a Lorentz invariant product of the derivatives of the respective fields. It turns out that the Λ-dependent coefficients exhibit finite order poles for positive rational values of ε the principal parts of which are independent of Λ. Actually, these poles have their origin in infra-red singularities! (In the limit $m \downarrow 0$ of the super-renormalizable ϕ^4 -theory in $(4-\varepsilon)$ dimensions the corresponding ε-poles have their origin in ultra-violet singularities!)

The problem has now been reduced to its hard core:
i) summation of the perturbation expansions of the Λ-dependent coefficients occurring in the renormalized effective Lagrangian, ii) proof of the existence of the limit $\Lambda \uparrow \infty$ giving an effective Lagrangian which for the limits of the Green's functions provides an asymptotic expansion for small values of the renormalized coupling constant iii) proof that the limits are the Green's functions of a respectable theory and iv) proof that the limits

are independent of the way in which the cut-off is introduced. The points i) - iv) have not as yet been solved. Point ii) is somewhat supported by the fact that there exist regularizations (Pais and Uhlenbeck [34]) and parametrizations (Zinn-Justin [30),31)]) such that the parametric functions β and γ occurring in the corresponding Callan-Symanzik equations, are non-trivial and Λ-independent and thus obviously have a non-trivial limit. Concerning point iv) all we can offer is a plausibility argument that the leading high energy behaviour of the limiting theory is independent of the way in which the cut-off is introduced: Consider a special class of regularizations in which we regularize the massless ϕ^4-theory in $(4+\varepsilon)$ dimensions à la Pais and Uhlenbeck by introducing into the kinetic part of the Lagrangian different sums of higher derivatives

$$\prod_{1}^{R} (1+a_i^2\Lambda^{-2}\Box)$$

characterized by $(a) = (a_1, \ldots, a_R)$. We then define for the resulting regularized theories two kinds of parametrizations a) an intrinsic parametrization e.g. by specifying the four point vertex function at zero momentum to be $-ig\mu^{-\varepsilon}$ where g is dimensionless and independent of Λ and where μ is a fixed mass and b) a parametrization à la Zinn-Justin in terms of a certain dimensionless parameter $\bar{g}$ also independent of Λ. The parameters g and $\bar{g}$ are related in a Λ-independent, but (a) -dependent fashion

$$\bar{g} = \bar{g}(g;(a),\varepsilon).$$

Assume that the Λ-independent β-functions, $\beta(\bar{g};(a),\varepsilon)$, in the Zinn-Justin parametrization all have a (first) zero for some finite positive value of $\bar{g}$: $\bar{g} = \bar{g}_\infty((a);\varepsilon)$. In addition assume that the regularized so renormalized theories all have a limit as the cut-off Λ tends to infinity. Obviously, the limit is characterized by (a), the parametrization and the corresponding parameter g or $\bar{g}$. The leading high energy behaviour of the limiting theory (a), g in the intrisic parametrization or (a), $\bar{g}$ $(g;(a),\varepsilon)$ in the Zinn-Justin parametrization is described by the conformal invariant [35)] theory corresponding to $\bar{g} = \bar{g}_\infty((a);\varepsilon)$ in the Zinn-Justin parametrization [30),31)]. A different regularization, say (a'), leads to a limiting theory (a'), g in the intrinsic parametrization whose leading high energy behaviour is described by the conformal invariant theory corresponding to $\bar{g} = \bar{g}_\infty((a');\varepsilon)$ in the Zinn-Justin parametrization. Actually, the two conformal invariant theories are the same since the set of all conformal invariant theories is believed to be discrete [36)]. Hence the leading high energy behaviour of the limiting theory is independent of the way in which the cut-off is introduced at least as long as the regularizations belong to the class specified above.

Instead of solving the points i) - iv) we shall explore the consequences of the assumption that the limits of the regularized renormalized Green's functions do exist. It will turn our that this assumption is equivalent to the condition that the Λ-dependent coefficients occurring in the renormalized regularized effective Lagrangian converge as Λ tends to infinity. We shall see that the limiting theory for non-rational values of ε will involve fractional powers of the renormalized (intrinsic) coupling constant. For $\varepsilon = 1,2$ we expect besides integer powers of the renormalized coupling constant also integer powers of its logarithms. This conforms with the old conjecture that for a non-renormalizable theory with non-trivial scattering the vertex functions are not infinite differentiable in the coupling constant at the origin. The arbitrariness encountered in perturbation theory reflects just this lack of infinite differentiability in the coupling constant.

In the second part of these lectures we shall be concerned with the exponential self-interaction of a scalar or pseudoscalar massless field $\phi(x)$ given by the non-renormalizable classical interaction Lagrangian $\mathcal{L}_I(x) = \exp(f\phi(x)) - 1$.

We aim to set up a systematic perturbation expansion in powers of $\mathcal{L}_I$. However as far as we are aware, only the first four terms of this expansion have been constructed so far. For their definition we apply the criterion of minimal singularity. The individual terms already involve logarithms of the coupling constant f. Thus the lack of infinite differentiability in the coupling constant is explicitly taken into account.

Actually, the cosine interaction seems a much easier object to study than the exponential one in view of its boundedness. Unfortunately, however, we are not able to exploit this appealing feature of the coupling. We are forced to contract the fields of the interaction Lagrangian and this introduces immediately the unbounded cosh-function of the free massless propagator D_F. Because of the simpler combinatorics we prefer to study the exponential self-interaction.

The superpropagator method indicated above for the exponential self-interaction has been applied to Lagrangians which realize the chiral symmetry groups SU(2) x SU(2) and SU(3) x SU(3) in a non-linear fashion and which provide a semi-realistic phenomenological description of the pion-nucleon and pseudoscalar mesonoctet-baryon-octet system respectively[37)-40)]. However, a comparison of the results of the computations, based on the superpropagator method, with the experimental results is not very conclusive as regards the validity of the method.

The ideas which lie at the bottom of the superpropagator approach date back to an article by Okubo in 1954. Yet only after they were rediscovered and slightly improved by M.K. Volkov in 1967, these ideas received a great deal of attention[41)-49)].

II. THE CUT-OFF DEPENDENCE OF THE REGULARIZED MASSLESS ϕ^4-THEORY IN $(4+\varepsilon)$ DIMENSIONS

A field theory in the non-integer space-time dimension $(4+\varepsilon)$ is defined through the perturbation expansion of its Green's functions using analytic (dimensional) integration[50)-52)]. Actually we prefer to work with the vertex functions arising from the Green's functions by omitting all contributions from one particle reducible graphs, amputating the remainder i.e. multiplying it in momentum space by the inverse two point Green's functions one for (and in) every external momentum and removing the factor $(2\pi)^{4+\varepsilon}\delta(\sum p)$. To deal with the non-renormalizable massless ϕ^4-theory in $(4+\varepsilon)$ dimensions, we introduce a cut-off Λ into the kinetic part of the Lagrangian[34)] to the effect that in the Feynman rules the propagator $i(k^2+i0)^{-1}$ is replaced by

$$\frac{i}{[k^2+i0][1-\Lambda^{-2}(k^2+i0)]} = \frac{i}{k^2+i0} - \frac{i}{k^2-\Lambda^2+i0}$$

This is a Pauli-Villars regularization entailing indefinite metric of which one has to get rid in the limit as the cut-off tends to infinity. Actually, even if we restrict the real part of ε to vary between zero and three, $0 \leq \mathrm{Re}\,\varepsilon < 3$, this modification of the kinetic part of the Lagrangian does not regularize all of the Feynman amplitudes: the self-energy part is not completely regularized in the interval $2 \leq \mathrm{Re}\,\varepsilon < 3$ and is completely regularized for $0 \leq \mathrm{Re}\,\varepsilon < 2$ only provided that Wick-ordering is employed. A mass counterterm with poles at positive rational values of ε takes care of all divergences which remain after the above Pais and Uhlenbeck regularization has been applied. The wave function and coupling constant renormalizations are finite in the interval $0 \leq \mathrm{Re}\,\varepsilon < 3$ whether Wick-ordering is employed or not.

The restriction to the massless theory is motivated by the simplicity of the dimensional analysis. It is not expected to cause infra-red problems. ε is taken to be generic, i.e. either complex or positive non-rational in order to avoid the appearance of logarithms in addition to fractional powers.

We propose to study the cut-off dependence of the regularized theory given by the Lagrangian

$$\mathcal{L}_{\Lambda B}(x) = \frac{1}{2}\partial_\mu\phi_B\partial^\mu\phi_B - \frac{\Lambda^{-2}}{2}\partial_\mu\partial_\nu\phi_B\partial^\mu\partial^\nu\phi_B - \frac{1}{2}m_{B0}^2\phi_B^2 - \frac{g_B}{4!}\phi_B^4$$

the r.h.s. also written as

$$-\frac{1}{2}\phi_B\Box(1+\Lambda^{-2}\Box)\phi_B - \frac{1}{2}m_{B0}^2\phi_B^2 - \frac{g_B}{4!}\phi_B^4 .$$

The bare field ϕ_B has dimension $1+\varepsilon/2$, the bare coupling constant g dimension $-\varepsilon$. m_{B0} is the bare mass for the physical mass zero theory.

Note that $(g_B\Lambda^\varepsilon)$ is the only dimensionless parameter in the theory. A first application of dimensional analysis (only integer powers of g_B occur) gives

$$m_{B0}^2 = m_{B0}(\Lambda, \varepsilon)^2 = \Lambda^2 \sum_{k=1}^{\infty} a_k(\varepsilon)(g_B \Lambda^{\varepsilon})^k$$

the coefficients $a_k(\varepsilon)$ being meromorphic functions in ε.

The vertex functions - here, the unrenormalized ones apart from mass renormalization -

$$\Gamma_{\Lambda B}(p_1, \ldots, p_{2n}; g_B, \varepsilon) = \Gamma_{\Lambda B}((2n); g_B, \varepsilon)$$

have dimension $4 - 2n - \varepsilon(n - 1)$.

After the wave- and coupling constant renormalizations have been performed:

$$\mathcal{L}_{\Lambda B}(x) \longrightarrow \mathcal{L}_{\Lambda}(x), \quad \Gamma_{\Lambda B}((2n); g_B, \varepsilon) \longrightarrow \Gamma_{\Lambda}((2n); g\mu^{-\varepsilon}, \varepsilon)$$

the large Λ-expansion of the renormalized vertex functions in the $(\phi^4)_{(4+\varepsilon)}$-theory is formally identical to the small m-expansion in the $(\phi^4)_{(4-\varepsilon)}$-theory when Λ^{-1} is replaced by m.

The cut-off dependence of the unrenormalized vertex functions is described in a compact form with the help of the effective unrenormalized Lagrangian

$$L_{\Lambda B}(x) = \{-\tfrac{1}{2}\phi_B \Box \phi_B - \tfrac{1}{4!} g_B \phi_B^4\} + \{-\tfrac{1}{2} m_{B0}^2 \phi_B^2 - \tfrac{1}{2}\Lambda^{-2}\phi_B \Box^2 \phi_B + \sum_{r=0}^{\infty}\sum_{s=1}^{\infty}\sum_{\nu=1}^{n_{rs}} f_{rs\nu}(g_B\Lambda^{\varepsilon}, \varepsilon)(\text{"}D^{2r}\phi_B^{2s}\text{"})_\nu\, g_B^{s-1}\Lambda^{4-2r-2s}\}$$

where

i) analytic integration is to be employed, supplemented by the convention that graphs with detached parts do not contribute i.e. parts which are joined to the rest of the graph by one vertex only and into which no external momentum enters,

ii) the terms in the second brackets are to be treated as repeated insertions into graphs defined by the first bracket,

iii) the sums $\sum_\nu$ extend only over those Lorentz invariant products of derivatives of the bare field ϕ_B which are linearly independent at zero momentum transfer i.e.

$$b_1(\text{"}D^{2r}\phi_B^{2s}\text{"})_{\nu_1} + \cdots + b_j(\text{"}D^{2r}\phi_B^{2s}\text{"})_{\nu_j} = \partial_\mu \sigma$$

implies

$$b_1 = \cdots = b_j = 0.$$

($\partial_\mu\phi_B\partial^\mu\phi_B$ and $\phi_B\Box\phi_B$ provide an example of operator products which are linearly dependent at zero momentum transfer

$$\partial_\mu\phi_B\partial^\mu\phi_B + \phi_B\Box\phi_B = \partial_\mu\{\phi_B\partial^\mu\phi_B\}),$$

iv) $$f_{rs\nu}(g_B\Lambda^{\varepsilon}, \varepsilon) = \sum_{h=1}^{\infty} f_{rs\nu h}(\varepsilon)(g_B\Lambda^{\varepsilon})^h$$

$f_{rs\nu h}(\varepsilon)$ being meromorphic functions of ε,

v) the infinite sums are to be interpreted as asymptotic expansions which for any finite order vertex graph based on the Lagrangian $\mathcal{L}_{\Lambda B}(x)$ produce the large cut-off behaviour of the corresponding amplitude to an arbitrary degree of accuracy

$$\Gamma_{\Lambda B}((2n); g_B, \varepsilon) = \sum_{j=0}^{\infty}\sum_{k=0}^{\infty} \Lambda^{-2j+\varepsilon k} F_{jk}((2n); g_B, \varepsilon)$$

where the F_{jk}'s are power series in g_B with meromorphic coefficients which receive contributions from the vertex graphs with k or more loops.

The effective unrenormalized Lagrangian is obtained from $\mathcal{L}_{\Lambda B}(x)$ by first subtracting the asymptotic series $\sum_r\sum_s\sum_v\sum_h$ the individual terms of which correspond to counterterms [11),12)] (oversubtraction at zero momentum) removing the finite contributions to $\Gamma_{\Lambda B}$

$$\sum_{j=0}^{\infty}\sum_{k=1}^{\infty} \Lambda^{-2j+\varepsilon k} F_{jk}((2n); g_B, \varepsilon)$$

caused by the ultra-violet poles of the multiple Mellin transform of

$$\prod_{\ell} \frac{1}{1-\Lambda^{-2}(q_\ell^2+i0)}$$

and then adding these series again. This yields the first and second bracket respectively.
Any vertex graph of finite order V contributes only to those f_{rsvh} with $h \leq V - (s-1)$.
Upon analytic integration all insertions apart from $-\frac{1}{2} m_{B0}^2 \phi_B^2$ and $\Lambda^2 f_{011}(g_B\Lambda^\varepsilon, \varepsilon)\phi_B^2$ give vanishing self-energy part at zero momentum for dimensional reasons. Thus we conclude

$$m_{B0}^2 = 2\Lambda^2 f_{011}(g_B\Lambda^\varepsilon, \varepsilon).$$

F_{00} is obtained from the Λ-independent part of the effective unrenormalized Lagrangian.
The coefficients f_{rsv} are closely related to the unrenormalized vertex functions $\Gamma_{\Lambda B}$. For instance, dimensional analysis for $2 < \mathrm{Re}\,\varepsilon < 3$ gives

$$f_{031}(g_B\Lambda^\varepsilon, \varepsilon) = -\frac{i}{6!} g_B^{-2}\Lambda^2\, \Gamma_{\Lambda B}(0,0,0,0,0,0; g_B, \varepsilon).$$

Out of all terms of $L_{\Lambda B}(x)$ only the once inserted "ϕ^6"-term contributes to the r.h.s.
The lowest order contribution to $\Gamma_{\Lambda B}(0,0,0,0,0,0; g_B, \varepsilon)$ comes from the third order triangle graph with propagators

$$\frac{i}{(k^2+i0)(1-\Lambda^{-2}(k^2+i0))} \quad .$$

The six point vertex function at zero momentum has infra-red singularities at positive rational values of ε in the interval $0 \leq \varepsilon \leq 2$.

The unrenormalized vertex function $\Gamma_{\Lambda B}((2n); g_B, \varepsilon)$ is a distribution - valued analytic function in ε in the complex plane cut along the real axis from $-\infty$ to 0 and from +3 to $+\infty$. In particular, $\Gamma_{\Lambda B}$ is regular in the interval $0 < \varepsilon < 3$ whereas the F_{jk}'s have poles there. Hence in this interval the poles of

F_{00} must cancel against what is left after partial mutual compensation of the poles of those $F_{jk}, j>0$, which are accompanied by Λ to the power zero: $-2j+\varepsilon k = 0$. Correspondingly, in this interval the singularities of F_{01} must cancel against the singularities which remain after partial mutual cancellation of the poles of F_{jk}, $j>0$, with $-2j+\varepsilon k = \varepsilon$ etc.

In order to pass to the renormalized quantities, we combine those terms of $L_{\Lambda B}(x)$ which involve the operator product $\phi_B\Box\phi_B$:

$$1 - 2f_{111}(g_B\Lambda^\varepsilon,\varepsilon) \doteq Z_3(g_B\Lambda^\varepsilon,\varepsilon)^{-1}$$

and the operator product ϕ_B^4 :

$$g_B[1-4!\,f_{021}(g_B\Lambda^\varepsilon,\varepsilon)] \doteq g\mu^{-\varepsilon}Z_3(g_B\Lambda^\varepsilon,\varepsilon)^{-2}$$

with g dimensionless and μ some fixed mass (g and μ occur only in the combination $g\mu^{-\varepsilon}$). By inverting the last equation:

$$g_B\Lambda^\varepsilon = \bar{g}(g(\Lambda/\mu)^\varepsilon,\varepsilon)$$

g_B is expressed in terms of $(\varepsilon,)\,g\mu^{-\varepsilon}$ and Λ:

$$g_B = g_B(g\mu^{-\varepsilon},\Lambda,\varepsilon).$$

Next, we let the bare coupling constant and the bare field depend on Λ in the following way

$$g_B = g_B(g\mu^{-\varepsilon},\Lambda,\varepsilon) = \Lambda^{-\varepsilon}\bar{g}(g(\Lambda/\mu)^\varepsilon,\varepsilon)$$

$$\phi_B = \phi_B(g\mu^{-\varepsilon},\Lambda,\varepsilon) = Z_3(g(\Lambda/\mu)^\varepsilon,\varepsilon)^{1/2}\phi$$

with the renormalized intrinsic coupling constant $g\mu^{-\varepsilon}$ and the renormalized field ϕ which are both independent of Λ.
Thereby the effective unrenormalized Lagrangian $L_{\Lambda B}(x)$ corresponding to the unrenormalized regularized Lagrangian $\mathcal{L}_{\Lambda B}(x)$ goes over into the effective renormalized Lagrangian $L_\Lambda(x)$

$$L_\Lambda(x) = \left\{-\tfrac{1}{2}\phi\Box\phi - \frac{g\mu^{-\varepsilon}}{4!}\phi^4\right\} + \left\{\sum_{\substack{r=0\\ r+s\geq 3}}^{\infty}\sum_{s=1}^{\infty}\sum_{\nu=1}^{n_{rs}} c_{rs\nu}(g(\Lambda/\mu)^\varepsilon,\varepsilon)\,(\text{"}D^{2r}\phi^{2s}\text{"})_\nu\,(g\mu^{-\varepsilon})^{s-1+(2r+2s-4)/\varepsilon}\right\}$$

with

$$c_{rs\nu}(z,\varepsilon) = z^{(4-2r-2s)/\varepsilon}\left\{-\tfrac{1}{2}\delta_{r1}\delta_{s1}Z_3 + f_{rs\nu}(\bar{g}(z,\varepsilon),\varepsilon)Z_3^s\left(\frac{\bar{g}(z,\varepsilon)}{z}\right)^{s-1}\right\}$$

$$Z_3 = Z_3(\bar{g}(z,\varepsilon),\varepsilon)$$

corresponding to the renormalized regularized Lagrangian

$$\mathcal{L}_\Lambda(x) = -\tfrac{1}{2}\phi_B\Box(1+\Lambda^{-2}\Box)\phi_B - \tfrac{1}{2}m_{B0}^2\phi_B^2 - \frac{g_B}{4!}\phi_B^4$$

where

$$\phi_B = \phi_B(g\mu^{-\varepsilon},\Lambda,\varepsilon),\quad g_B = g_B(g\mu^{-\varepsilon},\Lambda,\varepsilon).$$

We are entitled to call $g\mu^{-\varepsilon}$ the intrinsic renormalized coupling constant, ϕ the renormalized field etc. since

$$\Gamma_\Lambda(0,0;g\mu^{-\varepsilon},\varepsilon) = 0$$

$$\frac{\partial}{\partial p^2}\Gamma_\Lambda(p,-p;g\mu^{-\varepsilon},\varepsilon)\Big/_{p^2=0} = +i$$

$$\Gamma_\Lambda(0,0,0,0;g\mu^{-\varepsilon},\varepsilon) = -ig\mu^{-\varepsilon}.$$

Here $\Gamma_\Lambda(p_1,\cdots,p_{2n};g\mu^{-\varepsilon},\varepsilon) = \Gamma_\Lambda((2n);g\mu^{-\varepsilon},\varepsilon)$ denote the vertex functions corresponding to $\mathcal{L}_\Lambda(x)$. For dimensional reasons, the left hand sides of the above equations receive contributions only from the first bracket of $L_\Lambda(x)$. Moreover, these contributions are of lowest order in g.

The coefficients $c_{rs\nu}(g(\Lambda/\mu)^\varepsilon,\varepsilon)$ are closely related to the renormalized vertex functions Γ_Λ. For instance, for $2<\mathrm{Re}\,\varepsilon<4$

$$c_{031}(g(\Lambda/\mu)^\varepsilon,\varepsilon) = -\frac{i}{6!}[g\mu^{-\varepsilon}]^{-2-2/\varepsilon}\cdot\Gamma_\Lambda(0,0,0,0,0,0;g\mu^{-\varepsilon},\varepsilon).$$

In this interval, $\Gamma_0(0,\cdots,0;g\mu^{-\varepsilon},\varepsilon)$ is equal to zero since

$$\frac{i}{k^2+i0} - \frac{i}{k^2-\Lambda^2+i0} \xrightarrow[\Lambda\downarrow 0]{} 0 \qquad \text{sufficiently strongly.}$$

In the interval $0<\varepsilon\leq 2$, the renormalized six point function at zero momentum $\Gamma_\Lambda(0,\cdots,0;g\mu^{-\varepsilon},\varepsilon)$ has infra-red poles at positive rational values of ε whereas its Λ-derivative is regular there;

$$\Lambda\frac{\partial}{\partial\Lambda}\Gamma_\Lambda(0,\cdots,0;g\mu^{-\varepsilon},\varepsilon) = \Lambda^{-2-2\varepsilon}a(g(\Lambda/\mu)^\varepsilon,\varepsilon)$$

where

$$a(z,\varepsilon) = \sum^{\infty} a_k(\varepsilon)z^k, \quad z = g(\Lambda/\mu)^\varepsilon$$

and where the $a_k(\varepsilon)$[3] are analytic in ε in the complex plane cut along the real axis from $-\infty$ to 0 and from +4 to $+\infty$.
It is easy to make the ε-singularities of $c_{031}(g(\Lambda/\mu)^\varepsilon,\varepsilon)$ explicit by differentiation and reintegration: For $2<\mathrm{Re}\,\varepsilon<4$

$$c_{031}(g(\Lambda/\mu)^\varepsilon,\varepsilon) = -\frac{i}{6!}\sum_{k=3}^{\infty}\frac{a_k(\varepsilon)}{(k-2)\varepsilon-2} - \frac{i}{6!}[g\mu^{-\varepsilon}]^{-2-2/\varepsilon}\cdot$$
$$\cdot\{\Gamma_\Lambda(0,\cdots,0;g\mu^{-\varepsilon},\varepsilon) - \Gamma_{\mu g^{-1/\varepsilon}}(0,\cdots,0;g\mu^{-\varepsilon},\varepsilon)\}.$$

For $0<\mathrm{Re}\,\varepsilon\leq 2$, c_{031} is given by the analytic continuation of the r.h.s. The difference of the vertex functions in the last bracket is analytic for $0<\mathrm{Re}\,\varepsilon<4$.
We note that neither the positions nor the residues of the poles of $c_{031}(z,\varepsilon)$ depend on Λ. The singular parts of $c_{031}(z,\varepsilon)$ can be calculated exactly by finite order perturbation theory, whereas exact computation of the finite part of $c_{031}(z,\varepsilon)$ at the pole $\varepsilon = 2/(k-2)$ requires summation of the perturbation expansion.
A similar analysis can also be carried out for $c_{rs\nu}(z,\varepsilon)$ in the general case.

The renormalized vertex functions Γ_Λ have the following asymptotic expansion for large values of the cut-off

$$\Gamma_\Lambda((2n);g\mu^{-\varepsilon},\varepsilon) = h_{00}((2n);g\mu^{-\varepsilon},\varepsilon)+\sum_{j=1}^{\infty}\sum_{k=0}^{\infty}\Lambda^{-2j+\varepsilon k}h_{jk}((2n);g\mu^{-\varepsilon},\varepsilon)$$

where the $h_{jk}((2n);g\mu^{-\varepsilon},\varepsilon)$'s are power series in $(g\mu^{-\varepsilon})$ starting with $(g\mu^{-\varepsilon})^{n+k+j-1}$ the coefficients of which being meromorphic in ε. h_{00} is obtained from F_{00} by replacing g_B by $(g\mu^{-\varepsilon})$. Now, Γ_Λ is an analytic function of ε in the complex plane cut along the real axis from $-\infty$ to 0 and from +4 to $+\infty$. In particular, Γ_Λ is regular for real positive ε with $0<\varepsilon<4$ whereas in this interval the h_{jk}'s have poles. Again, the ε-singularities of h_{00} must cancel the singularities which are left after partial mutual compensation of the ε-poles of $h_{jk}, j>0$, with $-2j+\varepsilon k=0$ and the ε-singularities of h_{jk}, $j>0$, with $-2j+\varepsilon k = \pm\varepsilon, \pm 2\varepsilon, \dots$ must cancel each other. In both cases the cancellation must be identical in g and in the momenta.

The small momentum expansion of the renormalized regularized vertex function Γ_Λ is a double series in integer and in fractional powers of the momentum scale. After summation over the fractional powers it takes the form

$$\Gamma_\Lambda((2n); g\mu^{-\varepsilon},\varepsilon) = h_{00}((2n)\cdot g\mu^{-\varepsilon},\varepsilon) + \sum_{j=1}^{\infty} (g\mu^{-\varepsilon})^{2j/\varepsilon} h_j((2n)\cdot g\mu^{-\varepsilon},\varepsilon|z)$$

with

$$h_j((2n)\cdot g\mu^{-\varepsilon},\varepsilon|z) = \sum_{k=0}^{\infty} z^{k-2j/\varepsilon}\left[(g\mu^{-\varepsilon})^{-k} h_{jk}((2n); g\mu^{-\varepsilon},\varepsilon)\right],$$

$$h(\varkappa p_1,\dots,\varkappa p_{2n}; g\mu^{-\varepsilon},\varepsilon|z) = \varkappa^{4+2j-2n}[g\mu^{-\varepsilon}]^{n-1}\{h_j(p_1,\dots,p_{2n}; g_\varkappa\mu^{-\varepsilon},\varepsilon|z)/[g_\varkappa\mu^{-\varepsilon}]^{n-1}\},$$

$g_\varkappa = \varkappa^\varepsilon g$

Now, assume that with Λ going to infinity the renormalized regularized vertex functions converge sufficiently uniformly in ε (in the sense of distributions) to finite limits for all positive integers n

$$\lim_{\Lambda\uparrow\infty} \Gamma_\Lambda((2n); g\mu^{-\varepsilon},\varepsilon) = \Gamma_\infty((2n); g\mu^{-\varepsilon},\varepsilon).$$

Then, because of the close relation between the vertex functions Γ_Λ and the coefficients $c_{rs\nu}$, also the limits

$$\lim_{z\uparrow\infty} c_{rs\nu}(z,\varepsilon) = \bar{c}_{rs\nu}(\varepsilon)$$

exist for all r,s,ν and the $\bar{c}_{rs\nu}(\varepsilon)$ have the same singular parts as the $c_{rs\nu}(z,\varepsilon)$.

Conversely, assume that with z going to infinity the coefficients $c_{rs\nu}(z,\varepsilon)$ converge to well-defined limits $\bar{c}_{rs\nu}(\varepsilon)$ for all r,s,ν. Then the effective renormalized Lagrangians $L_\Lambda(x)$ converge to a well-defined effective Lagrangian $\bar{L}(x)$. Moreover, if the coefficients $c_{rs\nu}(z,\varepsilon)$ converge sufficiently uniformly in ε, the above cancellation mechanism persists to work in the limit when the cut-off is removed. Then the limit

$$h_{00}((2n); g\mu^{-\varepsilon},\varepsilon) + \lim_{z\uparrow\infty}\left\{\sum_{j=1}^{\infty}(g\mu^{-\varepsilon})^{2j/\varepsilon} h_j((2n); g\mu^{-\varepsilon},\varepsilon|z)\right.$$

exists and defines a small momentum expansion or, which is the same, a small g expansion which coincides with the expansion

$$h_{00}((2n); g\mu^{-\varepsilon},\varepsilon) + \sum_{j=1}^{\infty}(g\mu^{-\varepsilon})^{2j/\varepsilon} h_j((2n); g\mu^{-\varepsilon},\varepsilon)$$

furnished by $\bar{L}(x)$. (Note that the individual terms h_{00} and h_j still have singularities for $0<\mathrm{Re}\,\varepsilon<4$.) Provided that the mapping from the sequence of the renormalized regularized vertex functions $\{\Gamma_\Lambda((2n); g\mu^{-\varepsilon},\varepsilon)\}_{n=1,2,\dots}$ to the sequence of their small momentum expansions is invertible and (in a certain topology) continuous in both directions, we may conclude that the vertex functions Γ_Λ converge (sufficiently uniformly in ε) to well defined limits Γ_∞. As functions of ε the Γ_∞'s are analytic in the complex plane cut along the real axis from $-\infty$ to 0 and from +4 to $+\infty$. It is clear that the expansion

$$\Gamma_\infty((2n);g\mu^{-\varepsilon},\varepsilon) = h_{00}((2n);g\mu^{-\varepsilon},\varepsilon)+\sum_{j=1}^{\infty}(g\mu^{-\varepsilon})^{2j/\varepsilon}h_j((2n);g\mu^{-\varepsilon},\varepsilon)$$

is not very useful because its individual terms have singularities in ε for $0<\mathrm{Re}\,\varepsilon<4$ where Γ_∞ has none. Thus it seems mandatory to reorganize the expansion in such a way that there are no more spurious ε-singularities present. This is just another point to be added to the list of problems which was given in the introduction.

In the asymptotic expansion of Γ_∞ for small values of g besides integer powers of g also fractional powers of g occur. If ε is equal to one or two, along with the integer powers of g also integer powers of logarithms of g appear.

II. EXPONENTIAL LAGRANGIANS

Lagrangian densities of the type

$$\mathcal{L}_I(x)=\sum_n^\infty a_n[f\phi(x)]^n,\quad a_n\neq 0 \text{ for infinitely many } n$$

where $\sum a_n z^n$ is an entire function, $\phi(x)$ a scalar field and f a coupling constant, are very singular objects. This can already be seen from the behaviour of the vacuum expectation values (VEV) of $\mathcal{L}^0_I(x)\mathcal{L}^0_I(y)$ in momentum space

$$\mathcal{F}_x\{\langle|\mathcal{L}^0_I(x/2)\mathcal{L}^0_I(-x/2)|\rangle\}(k) = \sum^\infty a_n^2 n!\,\mathcal{F}_x\{[if^2\Delta^{(+)}(x)]^n\}(k)$$
$$\sim\Theta(k^2)\Theta(k_0)\sum_n\frac{n!\,a_n^2 f^{2n}}{(n-2)!(n-1)!}\left(\frac{k^2}{4\pi^2}\right)^n$$

where $\mathcal{L}^0_I(x)=\sum_n^\infty a_n f^n:\phi_0^n:(x)$, $\phi_0(x)$ the free scalar field.

Jaffe and Meimann [53),54)] have investigated the definition of such operators $\mathcal{L}^0_I$. In order to compensate for the steep increase of their vacuum expectation values for large momenta, the test functions have to decrease sufficiently rapidly. However, if they decrease too rapidly in momentum space they will be analytic in configuration space and no test function with compact support in configuration space will remain. Thus, strict locality in the form of vanishing commutators for space-like separation can no longer be formulated. If we insist on strict locality - and we certainly would like to - then test functions with less rapid decrease in momentum space must already kill the increase of the vacuum expectation values.

This leads to the condition on $\mathcal{L}_I$ that $\sum a_n z^n$ must be an entire function of order less than two [55)]. We restrict ourselves to this class of Lagrangians, the so-called localizable Lagrangians.

Now, Epstein and Glaser [56)] have shown that for any localizable interaction Lagrangian, in particular for the exponential one, the time ordered products

$$T(x_1,\cdots,x_n) = i^{n+1}\,T\,\mathcal{L}^0_I(x_1)\cdots\mathcal{L}^0_I(x_n)$$

entering into the Gell-Man Low formula for the Green's functions or into the perturbation expansion for the S-matrix

$$S = 1 + \sum_{n=1}^{\infty} \frac{1}{n!} \int \cdots \int dx_1 \cdots dx_n \, i^n \, T \, \mathcal{L}_I^0(x_1) \cdots \mathcal{L}_I^0(x_n)$$

can be defined such that the T's satisfy:
a) unitarity b) locality c) a bound on the type of singularities for partly or fully coinciding arguments d) Lorentz invariance e) symmetry under permutations of the arguments.
Point c) needs explanation. The anti-commutator of two $\mathcal{L}_I^0$'s enters into the corresponding time ordered product as its absorptive part thus giving a lower bound for the degree of singularity of $T(x_1, x_2)$ at $x_1 = x_2$. Similarly, in the successive construction of the T's , products of the form

$$T(x_{i_1}, \cdots, x_{i_r}) \, T(x_{i_{r+1}}, \cdots, x_{i_n})^{\dagger}$$

assumed to be already determined, appear as absorptive parts $T - T^{\dagger}$ for the T's and thus give a lower bound for the degree of singularity for the T's.
Now, one would like to define the dispersive part $T + T^{\dagger}$ such that the "degree" of the singularities does not exceed the degree already dictated by $T - T^{\dagger}$.
In particular, if the vacuum expectation value $\langle | T - T^{\dagger} | \rangle$ can be bounded in momentum space by an entire function in the invariant momenta of some exponential order and type, one would like to define the vacuum expectation value $\langle | T + T^{\dagger} | \rangle$ such that it is bounded by an entire function of the same exponential order and type. Epstein and Glaser constructed $\langle | T + T^{\dagger} | \rangle$'s which are bounded by entire functions in the invariant momenta the exponential order of which may increase by an infinitesimal amount as compared to the exponential order of the corresponding $\langle | T - T^{\dagger} | \rangle$'s.

These properties a) - e) still leave a great deal of arbitrariness in the definition of the T-products and the problem is exactly the removal of this arbitrariness.
In view of the polynomial boundedness of the cross-sections, proved by Epstein, Glaser and Martin [57] as well as by Mc Dowell, Roskies and Schroer [58] for arbitrary localizable interactions, we want the real part of the Green's functions - the only part we can dispose of - to have a high-energy behaviour which is as mild as possible. This will be our minimality criterion which singles out at least up to fourth order a unique definition. In a way this is a directionally dependent version of the minimality requirement of polynomial renormalizable theories.

We now turn to the special case of the exponential self-interaction of a scalar field

$$\mathcal{L}_I^0(x) = :e^{f\phi_0(x)} - 1:$$

For exponential interactions the combinatorics is particularly simple. We have the following formal relation between time - and normal ordered products

$$T:e^{f\phi_0(x_1)}:\ldots:e^{f\phi_0(x_n)}: = \prod_{1\leq i<j\leq n} e^{if^2\Delta_F(x_i-x_j)}:e^{f\phi_0(x_1)}\ldots e^{f\phi_0(x_n)}:$$

where the first factor on the r.h.s. is still to be defined. This formula allows us to pass with ease from time - ordered products

$$T(x_1,\ldots,x_n) = i^{n+1}\, T\, \mathcal{L}_I^0(x_1)\ldots\mathcal{L}_I^0(x_n)$$

to their connected vacuum expectation values

$$\tau(x_1,\ldots,x_n) = i^{n+1}\langle|T\,\mathcal{L}_I^0(x_1)\ldots\mathcal{L}_I^0(x_n)|\rangle_{conn.}$$

and vice versa.

In a successive construction of the time - ordered vacuum expectation values along Bogoliubov lines, $\tau(x_1,\ldots,x_n)$ is determined by $\tau(x_1,\ldots x_r)$, $r<n$, via locality and unitarity up to an arbitrary real Lorentz invariant localizable distribution (the latter with support in the points where all n arguments $x_1,\ldots,x_n$ coincide). In the following it is precisely the elaboration of the criterion for the removal of this arbitrariness that will concern us. We shall find it advantageous not to expand the exponentials

$$\exp(if^2\Delta_F(x_i-x_j)) - 1$$

in powers of f^2 which is just the superpropagator method.

To illustrate the general idea, let us consider for three particular cases the problem of extending a real symmetric distribution $\tau(\xi)$ of one real variable ξ supposed to be known and to be continuous outside the origin.

Case 1: The definition of τ can be extended so as to give an integrable function over all of $\mathbb{R}^1$.
Naturally we choose just this integrable extension. Any other definition would introduce δ-derivative type singularities previously absent which can be separated from the integrable background by a certain shrinking procedure.

Case 2: τ goes like $|\xi|^{-2k+3}$ for some finite integer $k \geq 2$; τ piles up a nonintegrable singularity at the origin.
Then the δ-function at the origin and its first k-2 even derivatives there, $\delta^{(2\nu)}(\xi)$ $\nu = 0,\ldots,k-2$, cannot be separated from the rest whereas the derivatives $\delta^{(2\nu)}(\xi)$ with $\nu \geq k-1$ can be separated.
In this case, we obtain a (k - 1)-parameter class of equally singular definitions. Extensions involving $\delta^{(2\nu)}(\xi)$ with $\nu \geq k-1$ would be more singular.

Case 3: τ goes like $\sum_1^\infty |C_n|\cdot|\xi|^{-n}$, $C_n \neq 0$ for infinitely many n.
The behaviour of τ on the reals near the origin does not suggest any finite parameter-class of extensions.

Next, we have to explain in what way this simple minded discussion is related to the problem of defining the time ordered vacuum expectation values $\tau(x_1, \cdots, x_n)$.

We first consider the problem for the time ordered vacuum expectation value $\tau(x_1, x_2)$ of two $\mathcal{L}_I^{\circ}$'s which is formally

$$-i\left\{e^{f^2 i\Delta_F(x_1-x_2)} - 1\right\}.$$

We confine our attention to its real part since it is only the latter which is not completely determined.
Moreover, since time ordering involves just multiplication of a well-defined object with step functions that depend only on the time variables, we average over the spatial variables with real, sufficiently smooth test functions i.e. with test functions whose Fourier transforms decrease faster than

$$\exp\left\{-A\left[\|\vec{p}_1\|^{2/3} + \|\vec{p}_2\|^{2/3}\right]\right\}$$

for any positive A. We study the resulting distribution in the time difference variable $\mathcal{R}e\,\tau_f(x_1^{\circ}-x_2^{\circ})$. It turns out that we are dealing with the situation of case 1. In fact, $\mathcal{R}e\,\tau_f$ cannot only be extended to an integrable function but even to an infinite differentiable function. This statement holds for all real, sufficiently smooth test functions. Consequently, we obtain unique least singular choices of $\mathcal{R}e\,\tau_f$ for all such f's.

It turns out that these $\mathcal{R}e\,\tau_f(x_1^{\circ}-x_2^{\circ})$ for various f's yield, when patched together, one Lorentz invariant definition for $\mathcal{R}e\,\tau(x_1-x_2)$ as a distribution in the time and space variables. If we add $\mathcal{I}m\,\tau(x_1-x_2)$ to it we are led to a unique least singular definition of $\tau(x_1-x_2)$, the superpropagator.

On the other hand, the distributions $(i\,\Delta_F)^n$, $n \geq 2$, subjected to the same treatment of spatial smearing, provide examples of case 2. We are thus led to an ever increasing (n - 1) - parameter class of definitions. If we would resolve the superpropagator into products of $i\Delta_F$'s

$$-i\left\{e^{f^2 i\Delta_F(x_1-x_2)} - 1\right\} = -i\sum_{n=1}^{\infty}\frac{\left(f^2 i\,\Delta_F(x_1-x_2)\right)^n}{n!}$$

we would then be left with an infinite parameter-class of definitions.

The time ordered vacuum expectation values of an arbitrary number of $\mathcal{L}_I^{\circ}$'s can be represented by sets of graphs in which each pair of points (x_i, x_j) is connected by at most one line representing the superpropagator. For our purposes , all vertices are to be considered external. For instance, the time ordered vacuum expectation value of three $\mathcal{L}_I^{\circ}$'s can be depicted graphically by

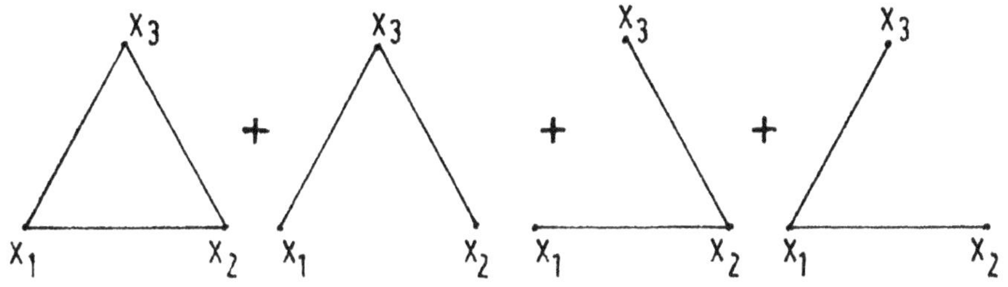

From here on we restrict ourselves to exponentially self-coupled massless fields. The restriction to massless fields is imposed for purely technical reasons

$$i\Delta_F(x) = -\frac{1}{4\pi^2}\,\frac{1}{x^2 - i0}$$

The results of our ultra-violet analysis are not expected to depend on this restriction.

We adopt the same procedure as before i.e. we confine our attention to $\mathcal{R}e\,\tau_f(x_1^\circ, x_2^\circ, x_3^\circ)$, where the subscript f again denotes averaging with respect to the spatial variables, since $\mathcal{I}m\,\tau(x_1,\cdots,x_3)$ is already completely defined by the first and second order. $\mathcal{R}e\,\tau_f(x_1^\circ,\cdots,x_3^\circ)$ is defined everywhere except at the points where $x_1^\circ = x_2^\circ = x_3^\circ$. A detailed analysis shows the following structure of $\mathcal{R}e\,\tau_f$ outside the points of totally coinciding time arguments. $\mathcal{R}e\,\tau_f$ consists of a background and of δ-derivative type singularities on top of it. The background is smooth - nothing piles up near the points where two times coincide, especially not near the points where all three times coincide. The δ-derivative type singularities

$$\sum_{\sigma\in\mathfrak{S}_3}\sum_{\nu=0}^{\infty} h^{\nu}_{\sigma,f}(x^\circ_{\sigma(1)} - x^\circ_{\sigma(2)})\,\delta^{(\nu)}(x^\circ_{\sigma(2)} - x^\circ_{\sigma(3)})$$

can be separated from the background. They are concentrated on the lines where two times coincide. The $h^{\nu}_{\sigma,f}$'s are smooth (C^∞) functions outside the origin and do not pile up non-integrable singularities (within the lines) as one approaches the origin. Consequently, we can extend the background as well as the $h^{\nu}_{\sigma,f}$'s to smooth functions over all of $\mathbb{R}^3$ and $\mathbb{R}^1$ respectively, a case 1 situation.

This provides us with a particular extension which is the least singular definition of $\mathcal{R}e\,\tau_f$ on all of $\mathbb{R}^3$, since any other choice would introduce singularities attached to the points of totally coinciding time arguments which could be separated from the rest. This statement holds for all real sufficiently smooth test functions f. When pieced together, the definitions of $\mathcal{R}e\,\tau_f$ for the various f's turn out to be the spatial averages of a single Lorentz invariant distribution $\mathcal{R}e\,\tau(x_1,\cdots,x_3)$. In this way we are led to a least singular choice of $\tau(x_1,\cdots,x_3)$.

Next, we consicer the fourth order graphs

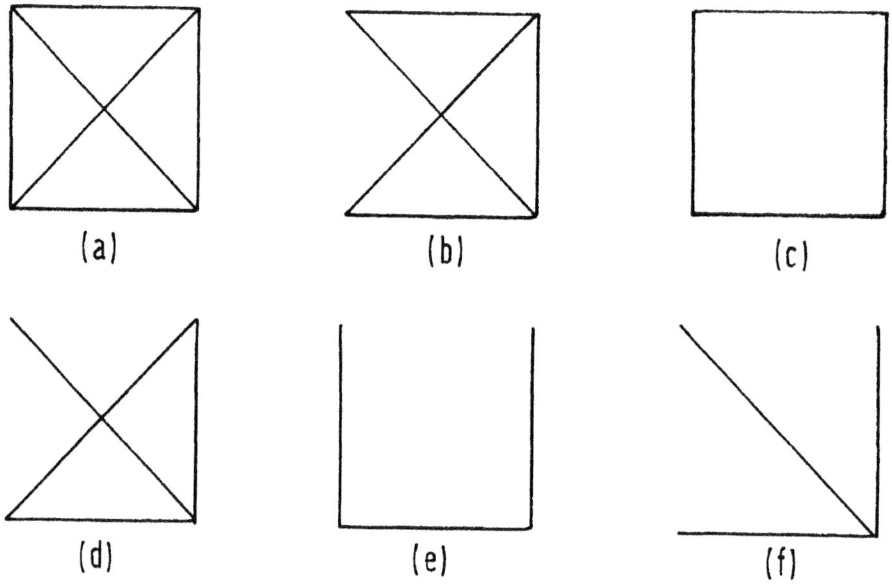

Here we encounter for the first time overlapping divergences. We again follow the same beaten track and confine our attention to $\mathcal{R}e\,\tau_f(x_1^0,\cdots,x_4^0)$, the subscript f denoting as usual space averages with a real sufficiently smooth test function, $\mathcal{I}m\,\tau(x_1,\cdots,x_4)$ being already defined.

The structure of a general choice of $\mathcal{R}e\,\tau_f(x_1^0,\cdots,x_4^0)$ is as follows :

$\mathcal{R}e\,\tau_f$ consists of a smooth background, -derivative type singularities (concentrated on planes which are tensorially multiplied by functions integrable over all of the respective planes

δ -derivative type singularities (concentrated on the lines $x_i^0 = x_j^0 = x_k^0$ -three coinciding times - or on the lines $x_i^0 = x_j^0$ $x_k^0 = x_l^0$ - two pairs of coinciding times -) which are tensorially multiplied by functions integrable over all of the respective lines

and a localizable distribution attached solely to the points where all four time arguments coincide.

It is possible to separate all the listed contributions from each other, in particular those singularities which are attached to the points where all time arguments coincide from all remaining singularities. The preferred choice of $\mathcal{R}e\,\tau_f$ to which we are led is the one which does not receive any singular contribution from the points where $x_1^0 = x_2^0 = x_3^0 = x_4^0$. This is the least singular extension which is also the most naive one. The preferred choices of $\mathcal{Re}\,\tau_f$ for the individual f's can be incorporated into one least singular definition of $\mathcal{Re}\,\tau(x_1,\cdots,x_4)$ leading to a unique least singular definition of $\tau(x_1,\cdots,x_4)$.

To establish the reported structure of $\mathcal{Re}\,\tau(x_1,\cdots,x_4)$ we had to prove that the following function

$$D_{\xi_1} D_{\xi_2} \exp\left\{-\sum_{j=\pm1}\sum_{k=\pm1} \frac{\lambda}{\left(\xi + \frac{(-1)^j\xi_1+(-1)^k\xi_2}{2}\right)^2 - i0}\right\}\Bigg/_{\xi_1=\xi_2=0}$$

with

$$D_x = \sum_{n=0}^{\infty} \frac{\left(-\frac{\lambda}{4}\Box_x\right)^n}{n!\,(n+1)!\,(n+2)!}\ , \quad \lambda = f^2/4\pi^2$$

is bounded for $\xi^2 > 0$. This turned out to be a very difficult problem.

For the Fourier transforms of $\mathcal{Re}\,\tau(x_1,\cdots,x_n)$, n = 2,3,4, the minimality property means the maximal possible decrease in the sector where all momenta and all partial sums of momenta are time - like. Summation of the perturbation theoretical contributions, say, to the two point vertex function preserves the maximal decrease of the real part in momentum space.

Any non-minimal choice of $\tau(x_1,\cdots,x_n)$ is expected to lead to a quantum theory which is even more singular than the one which we suggest to associate with the classical exponential Lagrangian.

The author is greatly indebted to Prof. Symanzik for detailed explanations of his new approach towards the non-renormalizability problem.

REFERENCES AND FOOTNOTES

1) The selection of the following references is to a large extent arbitrary

S. Okubo: Progr. Theor. Phys. 11, 80 (1954)

R. Arnowitt and S. Deser: Phys. Rev. 100, 349 (1955)

G. V. Efimov: Soviet. Phys. JETP 44, 2107 (1963)
TH. 1087 - CERN (1969)

E. S. Fradkin: Nucl. Phys. 49, 624 (1963)

B. Schroer: in "High-Energy Physics and Elementary Particles" IAEA, Vienna 1965

K. Bardakci and B. Schroer: J. Math. Phys. 7, 1016 (1966)

W. Güttinger: Fortschr. Phys. 14, 483 (1966)

W. Güttinger and E. Pfaffelhuber: Nuovo Cim. 52, 389 (1967)

O. W. Greenberg: Phys. Rev. 150, 1076 (1966)

A. Salam and J. Strathdee: Phys. Rev. D1, 3296 (1970)

C. J. Isham, A. Salam and J. Strathdee: Phys. Rev. D3, 1805 ('71)

N. Christ: Phys. Rev. D5, 2486 (1972)

G. Parisi: Note interne n. 573 (1974), n. 621 (1975), Istituto di Fisica "G. Marconi" I.N.F.N. - Sezione di Roma

F. Jegerlehner: preprint FUB HEP June 1975/11, Freie Universität Berlin

2) C. Feinberg and A. Pais: Phys. Rev. 131, 2724 (1963), Phys. Rev. 133 B, 477 (1964)

3) T. D. Lee: Nuovo Cim. 59, 579 (1968)

4) M. K. Volkov: Yadern. Fiz. 6, 1100 (1967), Yadern. Fiz. 7, 448 (1968), Fortschr. Phys. 22, 499 (1974)

5) K. Symanzik: Desy report 75/12, Desy report 75/24 (1975)

6) K. Symanzik: Proccedings of the Symposium on Mathematical Problems in Theoretical Physics, Kyoto (1975) Springer Lecture Notes in Physics 39 (1975)

7) N. N. Bogoliubov and O. S. Parasiuk: Acta Math. 97, 227 (1957)

8) K. Hepp: Commun. math. Phys. 2, 301 (1966)

9) F. J. Dyson: Phys. Rev. 75, 1736 (1949)

10) N. N. Bogoliubov and D. V. Shirkov: Introduction To The Theory Of Quantized Fields, § 28, Interscience Publ., New York 1959

11) N. N. Bogoliubov and D. V. Shirkov: Ibid. § 24 and §25

12) E. R. Speer: Generalized Feynman Amplitudes, p. 29 Princeton University Press, Princeton 1969

13) There are exactly soluble non-renormalizable models with trivial scattering, however, which can be quantized e.g. the s(v) coupling of bosons to fermions.

F. J. Dyson: Phys. Rev. 73, 929 (1948)

B. Klaiber: Nuovo Cim. 36, 165 (1965)

14) see e.g. L. D. Landau and E. M. Lifschitz: The Classical Theory of Fields, Pergamon Press, London 1962

15) see e.g. Marshak, Riazuddin and Ryan: Theory of Weak Interactions in Particle Physics, Wiley - Interscience, 1969

16) S. Weinberg: Phys. Rev. 166, 1568 (1968)

17) K. Hepp: Commun. math. Phys. 35, 265 (1974)

18) D. M. Capper, G. Leibbrandt and M. Ramón Medrano: Phys. Rev. D8, 4320 (1973)

19) G. t'Hooft and M. Veltman: Ann. Inst. H. Poincaré 1974

20) G. Kramer: Fortschr. Phys. 22, 633 (1974)

21) G. Ecker and J. Honerkamp: Nucl. Phys. B35, 481 (1971)

22) S. Weinberg: Phys. Rev. Lett. 19, 1264 (1967)

23) G. t'Hooft: Nucl. Phys. B33, 173 (1971)
Nucl. Phys. B35, 167 (1971)

24) B. W. Lee: Phys. Rev. D5, 823 (1972)

25) C. Becchi and R. Stora: Lectures at this school

26) K. Symanzik: Lett. Nuovo Cim. 8, 771 (1973), Cargèse Lectures 1973 (Ed. E. Brézin), Desy report 73/58

27) K. G. Wilson and J. B. Kogut: Phys. Rep. 12C, 75 (1974)

28) K. G. Wilson: Phys. Rev. D7, 2911 (1973)

29) G. Parisi: Cargèse Lectures 1973 (Ed. E. Brézin)

30) J. Zinn-Justin: Cargèse Lectures 1973 (Ed. E. Brézin)

31) E. Brézin, J. C. Le Guillou and J. Zinn-Justin: DPh-T/74/100 Saclay (1974)

32) G. Mack: International Summer Institute on Theoretical Physics, Kaiserslautern 1972, Springer Lecture Notes in Physics 17 (1973)

33) see also references 5) and 6)

34) A. Pais and G. Uhlenbeck: Phys. Rev. 79, 145 (1950)

35) B. Schroer: Lett. Nuovo Cim. 2, 867 (1971)

36) G. Mack: in "Renormalization and Invariance in Quantum Field Theory" (Ed. E. R. Caianiello) New York, Plenum Press (1974)

G. Mack and I. T. Todorov: Phys. Rev. D8, 1764 (1973)

37) H. Lehmann: Phys. Lett. 41B, 529 (1972), Lectures at "XII. Internationale Universitätswochen für Kernphysik, Schladming 1973

38) G. Ecker and J. Honerkamp: Nucl. Phys. B52, 211 (1973)

39) M. K. Volkov: Dubna preprint JINR E2-8869 (1975)

40) G. Ecker and J. Honerkamp: Nucl. Phys. B62, 509 (1973)

41) M. K. Volkov: ITP preprint 69-5, Kiev (1969)

42) H. Lehmann and K. Pohlmeyer: Commun. math. Phys. 20,101 (1971)

43) P. K. Mitter: Commun. math. Phys. 20, 251 (1971)

44) M. Karowski: Commun. math. Phys. 19, 289 (1970)

45) K. Pohlmeyer: On Superpropagators Involving Massive Fields IAS preprint 1971

46) K. Pohlmeyer: Commun. math. Phys.26,130 (1972)

47) M. Daniel and P. K. Mitter: J. Math. Phys. 13, 1026

48) M. K. Volkov: Dubna preprint JINR E2-6728 (1972)

49) K. Pohlmeyer: Commun. math. Phys. 35, 321 (1974)

50) C. G. Bollini and J. J. Giambiaggi: Phys. Lett. 40B, 566 (1972) Nuovo Cim 12B, 20 (1972)

51) G. t'Hooft and M. Veltman: Nucl. Phys. B44, 189 (1972)

52) E. R. Speer: Lectures at this schoool

53) A. Jaffe: Phys. Rev. Lett. 17, 661 (1966), Phys. Rev. 158, 1454 (1967) and unpublished manuscript "High Energy Behaviour of Strictly Localizable Fields II"

54) N. N. Meimann: Soviet Phys. JETP 20, 1320 (1965)

55) The correct more subtle condition is given in reference 53)

56) H. Epstein and V. Glaser: Commun. math. Phys. 27, 181 (1972)

57) H. Epstein, V. Glaser and A. Martin: Commun. math. Phys. 13, 257 (1969)

58) S. W. McDowell, R. Roskies and B. Schroer: Phys. Rev. 166, 1691 (1968)

GPSR Compliance

The European Union's (EU) General Product Safety Regulation (GPSR) is a set of rules that requires consumer products to be safe and our obligations to ensure this.

If you have any concerns about our products, you can contact us on ProductSafety@springernature.com

In case Publisher is established outside the EU, the EU authorized representative is:

Springer Nature Customer Service Center GmbH
Europaplatz 3
69115 Heidelberg, Germany

Batch number: 10762209

Printed by Printforce, the Netherlands